W9-AOM-126

CARL VON CLAUSEWITZ

On War

PREPARED UNDER THE AUSPICES OF
THE CENTER OF INTERNATIONAL STUDIES
PRINCETON UNIVERSITY

A LIST OF OTHER CENTER PUBLICATIONS
APPEARS AT THE BACK OF THE BOOK

CARL VON CLAUSEWITZ

ON WAR

Edited and Translated by

MICHAEL HOWARD *and* PETER PARET

Introductory Essays by PETER PARET,
MICHAEL HOWARD, *and* BERNARD BRODIE;
with a Commentary by BERNARD BRODIE

Index by ROSALIE WEST

PRINCETON UNIVERSITY PRESS

PRINCETON, NEW JERSEY

Published by Princeton University Press, 41 William Street,
Princeton, New Jersey 08540
In the United Kingdom: Princeton University Press, Chichester, West Sussex

Library of Congress Cataloging-in-Publication Data

Clausewitz, Carl von, 1780–1831.
On war.
Translation of: Vom Kriege.
Includes bibliographical references and index.
1. Military art and science. 2. War. I. Howard, Michael Eliot, 1922-
II. Paret, Peter. III. Title.
U102.C65 1984 355 84-3401
ISBN 0-691-05657-9
ISBN 0-691-01854-5 (pbk.)

First Princeton Paperback printing, 1989

Princeton University Press books are printed on acid-free paper
and meet the guidelines for permanence and durability of
the Committee on Production Guidelines for Book
Longevity of the Council on Library Resources

http://pup.princeton.edu

Printed in the United States of America

15 14 13 12 11 10
15 14 13 12 11 pbk.

ISBN-13: 978-0-691-1854-6 (pbk.)

ISBN-10: 0-691-01854-5 (pbk.)

CONTENTS

BOOK TWO
On the Theory of War

BOOK THREE
On Strategy in General

CONTENTS

BOOK FOUR
The Engagement

BOOK FIVE
Military Forces

BOOK SIX
Defense

BOOK SEVEN
The Attack

EDITORS' NOTE

The reader may wonder why another English translation of *Vom Kriege* is needed when two already exist. The first, made by Colonel J. J. Graham in 1874, was republished in London in 1909. The second, by Professor O. J. Matthijs Jolles, appeared in New York in 1943. But Graham's translation, apart from its dated style, contains a large number of inaccuracies and obscurities; and while Jolles' translation is more precise, both his version and Graham's were based on German texts that contained significant alterations from the first edition published in 1832.

The growing interest in Clausewitz's theoretical, political and historical writings in recent years suggested that the time had come for an entirely new translation. We have based our work on the first edition of 1832, supplemented by the annotated German text published by Professor Werner Hahlweg in 1952, except where obscurities in the original edition—which Clausewitz himself never reviewed—made it seem advisable to accept later emendations.

In all but one respect we have followed the original arrangement of the text. The first edition printed four notes by Clausewitz on his theories, dating from various periods between 1816 and 1830, as introductions to *On War* itself—a practice adopted by most subsequent German and foreign editions. We have abandoned the haphazard arrangement in which these have always appeared, and instead print them in the order in which we believe the notes to have been written. Read consecutively they help to indicate how *On War* took shape in Clausewitz's mind, and suggest how it might have further developed had he lived to complete it. We have also included Marie von Clausewitz's Preface to the first edition of Clausewitz's posthumous works, which adds information on the genesis of *On War*, and on the manner in which the manuscript was prepared for publication. A brief note she inserted at the beginning of the third volume of Clausewitz's *Works*, immediately preceding Book Seven of *On War*, has been deleted since its primary concern is not with *On War* but with other historical and theoretical writings.

We have attempted to present Clausewitz's ideas as accurately as possible, while remaining as close to his style and vocabulary as modern English usage would permit. But we have not hesitated to translate the same term in different ways if the context seemed to demand it. For instance, we have translated *Moral* and *moralische Kraft* variously as "morale," "moral," and "psychological." Clausewitz himself was far from consistent in his terminology, as might be expected of a writer who was less concerned with establishing a formal system or doctrine than with achieving understanding and clarity of expression. At times he writes *Geisteskräfte, Seelenkräfte,* even *Psychologie* instead of *moralische Kraft* or *moralische Grössen,* and a similar flexibility characterizes his use of such terms as "means," "purpose," "engagement," "battle," etc. As he writes in Book Five, Chapter Seven: "Strict adherence to terms would clearly result in little more than pedantic distinctions."

The task of translation was initially undertaken by Mr. Angus Malcolm, formerly of the British Foreign Office, who to the deep regret of his many friends died while he was still engaged on the project. He had however already done much valuable preliminary work, for which we are greatly in his debt. We should like to thank Mrs. Elsbeth Lewin, editor of *World Politics,* and Professor Bernard Brodie of the University of California at Los Angeles for checking the manuscript and helping us resolve many ambiguities, and Messrs. Herbert S. Bailey, Jr. and Lewis Bateman of Princeton University Press for the care they took in preparing the manuscript for publication. Financial assistance by the Center of International Studies of Princeton University facilitated the early phases of our work. Finally, it is a pleasure to express our gratitude to Professors Klaus Knorr of Princeton University and Gordon Craig of Stanford University, without whose interest and encouragement this task would never have been undertaken.

NOTE FOR THE 1984 EDITION

We have corrected some errors and attempted to remove a few infelicities in our translation of Clausewitz's text. As in the past, however, we believe that this work demands translators who combine a deep respect for the author with the willingness to seek equivalents whenever too close a correspondence with the original would lead to artificiality.

In the introductory essays, minor changes were made in "The Genesis of War," and two paragraphs on the Marxist interpretation of Clausewitz were added to "The Influence of Clausewitz." The only other change from our original edition is the inclusion of an index, which Mrs. Rosalie West has compiled on the model of the index in Professor Werner Hahlweg's 1952, 1972, and 1980 German editions of *On War.*

MICHAEL HOWARD
Oxford University

PETER PARET
Stanford University

INTRODUCTORY ESSAYS

By Peter Paret, Michael Howard,
and Bernard Brodie

PETER PARET

The Genesis of *On War*

Despite its comprehensiveness, systematic approach, and precise style, *On War* is not a finished work. That it was never completed to its author's satisfaction is largely explained by his ways of thinking and writing. Clausewitz was in his early twenties when he jotted down his first thoughts on the nature of military processes and on the place of war in social and political life. A pronounced sense of reality, skeptical of contemporary assumptions and theories, and an equally undoctrinaire fascination with the past, marked these observations and aphorisms and lent them a measure of internal consistency; but it would not be inappropriate to regard his writings before 1806 as essentially isolated insights—building-blocks for a structure that had not yet been designed.

The presence of a few of his earliest ideas in *On War* suggests the consequentiality with which his theories evolved, though in the mature work these ideas appear as components of a dialectical process that Clausewitz had mastered in the course of two decades and adapted to his own purposes. An example is his concept of the role genius plays in war, which lies near the source of his entire theoretical effort. Survivors of a somewhat different kind are his definitions of strategy and tactics, which he first formulated when he was twenty-four, or the characteristically unromantic comparison of war to commercial transactions, dating from the same time. Most of his early thoughts, however, expanded and acquired new facets in the years between Napoleon's defeat of Prussia and the Russian campaign. Clausewitz was a member of the loose alliance of reform-minded civilians and soldiers who attempted with some success to modernize Prussian institutions at this time, and his manifold activities as staff officer, administrator, and teacher further stimulated his intellectual interests and his creativity. Numerous passages from memoranda, lectures, and essays written during the reform era reappear, barely changed, in *On War*. After 1815, by which time his manuscripts on politics, history, philosophy, strategy, and tactics ran into thousands of pages, Clausewitz set to work on a collection of essays analyzing various aspects of war, which gradually coalesced into a comprehensive theory that sought to define universal, permanent elements in war on the basis of a realistic interpretation of the present and the past. In the course of a decade, he wrote six of eight planned parts, and drafted the remaining two. By 1827, however, he had developed a new hypothesis on what he called the "dual" nature of war, the systematic exploration of which demanded a far-reaching revision of the entire manuscript. He died before he could rewrite more than the first chapters of Book One.[1]

On War thus presents its author's thoughts in various stages of completion. They range from the magnificent opening sequence of logically unfolding propositions to the rich but at times one-sided or contradictory discussions of Books Two through Six, to the essayistic chapters of the last two books, which suggest with brilliant strokes what a final version might have contained. Nothing can take the place of this unwritten version; but we should remember that Clausewitz's decision in 1827 to revise his manuscript had not implied a rejection of earlier theories—he only meant to expand and refine them. As we read the present text of *On War*, we can at least approximate Clausewitz's intention by keeping his closely related hypotheses of the dual nature of war and of its political character clearly in mind. It will be useful, at the end of this discussion, to return to his ultimate hypotheses and outline their most significant aspects, the more so since he never fully developed their implications to theory.

That, despite the unevenness of its execution, *On War* offers an essentially consistent theory of conflict is indicative of the creative power of Clausewitz's method and ideas. Anyone prepared to enter into his man-

[1] Much of the older literature on the different phases of the writing of *On War* is based on inadequate sources and can be disregarded. Still valuable today is the short book by R. v. Caemmerer, *Clausewitz* (Berlin, 1905), and the suggestive article by H. Rosinski, "Die Entwicklung von Clausewitz' Werk 'Vom Kriege' im Lichte seiner 'Vorreden' und 'Nachrichten,'" *Historische Zeitschrift*, 151 (1935), pp. 278–293, which was amended in important respects by E. Kessel's response "Zur Entstehungsgeschichte von Clausewitz' Werk vom Kriege," *Historische Zeitschrift*, 152 (1935), pp. 97–100. W. M. Schering's speculations in his anthology of Clausewitz's writings, *Geist und Tat* (Stuttgart, 1941), are full of contradictions and factual errors; but since Schering was familiar with Clausewitz's unpublished drafts and seems to have been the last scholar to work on them before they disappeared at the end of the Second World War, his interpretations cannot be ignored. In a knowledgeable essay "Clausewitz," in *Makers of Modern Strategy*, ed. E. M. Earle (Princeton, 1943), pp. 93–113, H. Rothfels writes (p. 108, n. 65): "Clausewitz revised Book Eight and parts at least of Book One (probably Chapters One–Three) and of Book Two (certainly Chapter Two)." But he adds that Clausewitz regarded only Book One, Chapter One as complete. I believe that Rothfels considerably overstates the extent of Clausewitz's revisions after 1827. He gives no reason for his views other than internal evidence, but the passage from Book Eight he cites as proof of a late revision can be found in almost identical form in Clausewitz's manuscript on strategy of 1804. The best-informed evaluation of the entire question, incorporating the findings of a century of scholarship, is contained in E. Kessel's brilliant "Zur Genesis der modernen Kriegslehre," *Wehrwissenschaftliche Rundschau*, 3 (1953), no. 9, pp. 405–423.

ner of reasoning will grasp his thoughts on the timeless aspects of war. But our reading of *On War* can only benefit from an awareness of its genesis and intellectual context. What political and military experiences influenced its author? What were the assumptions and theories he reacted against? What, in his view, were the methodological requirements of sound analysis? Even a brief consideration of these questions will cast light on the development of Clausewitz's ideas and on the forms his ideas assumed in the various strata of *On War.*[2]

Clausewitz, the son of a retired lieutenant who held a minor post in the Prussian internal revenue service, first encountered war in 1793 as a twelve-year-old lance corporal. In the previous year the French legislative assembly had declared war on Austria, with whom Prussia had recently concluded a defensive alliance. The French action was caused less by considerations of national interest than by internal politics, but it opened twenty-three years of conflict between revolutionary and later imperial France and the rest of Europe. Aside from the Duke of Brunswick's initial invasion, which came to a halt at Valmy, the Prussians did reasonably well in a war to which they never committed more than part of their military resources. They defeated the French repeatedly in Alsace and the Saar, and captured thousands of prisoners; when the fighting ended in 1795, they controlled the line of the Rhine. But these achievements brought no political returns. As might be expected, the war with its exertions, bloodshed, and unspectacular outcome made a strong impression on the young Clausewitz; he himself later wrote of its impact on his emotions and thought. In the following years, while stationed in a small provincial garrison, he drew some tentative conclusions from these early experiences, three of which in particular were to have a lasting influence: There was no single standard of excellence in war. The rhetoric and policies of the French Republic, which proclaimed the coming of a new age, by no means overpowered the armies of the *ancien régime.* Mercenaries and forcibly enrolled peasants, led by officers whose effectiveness still rested as much on aristocratic self-esteem as on professional expertise, proved a match for the *levée en masse.* On the other

[2] Any interpretation of the genesis of Clausewitz's thought on war must rest not only on his works on military theory and history but also on his extensive writings on such subjects as education, politics, the theory of art, and on his correspondence. Especially valuable analyses of aspects of his broad intellectual development are H. Rothfels' *Carl von Clausewitz: Politik und Krieg* (Berlin, 1920), and E. Kessel's introduction to C. v. Clausewitz, *Strategie aus dem Jahr 1804* (Hamburg, 1937). Primary and secondary sources are discussed in detail in my book, *Clausewitz and the State* (New York, 1976), on which much of the following is based.

hand, Prussian drill failed to sweep away the revolutionary armies. As the Republic gained in stability and experience, it would have much to teach its opponents, whose ability to learn and to respond effectively remained in doubt. These events and his first readings in history suggested to Clausewitz that no one system was right to the exclusion of all others. Military institutions and the manner in which they employed violence depended on the economic, social, and political conditions of their respective states. Furthermore, political structures, like wars, could not be measured by a single standard. States were shaped by their particular past and present circumstances; very different forms had validity, and all were subject to continuing change.

Linked to this individualizing, antirationalist view of history and of social and military institutions was a second conclusion, which placed the young officer in opposition to prevailing opinion in Prussia and, indeed, Europe. He thought it was a mistake to believe that war could be mastered by observing this or that set of rules. The variety and constant change in war could never be fully caught by a system. Any dogmatic simplification—that victory depended on the control of key points, for instance, or on the disruption of the opponent's lines of communication —only falsified reality. Possibly Clausewitz already distrusted the conviction, held by most military theorists of his day, that the scope of chance in war should and could be reduced to a minimum by the employment of the correct operational and tactical doctrine. For someone who passionately wanted to understand war in a systematic and objectively verifiable manner it was particularly hard to accept the power of chance; but by the time he was in his mid-twenties his realism and the logic of his view of historical change had brought him to the point of regarding chance not only as inevitable but even as a positive element in war.

Finally, the campaigns of 1793 and 1794 set Clausewitz on the path of recognizing war as a political phenomenon. Wars, as everyone knew, were fought for a purpose that was political, or at least always had political consequences. Not as readily apparent was the implication that followed. If war was meant to achieve a political purpose, everything that entered into war—social and economic preparation, strategic planning, the conduct of operations, the use of violence on all levels—should be determined by this purpose, or at least accord with it. Even though soldiers had to acquire special expertise, and function in what in some respects was a separate world, it would be a denial of reality to allow them to carry on their bloody work undisturbed until an armistice brought their political employer back into the equation. Just as war and its institutions reflected their social environment, so every aspect of fighting should be suffused by its political impulse, whether this impulse was intense or moderate. The appropriate relationship between politics and war occupied Clausewitz throughout his life, but even his earliest manuscripts and letters show his awareness of their interaction.

The ease with which this link—always acknowledged in the abstract—can be forgotten in specific cases, and Clausewitz's insistence that it must never be overlooked, are illustrated by his polite rejection toward the end of his life of a strategic problem set by the chief of the Prussian General Staff, in which every military detail of the opposing sides was spelled out, but no mention made of their political purpose. To a friend who had sent him the problem for comment, Clausewitz replied that it was not possible to draft a sensible plan of operations without indicating the political condition of the states involved, and their relationship to each other: "War is not an independent phenomenon, but the continuation of politics by different means. Consequently, the main lines of every major strategic plan are *largely political in nature*, and their political character increases the more the plan applies to the entire campaign and to the whole state. A war plan results directly from the political conditions of the two warring states, as well as from their relations to third powers. A plan of campaign results from the war plan, and frequently—if there is only one theater of operations—may even be identical with it. But the political element even enters the separate components of a campaign; rarely will it be without influence on such major episodes of warfare as a battle, etc. According to this point of view, there can be no question of a *purely military* evaluation of a great strategic issue, nor of a purely military scheme to solve it."[3]

In the second half of the 1790's, the young Clausewitz had taken only the first steps on the intellectual journey that was to lead to this conclusion; but, as I suggested earlier, from the outset he traveled a straight road, with few tangents or interruptions. The five years he spent as subaltern in the small town of Neuruppin have commonly been dismissed as a time of stagnation, but it seems that biographers have been too literal in their interpretation of a characteristically critical and self-critical comment on the period that he made years later. In reality his situation was not without advantages. Far from serving in an undistinguished provincial unit, he belonged to a regiment that had a member of the royal family, Prince Ferdinand, as honorary colonel and patron. Near the town lay the residence of another Hohenzollern, Prince Henry, Frederick the Great's most gifted brother, whose library, opera, and theater were open

[3] C. v. Clausewitz to C. v. Roeder, 22 December 1827, in *Zwei Briefe des Generals von Clausewitz*, special issue of the *Militärwissenschaftliche Rundschau*, 2 (March 1937), p. 6. English edition forthcoming.

to the officer corps. Most important, the regiment was known throughout the army for its innovative educational policies, financed largely by the officers themselves. On its return from France the regiment had organized a primary and trade school for the children of the rank and file, and a more advanced school for its cadets and ensigns, which also admitted sons of the local gentry. It is probable though not certain that like other lieutenants, Clausewitz taught classes in the latter institution; and there can be no doubt that his exposure to a serious teaching program deepened the interest he already felt in education. As a fifteen-year-old, he later wrote, he had been captivated by the idea that the acquisition of knowledge could lead to human perfectibility. Soon the goal of improving society reinforced that of self-improvement, and his desire to learn was joined by concern with the methodology of education. The ways in which abstractions might accurately reflect and convey reality, the manner in which men can be taught to understand the truth, and the ultimate purpose of education—which, he held, consisted not in the transmission of technical expertise but in the development of independent judgment—all came to be major considerations in Clausewitz's theoretical work.

In 1801 Clausewitz gained admission to the new War College that Scharnhorst, recently transferred from the Hanoverian service, had organized in Berlin. Clausewitz graduated at the head of the class in 1803, and was appointed adjutant to a young prince, son of his former commander Prince Ferdinand, an assignment that enabled him to remain in the capital, in close contact with his teacher Scharnhorst. The impact that Scharnhorst exerted on Clausewitz's life and on the development of his ideas cannot be emphasized enough. Scharnhorst was an exceptionally energetic, daring soldier, as well as a scholar and a gifted politician—a harmonious combination of seeming opposites that his favorite pupil was never to equal. This is not the place to discuss his opinions on strategy, on conscription, and on command- and staff-organization, which constituted a pragmatic reconciliation of the old and the new; important for our purpose is the intellectual independence with which he approached the fundamental military issues of the age, as well as his sympathy with the aims of humanistic education, and his conviction that the study of history must be at the center of any advanced study of war. Clausewitz's tentative attitudes on military theory and on education were confirmed and guided further by Scharnhorst, who also deepened Clausewitz's awareness of the social forces that determined the military style and energies of states. Scharnhorst, the son of a free peasant who had risen to the rank of squadron sergeant-major, had had a difficult career in the Hanoverian army, where he had been repeatedly slighted in favor of well-connected noble comrades. The experience did not turn him into a democrat, nor—having achieved professional success, including a title of nobility—did he fall into facile acceptance of privilege. What mattered to him was not the particular structure of society or the form taken by its institutions, but the spirit that animated them. To give a specific example, in the regimental school for soldiers' children at Neuruppin Clausewitz had witnessed something of the humanitarian, paternalistic concern for the poor that was a pronounced feature of the late Enlightenment in Prussia. Scharnhorst taught him that this was adequate neither for the individual nor for the state. If the French Revolution had proved anything, it was that states wishing to preserve their independence must become more efficient in tapping the energies of their populations. Elites existed in every society, and were justified so long as they strengthened the community, remained open to talent, and rewarded merit. But nothing could justify the continuation of privilege that protected mediocrity while depriving the state of the abilities and enthusiasm of the common man. It was this attitude that a few years later was to determine the direction of the Prussian reform movement—less perhaps in civil matters than on the military side under the leadership of Scharnhorst and his close associates. In the genesis of Clausewitz's ideas, the essentially unideological view of social and political arrangements, which he had learned in part from Scharnhorst and which he expressed as early as 1804 and 1805, clearly parallels his undoctrinaire approach to war. Statesman and soldier must shed tradition, convenience, any influence that interferes with their achieving the major objective. Similarly, the theorist, wishing to understand the nature of the state and the nature of war, must never allow his thoughts to diverge far from the element central to each—power in politics, violence in war.

The most important task that faced Prussian soldiers in the opening years of the nineteenth century was to come to terms intellectually and institutionally with the new French way of warfare. Within one decade the resources that France mobilized for war had risen to unprecedented levels. The number of soldiers now available to her generals made possible campaigns that accepted greater risks, brought about battle more frequently, spread over more territory, and pursued political goals of greater magnitude than had been feasible for the armies of the *ancien régime*. This new technique was used by Napoleon with a brilliance that shocked as much as did his ruthlessness. For most Germans it was difficult enough to understand his system, which combined the gifts of an exceptional individual with social, administrative, and psychological achievements of the Revolution, which were necessarily alien to them. For theorists of any nationality it was even more difficult to recognize Napoleonic strategy and tactics as a historical phenomenon, inevitably subject to change, rather than as the ultimate in war, a permanent standard of excellence for war past, present, and future.

European military literature commented with considerable insight on separate elements of this system, but, as Clausewitz saw early on, failed in attempts at comprehensive analysis. The best work in this area was done by the Prussian theorist Heinrich von Bülow and the Swiss-French staff officer Antoine Jomini on whose writings Clausewitz sharpened his theoretical skills in the years preceding and immediately following the Prussian debacle of 1806. Bülow had grasped the value of such recent tactical developments as skirmishing in large numbers, rapidity of movement, and aimed fire; at the same time he discounted the effectiveness of battle in the new age, regarded it as a "resort of despair," and instead postulated a strategic system of points of domination and angles of approach, whose geometric patterns combined in a fantastic manner with his paeans on the natural, unfettered fighting man. In his first published work, a long essay on Bülow, Clausewitz acknowledged the usefulness of some of his terminology, as he was to find merit in some of Jomini's concepts, but pointed out that his method of analysis was flawed and that its conclusions were unrealistic. In his urge to rationalize war, turn it into a science, and make it predictive, Bülow attributed dominant roles to geographic features and the appropriate arrangement of the supply system, while largely ignoring the physical and psychological effects that might result from unexpected movements of the opponent, from violence, from the fortuitous. Strategy, Clausewitz objected, comprises "not only the forces that are susceptible to mathematical analysis; no, the realm of the military art extends wherever in psychology our intelligence discovers a resource that can serve the soldier."[4]

Jomini came closer to contemporary reality, but erred, Clausewitz thought, in taking one part of war—major armies seeking a decisive victory—for the whole. His claim that he had distilled general principles of war from the operations of Napoleon, and from supposedly similar, though inferior, operations of Frederick, Clausewitz dismissed as absurd. He wrote in 1808 that Jomini's principles would lose their absolute validity if it could be shown that earlier generations had good reasons to ignore them. Caesar or Eugene of Savoy, responding to the social, technological, and political realities of their times, were not inferior to Napoleon because they did not fight in a manner that the French Revolution

[4] [C. v. Clausewitz], "Bemerkungen über die reine und angewandte Strategie des Herrn von Bülow," *Neue Bellona*, 9 (1805), no. 3, p. 276.

had made possible. And just as the past could be understood only in its own terms, men, too, must be interpreted as individuals, not as abstractions. Jomini had unrealistically imposed one rational standard of behavior on men with such different personalities as Frederick and Napoleon, and besides had ignored the differences in their experiences, to which each naturally reacted in his own way.[5]

If the present did not provide the ideal against which war in the past could be measured, Clausewitz was equally insistent that Napoleonic war could not establish standards for the future.[6] What did this mean for theory? To Clausewitz the answer was obvious: The theory of any activity, even if it aimed at effective performance rather than comprehensive understanding, must discover the essential, timeless elements of this activity, and distinguish them from its temporary features. Violence and political impact were two of the permanent characteristics of war. Another was the free play of human intelligence, will, and emotions. These were the forces that dominated the chaos of warfare, not such schematic devices as Bülow's base of operations or Jomini's operating on interior lines.

There was nothing new about stressing the significance of psychological factors in war. But even those writers who attributed predominance to the emotions had little of substance to say about them; discussions of courage, fear, and morale figure only on the margin of the works of Maurice de Saxe or Henry Lloyd. By contrast the young Clausewitz placed the psychological at the center of his theoretical speculations. But since psychology was still a rudimentary discipline that offered him few of the taxonomic and interpretive tools he needed, he did so in a manner that modern readers may find puzzling: he subsumed a large part of his interpretation of emotional and moral characteristics under the concept of genius. It is essential to understand that by genius Clausewitz meant not only originality and creativity raised to their highest power but also, as he wrote in *On War*, gifts of mind and temperament in general. Genius served as his favorite analytic device to conceptualize the various abilities and feelings that affected the behavior of more ordinary as well as of exceptional men.

Even in his early writings Clausewitz had no difficulty in exposing the

[5] In his addition to his essay "On Abstract Principles of Strategy," a later addition to his manuscript on strategy of 1804, printed in *Strategie aus dem Jahr 1804*, pp. 71-73.

[6] See for instance his essay "On the State of Military Theory," written in his twenties, which opens with the declaration that contrary to the belief of some writers the art of war had not yet attained perfection: "Any scientific discipline—unless like logic it is complete unto itself—must always be capable of growth, of constant accretion. In any event it is not all that easy to set limits to the human intellect." *Geist und Tat*, p. 52.

inadequacy of prescriptive systems when faced with the infinite resources of the mind and spirit. In his essay on Bülow he wrote that there must be no conflict between common sense and sound theory since sound theory rested on common sense and genius, or gave them expression.[7] He was to hold fast to this thought; it occurs repeatedly in *On War*, not only in the chapter "On Military Genius," but elsewhere as well, for example in the chapter "On the Theory of War," where characteristically it is linked to a sarcastic attack on the surrender of such system-builders as Bülow and Jomini before the unpredictable riches of the spirit: "Anything that could not be reached by the meager wisdom of one-sided points of view was held to be beyond scientific control: it lay in the realm of genius, *which rises above all rules*. Pity the soldier who is supposed to crawl among these scraps of rules, which are not good enough for genius, which genius can ignore or laugh at. No; what genius does is the best rule, and theory can do no better than show how and why this should be the case. Pity the theory that conflicts with reason!"[8] Theory and its resultant doctrines are thus subordinate to the great creative talent, and to the universals of reason and feeling that it expresses.

Clausewitz himself was still far from formulating a theory that explained why and how the action of genius should be the finest rule. He needed to develop additional analytic devices before he could advance appreciably, and it might be added that he never fully overcame the difficulties inherent in the dual role that he ascribed to the concept of genius. The problems of theory, however, were not identical with the problems of historical understanding; here attention to the emotions of individuals and groups combined readily with belief in the particularity of past epochs. Clausewitz's history of Gustavus Adolphus in the Thirty Years War, written about 1805, constitutes his initial effort to integrate these two interpretive principles on a large scale.[9] It was a remarkably successful attempt, and only the first of many historical studies he was to write in the course of his life. Indeed, if quantity is the measure, Clausewitz was more of a historian than a theorist. That he was innovative in this discipline, too, tends to be forgotten—possibly because his most original historical writings were not published for decades, and because German historical scholarship soon developed and expanded the vein which he had been among the first to work, while as a theorist he remained without true successors. For a man of his time, he took an un-

[7] [C. v. Clausewitz], "Bemerkungen," *Neue Bellona*, 9 (1805), no. 3, pp. 276–277.

[8] "On the Theory of War," Book Two, Chapter 2, *On War*.

[9] The study, "Gustav Adolphs Feldzüge von 1630–1632," some hundred pages long, was published in 1837 in volume 9 of Clausewitz's collected works, *Hinterlassene Werke des Generals Carl von Clausewitz*, 10 vols. (Berlin, 1832–1837).

usually straightforward approach to the past. He did not hide an ironic interest in the passions and limitations of his characters, especially when writing about recent events; but he rarely showed ideological or patriotic prejudice. As well as he could, he tried to discover how and why things happened as they did. His urge to be objective was intensified by his belief, based on personal predilection and Scharnhorst's teaching, that military theory in a variety of ways was dependent on history. His mature conclusions on their appropriate relationship are best discussed when we come to the writing of *On War*.

The defeat of Prussia in 1806 confirmed Clausewitz in his view that war could not be considered in isolation, as an essentially military act. It was obvious to him that the politics of the previous decade had largely decided the issue before fighting began, while social conditions of long standing in the Prussian monarchy had created military institutions and attitudes that proved helpless against an opponent who was numerically superior and attuned to the new forms of fighting. For Clausewitz personally the campaign was once more an infantryman's war; he served with a grenadier battalion until his unit was compelled to surrender. After internment in France and a sojourn in Switzerland he returned to Prussia in the spring of 1808. For the next four years he acted as confidential assistant to his former teacher Scharnhorst, who employed him in a variety of assignments related to the modernization of the army: reorganizing and reequipping the troops, drafting new tactical and operational instructions, disseminating the new doctrine as teacher at the War College and military tutor to the crown prince. Finally, Clausewitz played a larger part than might be expected of a junior officer in the evolution of the political and strategic thinking of the reform party. The practical experience he acquired was unusually broad, and further strengthened the pragmatic note that ran through his theoretical as well as his historical writings. During these years he married. His wife, a sophisticated, intelligent woman, shared his literary and philosophic interests, and fully supported his growing political and professional independence; only the lack of children marred an otherwise exceptionally happy marriage. He also formed a lasting friendship with the second leader of the military reformers, Gneisenau, a relationship that was to shape much of his subsequent career. After Prussia was forced to contribute a corps to the army that Napoleon was assembling for the invasion of Russia he resigned his commission, and in the spring of 1812 accepted a staff appointment in the Russian army.

The richness as well as the volume of his writings during these very active years is astonishing. To outline only the major hypotheses that Clausewitz advanced in fields as seemingly diverse as grand strategy and national character would take up more space than is possible here; but even a brief introduction ought not to ignore the conclusions he reached on the nature and function of military theory, since they were to determine the approach he followed in *On War*. Something must also be said about the analytic method he was developing. Finally, his many advances in the content of theory can at least be suggested by discussing one representative conceptualization of this period—the concept of friction, with which he complemented earlier ideas and made them productive in scientific enquiry.

By 1808 Clausewitz firmly distinguished between the utilitarian, pedagogic, and cognitive functions of theory. The first—improving the soldier's effectiveness—was the major, often the only, aim of contemporary military theorists. Clausewitz shared their wish to define and respond to the practical issues of modern war, and never more so than in the years when he was passionately involved in rebuilding the Prussian army for the inevitable second contest with Napoleon. But on grounds of logic as well as realism he grew skeptical about the direct link between theory and performance that military theorists took for granted. His study of Kantian philosophy before 1806 gave him at least some of the intellectual tools he needed to resolve his doubts—his most significant borrowings being the view of theory held by late-Enlightenment writers on aesthetics, and their concepts of "means" and "purpose," which came to play a pervasive role in *On War*. An essay "Art and Theory of Art" illustrates his use of aesthetics to explore the violent art of defeating one's enemies. "Art," he wrote, "is a developed capacity. If it is to express itself it must have a *purpose*, like every application of existing forces, and to approach this purpose it is necessary to have means. . . . To combine purpose and means is to create. Art is the capacity to create; the theory of art *teaches* this combination [of purpose and means] to the extent that concepts can do so. Thus, we may say: *theory is the representation of art by way of concepts*. We can easily see that this constitutes the whole of art, with two exceptions: *talent*, which is fundamental to everything, and *practice*"—neither of which can be the product of theory.[10] In short, even the most realistic theory could never match reality. It followed that all attempts to establish rules with prescriptive power were pointless in an activity such as fighting, and that military theory could never be immediately utilitarian. As Clausewitz wrote in the same essay, "rules are not intended for individual cases, and action in the in-

[10] *Geist und Tat*, p. 159. The essay is undated, but was probably written during the reform period.

dividual case can be determined only by [applying the concepts of] purpose and means."[11] All that theory could do was to give the artist or soldier points of reference and standards of evaluation in specific areas of action, with the ultimate purpose not of telling him how to act but of developing his judgment.

It was this process of refining the judgment and "instinctive tact" of the acting individual that constituted the pedagogic function of theory, not drawing up rules to be learned by rote. (Another pedagogic aspect of theory, important to Clausewitz personally, had to do with the creative process. By developing an analytic framework for war, Clausewitz strengthened his intellectual capacities and implemented the program of self-education from which he had not swerved since adolescence.) But although only serious speculative inquiry could set the mind free, Clausewitz believed that most men were neither capable of achieving intellectual mastery over complex areas of human activity nor much interested in it. To help them through the confusion of war they demanded relatively firm guides. How were these to be provided? According to Clausewitz, experience went a long way, but in the end appropriate guides for conduct could only grow out of a comprehensive and scientific analysis.

This was the cognitive side of theory. Nonutilitarian analyses, concerned solely with gaining a deeper understanding, might bring about improvements in operational and strategic performance. But for Clausewitz scientific inquiry needed no justification. While he never lost interest in the military here and now, understanding as such was what eventually mattered most to him, and it was to this task that *On War* addressed itself.

When Clausewitz first began to think of writing a study that would explore the whole of war, not merely some of its parts, he chose as intellectual models such books as Montesquieu's *De l'Esprit des lois* and Kant's *Critique of Practical Reason*. If in its final version *On War* bore little resemblance to these works, they nevertheless indicate something about the method employed by its author. Earlier I characterized this method as dialectical. It was that, but in a special sense. Certainly, he did not proceed in a formal, highly structured manner. Hegel's thesis, antithesis, synthesis, to mention an approach that has often been read into *On War*, would have seemed inappropriate to Clausewitz, as did any system, the logical and intellectual symmetry of which were achieved at the expense of reality. But he frequently did develop his ideas in what may be called a modified form of thesis and antithesis, which permitted him to explore the specific characteristics of a particular phenomenon with a high degree of exactitude. Purpose and means, strategy and tac-

[11] *Ibid.*, p. 162.

tics, theory and reality, intent and execution, friend and enemy—these are some of the opposites he defines and compares not only to gain a truer understanding of each member of the pair but also to trace the dynamic links that connect all elements of war into a state of permanent interaction. One of the striking features of this way of thinking is that it defines each element as sharply as possible while insisting on the absence of discrete limits. War and politics, attack and defense, intelligence and courage—to mention some additional pairings—are never absolute opposites; rather one flows into the other.

Again German philosophy, together with certain analytic and structural assumptions of the natural sciences, provided Clausewitz with a fundamental attitude and with the intellectual tools to express it. Belief in the need to ascertain the essence of each phenomenon, or its regulative idea—as violence, according to Clausewitz, was the essential idea of the phenomenon "war"—combined with a universal view and with the sense that small details contained the key to large forces, as knowledge of one flower was basic to the understanding of nature, or knowing why and how a man fought was essential to understanding war.

It was in accord with this larger cultural outlook as well as with his personal tendencies that Clausewitz eschewed generalization and simultaneously rejected the anarchy of pure pragmatism. His aim was to achieve a logical structuring of reality. He believed this could be done if the search for regulative ideas and their elaboration were informed and controlled by the theorist's respect for present and past reality. Consequently, his method consisted in a permanent discourse between observation, historical interpretation, and speculative reasoning. As the analysis proceeds it tries to take account of every element of war in its present and past dimensions, accommodate itself to all, integrate all, and never emphasize one to the exclusion of others. We will see that this characteristic also holds true of the resulting theory, which floats, Clausewitz said, between the major phenomena of war, without stressing any one in particular. The dangers of exaggeration, of being blinded by contemporary conditions, let alone of one-sided advocacy, are thus largely avoided.

An example of the way in which Clausewitz's method transformed reality into analyzable form is provided by his development of the concept of friction. He first used the term during the campaign of 1806 to describe the difficulties Scharnhorst encountered in persuading the high command to reach decisions, and the further difficulties of having the decisions implemented. Uncertainty, ignorance, confusion, fatigue, error, countless other imponderables—all interfered with the effective application of force. During the reform era Clausewitz expanded the concept, and linked it with other ideas, until by 1812 he had fully grasped its theoretical implications. An essay he addressed to the crown prince at the end of his tutorials concluded with a section on friction, which both in content and wording became the basis for the chapter "Friction in War" in *On War*, and for the discussion of friction that runs through the entire work.[12] Waging war is very difficult, he wrote, "but the difficulty is not that erudition and great talent are needed . . . there is no great art to devising a good plan of operations. The entire difficulty lies in this: *To remain faithful in action to the principles we have laid down for ourselves*."

To explain why this should be so, Clausewitz resorted to a simile: "The conduct of war resembles the workings of an intricate machine with tremendous friction, so that combinations which are easily planned on paper can be executed only with great effort. Consequently the commander's free will and intelligence find themselves hampered at every turn, and remarkable strength of mind and spirit are needed to overcome this resistance. Even then many good ideas are destroyed by friction, and we must carry out more simply and modestly what in more complicated form would have given greater results."

Friction, he continued, even if it is created by physical forces—bad weather, for instance, or hunger—always has a psychologically inhibiting effect; psychic energy must therefore take a part in overcoming it: "In action our physical images and perceptions are more vivid than the impressions we gained beforehand by mature reflection. But they are only the outward appearances of things, which, as we know, rarely match their essence precisely. We therefore run the risk of sacrificing mature reflection to first impressions." In the face of these pressures, men must hold to their convictions and retain confidence in their knowledge and judgment; otherwise they will succumb to the force of friction. Friction, he was to conclude in *On War*, is the only notion that more or less comprises those matters that distinguish the real war from war on paper.[13]

By creating the concept of friction he rendered one of the most important elements in his image of war—chance—subject to theoretical analysis. Insofar as friction interfered with one's own actions, it stood only for the negative aspects of chance. The positive aspects of chance were represented by the equally pervasive force of friction on the enemy's side. To appreciate the significance of this development we must recall that the military writers of the Enlightenment, while often acknowledging the

[12] The essay, "Die wichtigsten Grundsätze des Kriegführens . . . ," has been brought out in English by H. Gatzke under the somewhat misleading title *Principles of War* (Harrisburg, Pa., 1942). For the following quotations, which are in my translation, compare Gatzke, pp. 60–61, 67.

[13] "Friction in War," Book One, Chapter Seven, *On War*.

power of the fortuitous, did their best to reduce the scope of chance. Their spiritual successors Bülow and Jomini strove for the same goal by means of systems that extended the enormously detailed rules of eighteenth-century march, camp, and tactical arrangements to strategy. Success could be assured by choosing "correct" techniques. Other writers claimed modern war to be anarchic, susceptible only to empirical treatment. Scharnhorst, on the contrary, held that the natural behavior of societies and individuals in war could be understood and thus to some extent guided, and Clausewitz gave this belief theoretical form. In their view, to exclude or deny chance was to go against nature; indeed, chance was to be welcomed because it was part of reality. It was not only a threat but also a positive force to be exploited. Napoleon expressed this idea perfectly in his operational dictum: Engage the enemy, and see what happens. The commander put himself in the way of chance; the power at his disposal and his will to use it enabled him to turn chance into a new reality.

The force that could most effectively create and exploit this reality was genius. Thus the concept of friction came to form the counterpart in external life to the result of Clausewitz's earlier analyses of the inner life of the individual. Observation and reflection had led him to elevate genius—the harmonious combination of exceptional gifts, and by extension, intellectual and emotional qualities in general—to a central position in his conceptualization of war. The concepts of genius, friction, chance, in their manifold interaction, now made it possible for the theorist to subject vast areas of military reality to logical, systematic analysis.

During the war of 1812 Clausewitz served as staff officer with several Russian commands, his ignorance of the language limiting him to the role of observer until the end of December, when he took part in talks between Russian authorities and the commander of the Prussian corps in the *Grande Armée* that led to the strategically and politically important separation of the Prussian forces from French control. As the fighting moved west he devised the plan for organizing the East Prussian militia, a further significant step in the process of detaching Prussia from French dominance. In the spring campaign of 1813, still in Russian uniform, he acted as adviser to Scharnhorst and Gneisenau until the former's death, and then became chief of staff of a small international army that covered the Baltic flank of the Allies. Though strict monarchists, including the king himself, continued to resent his earlier refusal to follow official policy and fight for the French, he eventually gained readmission to the Prussian service. During the Waterloo campaign he served as chief of staff of one of the four corps making up the Prussian field army, and fought at Ligny and Wavre, where his corps tied down Grouchy's superior force until the main French army was beyond help. In 1816 he became chief of staff of Gneisenau's new command, with headquarters at Coblenz on the Rhine; two years later he was transferred to Berlin as Superintendent of the War Academy. His new duties were neither onerous nor particularly rewarding. Several times he sought to exchange the army for diplomacy; but since his reformist politics made him unacceptable to the court, he remained in his administrative position for twelve years, not displeased, on balance, with the opportunity of giving much of his time to study and writing.

It was in the early years of peace, after the violent interval of the last campaigns against Napoleon, that Clausewitz returned seriously to theoretical work. A note found in his papers, which his wife quotes in her introduction to *On War*, indicates that while stationed in the Rhineland he began writing brief essays on strategy, addressed to the expert.[14] None of these pieces seems to have survived; but we possess at least one preliminary study from which Clausewitz hoped to distill the aphoristic essay he was aiming for: "On Progression and Pause in Military Activity." It provided the basis for Chapter Sixteen in Book Three of *On War*, which in turn elaborates one of the key arguments in the first chapter of the work: real war falls short of the total violence that is its essence in theory because, among other reasons, war does not consist in a single act or in a group of simultaneous actions, but extends over time with periods of action and inaction alternating. Another, much less significant, essay may have been the discussion on army organization that is usually printed as an appendix in German editions of *On War*. Its essential points can be found in Chapter Five of Book Five.

These essays, concise as they are, do not match the extreme brevity of the chapters in Montesquieu's works, which, Clausewitz writes, served him as a kind of general model at this time. Nor does the structure of his argument resemble Montesquieu's. But the character of *De l'Esprit des lois*, and the personality of its author, indicate clearly enough the basis for Clausewitz's sense of affinity. The introduction, to mention one

[14] M. v. Clausewitz, "Preface," *On War*. The quotation from her husband appears on p. 63. In time of writing this is the second of four introductory notes by Clausewitz to *On War*. The first, "Author's Preface," dates from 1816 to 1818, and refers to the essays Clausewitz was writing in those years (pp. 61–62). The second, included in his wife's preface, refers to the expansion of the original scheme. The third, dated 10 July 1827, constitutes the first half of the "Notice," and mentions Clausewitz's plan for a complete revision of Books One to Six, and of the sketches for Books Seven to Eight (pp. 69–70). The second part of the "Notice" was written later, possibly in 1830, and indicates that Clausewitz had not progressed very far with his revision (pp. 70–71).

example, contains sentences that Clausewitz himself might have written: "I ask for a favor that I fear will not be granted—do not judge the labor of twenty years at a moment's notice; approve or reject the entire work, not a few of its statements. If one wishes to seek the author's intention, it can be discovered only in the scheme of the work." A subsequent passage in which Montesquieu states that while writing he "knew neither rules nor exceptions" can hardly be improved on as a description of Clausewitz's attitude to the study of war.[15]

The essays, each singling out a particular phenomenon or concept, had the advantage of revealing the major features of each with great clarity, but their inevitably fragmented analysis left Clausewitz dissatisfied. As he added new sections and revised existing ones, the compressed aphoristic character of his work gave way to fuller treatment, which responded to his preference for the systematic development of ideas and the uniform application of concepts on a broad range of phenomena. Coincidentally, an expanded and more explicit analysis would be suitable, he felt, for a wider audience than he had had originally in mind. The result was *On War* essentially as we know it today, except for the limited revisions made from 1827 on.

Readers of this work and of the studies leading up to it may ask why Clausewitz felt it necessary to assert repeatedly that violence is the essence of war, and dismiss his reiteration as a pedantic insistence on the obvious. But Clausewitz stressed the point not only because experience and the study of the past had convinced him of its truth; he was also responding to the surprisingly numerous theorists who continued to claim that wars could be won by maneuver rather than bloodshed. What matters, in any case, are the deductions he drew from the self-evident. When he was twenty-four he had written that war must always be waged with the greatest possible amount of energy—that only "the most decisive operations accord with the nature of war."[16] Eight years later he instructed his pupil, the crown prince, that war always demanded the fullest mobilization of resources, and their most energetic exploitation.[17] Here were specific implications derived from the concept of *absolute* war, of war that ideally should be waged with the extreme of violence—ideally, because the extreme of violence accorded with its nature. If war was an act of force, Clausewitz could discern no logical "internal" or self-imposed limits on the use of force. His insistence on extremes during the Napoleonic era resulted, of course, not only from logic but also from the historical situation. Between 1792 and 1815 exceptional effort and

[15] Montesquieu, *De l'Esprit des lois*, Geneva, 1749, pp. iii, vi.
[16] "Plan of Operations," *Strategie aus dem Jahr* 1804, pp. 51–52.
[17] *Principles of War*, p. 46.

the willingness to take great risks were, in fact, needed to preserve Europe's independence, or to regain it. But even in the years of greatest challenge Clausewitz recognized that the demand of absolute or utmost violence, though logically valid, was rarely satisfied in reality. Absolute war was a fiction, an abstraction that served to unify all military phenomena and helped make their theoretical treatment possible. In practice the use of force tended to be limited. The power of friction reduced the abstract absolute to the modifications it assumed in reality. The major, unrevised part of *On War* is dominated by the mutually clarifying dialectical relationship between absolute and real war.

But was it actually true that real war always modified the abstract absolute? And, secondly, was it valid to deduce from the concept of the absolute that all wars, whatever their cause and purpose, must be waged with supreme effort? In 1804 Clausewitz already distinguished between wars fought "to exterminate the opponent, to destroy his political existence," and wars waged to weaken the opponent sufficiently so that one could "impose conditions [on him] at the peace conference."[18] Yet while drawing this distinction, Clausewitz denied that limited aims justified a limitation of effort. He argued that even if no more were intended than compelling the opponent to agree to terms, his power and will to resist must be broken. For political and social as well as for military reasons the preferred way of bringing about victory was the shortest, most direct way, and that meant using all possible force. In this view, as I have suggested, experience buttressed the demands of logic. It was not difficult to believe that from the first campaign of the Revolution to the wars of 1806 and 1809 France emerged victorious because her opponents would not exert themselves to the utmost. And it was in part because contemporary reality seemed to confirm that every war was a modification of the absolute and that every war should be waged without restrictions being placed on the rational application of force that these arguments retained what might be called a formal supremacy in Clausewitz's work even as he was coming to appreciate that they were one-sided.

His essay on "Progression and Pause" indicates that by 1817 he was no longer content to impute the modification of military activity wholly to the force of friction. Because war consisted in a series of interactions between opponents, it was proper both in logic and reality that not every minute should pass at the highest pitch of effort and violence. Numerous hints in Books One through Six of *On War* point in the same direction. By the middle of the 1820's Clausewitz fully recognized that the second type of war in actuality—a war fought for limited goals—was not necessarily a modification or corruption of the theoretical principle of absolute

[18] "Plan of Operations," *Strategie aus dem Jahr 1804*, p. 51.

war. As he stated in his "Notice" and in the last revision of Chapter One of Book One, a second type of war existed that was as valid as absolute war, not only in the field but also philosophically. Limited wars might be a modification of the absolute, but need not be, if the purpose for which they were waged was also limited. Violence continued to be the essence, the regulative idea, even of limited wars fought for limited ends; but in such cases the essence did not require its fullest possible expression. The concept of absolute war had by no means become invalid, it continued to perform decisive analytic functions; but it was now joined by the concept of limited war.

The dual nature of war, as Clausewitz formulated it in the last years of his life, is expressed in two pairs of possible conflicts, each defined according to the purpose involved: War waged with the aim of completely defeating the enemy, in order (1) to destroy him as a political organism, or (2) to force him to accept any terms whatever; and wars waged to acquire territory, in order (1) to retain the conquest, or (2) to bargain with the occupied land in the peace negotiations. In the "Notice" Clausewitz stated his intention of revising the entire text of *On War* to develop these different types systematically. But he went further. As a second major theme the revision would trace the political character of war. The distinction he drew between the two themes is puzzling since the previous paragraph declares that political motives determine whether a conflict is limited or unlimited. Clausewitz did not explain his separation of the dual nature of war and the political character of war, but Eberhard Kessel has suggested a reason based on arguments and observations that recur throughout Clausewitz's writings.[19] War is influenced by objective and by subjective political factors. The objective factors comprise the specific characteristics and strengths of the state in question, and the general characteristics of the age—political, economic, technological, intellectual, and social. The subjective factors consist in the free will of the leadership, which should conform to the objective realities, but often does not. Put differently, Clausewitz separated the political consequences of general conditions and those arising from individual intelligence, emotions, genius. He may have sought analytic clarity by linking his discussion of the objective political realities mainly to the concept of the dual nature of war, and the issues of leadership mainly to the concept of the political character of war. But however Clausewitz's programmatic statement is interpreted, the reader of *On War* will find himself in accord with its author if he gives the political motives and

[19] Kessel, "Zur Genesis der modernen Kriegslehre," pp. 415-417. See also the same author's "Die doppelte Art des Krieges," *Wehrwissenschaftliche Rundschau*, 4 (July 1954), no. 7.

character of war more prominence than they receive in much of the text, and, further, if he amends the unrevised sections to the effect that limited wars need not be a modification, but that theoretically and in reality two equally valid types of war exist.

Clausewitz came to the recognition of the dual nature of war largely by way of historical study, which convinced him that limited conflicts had often occurred not because the protagonists' means precluded greater effort or their leadership had faltered, but because their intentions were too limited to justify anything more. In the face of the historical evidence, theory had to be corrected. As Clausewitz insisted throughout his life, the present could claim no final superiority over the past; and to be valid at all, theory must be universally valid. From the first, as we know, history helped guide his ideas on war. It may be thought that this was hardly unusual. Just as few theorists failed to acknowledge the role psychological factors played in war, so most proclaimed the value of military history for a sound understanding of war. But what Clausewitz had in mind differed profoundly from the unreflective chronicles and utilitarian illustrations of strategic and tactical "laws" that passed for history in the military literature. He did not regard history as a book of examples from which soldiers could learn, directly or by analogy. His individualizing way of thinking, which enabled him to single out the force of character and intelligence in the clash of mass armies, and to interpret institutions, societies, and nations as larger personalities—separate and different from each other—extended to his view of the past. History, too, was marked by constant variety, not subject to patterns—the march of progress, for instance, or man's search for God—which to Clausewitz were simply assumptions created by fashion, themselves always changing. Each period existed for itself, not as part of a grand scheme, and could be understood only on its own terms. Certain large themes recurred throughout time; they derived from the elementary human desires for safety, strength, knowledge, but they expressed themselves in constantly changing forms. Like military theory, history had no lessons or rules to offer the student, it could only broaden his understanding and strengthen his critical judgment.

In Clausewitz's pedagogic and theoretical work, history had the additional function of expanding the student's or reader's experience, or substituting for it when experience was lacking. History depicted reality and stood for reality. The role of theory, on the contrary, Clausewitz once declared, was merely to help us comprehend history—a highly telling reversal of roles that few other theorists would have agreed with or even understood.[20]

[20] *Principles of War*, p. 67.

This conception imposed certain demands on the writing and study of history that constitute a further difference between Clausewitz and most of his contemporaries. Generalized accounts of the past he considered useless; far better, he said, to study one campaign in minute detail than to acquire vague knowledge of a dozen wars. His own historical writing shows a concern for specifics that in his day was exceptional, the more so since the mass of statistical, organizational, and cartographic data is combined with very extensive speculations on intentions and implications. *On War* is full of historical references. They have often been criticized—and sometimes deleted—as unnecessary, outdated detail; in fact they are depictions of reality that alone justify the theoretical superstructure, and that should stimulate the modern reader to reflect on his own experiences, to have recourse to his knowledge of events of his own time as well as of the past.

When Clausewitz decided that the text of *On War* must be recast to take sufficient account of the dual nature and political character of war, he did not complete the manuscript but instead turned to historical research. Between 1827 and 1830, when new duties interrupted his studies, he revised only a few chapters of *On War*; most of his time was given to writing histories of the campaign of 1815 and of two limited wars, the Italian campaigns of 1796 and 1799.[21] He needed to understand how his ideas worked in reality before he could proceed to their systematic theoretical treatment. When he might have felt ready to return to *On War*, external circumstances intervened. His transfer to the army's artillery inspectorate compelled him to familiarize himself with a branch of the service of which he knew relatively little. Hardly had he taken up his new assignment when the French Revolution of 1830 brought about another change. His friend Gneisenau was recalled to active service to command the army that Prussia mobilized, and he asked for Clausewitz as his chief of staff. As the cautious foreign policy of the new French regime and the Polish uprising against Russia shifted the crisis to the east, Gneisenau's forces were deployed along the East Prussian border to protect the country against Polish incursions and the cholera, which had spread from Russia to Poland. The epidemic could not be halted, however, and in August 1831 Gneisenau became one of its victims. On 16 November, shortly after he had returned to his regular duty as Inspector General of the Silesian artillery, Clausewitz died suddenly, probably of a heart attack brought about by a relatively mild case of the cholera.

Even after his ultimate revisions Clausewitz had known that his ideas

[21] The manuscripts, which were published in volumes 4 through 6, and 8 of his collected works, add up to 1,500 pages in print.

needed further development, and passages in *On War* and in the correspondence of his last years point to important additions to theory that he never worked out in detail. Book Six, for instance, explicitly states that the dual nature of war applies to defensive as well as offensive war, but the definition in the opening chapter of the work refers only to the side that initiates the conflict. Perhaps this was one reason why in the "Notice" he singles out his treatment of the defensive as being little more than a first attempt that must be completely reworked. Again, his definitions assume that ultimately political and military goals were parallel, even though he was aware that their relationship tended to be more complex and that goals might change in the course of fighting. Despite the remarkable invention of the concept of escalation, Clausewitz never sufficiently explored the various ways in which one side influences the other, particularly in the defensive. But these are comments not criticisms. They remind us once more of the manner in which Clausewitz formed and refined his ideas. They also suggest the vitality of these ideas, which never coalesced into a finite system, but led to hypotheses that over a century-and-a-half have shown the capacity for continuing growth that Clausewitz believed to be the mark of true theory.

MICHAEL HOWARD

The Influence of Clausewitz

When Clausewitz's widow published *On War* in 1832 a year after her husband's death, it was received with a respect which may have owed more to Clausewitz's reputation as one of the great generation of Prussian military reformers, a pupil of Scharnhorst and a close colleague of Gneisenau, than to any deep or widespread study of its contents. "The streams whose crystal floods pour over nuggets of pure gold," warned one tactful reviewer, "do not flow in any flat and accessible river bed but in a narrow rocky valley surrounded by gigantic Ideas, and over its entrance the mighty Spirit stands guard like a cherub with his sword, turning back all who expect to be admitted at the usual price for a play of ideas."[1] In other words he found it heavy going, and he was clearly not the only reader to do so. The first edition of 1500 copies was still not exhausted twenty years later when the publishers decided to issue a new one. This time many of the obscurities in the original text—obscurities perhaps inevitable in the posthumous publication of so large and complex a work by a devoted but inexpert widow—were clarified by the liberal revisions and corrections of the author's brother-in-law, the Count Friedrich von Brühl. No further edition then appeared until 1867. In that year the military writer Wilhelm Rüstow devoted a chapter to Clausewitz in his survey of *The Art of War in the Nineteenth Century* but said of him that he was "well-known but little read," an aphorism which has lost none of its accuracy with the passage of time. But even those who had not read him knew that his teachings embodied that freedom of thinking, that emphasis on the creative action of the individual and disdain for formalism which had lain at the root of Scharnhorst's reforms of the Prussian Army and which Scharnhorst's successor as War Minister, Hermann von Boyen, tried to keep alive during the sterile and reactionary period of the 1840's. The military conservatives preferred the teaching of General von Willisen, whose *Theorie des grossen Krieges* (1840) laid down positive rules and principles with Jominian dogmatism. Their domi-

[1] *Preussische Militair-Literatur Zeitung*, 1832. Quoted by Werner Hahlweg in his introduction to the sixteenth edition of *Vom Kriege* (Bonn, 1952). Hereafter referred to as *Hahlweg*.

nance in positions of influence at this time may have played some part in preventing Clausewitz's ideas from being more widely known.[2]

But the basic cause for Clausewitz's continuing obscurity must be sought in the text itself. So also must the wide diversity of interpretations to which it was to be subjected. Clausewitz himself had warned that if he did not live to complete his task he would leave behind "a shapeless mass of ideas" which would be endlessly misinterpreted and "made the target of much half-baked criticism." It was also a mass from which, because Clausewitz did not live to mold it into a finished and coherent shape, later writers were to quarry ideas and phrases to suit the needs of their own theories and their own times. Clausewitz, as it turned out, had less cause to fear his critics than to be wary of many of his professed admirers.

In the introductory note he wrote in 1827 Clausewitz made the position clear. He had completed six books. The seventh and eighth were still only rough drafts. When these were finished he would go through the entire work again and bring out the two great themes which would receive their final clarification in the last book. The first theme was the "dual nature" of war, as an instrument which could be used either to overthrow the enemy or to exact from him a limited concession. The second was the point "that must be made absolutely clear, namely that war is simply the continuation of policy by other means." This, he warned, "if firmly kept in mind throughout . . . will greatly facilitate the study of the subject and the whole will be easier to analyze" (see below, p. 69). But he had to rely on his readeıs to keep it in mind. His own revision went no further than the first chapter of Book One where he introduces us to the three elements in his theory: the intrinsic violence of war; the dominant role of rational policy in shaping and controlling it; and the all-important dimension of chance.

The above-quoted note makes it clear that had Clausewitz lived to finish the work it would have been the second of the above three elements which would have received the most emphasis: the dominance which the political end should exert over the military means. As it is, however, Clausewitz has very little to say about this even in Book Three on strategy. Strategy he defines baldly as "the use of the engagement for the purposes of the war" (see below, p. 177). It is here that we find the doctrine which was to be seized on so avidly by later writers: "The best strategy is to be very strong: first everywhere, and then at the decisive point" (see below, p. 204). The two types of war and the possibility that

[2] *Hahlweg*, pp. 12–13. Rüstow's work was published as *Die Feldherrnkunst des neunzehnten Jahrhunderts* (Zürich, 1867). See also Eugene Carrias, *La pensée militaire allemande* (Paris, 1948), pp. 224–228.

each might need to be conducted according to different principles receives here only the most glancing of references. In general the strategy dealt with in this book is simply the strategy, as Clausewitz saw it, of Napoleon; of war as "absolute" as the dictates of a powerful political motivation could make it.

The same limitation applies even more strongly to Book Four, "The Engagement." Here we find not a word about the two types of war or the supremacy of the political object. The centerpiece of this book is the major battle (*Hauptschlacht*) and its consequences; what Clausewitz termed "the true center of gravity of the war." Yet almost by definition "limited wars" are conflicts in which the issue is not brought to such a massive decision. It might be pleaded that this book emphasizes the central paradox of all war, the dialectic between the forces of violence and the forces of reason, and that the political requirement for rational control can no more mitigate the essentially violent nature of the means than the precise manipulation of an oxyacetylene flame diminishes its heat. Clausewitz was indeed at pains to reiterate this point in the revised first chapter of Book One, which must be taken as his considered view on the subject: "Kind-hearted people might think that there was some ingenious way to disarm or defeat an enemy without too much bloodshed and might imagine this is the true goal of the art of war. Pleasant as it sounds, this is a fallacy that must be exposed" (see below, p. 75). Thus, there is no reason to suppose that Clausewitz would in his revision have abandoned any of the beliefs expressed in Book Four, beliefs set out in phrases of dreadful vividness, the outcome of his own terrible experiences in 1806 and from 1812 to 1815. But perhaps he might have considered more deeply how this inexorably destructive flame could be moderated and controlled to serve the political ends he regarded as paramount.

As it was Clausewitz did not survive to make these revisions. In the text which he left behind him we find that, of the three elements in his theory, the political element, that by which he came to set the most store, is dealt with only in the last book and the first chapter of the first. It is the other two elements, the intrinsic violence of war and the omnipresence of chance, together with the demands which both make upon moral qualities, which are stressed throughout the rest of the work—except indeed in the long, rich, and complex Book Six on defense, which badly needed revision if its lessons were to be clearly brought out.

These were certainly the aspects of Clausewitz's work which most forcibly impressed posterity, not least the great Helmuth von Moltke, who became Chief of the Prussian General Staff in 1857, and who played the leading part in bringing Clausewitz's work to the attention of his countrymen. Moltke was to cite *On War* together with Homer and the Bible as one of the truly seminal works which had molded his own thinking.[3] He had been at the War Academy when Clausewitz was Superintendent but since Clausewitz had had virtually no contact with the students, there could have been no question of direct influence. Moreover, as Moltke's most recent biographer has pointed out,[4] many of the ideas we now think of as peculiarly Clausewitzian and to which Moltke was so signally to give effect in his campaigns—the annihilation of the enemy's main force, the concentration of effort at the decisive point, the overriding importance of moral forces, the need for self-reliance in the commander and flexibility in tactical method—were commonplaces among young Prussian officers who had shared the Napoleonic experience. As with so many thinkers, many of the ideas which Clausewitz codified and transmitted to posterity may have been shared generally, if unconsciously, among his contemporaries, and so intelligent and sensitive a pupil as Moltke would have quickly picked them up from his environment. Moltke's thinking should perhaps be seen rather as reinforcing and demonstrating Clausewitz's ideas than as simply deriving from them.

But it was Moltke's achievement in the German wars of unification in 1866 and 1870 which drew attention to the abiding validity of Clausewitz's work. In Moltke's own writings we find over and over again passages which echo Clausewitz. "Victory through the application of armed force is the decisive factor in war . . . it is not the occupation of a slice of territory or the capture of a fortress but the destruction of the enemy forces which will decide the outcome of the war. This destruction thus constitutes the principal object of operations." And again "Strategy is a system of expedients. It is more than a science, it is a science applied to everyday life . . . the art of acting under the pressure of most arduous circumstances."[5] And perhaps the most influential reflection of all: "In war it is often less important what one does than how one does it. Strong determination and perseverance in carrying through a simple idea are the surest routes to one's objective."[6] This was the kind of thing that appealed to practical soldiers.

But what one does not find in Moltke, or indeed in any of his disciples or successors, is any reflection of Clausewitz's insistence on the need for military means to be subordinate to political ends. He showed no signs, either in his writings or his work as Chief of Staff, of understanding

[3] Eberhard Kessel, *Moltke* (Stuttgart, 1957), p. 108.
[4] *Ibid.*
[5] See the passages quoted in Carrias, *La pensée militaire allemande*, pp. 238–241.
[6] Kessel, *Moltke*, p. 511.

Clausewitz's requirement for war to be versatile if it were to serve the political object. For Moltke war was not so much an instrument of policy as the inevitable fate of mankind, to be stoically endured and efficiently conducted. Certainly, he accepted the supremacy of the political authority so long as it was the King himself, the War Lord whose uniform he wore and to whom he had sworn allegiance. But it did not extend to the King's political advisers, who had no business whatever, in Moltke's view, to meddle in the matters which the King had delegated to him. When war broke out, he considered, "at the moment of mobilization the political adviser should fall silent, and should take the lead again only when the Strategist has informed the King, after the complete defeat of the enemy, that his task has been fulfilled."[7]

All this of course was totally at variance with Clausewitz's teaching about the relationship between the military and the political authorities as set out in Book Eight of *On War*. This assumed the continuing direction of the campaign by the cabinet as a whole, suggesting indeed that the military commander should be made a member of that body so that it could take part in making crucial strategic decisions.[8] But it was Moltke's view of the matter, not that of Clausewitz, which became dominant in Imperial Germany toward the end of the nineteenth century, even though it was during these years that Clausewitz was being most widely acclaimed. A fourth edition of *On War* appeared in 1880. The esteem in which Clausewitz was now held in Germany can be judged by the words with which, in 1883, Colmar von der Goltz opened his famous work *Das Volk im Waffen*, which ran into many editions and was translated into English as *The Nation in Arms*: "A military writer who, after Clausewitz, writes upon war, runs the risk of being likened to the poet who, after Goethe, attempts a Faust, or, after Shakespeare, a Hamlet. Everything of any importance to be said about the nature of war can be found stereotyped in the works left behind by that greatest of military thinkers."[9]

In 1905 a fifth edition was published with a commendatory introduction by the then Chief of the General Staff, Count Alfred von Schlieffen. Three more were to follow before the outbreak of the Great War, and no less than five during the war itself.

During this period Clausewitz's teaching about the relationship between war and policy was not ignored. Indeed it caused a certain bewilderment among his admirers. General von Caemmerer, the most distinguished Clausewitz scholar of his generation, drew attention to the

[7] Rudolph Stadelmann, *Moltke und der Staat* (Krefeld, 1950), p. 206.

[8] For the significance of this passage see below, p. 608.

[9] Colmar von der Goltz, *The Nation in Arms* (London, 1913), p. 1.

divergence on this subject between Clausewitz and Moltke, and expressed himself convinced "of the correctness in every respect of Clausewitz's views."[10] But von der Goltz expressed the majority view in refusing to accept this. It was not that he ignored the political element in the Clausewitzian trinity. He considered it to be no longer relevant. The situation, he claimed, had changed since Clausewitz had written. Now "if two European Powers of the first order collide, their whole organized forces will at once be set in motion to decide the quarrel. All political considerations bred of the half-heartedness of wars of alliance fall to the ground." War was certainly an instrument of policy in that war arose from a political situation; but, he continued: "War will on that account be in no way lowered in importance nor restricted in its independence, if only the commander in chief and the leading statesmen are agreed that, under all circumstances, war serves the end of politics *best by a complete defeat of the enemy*. By attention to this maxim the widest scope is allowed in the employment of the fighting forces."[11]

Thus, ingeniously, von der Goltz made the best of both worlds, reconciling the supremacy of policy with the independence of the commander. Given the increasing difficulties of military planning as armies grew ever larger, less flexible, at the mercy of railway timetables, it was probably the best that could be done to adapt Clausewitz's teaching to the exigencies of military policy in the Wilhelmine Era.

For it was not only in Germany that Clausewitz's views on this matter were seen, at the beginning of the twentieth century, to be increasingly anachronistic. In France Colonel Foch wrote in 1903 in his *Principles of War*, "You must henceforth go to the very limits to find the aim of war. Since the vanquished party now never yields before it has been deprived of all means of reply, what you have to aim at is the destruction of those very means of reply."[12] Even a very much more balanced French authority, Colonel Colin, whose *The Transformation of War* still ranks as the outstanding summary of the military situation in Europe on the eve of the Great War, wrote in exactly the same terms:

> Without speaking of the passions that would animate most of the belligerents, the material conditions of modern war no longer admit of the avoidance of the radical decision by battle. The two armies occupying the whole area of the theatre of operations march towards each other, and there is no issue but victory. . . . Therefore the indications which a government should give to a general on the political

[10] Rudolph von Caemmerer, *The Development of Strategical Sciences during the Nineteenth Century* (London, 1905), p. 86.

[11] Von der Goltz, *Nation in Arms*, p. 143.

[12] Ferdinand Foch, *The Principles of War* (London, 1918), p. 37.

> object is reduced to a very small affair. Once the war is decided on, it is absolutely necessary that a general should be left free to conduct it at his own discretion, subject to seeing himself relieved of his command if he uses his discretion with but little energy or competence.[13]

Such was the philosophy which was to animate, not only the Germans, but all the belligerent powers at the outset of the Great War. But it was not the strategic thinkers, nor even the requirements of military technology, that molded the experience of 1914 and the terrible years that followed. Clausewitz himself had written: "The aims a belligerent adopts and the resources he employs must be governed by the particular characteristics of his own position; but they will also conform to the spirit of the age and its general character."[14] It is in the light of this aphorism that the experience of the First World War has to be understood.

If there were understandable reasons, technical, political, and psychological, why Clausewitz's teaching about the subordination of the military means to the political end were no longer seen to be relevant in 1914, the same can hardly be said about his views on defense as the stronger form of war; views which had become more significant with practically every new development in military technology since 1870. Moltke himself had endorsed them, as had Caemmerer,[15] but very few other German writers did, and even fewer French. Von der Goltz was typical of the former.[16] "To make war means to attack," he wrote, and "happy the soldier to whom fate assigns the part of the assailant." In concluding an account of an imaginary battle he summed up: "In describing a battle we have been led, in spite of ourselves, to depict an offensive battle. What German soldier could do otherwise?" In fact, one German soldier *had* done otherwise, and that was Clausewitz himself. His picture of a battle in Chapter Two of Book Four was that of a grueling *bataille d'usure*; not all that different, on a very much smaller scale, from the battles which developed on the Western Front between 1915 and 1917. But, suggested Goltz, Clausewitz might have "changed his views as to the superiority of the defense over the attack if he had had the opportunity of carefully revising his original text"; an argument frequently used by admirers of Clausewitz—the present writer not excepted—who find aspects of his work not wholly to their liking.

If Clausewitz's teaching about the defensive was ignored and his views on the relations between war and policy were considered anachronistic, why then was he so admired in the German army? Primarily for his

[13] J. Colin, *The Transformation of War* (London, 1912), p. 343.

[14] See below, p. 594.

[15] Caemmerer, *Development of Strategical Sciences*, p. 95.

[16] Von der Goltz, *Nation in Arms*, pp. 262–263, 345.

achievement, as Hans Rothfels has stated,[17] in turning strategic thought away from a mechanistic concern with geometrical relations "to man and man's actions in the midst of all the uncertainties which are the proper element of war." Caemmerer singled out the chapters on "The Genius for War," "Friction in War," and "Tension and Rest in War," together with the first and third books generally as being the most influential in educating the Prussian army. "They freed us from all that artificiality which gave itself such airs in the theory of war and have shown us what, after all, is the real point."[18] Wilhelm Blume, in his widely read *Strategie* (first published in 1884) echoes Clausewitz when he wrote that "Every theory . . . which seeks to bind the reciprocal action of living forces by dead theories, must be rejected, for it will in practice have disastrous results."[19] And it was to this aspect of Clausewitz's teaching that von Schlieffen drew attention in his introduction to the fifth edition of *On War* in 1905. Clausewitz, he wrote, had taught "that every case in war must be considered and thought through in its own right (*nach seiner Eigenart*)." It was, he said, "the awakening of this recognition as a reality for which the Prussian and now the entire German army owes the great thinker everlasting thanks."[20]

Second, Clausewitz was quoted for his emphasis on the preeminence of moral forces in war. Clausewitzian phrases about the will-power of the commander, his need for resolution, self-confidence and *coup d'oeil* resound through German military writings; though perhaps it was the influence of Moltke rather than of Clausewitz which stressed the need for these qualities at every level of command and not simply in the commander-in-chief. The emphasis on simplicity and directness rather than on ingenious maneuver, on resolution rather than on subtlety, on bold initiative rather than on elaborate calculation was to be found in every German textbook between 1870 and 1914;[21] all the more so since the conditions of twentieth-century warfare clearly made these qualities even more vital to military success than they had been in the Napoleonic era. In the enormous armies of 1900, their communications dependent at best on fragile field-telephones, their size and complexity rendering elaborate maneuver out of the question, commanders-in-chief could give only the broadest of directives to their subordinates and rely on their intelligence

[17] Hans Rothfels, "Clausewitz," in E. M. Earle, ed., *Makers of Modern Strategy* (Princeton, 1943), p. 100. "In a way," went on Rothfels, "this was a Copernican revolution." This is an unhappy analogy. Copernicus dethroned man as the center of the universe: Clausewitz and his contemporaries restored him.

[18] Caemmerer, *Development of Strategical Sciences*, p. 82.

[19] Quoted in Carrias, *La pensée militaire allemande*, p. 263.

[20] Clausewitz, *Vom Kriege*, 5th ed. (Berlin, 1905), pp. iii–vi.

[21] See the examples in Carrias, *La pensée militaire allemande*, pp. 268 ff.

and initiative to carry them out in detail. Junior officers were likely to find themselves isolated on vast battlefields in a strange, sometimes barely endurable environment, with no recourse save their inner strength to keep them going and their common sense to tell them what to do. For these circumstances Clausewitz's teaching was marvelously appropriate, and his disciples did very well to stress it.

And what was the object of all this resolution, all this common sense? Again Clausewitz seemed to provide a clear and simple answer; the *annihilation* of the enemy; the aspect of his teaching which Schlieffen stressed in his introduction to the fifth edition of *On War*. As to whether this was always and necessarily so, whether Clausewitz's doctrine of the two kinds of war did not imply a valid alternative objective in the *attrition* of the enemy, a scholarly controversy raged in military and historical periodicals for thirty years, a controversy centering on the military historian Hans Delbrück, who as early as 1881 put forward the thesis that if Clausewitz had lived to revise his work he would have devoted far more appreciation and attention to this strategy of attrition (*Ermattungsstrategie*) which had distinguished eighteenth-century warfare and the campaigns of Frederick the Great, as against the strategy of annihilation (*Vernichtungsstrategie*) characteristic of Napoleon.[22] The experience of the First World War was most gloomily to justify Delbrück's view that the former strategy was no less valid than the latter; but since no echo of that view found its way into German military textbooks before 1914, the question remained academic. The object of strategy, German soldiers were taught, was the destruction of the enemy armed forces by battle; and the greater the battle, the more effectively could that object be achieved.

Of course, all this was to be found in Clausewitz. Not only was the concept of battle central to his strategic thought, but he wrote about it with a vigor and a vivacity which make those chapters leap from the pages like a splash of scarlet against a background of scholarly gray.[23] The famous phrases about the inevitability of bloody slaughter in a successful battle and of bloody battles in a successful strategy were taken up and quoted with grisly relish; not so much in serious military text-

[22] Delbrück first set out his arguments in the *Zeitschrift für preussische Geschichte und Landeskunde*, vols. 11, 12 (1881), and repeated them in his *Geschichte der Kriegskunst im Rahmen der politischen Geschichte* (Berlin, 1920), 4: 439–444. The controversy was concluded only in 1920 with an exchange with Otto Hintze in *Forschungen zur brandenburgischen und preussischen Geschichte*, vol. 33. See also the essay on Delbrück by Gordon Craig in Earle, *Makers of Modern Strategy*.

[23] For some interesting speculations as to the psychological reasons for this see Bernard Brodie, "Clausewitz: a Passion for War," *World Politics* 25, 2 (January, 1973).

books as in the popular militaristic writings of von der Goltz, Bernhardi, and their innumerable imitators in the Wilhelmine Reich. The name of Clausewitz became associated in the popular mind with battle and blood. As for the military specialists, the concept of annihilation as the object of strategy was no less dominant—if only because they could not see how else, under the conditions of early twentieth-century warfare wars, especially wars conducted by Germany on two fronts, could conceivably be won. As Schlieffen himself wrote: "It is not possible to conduct a strategy of attrition when the support of millions of combattants runs into milliards of marks."[24] If Germany could not destroy one or other of her principal adversaries as rapidly and as totally as she had destroyed French military power in 1870, she was likely eventually to be squeezed to death between them. To that extent the strategy of annihilation appeared inevitable. What Schlieffen and his successors underrated were the limitations of a strategic plan which, in concentrating on the destruction of a major land power, provoked the antagonism of a major sea power. But then Clausewitz himself never considered the significance of sea power in the Napoleonic wars. With all its depth and genius, his strategic thinking was intensely parochial, conceived as it was within the framework of a land-locked Prussia. By his own definition strategy was concerned with the movement of armies. None of his disciples considered how his teaching might be adapted to the requirements of a German Empire with aspirations to world power.

So far we have considered the influence of Clausewitz on the German army alone. But the German army was, at the beginning of the twentieth century, the model for all others, and in imitating its training methods and tactical doctrines foreign armies absorbed the doctrines of Clausewitz as much unconsciously as consciously. But the French did so quite consciously. A French translation of *On War* had been published as early as 1849 and four years later a commentary on it was published by a professor at Saint-Cyr.[25] Neither work appears to have made much impression on a French army which assumed inborn genius in its commander and required of its junior officers only obedience to orders, a good seat on a horse, and unflinching bravery under fire. The events of 1870 proved these to be insufficient equipment for the responsibilities of modern warfare, but as a result most French military writers set about studying, not why the Prussian army had won, but what the great Napoleon would have done had he been presented with the same problem. Clausewitz was considered, insofar as he was considered at all, as one interpreter of the true Napoleonic doctrine among many, and one who

[24] Graf von Schlieffen, *Gesammelte Schriften*, 2 vols. (Berlin, 1913), 1: 17.
[25] Eugene Carrias, *La pensée militaire française* (Paris, 1960), p. 252.

had as often as not distorted the pure beams of that sacred truth.[26] But one French writer had begun independently to explore the relationship of moral to physical forces in war—Colonel Ardent du Picq, who was killed at Metz in 1870 and whose *Etudes sur le Combat*, when it was published ten years later, won immediate acclaim. Attention was thus already becoming focused on the question of *morale* when an instructor at the Ecole de Guerre, Lucien Cardot, was inspired by reading von der Goltz to give a course of lectures on Clausewitz in 1884 which was to influence an entire generation of French officers; the generation which was to mold the thinking of the French army at the turn of the century and to lead it during the Great War.[27]

These men seized upon the Clausewitzian teaching about morale, about the battle, and about the offensive spirit, and preached it with an enthusiasm which outmatched even the Germans. Their ardor was fanned by a national mystique about the *furia francese* and by the contemporary popular philosophy of Henry Bergson, with all its emphasis on *l'élan vital*. It was a mood rather than a doctrine, and one which found its greatest exponent in one of the officers who had attended Cardot's lectures: Ferdinand Foch. In fact, by the end of the century the French army had become as totally imbued with these oversimplified neo-Clausewitzian ideas as had its German adversaries. Witness the Field Service Regulations of 1895: "Combat may be offensive or defensive, but it always has for its end to break by force the will of the enemy and to impose on him our own. Only the offensive permits the obtaining of decisive results. The passive defense is doomed to certain defeat; it is to be rejected absolutely."[28]

And this teaching seemed justified by the lessons of the next major conflict between great powers, the Russo-Japanese War of 1904. In their conduct of operations, the Japanese army displayed all the qualities so lauded by Clausewitz: the offensive spirit, the simple and direct strategy, the initiative at every level, as against the passivity of their Russian adversaries. Whether the Japanese army would have performed in the same manner if they had not been trained by an enthusiastic disciple of Clausewitz, General von Meckel, is a reasonable matter for speculation; but certainly *On War* had been translated into Japanese, and the Japanese commanders politely acknowledged their debt to it.[29]

[26] *Ibid.* See especially the work of General Bonnal and Colonel Camon for grotesque examples of this kind of chauvinism.

[27] Dallas D. Irvine, "The French Discovery of Clausewitz and Napoleon," *Journal of the American Military Institute*, 4 (1940): 143.

[28] Quoted by Irvine, *ibid.*

[29] In 1904 the publishers of *Vom Kriege*, Dümmler Verlag, sent advance copies of the forthcoming fifth edition to the Japanese commander Count Kuroki. He replied

The debt was noted by the military observers of that other island Empire, Great Britain, who observed the performance of their new ally in the Far East with particular interest.[30] After the humiliating performance of the British army against the Boer Republics in South Africa, 1899-1902, a revival of military thinking was under way in Britain—a revival hastened by a growing perception of the possibility that in the not far distant future Britain might become involved in land warfare against the German army. An English translation of *On War* by Colonel J. J. Graham had appeared in 1874, but it had long been out of print. The general contemptuous ignorance in which Clausewitz was held by British soldiers was probably well summed up by the most admired of all the teachers at the Staff College, Colonel G.F.R. Henderson, who in a lecture at the Royal United Services Institution in 1894 on "Lessons of the Past for the Present" had mentioned Clausewitz only to say, sarcastically and inaccurately: "Clausewitz, the most profound of all writers on war, says that everyone understands what moral force is and how it is applied. But Clausewitz was a genius, and geniuses and clever men have a distressing habit of assuming that everyone understands what is perfectly clear to themselves."[31]

This complacent anti-intellectualism, which has long been the predominant characteristic of a British army which takes a perverse delight in learning all its lessons the hard way, was temporarily shaken in the ten years before 1914.[32] Clausewitz began to receive almost as much attention at Camberley as he did in Continental Staff Colleges. A new, truncated translation was published by T. M. Maguire in 1909, but simultaneously Colonel Graham's translation was reprinted, with an introduction by Colonel F. N. Maude drawing attention to its relevance for an army likely to have to encounter the Germans, in the three red volumes through which so many British and American readers have come to know *On War* for the past seventy years. But, more interestingly, Clausewitz was studied also by Britain's leading naval historian, Sir Julian Corbett, who in his *Principles of Maritime Strategy* (1911) based on the lectures he had given at the Royal Naval College Greenwich, both indicated Clausewitz's relevance to the problems of naval warfare, and added a new and significant dimension to his concept of

that the work had been already translated into Japanese and had been influential in the conduct of the campaign. *Hahlweg*, p. 52.

[30] See especially the articles by *The Times* military correspondent, Colonel Repington, reprinted in *The Times History of the War in the Far East* (London, 1905), pp. 548-553.

[31] Reprinted in G.F.R. Henderson, *The Science of War* (London, 1905), p. 173.

[32] See John Gooch, *The Plans of War: The General Staff and British Military Strategy 1900-1916* (London, 1974), *passim*.

Limited War. Corbett was one of the very few thinkers who not only interpreted Clausewitz for his own generation but constructively developed his ideas.[33]

Hardly had Clausewitz become known in the Anglo-Saxon world than a powerful reaction set in against him. After 1914 British readers saw him, especially as interpreted by Bernhardi and von der Goltz, as a prophet of that bloodthirsty "Prussianism" against which they had taken up arms. Liberals on both sides of the Atlantic seized on garbled versions of his teaching about the relationship of war to policy as evidence of naked and unashamed militarism. Clausewitz's popularity in Germany during the Great War was reason enough for him to be unpopular with Germany's enemies. And this unpopularity was to be inherited by a postwar generation which saw in the holocaust of the First World War the direct result of Clausewitz's teaching. Nor were they entirely wrong. Clausewitz can hardly be blamed for those distorted notions of the offensive which sent nearly a million young Frenchmen to their deaths in 1914 and 1915. But in the grinding battles of attrition of 1916 and 1917 and the arguments used to justify them, one can clearly trace a Clausewitzian philosophy both of tactics and of strategy. The skepticism for strategic maneuver; the accumulation of maximum force at the decisive point in order to defeat the enemy main force in battle; the conduct of operations so as to inflict the greatest possible number of losses on the enemy and compel him to use up his reserves at a greater rate than one was expending one's own; the dogged refusal to be put off by heavy casualties; all these familiar Clausewitzian principles were deployed to justify the continuation of attacks on the Western Front by British commanders who almost self-consciously embodied those qualities of calm, determination and perseverance which Clausewitz had praised so highly.

It is not surprising therefore that the leading British postwar critic of the Western Front strategy, Captain B. H. Liddell Hart, should have extended his censures to include Clausewitz himself as well as his disciples. Liddell Hart in his many writings admitted that those disciples had often misunderstood Clausewitz because of the "obscurity" of his writings, but his own comments often revealed a comparable degree of misunderstanding. Clausewitz, he wrote, "had proclaimed the sovereign virtues of the will to conquer, the unique value of the offensive carried out with unlimited violence by a nation in arms *and the power of military action to override everything else*"[34] (my italics). In view of Clausewitz's explicit and repeated insistence on the need to subordinate military

[33] On Corbett see the essay in Donald M. Schurman, *The Education of a Navy: the Development of British Naval Strategic Thought* 1867–1914 (London, 1965), pp. 147–184.

[34] B. H. Liddell Hart, *Foch: Man of Orleans* (London, 1931), p. 22.

means to political ends this final assertion is puzzling. Even more curious was Liddell Hart's anathema of Clausewitz in his book *The Ghost of Napoleon* published at the height of the reaction against the Western Front strategy in 1933: "He was the source of the doctrine of 'absolute war,' the fight to a finish theory which, beginning with the argument that 'war is only a continuation of state policy by other means,' ended by making policy the slave of strategy . . . Clausewitz looked only to the end of war, not beyond war to the subsequent peace."[35]

Clausewitz did, Liddell Hart agreed, recognize "a modification in the reality" and taught that "the political object should determine the effort made"; but "unfortunately his qualifications came on later pages and were conveyed in a philosophical language that befogged the plain soldier, essentially concrete minded."[36] Of course the "qualifications" are very emphatically set out, as an intrinsic part of Clausewitz's argument, not "on later pages," but the very first chapter; but of this chapter Liddell Hart was to write "Not one reader in a hundred was likely to follow the subtlety of his logic or to preserve a true balance among such philosophical jugglery."[37] The reader will have to judge the correctness of this verdict for himself.

Finally Liddell Hart, having castigated the Clausewitzian concept of "absolute war," proposed in its place a "strategy of limited aim." A government, he suggested,

> may desire to wait, or even to limit its military effort permanently, while economic or naval action decides the issue. It may calculate that the overthrow of the enemy's military power is a task definitely beyond its capacity, or not worth the effort—and that the object of its war policy can be assured by seizing territory which it can either retain or use as bargaining counters when peace is negotiated. . . . There is ground for inquiry whether this "conservative" military policy does not deserve to be accorded a place in the theory of the conduct of war.[38]

One does not need to read very far in Clausewitz—no further, indeed, than the first paragraph of the "Notice" of 10 July 1827—before finding the most explicit and lucid exposition of precisely this doctrine of "limited aim."

In many respects Liddell Hart's critique of Clausewitz was quite justi-

[35] B. H. Liddell Hart, *The Ghost of Napoleon* (London, 1933), p. 121.

[36] *Ibid.*, p. 123.

[37] B. H. Liddell Hart, *Strategy: The Indirect Approach*, 3d ed. (London, 1954), p. 355. This work contains verbatim many other passages critical of Clausewitz which first appeared in *The Ghost of Napoleon*.

[38] *Strategy*, p. 334.

fied: the reiterated emphasis on battle, the small concern for maneuver, a definition of "strategy" which ignored all save purely military means, the ignoring of naval and economic factors, the tortuous and self-contradictory quality of much of the writing; all these were shortcomings which, however understandable in their context, needed and still need to be pointed out. But the final picture Liddell Hart painted of Clausewitz's teaching was distorted, inaccurate, and unfair. And since Liddell Hart was in his time probably the most widely read military writer in the English-speaking world, this picture was by the Second World War very generally accepted as true.

In Germany however Clausewitz lost none of his popularity. During the 1920's General von Seeckt continued to hammer home to the Reichswehr the lessons Schlieffen had derived from him: the importance of the initiative, of moral forces, of flexibility and self-reliance, of the study of historical examples as guides to action.[39] A fourteenth edition of *On War* was published in 1933 to celebrate the one hundredth anniversary of Schlieffen's birth, and the commander-in-chief of the new Wehrmacht, General von Blomberg, declared: "In spite of the fundamental transformation of all technical modalities, Clausewitz's book *On War* remains for all time the basis for any rational development in the Art of War."[40] Further popular editions followed during the 1930's as well as another complete edition in 1937, and articles by such scholars as Karl Linnebach, Hans Rothfels, Herbert Rosinski, Walther Schering, and Eberhart Kessel frequently appeared in historical and military periodicals.[41] Since he was seen, incorrectly, as a pioneer of German nationalism as well as the greatest of all writers on war, Clausewitz enjoyed high esteem in the Nazi pantheon.

The same could not be said of all his students, and two of the most eminent of them, Hans Rothfels and Herbert Rosinski, had to seek refuge from racial persecution in the United States. Apart from some perceptive comments in his book on *The German Army* (1940) most of Rosinski's studies on Clausewitz remained unpublished, but his lectures at universities and service colleges introduced his work to a wide range of audiences. Hans Rothfels published a seminal article in E. M. Earle's collection of studies, *Makers of Modern Strategy* (1943) which revealed

[39] See the collection of von Seeckt's speeches collected in *Thoughts of a Soldier* (London, 1930).

[40] *Wissen und Wehr*, 1933, p. 477.

[41] See particularly *Historische Zeitschrift* for 1935 and 1943, and *Wissen und Wehr* for 1931, 1933, and 1936. In the *Historische Zeitschrift*, 167 (1943): 41, the historian Gerhard Ritter published an acute analysis of Clausewitz's doctrine of the political sense of war and the difficulty of applying it in the conditions of the twentieth century. It was reprinted in his *Staatskunst und Kriegshandwerk* (Munich, 1954), 1: 67–96.

to a new generation the Clausewitz who had for so long been studied and admired by German scholars as well as German soldiers, and went a long way to dissipate the false image which had dominated the English-speaking world since 1914. A new translation by O. J. Mattijs Jolles was published in the United States in 1943, clarifying many of the obscurities in the Graham version of *On War*.

There is little to suggest that Clausewitz was intensively studied in American military schools between the wars. As Bernard Brodie has made clear, Jomini's influence there had remained almost unchallenged since the days of the Civil War. But those aspects of Clausewitz's thought which had dominated European strategic thinking in 1914 had certainly crossed the Atlantic by the end of the First World War. U.S. Army Field Service Regulations for 1923 stated: "The ultimate objective of all military operations is the destruction of the enemy's armed forces by battle. Decisive defeat in battle breaks the enemy's will to war and forces him to sue for peace."[42]

This was certainly the attitude that led General Marshall, when plans had to be made for the defeat of Germany in the Second World War, to hold so staunchly to his plan for concentrating the bulk of American forces at the decisive point, Northwest Europe, the only place where the strength of the German Wehrmacht could be broken in battle.[43]

But it was the next great conflict in which the United States took part, that in Korea in 1950–1953, that led to a serious renewal of Clausewitzian studies on both sides of the Atlantic. That war compelled the American government to grapple with two of the problems which Clausewitz had studied most deeply: the relationship between the civil and the military powers in the conduct of a war, and the conduct of a war for a limited aim—that is, one not aimed at the total overthrow of the enemy. The commander of the United States forces in the Far East, General Douglas MacArthur, was a firm believer in the creed which had animated European military thinkers in 1914, and he set it out in terms reminiscent of Moltke himself:

> A theater commander (he informed the Senate after his dismissal) is not merely limited to the handling of his troops; he commands the whole area, politically, economically and militarily. At that stage of the game when politics fails and the military takes over, you must trust

[42] Quoted by Maurice Matloff, "The American Approach to War 1919–1945," in Michael Howard, ed., *The Theory and Practice of War* (London, 1965), p. 223.

[43] See especially Maurice Matloff and Edwin M. Snell, *Strategic Planning for Coalition Warfare 1941–1942* (Washington, D.C., 1953), pp. 174–197, and Forrest C. Pogue, *George C. Marshall: Ordeal and Hope 1939–42* (New York, 1966), pp. 303–320.

> the military. . . . I do unquestionably state that when men become locked in battle, that there should be no artifice under the name of politics which should handicap your own men, decrease their chances for winning, and increase their losses.[44]

This statement and the attitude which it expressed caused deep concern both in government circles in the United States and among the growing community of strategic thinkers on both sides of the Atlantic. The development of atomic weapons by both sides already made it likely that the kind of military solution advocated by General MacArthur might involve a quite unacceptable degree of reciprocal destruction, which the advent of thermonuclear weapons would soon raise to an inconceivable order of magnitude. It became almost impossible to visualize any political objective for which the use of such weapons would be appropriate. It was hardly necessary to read Clausewitz in order to rediscover the concept of "limited war." Like Molière's M. Jourdain, who had never realized that what he had been talking all the time was prose, the American forces and their allies in Korea had been fighting a Clausewitzian "limited war" without knowing it.

Few of the numerous writers who during the 1950's wrote about "limited war" needed to acknowledge any debt to Clausewitz.[45] They thought the idea through for themselves. But some, notably Robert Osgood and Bernard Brodie, found in Clausewitz a pattern of thought whose contribution to their own they generously acknowledged, and through the influence of these and other writers Clausewitz began to be studied again, and by a wider readership than ever before.[46] This time he was not only read by soldiers concerned with the conduct of war but by students of international politics concerned with the preservation of peace. If the nineteenth century had laid stress on Clausewitz's teaching about moral forces, readers in the mid-twentieth were to concentrate—perhaps almost equally to excess—on his emphasis on the primacy of the political aim.

This has certainly been the case with the Marxist disciples of Clausewitz. When Friedrich Engels first came across *On War* it was not, in fact, this aspect of Clausewitz's writing that struck him most forcefully. It was the analogy between war and commerce to which he drew Marx's attention;

[44] Quoted in Walter Millis, *Arms and the State* (New York, 1958), p. 325.

[45] Writings on "limited war" in this period are conveniently cataloged in Morton H. Halperin, *Limited War in the Nuclear Age* (New York, 1963).

[46] Robert E. Osgood, *Limited War: the Challenge to American Strategy* (Chicago, 1957); Bernard F. Brodie, *Strategy in the Missile Age* (Princeton, N.J., 1959).

"A remarkable way of philosophizing about the question," he commented, "but very good." Marx expressed equal approval: "the rascal has a 'common sense' bordering on wit" he replied.[47] However, it was on the conçept of "War as the Continuation of Politics by other (i.e. violent) means" that Lenin focussed in the study *Socialism and War* (1915). "This dictum," he wrote, "was uttered by one of the profoundest writers on the problems of war. Marxists have always rightly regarded this thesis as the theoretical basis of views on the significance of any war.[48] Every war, he explained, in this and succeeding pamphlets, was indissolubly linked with the political order whence it derived, and with the policy pursued by the ruling class. Its character was "not determined at the point where the opposing armies take their stand, [but by] what policy is carried on by the war, what class is conducting the war and what objectives it is pursuing in the course of it.[49]

This generous homage legitimized the bourgeois philosopher in the eyes of Marxist-Leninists. Stalin, to be sure, dismissed him as "representative of the hand-workers era" who had nothing to teach the industrial age. But respectful references continued to be made in Soviet military literature, until a new generation felt it necessary to purge Soviet thought of such alien intrusions. A comprehensive study by B. Byely and others declared that by denying the class basis of politics, Clausewitz "propounded a false, idealistic view of politics, which he called the mind of the personified state. . . . [He] completely ignored the fact that politics is conditioned by deep causes rooted in the economic system of society." Instead they propounded their own definition: "war is the continuation of the politics of definite classes and states by other means."[50] Thus modified, Clausewitz is allowed to retain his place in the communist pantheon, and few Marxist textbooks on strategy fail to pay at least lip-service to him.

It is thus entirely appropriate that *On War* should today be studied as much in universities as in military academies. But it must not be forgotten that Clausewitz was a soldier writing primarily for soldiers; that he looked forward to the continuation of war as something natural and inevitable; and that his teaching was intended for successive generations of patriotic Germans fighting for their Fatherland—not for world statesmen conducting international politics in an age of nuclear plenty. Too much should not be read in Clausewitz, nor should more be expected of him than he intended to give. It remains the measure of his genius that, although the age for which he wrote is long since past, he can still provide so many insights relevant to a generation, the nature of whose problems he could not possibly have foreseen.

[47] Marx-Engels, *Werke*, Bd. 29 (Berlin, 1963), pp. 252, 236.

[48] Quoted in Bernard Semmel (ed.): *Marxism and the Science of War* (Oxford, 1981), p. 67.

[49] Werner Hahlweg, introduction to *Vom Kriege* (Bonn, 1980), p. 98.

[50] *Marxism-Leninism on War and Army* (Moscow, 1972) pp. 17–19.

BERNARD BRODIE

The Continuing Relevance of *On War*

The late Herbert Rosinski, in his classic study *The German Army* called *On War* "the most profound, comprehensive, and systematic examination of war that has appeared to the present day." However, Rosinski also had some concern about its effectiveness, because elsewhere he wrote: "The fact that it towers above the rest of military and naval literature, penetrating into regions no other military thinker has ever approached, has been the cause of its being misunderstood."[1]

Misunderstood indeed it often has been, but Rosinski's explanation somewhat misses the mark. He was a close student of Clausewitz and of war, and his characterization of the book is sound enough, but when *On War* is misunderstood the reason lies not in any inherent difficulty in comprehending its ideas. Clausewitz's ideas, though densely packed in, are generally simple and are for the most part clearly expressed in jargon-free language, both in the original and in the present translation. However, these qualities may deceive the casual reader into thinking he is reading mere commonplaces. That may have been why a retired British officer of exalted rank, who was certainly not lacking in intelligence, remarked to this writer some years ago: "I once tried reading Clausewitz, but got nothing out of it." If he had encountered strange new ideas requiring some effort to comprehend them—like some recent strategic essays that use mathematics, game theory, and the like—he might well have made that effort and perhaps carried away a feeling of being suitably rewarded. Instead he encountered wisdom, and thought it was nothing new. Perhaps he also found some of the ideas not quite to his liking, and not liking ideas is a common reason for misunderstanding them.

Inasmuch as an introductory essay should have a purpose that warrants interposing it between the reader and his object, the purpose of this one is mostly to help him avoid the experience of my distinguished military acquaintance. One way to avoid it, naturally, is to avoid reading Clausewitz, which has been the way chosen by all but a minute proportion of literate people, including the great majority of those who have not hesitated to cite or quote him. Civilians have not read his work because

[1] The first of the two Rosinski quotations is from the revised edition of *The German Army* (Washington, 1944), p. 73, and the second is from the original edition (London, 1940), p. 122.

they have erroneously deemed the field recondite, or perhaps too remote from their interests; and the military, except for some special few, have had other reasons for ignoring it. The present reader, however, having the book in hand, obviously has the best of intentions. Let him, therefore, be assured at once that he will not be hindered by abstruse language or difficult-to-fathom ideas. Books on strategy are anyway not often of that character. They may be dull or they may be wild, but they are rarely difficult.

There are indeed some problems in reading Clausewitz, which we shall attempt to explore, because directly confronting these problems helps to diminish them. For one thing, large parts of the work are truly dated, and other large parts sound more dated than they really are because the historical examples cited to illustrate them are inevitably of older times. Also, *On War* is a work in which one easily loses the forest for the trees. Its very length, stretched by innumerable qualifications to its propositions, contributes to this quality, and it is certainly not on the same high level throughout.

Clausewitz himself made clear, in the "Notice" he left with his manuscript, that the revision he planned for the work would be drastic and would "rid the first six books of a good deal of superfluous material, fill in various gaps, large and small, and make a number of generalities more precise in thought and form." In expressing his dissatisfaction with the manuscript as it stood he meant what he said, though many of his most dedicated interpreters seem to forget that. The contrast is often striking between the one chapter he regarded as being revised and completed to his satisfaction, which is the opening chapter of the work, and many of the other chapters. We must, in short, be prepared for a work which is unfinished and therefore on the whole imperfectly organized, often repetitious, and at times even rambling. Sometimes, on the other hand, it is too spare. Occasionally the exact meaning of one or more points is obscure, not from any inherent difficulty of comprehension but because the author did not clearly set down his meaning. What exactly does he mean, for example, by his important concept of "the culminating point of victory" when he seems, not by accident, to exclude Napoleon's march to Moscow as an example of it? Actually, his omitting it is a clue to his meaning, though the casual reader will not notice it.

Although we shall hold that Clausewitz is worth reading today because he is basically timeless, everyone is a child of his age and his culture, and he whose mind eagerly absorbs new ideas will be such in a quite special way. We have already noted and will say more about the datedness of much of Clausewitz's writing, but we encounter also a special idiom, not only of language but occasionally also of thought. A man who is a young German at the beginning of the nineteenth century (and whose life will be over before a third of that century is run), who is intensely intellectual but also with limited formal education, who is deeply sensitive and passionate and yet lives in an age and is committed to a profession which together expose him to an extraordinary experience with war, and who, like all of us, has certain peculiarities of personality and character, will write in a manner that somehow reflects these things. With Clausewitz no more than with any other great thinker and writer are we dealing with a disembodied intellect.

It would be space-consuming, probably tiresome, and in any case beyond our purpose to attempt the always hazardous business of linking some special thought Clausewitz expresses to what we know of his experience or think we can guess of his character, but sometimes it cannot be avoided. Many readers, for example, have been thrown into confusion at the very outset of *On War* by the author's notion of "absolute war" (a term less used in this translation than in others), and by the metamorphosis that occurs within a few pages from concentration on the needs and properties of the absolute or "pure concept" of war to discussion of something much more practical. Yet what could be more natural to an author living in the age and land of Kant and Hegel who is determined to write what readers will recognize as the most deeply penetrating as well as comprehensive treatise of war that has ever been written? Actually, Clausewitz's very slight infusion of metaphysics into this work poses no problems that cannot be explained with relatively few words, and it virtually disappears after those early pages. The greatest misfortune that has resulted from it has been the reputation accorded Clausewitz even by those who are supposed to know him well as someone who is deeply philosophical, in the metaphysical meaning of that term. His contemporary and rival, Antoine Henri Jomini, was already making such remarks about him—besides calling his work "excessive and arrogant"—and such appraisals have continued to the present day.

The mood with which we approach Clausewitz is bound to be affected by all the extravagant things that have been written about him and his chief work. Rosinski, whom we have already quoted, also says the following: "Out of Scharnhorst's fragmentary and aphoristic heritage he developed the systematic, closely knit, perfectly balanced theory, in which every factor, every aspect, every argument had its place from which it could not be removed without fatally endangering the delicate balance of the whole. From the deep appreciation of the revolution wrought by Napoleon in the art of war, he reached an infinitely broader conception embracing within its elastic framework and majestic sweep every conceivable form of warfare and strategy."[2] This hyperbole is clearly denied by Clausewitz himself. A work which has literally not one word to say about naval warfare can hardly be covering "every conceivable form of warfare and strategy" even for its own time, and we have already noted that Clausewitz was planning a revision which would clearly have removed some "factors" and "arguments."

A French scholar of an earlier generation who wrote a book about Clausewitz speaks of him as *le plus Allemand des Allemands . . . A tout instant chez lui on a la sensation d'être dans le brouillard métaphysique* ("The most German of Germans. . . . In reading him one constantly has the feeling of being in a metaphysical fog").[3] This is simply nonsense. Such quotations could be piled up, and they are by people who knew or professed to know Clausewitz's work intimately, as Rosinski certainly did. Awe may be an appropriate mood for some occasions, particularly religious ones, but it is not conducive to calm, perceptive, and therefore critical study.

We have already said something about readers not liking all the Clausewitzian ideas. Both soldiers and civilians have disliked some of them, often for contrary reasons. The soldier trained to revere offensive spirit does not feel comfortable with the argument that the defensive is obviously the stronger form of war, and he especially does not like being told that the military aim must always be subordinated to the political objectives laid down by the civilian leaders. Among civilians there may be some who feel that there is more than a shade too much ruthlessness in Clausewitz, though this attitude is likely to characterize nonreaders who have formed their opinions by hearsay rather than those who have actually read the book. Clausewitz knew that war is not a pleasant affair, and he enjoins the reader to be immediately clear about that too, so that they can proceed jointly to consider the business in hand—which is to understand basically what war is all about, on its various levels of commitment and of violence. One purpose of such understanding is to increase chances of success in this most demanding of pursuits.

By his time the idea was already ancient that war was inherently evil and often also foolish. His countryman and older contemporary, Immanuel Kant, whose work he knew and respected, had written a tract, *Perpetual Peace* (1795), reasserting that idea within the framework of the new knowledge of his time. But this view enjoyed immeasurably less

[2] *The German Army*, 2nd ed., 1944, p. 73.

[3] Hubert Camon, *Clausewitz* (Paris, 1911), p. vii, quoted by H. Rothfels in "Clausewitz," *Makers of Modern Strategy*, E. M. Earle, ed. (Princeton: Princeton University Press, 1943), p. 93.

acceptance then than it does today, which is not to say that it is altogether commonplace now. Anyway, here was a man whose military career was begun when he was age twelve, in an army still imbued with the traditions of Frederick the Great, at a moment that marked the onset of nearly a quarter-century of wars with Revolutionary and Napoleonic France. Also, from what clues he left in his letters and his personal behavior about the nature of his inner life, he seems to have had something more than the usual psychological need for recognition, which for him could come only through some mode of excellence in the profession in which he found himself. Thus there is no reason to wonder at his dedication to his grim subject. He was sufficiently sensitive to the extreme costs and hazards of war, of which he had had no lack of personal experience, to place a high value on competence in conducting it expertly and thus with optimum chance of success. Also, what is much rarer, he attached comparable importance to understanding its purpose.

But the reader may have another and more stubborn concern. Can it be, he will ask, that a book written a century and a half ago, and on war of all things, is really worth his time? That question would arise even if nuclear weapons had never been invented, but those weapons do indeed seem to make a totally new universe. Or do they? There has been a good deal of fighting without nuclear weapons since the two were used on Japan in 1945, including wars which for some of the participants represented total commitment. Still, if it is not yet an established fact it is at minimum a strong possibility that, at least between the great powers who possess nuclear weapons, the whole character of war as a means of settling differences has been transformed beyond all recognition. Why then read Clausewitz?

In our crowded times, it is not enough to argue that a book has exceptional merit. Too many books do which we have no time to read. The commitment to the reading of a substantial book like the present one represents in striking fashion the play of what the economist calls "opportunity cost" (that relinquished object or benefit which might have been gained for the same units of value). Reading time, even for the most favored, is a sharply limited commodity. The reading of a serious book is therefore always a serious undertaking, rationally considered in the form of the following question: Is the reading of this book at this time worth more to me than the reading of any other works that I could read with the same time?

It is well that we do not keep this question at the forefronts of our minds, or we should worry so much each time whether we have made the supremely right choice that we should get nothing read. Still, except for some circumstances where the choices are made for us, as in undergraduate courses, we do in fact tend to keep this question somewhere in the backs of our minds. We pick and choose among books to read and put many of them down unfinished. Among the books we skip are usually the classics, especially those which are not purely literary, because we tend to assume, first, that however great they were in their own times they are not particularly pertinent to ours, and second, that whatever wisdom they do contain which is relevant to our times has no doubt been absorbed and exploited by later writers.

Clausewitz's *On War* does not conform with either assumption. Other classics may sometimes be worth reading because they have a distinctive flavor not fully represented or captured even by those later writers who have fully absorbed and refined their thought—Darwin's *Origin of Species* comes to mind, and there are others. But Clausewitz's work stands out among those very few older books which have presented profound and original insights that have *not* been adequately absorbed in later literature. Naturally, it will be read only by those who have some strong interest, professional or otherwise, in the subject of its title, but for them it is quite indispensable. Of course, there are other books in the field much worth reading in addition to Clausewitz, certainly including some dealing with contemporary and especially nuclear weapons issues, but none can equal it in importance or displace it in its timelessness.

Clausewitz's work was, for example, far more pertinent to the problems and issues of World War I than was Ferdinand Foch's *Principles of War*, published in 1903, only eleven years before that war began. For Foch and his followers the idea of the dominance of the political objective, of which Clausewitz made so much, simply did not apply to modern times. In addition, they romanticized the role of the commander and apotheosized the offensive to a fantastic degree that also proved, insofar as it was followed in action, inordinately costly. Foch gave lip service to the name of Clausewitz, whose work he claimed to have read and absorbed, but his own writing is of totally different character.

Clausewitz indeed made much of the role and talent of the commanding general, but altogether more soberly than Foch. He weighed with great care the relationship of the offensive to the defensive, concluding that the latter was the stronger form of war. If it was so in his time, it was much more so in Foch's, though the latter took the opposite view. For the war of 1914–1918 Foch's highly influential book may not have been dated but it was dreadfully wrong, and it took a whole sea of blood to prove it. There is *no* utility in reading Foch today, except to observe to what aberrant extremes thinking in this field can go and how ill-considered can be the slogans that guide the military policies of great nations.

And, of course, reading it helps one to understand the stupendous catastrophe that was the First World War.

Another work written after that war which had enormous influence on the organization especially of American forces and on the waging of great campaigns in World War II, that of Giulio Douhet, is also today a museum piece. The several essays usually gathered together under the name of the most famous of them—*Command of the Air*—are brilliant, but they are also narrow in outlook, dogmatic, and, as the Second World War proved, in all their specific prescriptions altogether wrong. Air power enthusiasts refer to Douhet reverently as the "prophet of air power," and they will therefore reject this appraisal, perhaps indignantly. But all they need do is read him carefully, checking his detailed predictions against the experience of the Second World War, which was the "war of the future" he was writing about. He argued that battle lines on the ground would remain static and that the decision would anyway be won by the nation's bombers in just a few days. Certainly that is not what happened. It is no doubt true that his ideas would be more appropriate to nuclear weapons than they were to the bombs he had in mind, but it is also true that the nuclear age hardly needs a Douhet to tell it what havoc and terror can be achieved by these weapons. In any case, his specific prescriptions would now be out of date. As with Foch's book we again have a body of work that has no utility today.

Coming down to our own era, Clausewitz is probably as pertinent to our times as most of the literature specifically written about nuclear war. Among works of the latter genre we pick up a good deal of useful technological and other lore, but we usually sense also the absence of that depth and scope which are particularly the hallmark of Clausewitz. We miss especially his tough-minded pursuit of the idea that war in all its phases must be rationally guided by meaningful political purposes. That insight is quite lost in most of the contemporary books, including one which bears a title that boldly invites comparison with the earlier classic, Herman Kahn's *On Thermonuclear War*. Kahn incidentally based his main argument—that the United States could survive and therefore ought not too much fear a thermonuclear war with its chief rival—on technical premises which are certainly obsolete today, whether or not they were realistic when his book was published in the not-so-distant year of 1960. Also, Kahn's book does not, as Clausewitz's does, have much to say of relevance to the Vietnam War which has intervened since and which caused the United States so much soul-searching and agony, though far less of the latter than that borne by the nation it set out to save. Kahn may still usefully supplement Clausewitz, but only in a limited sense is he more timely, and he does not in any way help to supplant him.

From all this we deduce that there must be something about the field of strategic thinking and writing that makes it different from other fields of intellectual endeavor. In most other fields the works of older writers tend to become outmoded because they are either absorbed or disproved. They are sometimes interesting to read for historical reasons and often also for various intrinsic qualities, but they are easily skipped without significant penalty. I have mentioned the name of Darwin, who represents (as does Freud in another field) the great discoverer, whose contribution is never quite matched by any successor. But there is also the great innovator rather than the discoverer, like Adam Smith, whose life span overlapped that of Clausewitz and who wrote in a field remarkably akin to strategy in several respects, including a preoccupation with efficiency in the use of resources for specified goals and with solutions which are at least pragmatic, whether or not they confirm laws describing invariable behavior.

His great seminal work *The Wealth of Nations* (1776) is generally acknowledged to be the fountainhead of modern economics, owing something to others but nevertheless marking a distinctive break with the mercantilist tradition which preceded it and to which no economist worthy of the name would thenceforward return. But this great work has had mighty successors in the two centuries since its publication, and work in this field continues today a most flourishing business, easily attracting its due proportion of gifted minds. All of Smith's essential contribution was fully absorbed and further developed by later writers, who recognized their indebtedness to him. Clausewitz, on the other hand, though he easily compares in talent and innovation with an Adam Smith, has had no comparable cavalcade of brilliant successors.

Thus, in the more noteworthy writings on strategy there is a discontinuity that is not observed in other fields, partly because those fields are far more densely populated with eager workers and partly because of the discontinuity of war itself. Also, while genius has scarcity value in every field of human endeavor, in the field of strategic writing it has a special rarity. The reason is that soldiers are rarely scholars, and civilians are rarely students of strategy. Clausewitz's genius is indisputable, and also in his field unique.

We therefore find ourselves with at least two reasons why Clausewitz continues to be worth the most careful study: first, he was striving always, with a success that derived from his great gifts as well as his intense capacity for work, to get to the fundamentals of each issue he examined, beginning with the fundamental nature of war itself; and second, he is virtually alone in his accomplishment. His is not simply the greatest but the only truly great book on war. Where various other writers on that subject seek to be analytical rather than simply historical, they may be highly respectable in their achievements but as compared with Clausewitz the invariable conclusion has to be that they do not come close.

One must so judge, for example, the work of Alfred Thayer Mahan, who of course limited himself to the naval side of warfare and whose writing is mostly historical. His dimensions and characteristics as a thinker are reflected in his owning himself greatly indebted to Jomini but hardly at all to the latter's greater contemporary, Clausewitz. Another naval historian and analyst contemporary with Mahan, Julian S. Corbett, did indeed pay attention to the work of Clausewitz at great benefit to his own. We might incidentally note, insofar as we are thinking of books being dated, that though Mahan and Corbett lived and wrote in an age of essentially modern steam warships, their writings, so tremendously influential especially in the case of Mahan, develop doctrines deriving almost exclusively from naval warfare in the days of sail.

Nevertheless, we must confront the issue of datedness and consider to what extent this factor detracts from the utility of reading Clausewitz today. Obviously, to the military historian it will detract nothing at all but will make the work, on the contrary, advantageous and indeed necessary to read. If he wonders, for example, why the armies of Wellington and Blücher were spread over so much territory when Napoleon came up to fight them in June of 1815, he will gain some enlightenment from Chapter Thirteen of Book Five, which happens to be on the subject of billets, and the description there of that situation gains authority as well as clarity from the fact that Clausewitz was then with the Prussian army and fought in two of the ensuing battles. But much more than that, Clausewitz was himself a keen military historian—*On War* represents less than a quarter of all his work that ultimately found its way into print, and much of the rest is historical in nature—and he was finely alert to the changes in military practices which separated his own time from the generations preceding. Much of his shrewd observation on these matters finds its way in highly condensed fashion into the present work.

Naturally, military historians comprise a very small proportion of the human race, and a small proportion even of those who might want to read Clausewitz. However, anyone interested enough in what Clausewitz represents to wish to read his book ought surely not be deterred by the fact that one will in the process gain some insights into how war was fought in his day. Our own generation is unique, but sadly so, in producing a school of thinkers who are allegedly experts in military strategy and who are certainly specialists in military studies but who know virtually nothing of military history, including the history of our most recent wars, and who seem not to care about their ignorance. Their skill in systems analysis and related esoteric disciplines is undoubtedly of enormous value in helping them to tread their way among the conflicting claims of the sellers and proponents of different varieties of our extraordinarily complicated modern weaponry. Yet the only empirical data we have about how people conduct war and behave under its stresses is our experience with it in the past, however much we have to make adjustments for subsequent changes in conditions.

Until this new school developed in the period following the Second World War, it was axiomatic that an intimate knowledge of its history was indispensable to an understanding of war. Clausewitz believed it devoutly. "Undoubtedly, the knowledge which is basic to the art of war," he says (in Chapter Six of Book Two), "is empirical." And also, "Historical examples make everything clear, and also provide the best kind of proof in the empirical sciences." Nor is he content with such generalizations, but instead goes on to a careful and characteristically penetrating analysis of the ways in which military history should be used to build up theory.

Still, we cannot avoid considering the drawbacks which attend the fact that Clausewitz died almost a century and a half ago as these lines are written. This condition affects the present utility of his work in various ways, the most obvious of which we have already mentioned. Clausewitz himself asserts that the utility of an historical illustration is generally inversely proportional to its age, and he announces that he will avoid in this work examples antedating the War of the Austrian Succession, the beginning of which in 1740 coincides with the beginning of the First Silesian War and, more significantly, the accession of Frederick II, later "the Great," to the throne of Prussia. Thus, one finds throughout *On War* hardly more than mention of another undoubted genius of war, the Duke of Marlborough, and of his talented colleague, Prince Eugene of Savoy, who collaborated in the brilliant campaign that ended with Blenheim only thirty-six years before Frederick's accession.

Thus, Clausewitz, who as Peter Paret points out wrote also a study on Gustavus Adolphus, confines himself in this work, with but rare exceptions, to historical examples deriving from the seventy-five years ending with Waterloo, which was the last battle he knew and which occurred sixteen years before his death. He dwells on the extraordinarily significant changes in the art of war that occurred during that period, and we cannot help observing that those changes have to be compared to those which have occurred since because of the vast technological revolution in warfare that began about when he died. After all, the weaponry used in Frederick's time differed only slightly from that used in Napoleon's, and to us it is remarkable that very significant changes in practice could occur despite insignificant changes in arms, not to mention transportation or communication.

Anyway, the illustrative historical material Clausewitz uses has the double disadvantage for us, first, that even the latest of it is so far removed from us in time and condition that Clausewitz by his own standards of acceptance would not have considered it, and second, largely because of that remoteness, exceedingly few of his readers will have any prior knowledge at all of the many campaigns or battles he refers to. One supposes that everyone knows something about Napoleon's invasion of Russia in 1812. Tschaikovsky wrote a popular overture about it and Tolstoy a very great novel, and the latter work has been made into several movies and a television series. But who today besides a few specialists knows anything about the campaigns of Frederick, or for that matter about most of the other campaigns of Napoleon?

Fortunately, in his use of an historical example Clausewitz often reconstructs enough about it to give us an adequate picture of what went on and its relevance to the point he is making. But often he does not. We must, therefore, admit that simply from the expository point of view, we lose a great deal of the richness of his analysis as it must have presented itself to his contemporaries. Naturally, we can take steps to amend this shortcoming by learning something of the history he utilizes—a far less severe burden than, say, learning Greek in order to enjoy the poems of Sappho—but in the net we have to write this factor off as a debit.

There is indeed an obverse side to this issue. War, as Clausewitz asserts in one place, *is different from anything else*. Thus, however much it may change within itself from one era to another, its essential character remains distinct from every other pursuit of man. By the same token, it is not in vain that we seek certain elemental qualities which change very little if at all. We are not speaking of the "unchanging principles" of Jomini but of something more fundamental. This element touches basically on why we read Clausewitz, who comes closer to revealing those fundamentals to us than does anyone else, but it also affects the matter of his historical examples.

The reader can himself provide from whatever store of historical knowledge and personal experience he may possess an example to test whether the point in question remains valid, or at least whether it still applied to a much later time than Clausewitz's. Thus where the latter, drawing examples from the campaigns of Frederick and of Napoleon, concedes that there are exceptions to the rule of concentration (which otherwise he most strongly supports) and suggests that there are times when a commander *should* divide his forces in the presence of the enemy, one can reflect on how brilliantly Lee did so at Chancellorsville or how Admiral William F. Halsey foolishly failed to do so at Leyte Gulf. And there is a thrill of discovery in seeing suddenly leaping out at one, in the very last chapter of the work, a pattern of ideas which must certainly have provided the conceptual inspiration for the military aspects of the famous Schlieffen Plan. One may remember too that Count von Schlieffen was sufficiently a student of Clausewitz to have absorbed the latter's repeated maxims about not letting the political end be dominated by the military objective, for he is on record as having urged in writing that if the Plan should fail, as of course it did in 1914, Germany should at once seek a negotiated peace. Unfortunately for Germany and for the world, this most basic of the Clausewitzian ideas was discarded by the successors of Schlieffen and of the younger von Moltke. The Schlieffen Plan, to be sure, had an enormous built-in defect of its own which was basically anti-Clausewitzian—that is, the requirement for the invasion of Belgium (originally also Holland), which was bound to bring Britain into the war.

It is for any student of war, or of politics, a never-ending exercise, in reading of some older problems and how they were handled, to make adjustments to later times. It soon becomes automatic, because it really presents few serious intellectual difficulties. Some ideas and admonitions are immediately recognized as still pertaining today, others as being useful only for the better understanding of military or political history.

More troublesome, admittedly, than the question of mere examples are those quite long stretches where Clausewitz discusses methods of marching, provisioning, and the like that belong to a vanished past. This does not apply to all of Books Four to Seven inclusive, but it applies to much that is found within those books. Over these stretches one's reading may tend to speed up somewhat—perhaps a little warning will help one in accomplishing the acceleration—but one would have to be very pressed for time to want to skip them entirely. In these passages the author shares with us his great knowledge of the conduct of campaigns in his own time, and he is at some pains to make us aware of certain important changes from earlier times. Various abbreviated editions have omitted some of these sections, but it is no doubt best to let the reader decide for himself whether or not he desires to accompany so great a master into these areas. All too little of what Clausewitz wrote has been published in English translation, and few of us would want to see his masterwork truncated. Besides, the reader will find that even in the most unlikely pages he will encounter some sage and penetrating observations that are characteristically Clausewitzian and that apply as much to our own day as they did to his.

Apart from the question of obsolescence or datedness, there are other characteristics of Clausewitz which, though they may be virtues rather than defects, militate against recognition of his genius and accomplishment. Chief among these is his pronounced disinclination to provide formulas or axioms as guides to action. He is often intent upon demonstrating the pitfalls of such axioms, which is the quality chiefly distinguishing him from Jomini, as well as from virtually all his successors. That is one of the chief reasons why military people are so often disappointed with Clausewitz, for they are particularly accustomed in their training to absorbing against a tight schedule of time specific rules for conduct, a practice reflected in their broad use of the term "indoctrination." Clausewitz, on the contrary, invites his readers to ruminate with him on the complex nature of war, where any rule that admits of no exceptions is usually too obvious to be worth much discourse.

This quality is seen especially in his attitude toward such notions as were already beginning to be called "principles of war." Though he could hardly avoid establishing certain generalizations, which is inevitably the result and the purpose of analytical study, he specifically and vehemently rejected the notion that the conduct of war can reasonably be guided by a small number of pithy axioms. It was Jomini, not Clausewitz, who has been responsible for the endlessly quoted remark that "methods change but principles are unchanging," and it is largely for that reason that Jomini had far the greater influence on military thinking in his own and later times, at least among non-Germans. It was Jomini who was looked to for guidance by both sides in the American Civil War, which in his very long life he lived to see concluded. And, as we have seen, it was Jomini whom Mahan called "my best military friend."

It was only following the First World War that various (initially American) field manuals began attempting to encapsulate centuries of experience and volumes of reflection into a few tersely worded and usually numbered "principles of war," such as the "principle of concentration," the "principle of economy of force," the "principle of surprise," and so forth. Though books have been written to explain and elaborate these principles, the main emphasis has been on keeping them lean and taut—to make them more readily communicated in a few days of a war college course by whoever happened to be assigned as an instructor for the purpose, and also more easily carried into combat situations. Clausewitz would have been appalled at such attempts, and not surprised at some of the terrible blunders that have been perpetrated in the name of those "principles." Some who in his own time attempted comparable things he called "the scribblers of systems and compendia."

The price of admission to the Clausewitzian alternative of intensive rumination, sometimes in pages most densely packed with sharp insights, is a commitment to be responsive. This requires a different kind of reading from what we are normally accustomed to. In our time training courses are given to increase reading speed, and no one doubts the benefit of speedy reading with respect to the great masses of stuff that almost any professional person has to cover. With Clausewitz, however, one should be prepared to tarry, to pause frequently for reflection. Clausewitz's basic wish about his book, though no modest one, will reward anyone who does so.

"It was my ambition," he said in a note found among his papers, "to write a book that would not be forgotten after two or three years, and that possibly might be picked up more than once by those who are interested in the subject."

Thinking that we have interposed enough comment between the title page and the text, we have placed at the end of the book "A Guide to the Reading of *On War*." Eds.

CARL VON CLAUSEWITZ

On War

To an Unpublished Manuscript on the Theory of War, Written between 1816 and 1818

There is no need today to labor the point that a scientific approach does not consist solely, or even mainly, in a complete system and a comprehensive doctrine. In the formal sense the present work contains no such system; instead of a complete theory it offers only material for one.

Its scientific character consists in an attempt to investigate the essence of the phenomena of war and to indicate the links between these phenomena and the nature of their component parts. No logical conclusion has been avoided; but whenever the thread became too thin I have preferred to break it off and go back to the relevant phenomena of experience. Just as some plants bear fruit only if they don't shoot up too high, so in the practical arts the leaves and flowers of theory must be pruned and the plant kept close to its proper soil—experience.

It would obviously be a mistake to determine the form of an ear of wheat by analyzing the chemical elements of its kernel, since all one needs to do is to go to a wheat field to see the grown ears. Analysis and observation, theory and experience must never disdain or exclude each other; on the contrary, they support each other. The propositions of this book therefore, like short spans of an arch, base their axioms on the secure foundation either of experience or the nature of war as such, and are thus adequately buttressed.[1]

Perhaps it would not be impossible to write a systematic theory of war, full of intelligence and substance; but the theories we presently possess are very different. Quite apart from their unscientific spirit, they try so hard to make their systems coherent and complete that they are stuffed with commonplaces, truisms, and nonsense of every kind. An accurate impression of their character can be gained by reading Lichtenberg's "Extract from a Fire Regulation":

"If a house is on fire, one must above all seek to save the right wall of the house on the left, and on the other hand the left wall of the house on the right. For if, for example, one were to try and protect the left wall of the house on the left, one must remember that the right wall of the house is on the right of its left wall, and thus, since the fire is also to the right of this wall and of the right wall (for we have assumed that the house is on

[1] That this is not the case with many military writers, particularly those who have tried to deal with war scientifically, is shown by the many instances when the pros and cons of their reasoning dispose of one another so completely that, unlike the two lions in the fable, not even their tails are left. Clausewitz, hereafter cited Cl.

the left of the fire) the right wall is closer to the fire than the left, and the right wall of the house could burn down if it were not protected before the fire would reach the left, protected wall; consequently something could burn down that was not protected, and sooner than something else, even if this something else was also unprotected; consequently the latter must be left alone and the former must be protected. To fix the point firmly in mind, one need only remember: if the house is to the right of the fire, it is the left wall that matters; and if the house is on the left, it is the right wall."

The author does not want to put off the intelligent reader with such trite wisdom, or spoil the taste of the little good he has to offer by watering it down. Years of thinking about war, much association with able men who knew war, and a good deal of personal experience with it, have left him with certain ideas and convictions, and these he has preferred to present in compressed form, like small nuggets of pure metal. That is how the chapters of this book took shape, only tentatively linked on the surface, but I hope not without internal consistency. Perhaps a greater mind will soon appear to replace these individual nuggets with a single whole, cast of solid metal, free from all impurity.

On the Genesis of his Early Manuscript on the Theory of War, Written around 1818[1]

The statements set down here deal with what in my opinion are the major elements of strategy. I regarded them as early drafts, and had more or less reached the point of fusing them into a single work.

These drafts did not follow any preliminary plan. My original intention was to set down my conclusions on the principal elements of this topic in short, precise, compact statements, without concern for system or formal connection. The manner in which Montesquieu dealt with his subject was vaguely in my mind. I thought that such concise, aphoristic chapters, which at the outset I simply wanted to call kernels, would attract the intelligent reader by what they suggested as much as by what they expressed; in other words, I had an intelligent reader in mind, who was already familiar with the subject. But my nature, which always drives me to develop and systematize, at last asserted itself here as well. From the studies I wrote on various topics in order to gain a clear and complete understanding of them, I managed for a time to lift only the most important conclusions and thus concentrate their essence in smaller compass. But eventually my tendency completely ran away with me; I elaborated as much as I could, and of course now had in mind a reader who was not yet acquainted with the subject.

The more I wrote and surrendered to the spirit of analysis, the more I reverted to a systematic approach, and so one chapter after another was added.

In the end I intended to revise it all again, strengthen the causal connections in the earlier essays, perhaps in the later ones draw together several analyses into a single conclusion, and thus produce a reasonable whole, which would form a small volume in octavo. But here too, I wanted at all costs to avoid every commonplace, everything obvious that has been stated a hundred times and is generally believed. It was my ambition to write a book that would not be forgotten after two or three years, and that possibly might be picked up more than once by those who are interested in the subject.

[1] See the Preface by Marie von Clausewitz on p. 66 below. Editors, hereafter cited Eds.

By Marie von Clausewitz to the Posthumous Edition of Her Husband's Works, Including *On War*

Readers will be rightly surprised that a woman should dare to write a preface for such a work as this. My friends will need no explanation; but I hope that a simple account of the circumstances that caused me to take this step will remove any impression of presumptuousness in the minds of those who do not know me.

The work which these lines precede occupied my inexpressively beloved husband almost completely for the last twelve years of his life. His fatherland and I unfortunately lost him far too early. To complete his work was his dearest wish, but it was not his intention to communicate it to the world during his lifetime. When I would try to dissuade him from this decision, he often responded, half jokingly, but perhaps also with a presentiment of his early death: "*You* shall publish it." These words (which in happier days often caused me tears, even though I scarcely took them seriously) oblige me in the view of my friends to introduce the posthumous works of my beloved husband with a few lines. Even though readers may have differing opinions on this point, they will surely not misinterpret the emotion that has caused me to overcome the timidity which makes it so difficult for a woman to appear before the reading public even in the most subordinate manner.

It goes without saying that I have no intention whatever of regarding myself as the true editor of a work that is far beyond my intellectual horizon. Only as a sympathetic companion do I want to help its entry into the world. I may claim this role since I was granted a similar function in the creation and development of the work. Those who knew of our happy marriage and knew that we shared *everything*, not only joy and pain but also every occupation, every concern of daily life, will realize that a task of this kind could not occupy my beloved husband without at the same time becoming thoroughly familiar to me. For the same reason no one can testify as well as I to the energy and love with which he dedicated himself to the task, the hopes he associated with it, and the manner and time of its creation. From early youth his richly endowed mind had felt the need for light and truth, and while he was broadly educated, his reflections were directed primarily toward military affairs, which are of such great importance to the well-being of nations and which constituted his profession. Scharnhorst first showed him the right course; his appointment as teacher at the General War College, as well as the simultaneous honor of being chosen to introduce His

Royal Highness the Crown Prince to the study of war, gave him additional reasons for directing his research and efforts toward these matters, as well as to set down his findings in writing. An essay with which he concluded the instruction of His Royal Highness the Crown Prince in 1812 already contains the seeds of his later works. But it was not until 1816, in Coblenz, that he again took up his scholarly work and began to gather the fruit that had ripened in the course of his rich experiences during four significant years of warfare. To begin with he developed his views in brief essays only loosely connected with each other. The following undated note, which was among his papers, seems to belong to that early stage: [See "Author's Comment," p. 63 above].

In Coblenz, where he had many duties, he could devote only a few hours now and then to his private studies. Not until 1818 when he was appointed Director of the General War College in Berlin did he acquire enough time to expand his work, and further enrich it with the historical interpretation of the more recent wars. This new leisure also reconciled him to his assignment, which in other respects could not quite satisfy him, since under the present arrangements of the college its educational program is not the Director's responsibility but is guided by a separate commission of studies. Free as he was of any petty vanity, of restless egotism and ambition, he nevertheless felt the need to be truly useful, and not let his God-given abilities go to waste. In his professional life he did not occupy a position that could satisfy this need, and he had little hope that he would ever reach such a position. Consequently, all his efforts were directed toward the realm of scientific understanding, and the benefits that he hoped would result from his work became his purpose in life. If in spite of this he was ever more determined not to have his work published until after his death, it must be the best proof that no vain desire for praise and recognition, no trace of egotistic motive, mingled with this noble urge for great and lasting influence.

He continued to work intensively until the spring of 1830, when he was transferred to the artillery. His energies were now taken up for a different purpose, and to such an extent that at least for the moment he had to renounce all literary work. He arranged his papers, sealed and labeled the individual packages, and sadly bade farewell to an activity that had come to mean so much to him. In August of that year he was transferred to Breslau, where he was assigned to head the 2d Artillery Inspection; but already in December he was recalled to Berlin and appointed chief of staff to Field Marshal Count Gneisenau (for the duration of the latter's command [in the East]). In March 1831 he accompanied his admired commander-in-chief to Posen. When he returned to Breslau in November, having suffered the most painful loss [in Gneisenau's death], he was cheered by the hope of resuming his work, and possibly completing it in the course of the winter. God decided otherwise. He returned to Breslau on 7 November, on the sixteenth he died, and the packages that his hand had sealed were not opened until after his death!

These literary remains are published in the following volumes, exactly as they were found, without one word being added or deleted. Nevertheless their publication called for a good deal of work, arranging of material, and consultation, and I am profoundly grateful to several loyal friends for their assistance in these tasks. Above all I must name Major O'Etzel, who was kind enough to read the proofs and to draw the maps that will accompany the historical sections of the edition. I also take the liberty of mentioning my beloved brother, my support in times of trial, who rendered so many different services in preparing the manuscripts for publication. Among others, in the course of careful checking and sorting the material he found the beginnings of the revision that my beloved husband mentions as a future project in the *Note of 1827*, printed below. The revisions have been inserted in those parts of Book I for which they were intended (they did not go further).

I want to thank many other friends for their advice and for the sympathy and affection they have shown me; though I cannot name them all, they will surely not doubt my warmest gratitude. This gratitude is the greater since I am firmly convinced that everything they have done was done not only for me but also for the friend whom God took from them so prematurely.

For twenty-one years I was profoundly happy at the side of *such* a man. Treasured memories, hopes, the rich inheritance of sympathy and friendship that I owe to the beloved departed, and the elevating sense that his rare distinction is so generally and nobly recognized sustain this happiness despite my irreplaceable loss.

The trust that led a noble prince and princess to call me to their side is a new favor for which I thank God.[1] It has given me a new and valued task, to which I dedicate myself gladly. May this task be blessed, and may the cherished little prince, who is presently entrusted to my care, someday read this book, and be inspired by it to deeds similar to the deeds of his glorious ancestors!

Written in the Marble Palace at Potsdam, 30 June 1832.

MARIE VON CLAUSEWITZ
Born Countess Brühl

First Lady in Waiting
of Her Royal Highness
Princess Wilhelm

[1] Marie von Clausewitz had been appointed Governess of Prince Friedrich Wilhelm, the later Emperor Frederick III. Eds.

Two Notes by the Author on His Plans for Revising *On War*

Note of 10 July 1827

I regard the first six books, which are already in a clean copy, merely as a rather formless mass that must be thoroughly reworked once more. The revision will bring out the two types of war with greater clarity at every point. All ideas will then become plainer, their general trend will be more clearly marked, their application shown in greater detail.

War can be of two kinds, in the sense that either the objective is to *overthrow the enemy*—to render him politically helpless or militarily impotent, thus forcing him to sign whatever peace we please; or *merely to occupy some of his frontier-districts* so that we can annex them or use them for bargaining at the peace negotiations. Transitions from one type to the other will of course recur in my treatment; but the fact that the aims of the two types are quite different must be clear at all times, and their points of irreconcilability brought out.

This distinction between the two kinds of war is a matter of actual fact. But no less practical is the importance of another point that must be made absolutely clear, namely that *war is nothing but the continuation of policy with other means.* If this is firmly kept in mind throughout it will greatly facilitate the study of the subject and the whole will be easier to analyze. Although the main application of this point will not be made until Book Eight, it must be developed in Book One and will play its part in the revision of the first six books. That revision will also rid the first six books of a good deal of superfluous material, fill in various gaps, large and small, and make a number of generalities more precise in thought and form.

Book Seven, "On Attack" (various chapters of which are already in rough draft) should be regarded as the counterpart of Book Six, "On Defense," and is the next to be revised in accordance with the clear insights indicated above. Thereafter it will need no further revision; indeed, it will then provide a standard for revising the first six books.

Book Eight, "War-Plans," will deal with the organization of a war as a whole. Several chapters of it have already been drafted, but they must not in any sense be taken as being in final form. They are really no more than a rough working over of the raw material, done with the idea that the labor itself would show what the real problems were. That in fact is what happened, and when I have finished Book Seven I shall go on at once and work out Book Eight in full. My main concern will be to apply the two principles mentioned above, with the idea of refining and simplifying everything. In Book Eight I also hope to iron out a good many kinks in the minds of strategists and statesmen and at all events to show what the whole thing is about and what the real problems are that have to be taken into account in actual warfare.

If the working out of Book Eight results in clearing my own mind and in really establishing the main features of war it will be all the easier for me to apply the same criteria to the first six books and make those features evident throughout them. Only when I have reached that point, therefore, shall I take the revision of the first six books in hand.

If an early death should terminate my work, what I have written so far would, of course only deserve to be called a shapeless mass of ideas. Being liable to endless misinterpretation it would be the target of much half-baked criticism, for in matters of this kind everyone feels he is justified in writing and publishing the first thing that comes into his head when he picks up a pen, and thinks his own ideas as axiomatic as the fact that two and two make four. If critics would go to the trouble of thinking about the subject for years on end and testing each conclusion against the actual history of war, as I have done, they would undoubtedly be more careful of what they said.

Nonetheless, I believe an unprejudiced reader in search of truth and understanding will recognize the fact that the first six books, for all their imperfection of form, contain the fruit of years of reflection on war and diligent study of it. He may even find they contain the basic ideas that might bring about a revolution in the theory of war.

Unfinished Note, Presumably Written in 1830

The manuscript on the conduct of major operations that will be found after my death can, in its present state, be regarded as nothing but a collection of materials from which a theory of war was to have been distilled. I am still dissatisfied with most of it, and can call Book Six only a sketch. I intended to rewrite it entirely and to try and find a solution along other lines.

Nevertheless I believe the main ideas which will be seen to govern this material are the right ones, looked at in the light of actual warfare. They are the outcome of wide-ranging study: I have thoroughly checked them against real life and have constantly kept in mind the lessons derived from my experience and from association with distinguished soldiers.

Book Seven, which I have sketched in outline, was meant to deal with "Attack," and Book Eight with "War-Plans," in which I intended to concern myself particularly with war in its political and human aspects.

The first chapter of Book One alone I regard as finished. It will at least serve the whole by indicating the direction I meant to follow everywhere.

The theory of major operations (strategy, as it is called) presents extraordinary difficulties, and it is fair to say that very few people have clear ideas about its details—that is, ideas which logically derive from basic necessities.

Most men merely act on instinct, and the amount of success they achieve depends on the amount of talent they were born with.

All great commanders have acted on instinct, and the fact that their instinct was always sound is partly the measure of their innate greatness and genius. So far as action is concerned this will always be the case and nothing more is needed. Yet when it is not a question of acting oneself but of persuading others in discussion, the need is for clear ideas and the ability to show their connection with each other. So few people have yet acquired the necessary skill at this that most discussions are a futile bandying of words; either they leave each man sticking to his own ideas or they end with everyone agreeing, for the sake of agreement, on a compromise with nothing to be said for it.

Clear ideas on these matters do, therefore, have some practical value. The human mind, moreover, has a universal thirst for clarity, and longs to feel itself part of an orderly scheme of things.

It is a very difficult task to construct a scientific theory for the art of war, and so many attempts have failed that most people say it is impossible, since it deals with matters that no permanent law can provide for. One would agree, and abandon the attempt, were it not for the obvious fact that a whole range of propositions can be demonstrated without difficulty: that defense is the stronger form of fighting with the negative purpose, attack the weaker form with the positive purpose; that major successes help bring about minor ones, so that strategic results can be traced back to certain turning-points; that a demonstration is a weaker use of force than a real attack, and that it must therefore be clearly justified; that victory consists not only in the occupation of the battlefield, but in the destruction of the enemy's physical and psychic forces, which is usually not attained until the enemy is pursued after a victorious battle; that success is always greatest at the point where the victory was gained, and that consequently changing from one line of operations, one direction, to another can at best be regarded as a necessary evil; that a turning movement can only be justified by general superiority or by having better lines of communication or retreat than the enemy's; that flank-positions are governed by the same consideration; that every attack loses impetus as it progresses.

BOOK ONE

On the Nature of War

CHAPTER ONE

What Is War?

1. Introduction

I propose to consider first the various *elements* of the subject, next its *various parts* or *sections*, and finally *the whole* in its internal structure. In other words, I shall proceed from the simple to the complex. But in war more than in any other subject we must begin by looking at the nature of the whole; for here more than elsewhere the part and the whole must always be thought of together.

2. Definition

I shall not begin by expounding a pedantic, literary definition of war, but go straight to the heart of the matter, to the duel. War is nothing but a duel on a larger scale. Countless duels go to make up war, but a picture of it as a whole can be formed by imagining a pair of wrestlers. Each tries through physical force to compel the other to do his will; his *immediate* aim is to *throw* his opponent in order to make him incapable of further resistance.

War is thus an act of force to compel our enemy to do our will.

Force, to counter opposing force, equips itself with the inventions of art and science. Attached to force are certain self-imposed, imperceptible limitations hardly worth mentioning, known as international law and custom, but they scarcely weaken it. Force—that is, physical force, for moral force has no existence save as expressed in the state and the law—is thus the *means* of war; to impose our will on the enemy is its *object*. To secure that object we must render the enemy powerless; and that, in theory, is the true aim of warfare. That aim takes the place of the object, discarding it as something not actually part of war itself.

3. The Maximum Use of Force

Kind-hearted people might of course think there was some ingenious way to disarm or defeat an enemy without too much bloodshed, and might imagine this is the true goal of the art of war. Pleasant as it sounds, it is a fallacy that must be exposed: war is such a dangerous business that the mistakes which come from kindness are the very worst. The maximum use of force is in no way incompatible with the simultaneous use of the intellect. If one side uses force without compunction, undeterred by the bloodshed it involves, while the other side refrains, the first will gain the upper hand. That side will force the other to follow suit; each will drive its opponent toward extremes, and the only limiting factors are the counterpoises inherent in war.

This is how the matter must be seen. It would be futile—even wrong—to try and shut one's eyes to what war really is from sheer distress at its brutality.

If wars between civilized nations are far less cruel and destructive than wars between savages, the reason lies in the social conditions of the states themselves and in their relationships to one another. These are the forces that give rise to war; the same forces circumscribe and moderate it. They themselves however are not part of war; they already exist before fighting starts. To introduce the principle of moderation into the theory of war itself would always lead to logical absurdity.

Two different motives make men fight one another: *hostile feelings* and *hostile intentions*. Our definition is based on the latter, since it is the universal element. Even the most savage, almost instinctive, passion of hatred cannot be conceived as existing without hostile intent; but hostile intentions are often unaccompanied by any sort of hostile feelings—at least by none that predominate. Savage peoples are ruled by passion, civilized peoples by the mind. The difference, however, lies not in the respective natures of savagery and civilization, but in their attendant circumstances, institutions, and so forth. The difference, therefore, does not operate in every case, but it does in most of them. Even the most civilized of peoples, in short, can be fired with passionate hatred for each other.

Consequently, it would be an obvious fallacy to imagine war between civilized peoples as resulting merely from a rational act on the part of their governments and to conceive of war as gradually ridding itself of passion, so that in the end one would never really need to use the physical impact of the fighting forces—comparative figures of their strength would be enough. That would be a kind of war by algebra.

Theorists were already beginning to think along such lines when the recent wars taught them a lesson. If war is an act of force, the emotions cannot fail to be involved. War may not spring from them, but they will still affect it to some degree, and the extent to which they do so will depend not on the level of civilization but on how important the conflicting interests are and on how long their conflict lasts.

If, then, civilized nations do not put their prisoners to death or devastate cities and countries, it is because intelligence plays a larger part in their methods of warfare and has taught them more effective ways of using force than the crude expression of instinct.

The invention of gunpowder and the constant improvement of firearms are enough in themselves to show that the advance of civilization has done nothing practical to alter or deflect the impulse to destroy the enemy, which is central to the very idea of war.

The thesis, then, must be repeated: war is an act of force, and there is no logical limit to the application of that force. Each side, therefore, compels its opponent to follow suit; a reciprocal action is started which must lead, in theory, to extremes. This is the *first case of interaction and the first "extreme"* we meet with.

4. The Aim Is To Disarm the Enemy

I have already said that the aim of warfare is to disarm the enemy and it is time to show that, at least in theory, this is bound to be so. If the enemy is to be coerced you must put him in a situation that is even more unpleasant than the sacrifice you call on him to make. The hardships of that situation must not of course be merely transient—at least not in appearance. Otherwise the enemy would not give in but would wait for things to improve. Any change that might be brought about by continuing hostilities must then, at least in theory, be of a kind to bring the enemy still greater disadvantages. The worst of all conditions in which a belligerent can find himself is to be utterly defenseless. Consequently, if you are to force the enemy, by making war on him, to do your bidding, you must either make him literally defenseless or at least put him in a position that makes this danger probable. It follows, then, that to overcome the enemy, or disarm him—call it what you will—must always be the aim of warfare.

War, however, is not the action of a living force upon a lifeless mass (total nonresistance would be no war at all) but always the collision of two living forces. The ultimate aim of waging war, as formulated here, must be taken as applying to both sides. Once again, there is interaction. So long as I have not overthrown my opponent I am bound to fear he may overthrow me. Thus I am not in control: he dictates to me as much as I dictate to him. This is the *second case of interaction and it leads to the second "extreme."*

5. The Maximum Exertion of Strength

If you want to overcome your enemy you must match your effort against his power of resistance, which can be expressed as the product of two inseparable factors, viz. *the total means at his disposal* and *the strength of his will.* The extent of the means at his disposal is a matter—though not exclusively—of figures, and should be measurable. But the strength of his will is much less easy to determine and can only be gauged approximately by the strength of the motive animating it. Assuming you arrive in this way at a reasonably accurate estimate of the enemy's power of resistance, you can adjust your own efforts accordingly; that is, you can either increase them until they surpass the enemy's or, if this is beyond your means, you can make your efforts as great as possible. But the enemy will do the same; competition will again result and, in pure theory, it must again force you both to extremes. This is *the third case of interaction and the third "extreme."*

6. Modifications in Practice

Thus in the field of abstract thought the inquiring mind can never rest until it reaches the extreme, for here it is dealing with an extreme: a clash of forces freely operating and obedient to no law but their own. From a pure concept of war you might try to deduce absolute terms for the objective you should aim at and for the means of achieving it; but if you did so the continuous interaction would land you in extremes that represented nothing but a play of the imagination issuing from an almost invisible sequence of logical subtleties. If we were to think purely in absolute terms, we could avoid every difficulty by a stroke of the pen and proclaim with inflexible logic that, since the extreme must always be the goal, the greatest effort must always be exerted. Any such pronouncement would be an abstraction and would leave the real world quite unaffected.

Even assuming this extreme effort to be an absolute quantity that could easily be calculated, one must admit that the human mind is unlikely to consent to being ruled by such a logical fantasy. It would often result in strength being wasted, which is contrary to other principles of statecraft. An effort of will out of all proportion to the object in view would be needed but would not in fact be realized, since subtleties of logic do not motivate the human will.

But move from the abstract to the real world, and the whole thing looks quite different. In the abstract world, optimism was all-powerful and forced us to assume that both parties to the conflict not only sought perfection but attained it. Would this ever be the case in practice? Yes, it would if: (a) war were a wholly isolated act, occurring suddenly and not produced by previous events in the political world; (b) it consisted of a single decisive act or a set of simultaneous ones; (c) the decision achieved was complete and perfect in itself, uninfluenced by any previous estimate of the political situation it would bring about.

7. War Is Never an Isolated Act

As to the first of these conditions, it must be remembered that neither opponent is an abstract person to the other, not even to the extent of that factor in the power of resistance, namely the will, which is dependent on externals. The will is not a wholly unknown factor; we can base a forecast of its state tomorrow on what it is today. War never breaks out wholly unexpectedly, nor can it be spread instantaneously. Each side can therefore gauge the other to a large extent by what he is and does, instead of judging him by what he, strictly speaking, ought to be or do. Man and his affairs, however, are always something short of perfect and will never quite achieve the absolute best. Such shortcomings affect both sides alike and therefore constitute a moderating force.

8. War Does Not Consist of a Single Short Blow

The second condition calls for the following remarks:

If war consisted of one decisive act, or of a set of simultaneous decisions, preparations would tend toward totality, for no omission could ever be rectified. The sole criterion for preparations which the world of reality could provide would be the measures taken by the adversary—so far as they are known; the rest would once more be reduced to abstract calculations. But if the decision in war consists of several successive acts, then each of them, seen in context, will provide a gauge for those that follow. Here again, the abstract world is ousted by the real one and the trend to the extreme is thereby moderated.

But, of course, if all the means available were, or could be, simultaneously employed, all wars would automatically be confined to a single decisive act or a set of simultaneous ones—the reason being that any *adverse* decision must reduce the sum of the means available, and if *all* had been committed in the first act there could really be no question of a second. Any subsequent military operation would virtually be part of the first—in other words, merely an extension of it.

Yet, as I showed above, as soon as preparations for a war begin, the world of reality takes over from the world of abstract thought; material calculations take the place of hypothetical extremes and, if for no other reason, the interaction of the two sides tends to fall short of maximum effort. Their full resources will therefore not be mobilized immediately.

Besides, the very nature of those resources and of their employment means they cannot all be deployed at the same moment. The resources in question are *the fighting forces proper, the country,* with its physical features and population, and its *allies.*

The country—its physical features and population—is more than just the source of all armed forces proper; it is in itself an integral element among the factors at work in war—though only that part which is the actual theater of operations or has a notable influence on it.

It is possible, no doubt, to use all mobile fighting forces simultaneously; but with fortresses, rivers, mountains, inhabitants, and so forth, that cannot be done; not, in short, with the country as a whole, unless it is so small that the opening action of the war completely engulfs it. Furthermore, allies do not cooperate at the mere desire of those who are actively engaged in fighting; international relations being what they are, such cooperation is often furnished only at some later stage or increased only when a balance has been disturbed and needs correction.

In many cases, the proportion of the means of resistance that cannot immediately be brought to bear is much higher than might at first be thought. Even when great strength has been expended on the first decision and the balance has been badly upset, equilibrium can be restored. The point will be more fully treated in due course. At this stage it is enough to show that the very nature of war impedes the *simultaneous concentration of all forces.* To be sure, that fact in itself cannot be grounds for making any but a maximum effort to obtain the first decision, for a defeat is always a disadvantage no one would deliberately risk. And even if the first clash is not the only one, the influence it has on subsequent actions will be on a scale proportionate to its own. But it is contrary to human nature to make an extreme effort, and the tendency therefore is always to plead that a decision may be possible later on. As a result, for the first decision, effort and concentration of forces are not all they might be. Anything omitted out of weakness by one side becomes a real, *objective* reason for the other to reduce its efforts, and the tendency toward extremes is once again reduced by this interaction.

9. In War the Result Is Never Final

Lastly, even the ultimate outcome of a war is not always to be regarded as final. The defeated state often considers the outcome merely as a transitory evil, for which a remedy may still be found in political conditions at some later date. It is obvious how this, too, can slacken tension and reduce the vigor of the effort.

10. The Probabilities of Real Life Replace the Extreme and the Absolute Required by Theory

Warfare thus eludes the strict theoretical requirement that extremes of force be applied. Once the extreme is no longer feared or aimed at, it becomes a matter of judgment what degree of effort should be made; and this can only be based on the phenomena of the real world and the *laws of probability.* Once the antagonists have ceased to be mere figments of a theory and become actual states and governments, when war is no longer a theoretical affair but a series of actions obeying its own peculiar laws, reality supplies the data from which we can deduce the unknown that lies ahead.

From the enemy's character, from his institutions, the state of his affairs and his general situation, each side, using the *laws of probability,* forms an estimate of its opponent's likely course and acts accordingly.

11. The Political Object Now Comes to the Fore Again

A subject which we last considered in Section 2 now forces itself on us again, namely the *political object of the war.* Hitherto it had been rather overshadowed by the law of extremes, the will to overcome the enemy and make him powerless. But as this law begins to lose its force and as this determination wanes, the political aim will reassert itself. If it is all a calculation of probabilities based on given individuals and conditions, the *political object,* which was the *original motive,* must become an essential factor in the equation. The smaller the penalty you demand from your opponent, the less you can expect him to try and deny it to you; the smaller the effort he makes, the less you need make yourself. Moreover, the more modest your own political aim, the less importance you attach to it and the less reluctantly you will abandon it if you must. *This is another reason why your effort will be modified.*

The political object—the original motive for the war—will thus determine both the military objective to be reached and the amount of effort it requires. The political object cannot, however, *in itself* provide the standard of measurement. Since we are dealing with realities, not with abstractions, it can do so only in the context of the two states at war. The same political object can elicit *differing* reactions from different peoples, and even from the same people at different times. We can therefore take the political object as a standard only if we think of *the influence it can exert upon the forces it is meant to move.* The nature of those forces therefore calls for study. Depending on whether their characteristics increase or diminish the drive toward a particular action, the outcome will vary. Between two peoples and two states there can be such tensions, such a mass of inflammable material, that the slightest quarrel can produce a wholly disproportionate effect—a real explosion.

This is equally true of the efforts a political object is expected to arouse in either state, and of the military objectives which their policies require. Sometimes the *political and military objective is the same*—for example, the conquest of a province. In other cases the political object will not provide a suitable military objective. In that event, another military objective must be adopted that will serve the political purpose and symbolize it in the peace negotiations. But here, too, attention must be paid to the character of each state involved. There are times when, if the political object is to be achieved, the substitute must be a good deal more important. The less involved the population and the less serious the strains within states and between them, the more political requirements in themselves will dominate and tend to be decisive. Situations can thus exist in which the political object will almost be the sole determinant.

Generally speaking, a military objective that matches the political object in scale will, if the latter is reduced, be reduced in proportion; this will be all the more so as the political object increases its predominance. Thus it follows that without any inconsistency wars can have all degrees of importance and intensity, ranging from a war of extermination down to simple armed observation. This brings us to a different question, which now needs to be analyzed and answered.

12. An Interruption of Military Activity Is Not Explained by Anything Yet Said

However modest the political demands may be on either side, however small the means employed, however limited the military objective, can the process of war ever be interrupted, even for a moment? The question reaches deep into the heart of the matter.

Every action needs a certain time to be completed. That period is called its duration, and its length will depend on the speed with which the person acting works. We need not concern ourselves with the difference here. Everyone performs a task in his own way; a slow man, however, does not do it more slowly because he wants to spend more time over it, but because his nature causes him to need more time. If he made more haste he would do the job less well. His speed, then, is determined by subjective causes and is a factor in the actual duration of the task.

Now if every action in war is allowed its appropriate duration, we would agree that, at least at first sight, any additional expenditure of time—any suspension of military action—seems absurd. In this connection it must be remembered that what we are talking about is not the progress made by one side or the other but the progress of military interaction as a whole.

13. Only One Consideration Can Suspend Military Action, and It Seems That It Can Never Be Present on More Than One Side

If two parties have prepared for war, some motive of hostility must have brought them to that point. Moreover so long as they remain under arms (do not negotiate a settlement) that motive of hostility must still be active. Only one consideration can restrain it: *a desire to wait for a better moment before acting.* At first sight one would think this desire could never operate on more than one side since its opposite must automatically be working on the other. If action would bring an advantage to one side, the other's interest must be to wait.

But an absolute balance of forces cannot bring about a standstill, for if such a balance should exist the initiative would necessarily belong to the side with the positive purpose—the attacker.

One could, however, conceive of a state of balance in which the side with the positive aim (the side with the stronger grounds for action) was the one that had the weaker forces. The balance would then result from the combined effects of aim and strength. Were that the case, one would have to say that unless some shift in the balance were in prospect the two sides should make peace. If, however, some alteration were to be foreseen, only one side could expect to gain by it—a fact which ought to stimulate the other into action. Inaction clearly cannot be explained by the concept of balance. The only explanation is that both are waiting for a better time to act. Let us suppose, therefore, that one of the two states has a positive aim—say, the conquest of a part of the other's territory, to use for bargaining at the peace table. Once the prize is in its hands, the political object has been achieved; there is no need to do more, and it can let matters rest. If the other state is ready to accept the situation, it should sue for peace. If not, it must do something; and if it thinks it will be better organized for action in four weeks' time it clearly has an adequate reason for not taking action at once.

But from that moment on, logic would seem to call for action by the other side—the object being to deny the enemy the time he needs for getting ready. Throughout all this I have assumed, of course, that both sides understand the situation perfectly.

14. Continuity Would Thus Be Brought About in Military Action and Would Again Intensify Everything

If this continuity were really to exist in the campaign its effect would again be to drive everything to extremes. Not only would such ceaseless activity arouse men's feelings and inject them with more passion and elemental strength, but events would follow more closely on each other and be governed by a stricter causal chain. Each individual action would be more important, and consequently more dangerous.

But war, of course, seldom if ever shows such continuity. In numerous conflicts only a very small part of the time is occupied by action, while the rest is spent in inactivity. This cannot always be an anomaly. Suspension of action in war must be possible; in other words, it is not a contradiction in terms. Let me demonstrate this point, and explain the reasons for it.

15. Here a Principle of Polarity Is Proposed

By thinking that the interests of the two commanders are opposed in equal measure to each other, we have assumed a genuine *polarity*. A whole chapter will be devoted to the subject further on, but the following must be said about it here.

The principle of polarity is valid only in relation to one and the same object, in which positive and negative interests exactly cancel one another out. In a battle each side aims at victory; that is a case of true polarity, since the victory of one side excludes the victory of the other. When, however, we are dealing with two different things that have a common relation external to themselves, the polarity lies not in the *things* but in their relationship.

16. Attack and Defense Being Things Different in Kind and Unequal in Strength, Polarity Cannot Be Applied to Them

If war assumed only a single form, namely, attacking the enemy, and defense were nonexistent; or, to put it in another way, if the only differences between attack and defense lay in the fact that attack has a positive aim whereas defense has not, and the forms of fighting were identical; then every advantage gained by one side would be a precisely equal disadvantage to the other—true polarity would exist.

But there are two distinct forms of action in war: attack and defense. As will be shown in detail later, the two are very different and unequal in strength. Polarity, then, does not lie in attack or defense, but in the object both seek to achieve: the decision. If one commander wants to postpone the decision, the other must want to hasten it, always assuming that both are engaged in the same kind of fighting. If it is in A's interest not to attack B now but to attack him in four weeks, then it is in B's interest not to be attacked in four weeks' time, but now. This is an immediate and direct conflict of interest; but it does not follow from this that it would also be to B's advantage to make an immediate attack on A. That would obviously be quite another matter.

17. The Superiority of Defense over Attack Often Destroys the Effect of Polarity, and This Explains the Suspension of Military Action

As we shall show, defense is a stronger form of fighting than attack. Consequently we must ask whether the advantage of *postponing a decision* is as great for one side as the advantage of *defense* is for the other. Whenever it is not, it cannot balance the advantage of defense and in this way influence the progress of the war. It is clear, then, that the impulse created by the polarity of interests may be exhausted in the difference between the strength of attack and defense, and may thus become inoperative.

Consequently, if the side favored by present conditions is not sufficiently strong to do without the added advantages of the defense, it will have to accept the prospect of acting under unfavorable conditions in the future. To fight a defensive battle under these less favorable conditions may still be better than to attack immediately or to make peace. I am convinced that the superiority of the defensive (if rightly understood) is very great, far greater than appears at first sight. It is this which explains without any inconsistency most periods of inaction that occur in war. The weaker the motives for action, the more will they be overlaid and neutralized by this disparity between attack and defense, and the more frequently will action be suspended—as indeed experience shows.

18. A Second Cause Is Imperfect Knowledge of the Situation

There is still another factor that can bring military action to a standstill: imperfect knowledge of the situation. The only situation a commander can know fully is his own; his opponent's he can know only from unreliable intelligence. His evaluation, therefore, may be mistaken and can lead him to suppose that the initiative lies with the enemy when in fact it remains with him. Of course such faulty appreciation is as likely to lead to ill-timed action as to ill-timed inaction, and is no more conducive to slowing down operations than it is to speeding them up. Nevertheless, it must rank among the natural causes which, *without entailing inconsistency, can bring military activity to a halt*. Men are always more inclined to pitch their estimate of the enemy's strength too high than too low, such is human nature. Bearing this in mind, one must admit that partial ignorance of the situation is, generally speaking, a major factor in delaying the progress of military action and in moderating the principle that underlies it.

The possibility of inaction has a further moderating effect on the progress of the war by diluting it, so to speak, in time by delaying danger, and by increasing the means of restoring a balance between the two sides. The greater the tensions that have led to war, and the greater the consequent war effort, the shorter these periods of inaction. Inversely, the weaker the motive for conflict, the longer the intervals between actions. For the stronger motive increases willpower, and willpower, as we know, is always both an element in and the product of strength.

19. Frequent Periods of Inaction Remove War Still Further from the Realm of the Absolute and Make It Even More a Matter of Assessing Probabilities

The slower the progress and the more frequent the interruptions of military action the easier it is to retrieve a mistake, the bolder will be the general's assessments, and the more likely he will be to avoid theoretical extremes and to base his plans on probability and inference. Any given situation requires that probabilities be calculated in the light of circumstances, and the amount of time available for such calculation will depend on the pace with which operations are taking place.

20. Therefore Only the Element of Chance Is Needed To Make War a Gamble, And That Element Is Never Absent

It is now quite clear how greatly the objective nature of war makes it a matter of assessing probabilities. Only one more element is needed to make war a gamble—chance: the very last thing that war lacks. No other human activity is so continuously or universally bound up with chance. And through the element of chance, guesswork and luck come to play a great part in war.

21. Not Only Its Objective But Also Its Subjective Nature Makes War a Gamble

If we now consider briefly the *subjective nature* of war—the means by which war has to be fought—it will look more than ever like a gamble. The element in which war exists is danger. The highest of all moral qualities in time of danger is certainly *courage*. Now courage is perfectly compatible with prudent calculation but the two differ nonetheless, and pertain to different psychological forces. Daring, on the other hand, boldness, rashness, trusting in luck are only variants of courage, and all these traits of character seek their proper element—chance.

In short, absolute, so-called mathematical, factors never find a firm basis in military calculations. From the very start there is an interplay of possibilities, probabilities, good luck and bad that weaves its way throughout the length and breadth of the tapestry. In the whole range of human activities, war most closely resembles a game of cards.

22. How in General This Best Suits Human Nature

Although our intellect always longs for clarity and certainty, our nature often finds uncertainty fascinating. It prefers to day-dream in the realms of chance and luck rather than accompany the intellect on its narrow and tortuous path of philosophical enquiry and logical deduction only to arrive—hardly knowing how—in unfamiliar surroundings where all the usual landmarks seem to have disappeared. Unconfined by narrow necessity, it can revel in a wealth of possibilities; which inspire courage to take wing and dive into the element of daring and danger like a fearless swimmer into the current.

Should theory leave us here, and cheerfully go on elaborating absolute conclusions and prescriptions? Then it would be no use at all in real life. No, it must also take the human factor into account, and find room for courage, boldness, even foolhardiness. The art of war deals with living and with moral forces. Consequently, it cannot attain the absolute, or certainty; it must always leave a margin for uncertainty, in the greatest things as much as in the smallest. With uncertainty in one scale, courage and self-confidence must be thrown into the other to correct the balance. The greater they are, the greater the margin that can be left for accidents. Thus courage and self-confidence are essential in war, and theory should propose only rules that give ample scope to these finest and least dispensable of military virtues, in all their degrees and variations. Even in daring there can be method and caution; but here they are measured by a different standard.

23. But War Is Nonetheless a Serious Means to a Serious End: A More Precise Definition of War

Such is war, such is the commander who directs it, and such the theory that governs it. War is no pastime; it is no mere joy in daring and winning, no place for irresponsible enthusiasts. It is a serious means to a serious end, and all its colorful resemblance to a game of chance, all the vicissitudes of passion, courage, imagination, and enthusiasm it includes are merely its special characteristics.

When whole communities go to war—whole peoples, and especially *civilized* peoples—the reason always lies in some political situation, and the occasion is always due to some political object. War, therefore, is an act of policy. Were it a complete, untrammeled, absolute manifestation of violence (as the pure concept would require), war would of its own independent will usurp the place of policy the moment policy had brought it into being; it would then drive policy out of office and rule by the laws of its own nature, very much like a mine that can explode only in the manner or direction predetermined by the setting. This, in fact, is the view that has been taken of the matter whenever some discord between policy and the conduct of war has stimulated theoretical distinctions of this kind. But in reality things are different, and this view is thoroughly mistaken. In reality war, as has been shown, is not like that. Its violence is not of the kind that explodes in a single discharge, but is the effect of forces that do not always develop in exactly the same manner or to the same degree. At times they will expand sufficiently to overcome the resistance of inertia or friction; at others they are too weak to have any effect. War is a pulsation of violence, variable in strength and therefore variable in the speed with which it explodes and discharges its energy. War moves on its goal with varying speeds; but it always lasts long enough for influence to be exerted on the goal and for its own course to be changed in one way or another—long enough, in other words, to remain subject to the action of a superior intelligence. If we keep in mind that war springs from some political purpose, it is natural that the prime cause of its existence will remain the supreme consideration in conducting it. That, however, does not imply that the political aim is a tyrant. It must adapt itself to its chosen means, a process which can radically change it; yet the political aim remains the first consideration. Policy, then, will permeate all military operations, and, in so far as their violent nature will admit, it will have a continuous influence on them.

24. War Is Merely the Continuation of Policy by Other Means

We see, therefore, that war is not merely an act of policy but a true political instrument, a continuation of political intercourse, carried on with other means. What remains peculiar to war is simply the peculiar nature of its means. War in general, and the commander in any specific instance, is entitled to require that the trend and designs of policy shall not be inconsistent with these means. That, of course, is no small demand; but however much it may affect political aims in a given case, it will never do more than modify them. The political object is the goal, war is the means of reaching it, and means can never be considered in isolation from their purpose.

25. The Diverse Nature of War

The more powerful and inspiring the motives for war, the more they affect the belligerent nations and the fiercer the tensions that precede the outbreak, the closer will war approach its abstract concept, the more important will be the destruction of the enemy, the more closely will the military aims and the political objects of war coincide, and the more military and less political will war appear to be. On the other hand, the less intense the motives, the less will the military element's natural tendency to violence coincide with political directives. As a result, war will be driven further from its natural course, the political object will be more and more at variance with the aim of ideal war, and the conflict will seem increasingly *political* in character.

At this point, to prevent the reader from going astray, it must be observed that the phrase, the *natural tendency* of war, is used in its philosophical, strictly *logical* sense alone and does not refer to the tendencies of the forces that are actually engaged in fighting—including, for instance, the morale and emotions of the combatants. At times, it is true, these might be so aroused that the political factor would be hard put to control them. Yet such a conflict will not occur very often, for if the motivations are so powerful there must be a policy of proportionate magnitude. On the other hand, if policy is directed only toward minor objectives, the emotions of the masses will be little stirred and they will have to be stimulated rather than held back.

26. All Wars Can Be Considered Acts of Policy

It is time to return to the main theme and observe that while policy is apparently effaced in the one kind of war and yet is strongly evident in the other, both kinds are equally political. If the state is thought of as a person, and policy as the product of its brain, then among the contingencies for which the state must be prepared is a war in which every element calls for policy to be eclipsed by violence. Only if politics is regarded not as resulting from a just appreciation of affairs, but—as it conventionally is—as cautious, devious, even dishonest, shying away from force, could the second type of war appear to be more "political" than the first.

27. The Effects of This Point of View on the Understanding of Military History and the Foundations of Theory

First, therefore, it is clear that war should never be thought of as *something autonomous* but always as an *instrument of policy*; otherwise the entire history of war would contradict us. Only this approach will enable us to penetrate the problem intelligently. *Second,* this way of looking at it will show us how wars must vary with the nature of their motives and of the situations which give rise to them.

The first, the supreme, the most far-reaching act of judgment that the statesman and commander have to make is to establish by that test the kind of war on which they are embarking; neither mistaking it for, nor trying to turn it into, something that is alien to its nature. This is the first of all strategic questions and the most comprehensive. It will be given detailed study later, in the chapter on war plans.

It is enough, for the moment, to have reached this stage and to have established the cardinal point of view from which war and the theory of war have to be examined.

28. The Consequences for Theory

War is more than a true chameleon that slightly adapts its characteristics to the given case. As a total phenomenon its dominant tendencies always make war a paradoxical trinity—composed of primordial violence, hatred, and enmity, which are to be regarded as a blind natural force; of the play of chance and probability within which the creative spirit is free to roam; and of its element of subordination, as an instrument of policy, which makes it subject to reason alone.

The first of these three aspects mainly concerns the people; the second the commander and his army; the third the government. The passions that are to be kindled in war must already be inherent in the people; the scope which the play of courage and talent will enjoy in the realm of probability and chance depends on the particular character of the commander and the army; but the political aims are the business of government alone.

These three tendencies are like three different codes of law, deep-rooted in their subject and yet variable in their relationship to one another. A theory that ignores any one of them or seeks to fix an arbitrary relationship between them would conflict with reality to such an extent that for this reason alone it would be totally useless.

Our task therefore is to develop a theory that maintains a balance between these three tendencies, like an object suspended between three magnets.

What lines might best be followed to achieve this difficult task will be explored in the book on the theory of war [Book Two]. At any rate, the preliminary concept of war which we have formulated casts a first ray of light on the basic structure of theory, and enables us to make an initial differentiation and identification of its major components.

CHAPTER TWO

Purpose and Means in War

The preceding chapter showed that the nature of war is complex and changeable. I now propose to inquire how its nature influences its purpose and its means.

If for a start we inquire into the objective of any particular war, which must guide military action if the political purpose is to be properly served, we find that the object of any war can vary just as much as its political purpose and its actual circumstances.

If for the moment we consider the pure concept of war, we should have to say that the political purpose of war had no connection with war itself; for if war is an act of violence meant to force the enemy to do our will its aim would have *always* and *solely* to be to overcome the enemy and disarm him. That aim is derived from the theoretical concept of war; but since many wars do actually come very close to fulfilling it, let us examine this kind of war first of all.

Later, when we are dealing with the subject of war plans, we shall investigate in greater detail what is meant by *disarming* a country. But we should at once distinguish between three things, three broad objectives, which between them cover everything: the *armed forces*, the *country*, and the *enemy's will*.

The fighting forces must be *destroyed*: that is, they must be *put in such a condition that they can no longer carry on the fight*. Whenever we use the phrase "destruction of the enemy's forces" this alone is what we mean.

The country must be occupied; otherwise the enemy could raise fresh military forces.

Yet both these things may be done and the war, that is the animosity and the reciprocal effects of hostile elements, cannot be considered to have ended so long as the enemy's *will* has not been broken: in other words, so long as the enemy government and its allies have not been driven to ask for peace, or the population made to submit.

We may occupy a country completely, but hostilities can be renewed again in the interior, or perhaps with allied help. This of course can also happen *after* the peace treaty, but this only shows that not every war necessarily leads to a final decision and settlement. But even if hostilities should occur again, a peace treaty will always extinguish a mass of sparks that might have gone on quietly smoldering. Further, tensions are slackened, for lovers of peace (and they abound among every people under all circumstances) will then abandon any thought of further action. Be that as it may, we must always consider that with the conclusion of peace the purpose of the war has been achieved and its business is at an end.

Since of the three objectives named, it is the fighting forces that assure the safety of the country, the natural sequence would be to destroy them first, and then subdue the country. Having achieved these two goals and exploiting our own position of strength, we can bring the enemy to the peace table. As a rule, destroying the enemy's forces tends to be a gradual process, as does the ensuing subjugation of the country. Normally the one reacts on the other, in that loss of territory weakens the fighting forces; but that particular sequence of events is not essential and therefore does not always take place. Before they suffer seriously, the enemy's forces may retire to remote areas, or even withdraw to other countries. In that event, of course, most or all of the country will be occupied.

But the aim of *disarming the enemy* (the object of *war in the abstract*, the ultimate means of accomplishing the war's political purpose, which should incorporate all the rest) is in fact not always encountered in reality, and need not be fully achieved as a condition of peace. On no account should theory raise it to the level of a law. Many treaties have been concluded before one of the antagonists could be called powerless—even before the balance of power had been seriously altered. What is more, a review of actual cases shows a whole category of wars in which the very idea of *defeating the enemy* is unreal: those in which the enemy is substantially the stronger power.

The reason why the object of war that emerges in theory is sometimes inappropriate to actual conflict is that war can be of two very different kinds, a point we discussed in the first chapter. If war were what pure theory postulates, a war between states of markedly unequal strength would be absurd, and so impossible. At most, material disparity could not go beyond the amount that moral factors could replace; and social conditions being what they are in Europe today, moral forces would not go far. But wars have in fact been fought between states of *very unequal strength, for actual war is often far removed from the pure concept postulated by theory.* Inability to carry on the struggle can, in practice, be replaced by two other grounds for making peace: the first is the improbability of victory; the second is its unacceptable cost.

As we saw in the first chapter, war, if taken as a whole, is bound to move from the strict law of inherent necessity toward probabilities. The more the circumstances that gave rise to the conflict cause it to do so, the slighter will be its motives and the tensions which it occasions. And this makes it understandable how an analysis of probabilities may lead to peace itself. Not every war need be fought until one side collapses. When the motives and tensions of war are slight we can imagine that the very faintest prospect of defeat might be enough to cause one side to yield. If from the very start the other side feels that this is probable, it will obviously concentrate on bringing about *this probability* rather than take the long way round and totally defeat the enemy.

Of even greater influence on the decision to make peace is the consciousness of all the effort that has already been made and of the efforts yet to come. Since war is not an act of senseless passion but is controlled by its political object, the value of this object must determine the sacrifices to be made for it in *magnitude* and also in *duration*. Once the expenditure of effort exceeds the value of the political object, the object must be renounced and peace must follow.

We see then that if one side cannot completely disarm the other, the desire for peace on either side will rise and fall with the probability of further successes and the amount of effort these would require. If such incentives were of equal strength on both sides, the two would resolve their political disputes by meeting half way. If the incentive grows on one side, it should diminish on the other. Peace will result so long as their sum total is sufficient—though the side that feels the lesser urge for peace will naturally get the better bargain.

One point is purposely ignored for the moment—the difference that the *positive* or *negative* character of the political ends is bound to produce in practice. As we shall see, the difference is important, but at this stage we must take a broader view because the original political objects can greatly alter during the course of the war and may finally change entirely *since they are influenced by events and their probable consequences.*

The question now arises how success can be made more likely. One way, of course, is to choose objectives that will incidentally bring about the enemy's collapse—*the destruction of his armed forces and the conquest of his territory*; but neither is quite what it would be if our real object were the total defeat of the enemy. When we attack the enemy, it is one thing if we mean our first operation to be followed by others until all resistance has been broken; it is quite another if our aim is only to obtain a single victory, in order to make the enemy insecure, to impress our greater strength upon him, and to give him doubts about his future. If that is the extent of our aim, we will employ no more strength than is absolutely necessary. In the same way, conquest of territory is a different matter if the enemy's collapse is not the object. If we wish to gain total victory, then the destruction of his armed forces is the most appropriate action and the occupation of his territory only a consequence. To occupy land before his armies are defeated should be considered at best a necessary evil. If on the other hand we do not aim at destroying the opposing army, and if we are convinced that the enemy does not seek a brutal decision, but rather *fears* it, then the seizure of a lightly held or undefended province is *an advantage in itself*; and should this advantage be enough to make the enemy fear for the final outcome, it can be considered as a short cut on the road to peace.

But there is another way. It is possible to increase the likelihood of success without defeating the enemy's forces. I refer to operations that have *direct political repercussions,* that are designed in the first place to disrupt the opposing alliance, or to paralyze it, that gain us new allies, favorably affect the political scene, etc. If such operations are possible it is obvious that they can greatly improve our prospects and that they can form a much shorter route to the goal than the destruction of the opposing armies.

The second question is how to influence the enemy's expenditure of effort; in other words, how to make the war more costly to him.

The enemy's expenditure of effort consists in the *wastage of his forces*—our *destruction* of them; and in his *loss of territory*—our *conquest*.

Closer study will make it obvious that both of these factors can vary in their significance with the variation in objectives. As a rule the differences will be slight, but that should not mislead us, for in practice, when strong motives are not present, the slightest nuances often decide between the different uses of force. For the moment all that matters is to show that, given certain conditions, different ways of reaching the objective are *possible* and that they are neither *inconsistent, absurd,* nor even *mistaken.*

In addition, there are three other methods directly aimed at increasing the enemy's expenditure of effort. The first of these is *invasion,* that is *the seizure of enemy territory; not with the object of retaining it* but in order to exact financial contributions, or even to lay it waste. The immediate object here is neither to conquer the enemy country nor to destroy its army, but simply *to cause general damage.* The second method is to give priority to operations that will increase the enemy's suffering. It is easy to imagine two alternatives: one operation is far more advantageous if the purpose is to defeat the enemy; the other is more profitable if that cannot be done. The first tends to be described as the more military, the second the more political alternative. From the highest point of view, however, one is as military as the other, and neither is appropriate unless it suits the particular conditions. The third, and far the most important method, judging from the frequency of its use, is *to wear down* the enemy. That expression is more than a label; it describes the process precisely, and is not so metaphorical as it may seem at first. Wearing down the enemy in a conflict means using *the duration of the war to bring about a gradual exhaustion of his physical and moral resistance.*

If we intend to hold out longer than our opponent we must be content with the smallest possible objects, for obviously a major object requires more effort than a minor one. The minimum object is *pure self-defense;* in other words, fighting without a positive purpose. With such a policy our relative strength will be at its height, and thus the prospects for a favorable outcome will be greatest. But how far can this negativity be pushed? Obviously not to the point of absolute passivity, for sheer endurance would not be fighting at all. But resistance is a form of action, aimed at destroying enough of the enemy's power to force him to renounce his intentions. Every single act of our resistance is directed to that act alone, and that is what makes our policy negative.

Undoubtedly a single action, assuming it succeeds, would do less for a negative aim than it would for a positive one. But that is just the difference: the former is more likely to succeed and so to give you more security. What it lacks in immediate effectiveness it must make up for in its use of time, that is by prolonging the war. Thus the negative aim, which lies at the heart of pure resistance, is also the natural formula for outlasting the enemy, for wearing him down.

Here lies the origin of the distinction that dominates the whole of war: the difference between *attack* and *defense*. We shall not pursue the matter now, but let us just say this: that from the negative purpose derive all the advantages, all the more effective forms, of fighting, and that in it is expressed the dynamic relationship between the magnitude and the likelihood of success. All this will be gone into later.

If a negative aim—that is, the use of every means available for pure resistance—gives an advantage in war, the advantage need only be enough to *balance* any superiority the opponent may possess: in the end his political object will not seem worth the effort it costs. He must then renounce his policy. It is evident that this method, wearing down the enemy, applies to the great number of cases where the weak endeavor to resist the strong.

Frederick the Great would never have been able to defeat Austria in the Seven Years War: and had he tried to fight in the manner of Charles XII he would unfailingly have been destroyed himself. But for seven years he skillfully husbanded his strength and finally convinced the allies that far greater efforts were needed than they had foreseen. Consequently they made peace.

We can now see that in war many roads lead to success, and that they do not all involve the opponent's outright defeat. They range from *the destruction of the enemy's forces, the conquest of his territory, to a temporary occupation or invasion, to projects with an immediate political purpose, and finally to passively awaiting the enemy's attacks.* Any one of these may be used to overcome the enemy's will: the choice depends on circumstances. One further kind of action, of shortcuts to the goal, needs mention: one could call them arguments *ad hominem.* Is there a field of human affairs where personal relations do not count, where the sparks they strike do not leap across all practical considerations? The personalities of statesmen and soldiers are such important factors that in war above all it is vital not to underrate them. It is enough to mention this point: it would be pedantic to attempt a systematic classification. It can be said, however, that these questions of personality and personal relations raise the number of possible ways of achieving the goal of policy to infinity.

To think of these shortcuts as rare exceptions, or to minimize the difference they can make to the conduct of war, would be to underrate them. To avoid that error we need only bear in mind how wide a range of political interests can lead to war, or think for a moment of the gulf that separates a war of annihilation, a struggle for political existence, from a war reluctantly declared in consequence of political pressure or of an alliance that no longer seems to reflect the state's true interests. Between these two extremes lie numerous gradations. If we reject a single one of them on theoretical grounds, we may as well reject them all, and lose contact with the real world.

CHAPTER TWO

So much then for the ends to be pursued in war; let us now turn to the means.

There is only one: *combat.* However many forms combat takes, however far it may be removed from the brute discharge of hatred and enmity of a physical encounter, however many forces may intrude which themselves are not part of fighting, it is inherent in the very concept of war that everything that occurs *must originally derive from combat.*

It is easy to show that this is always so, however many forms reality takes. Everything that occurs in war results from the existence of armed forces; *but whenever armed forces, that is armed individuals,* are used, the idea of combat must be present.

Warfare comprises everything related to the fighting forces—everything to do with their creation, maintenance, and use.

Creation and maintenance are obviously only means; their use constitutes the end.

Combat in war is not a contest between individuals. It is a whole made up of many parts, and in that whole two elements may be distinguished, one determined by the subject, the other by the objective. The mass of combatants in an army endlessly forms fresh elements, which themselves are parts of a greater structure. The fighting activity of each of these parts constitutes a more or less clearly defined element. Moreover, combat itself is made an element of war by its very purpose, by its *objective.*

Each of these elements which become distinct in the course of fighting is named an *engagement.*

If the idea of fighting underlies every use of the fighting forces, then their employment means simply the planning and organizing of a series of engagements.

The whole of military activity must therefore relate directly or indirectly to the engagement. The end for which a soldier is recruited, clothed, armed, and trained, the whole object of his sleeping, eating, drinking, and marching *is simply that he should fight at the right place and the right time.*

If all threads of military activity lead to the engagement, then if we control the engagement, we comprehend them all. Their results are produced by our orders and by the execution of these orders, never directly by other conditions. Since in the engagement everything is concentrated on the destruction of the enemy, or rather of *his armed forces,* which is inherent in its very concept, it follows that the destruction of the enemy's forces is always the means by which the purpose of the engagement is achieved.

The purpose in question may be the destruction of the enemy's forces, but not necessarily so; it may be quite different. As we have shown, the destruction of the enemy is not the only means of attaining the political object, when there are other objectives for which the war is waged. It follows that those other objectives can also become the purpose of particular military operations, and thus also the purpose of engagements.

Even when subordinate engagements are directly intended to destroy the opposing forces, that destruction still need not be their first, immediate concern.

Bearing in mind the elaborate structure of an army, and the numerous factors that determine its employment, one can see that the fighting activity of such a force is also subject to complex organization, division of functions, and combinations. The separate units obviously must often be assigned tasks that are not in themselves concerned with the destruction of the enemy's forces, which may indeed increase their losses but do so only indirectly. If a battalion is ordered to drive the enemy from a hill, a bridge, etc., the true purpose is normally to occupy that point. Destruction of the enemy's force is only a means to an end, a secondary matter. If a mere demonstration is enough to cause the enemy to abandon his position, the objective has been achieved; but as a rule the hill or bridge is captured only so that even more damage can be inflicted on the enemy. If this is the case on the battlefield, it will be even more so in the theater of operations, where it is not merely two armies that are facing each other, but two states, two peoples, two nations. The range of possible circumstances, and therefore of options, is greatly increased, as is the variety of dispositions; and the gradation of objects at various levels of command will further separate the first means from the ultimate purpose.

Thus there are many reasons why the purpose of an engagement may not be the destruction of the enemy's forces, the forces immediately confronting us. Destruction may be merely a means to some other end. In such a case, total destruction has ceased to be the point; the engagement is nothing but a *trial of strength*. In itself it is of no value; its significance lies in the outcome of the trial.

When one force is a great deal stronger than the other, an estimate may be enough. There will be no fighting: the weaker side will yield at once.

The fact that engagements do not always aim at the destruction of the opposing forces, that their objectives can often be attained without any fighting at all but merely by an evaluation of the situation, explains why entire campaigns can be conducted with great energy even though actual fighting plays an unimportant part in them.

This is demonstrated by hundreds of examples in the history of war. Here we are only concerned to show that it is *possible*; we need not ask how often it was appropriate, in other words *consistent with the overall purpose*, to avoid the test of battle, or whether all the reputations made in such campaigns would stand the test of critical examination.

There is only one means in war: combat. But the multiplicity of forms that combat assumes leads us in as many different directions as are created by the multiplicity of aims, so that our analysis does not seem to have made any progress. But that is not so: the fact that only one means exists constitutes a strand that runs through the entire web of military activity and really holds it together.

We have shown that the destruction of the enemy's forces is one of the many objects that can be pursued in war, and we have left aside the question of its importance relative to other purposes. In any given case the answer will depend on circumstances; its importance to war in general remains to be clarified. We shall now go into this question, and we shall see what value must necessarily be attributed to this object of destruction.

Combat is the only effective force in war; its aim is to destroy the enemy's forces as a means to a further end. That holds good even if no actual fighting occurs, because the outcome rests on the assumption that if it came to fighting, the enemy would be destroyed. It follows that the destruction of the enemy's force underlies all military actions; all plans are ultimately based on it, resting on it like an arch on its abutment. Consequently, all action is undertaken in the belief that if the ultimate test of arms should actually occur, the outcome would be *favorable.* The decision by arms is for all major and minor operations in war what cash payment is in commerce. Regardless how complex the relationship between the two parties, regardless how rarely settlements actually occur, they can never be entirely absent.

If a decision by fighting is the basis of all plans and operations, it follows that the enemy *can frustrate everything through a successful battle.* This occurs not only when the encounter affects an essential factor in our plans, but when any victory that is won is of sufficient scope. For every important victory—that is, destruction of opposing forces—reacts on all other possibilities. Like liquid, they will settle at a new level.

Thus it is evident that destruction of the enemy forces is always the superior, more effective means, with which others cannot compete.

But of course, we can only say destruction of the enemy is more effective if we can assume that all other conditions are equal. It would be a great mistake to deduce from this argument that a headlong rush must always triumph over skillful caution. Blind aggressiveness would destroy the attack itself, not the defense, and this is not what we are talking about. Greater effectiveness relates not to the *means* but to the *end;* we are simply comparing the effect of different outcomes.

When we speak of destroying the enemy's forces we must emphasize that nothing obliges us to limit this idea to physical forces: the moral element must also be considered. The two interact throughout: they are inseparable. We have just mentioned the effect that a great destructive act—a major victory—inevitably exerts on all other actions, and it is exactly at such times that the moral factor is, so to speak, the most fluid element of all, and therefore spreads most easily to affect everything else. The advantage that the destruction of the enemy possesses over all other means is balanced by its cost and danger; and it is only in order to avoid these risks that other policies are employed.

That the method of destruction cannot fail to be expensive is understandable; other things being equal, the more intent we are on destroying the enemy's forces, the greater our own efforts must be.

The danger of this method is that the greater the success we seek, the greater will be the damage if we fail.

Other methods, therefore, are less costly if they succeed and less damaging if they fail, though this holds true only if both sides act identically, if the enemy pursues the same course as we do. If he were to seek the decision through a major battle, *his choice would force us against our will to do likewise*. Then the outcome of the battle would be decisive; but it is clear—other things again being equal—that we would be at an overall disadvantage, since our plans and resources had been in part intended to achieve other goals, whereas the enemy's were not. Two objectives, neither of which is part of the other, are mutually exclusive: one force cannot simultaneously be used for both. If, therefore, one of the two commanders is resolved to seek a decision through major battles, he will have an excellent chance of success if he is certain that his opponent is pursuing a different policy. Conversely, the commander who wishes to adopt different means can reasonably do so only if he assumes his opponent to be equally unwilling to resort to major battles.

What has been said about plans and forces being directed to other uses refers only to the *positive* purposes, other than the destruction of enemy forces, that can be pursued in war. It pertains *in no way to pure resistance*, which seeks to wear down the opponent's strength. Pure resistance has no *positive* intention; we can use our forces only to frustrate the enemy's intentions, and not divert them to other objectives.

Here we must consider the negative side of destroying the enemy's forces—that is, the preservation of our own. These two efforts always go together; they interact. They are integral parts of a single purpose, and we only need to consider the result if one or the other dominates. The effort to destroy the enemy's forces has a positive purpose and leads to positive results, whose final aim is the enemy's collapse. Preserving our own forces has a negative purpose; it frustrates the enemy's intentions—that is, it amounts to pure resistance, whose ultimate aim can only be to prolong the war until the enemy is exhausted.

The policy with a positive purpose calls the act of destruction into being; the policy with a negative purpose waits for it.

How far such a waiting attitude may or should be maintained is a question we shall study in connection with the theory of attack and defense, whose basic element is here involved. For the moment we need only say that a policy of waiting must never become passive endurance, that any action involved in it may just as well seek the destruction of the opposing forces as any other objective. It would be a fundamental error to imagine that a negative aim implies a preference for a bloodless decision over the destruction of the enemy. A preponderantly negative effort may of course lead to such a choice, but always at the risk that it is not the appropriate course: that depends on factors that are determined not by us but by the opponent. Avoidance of bloodshed, then, should not be taken as an act of policy if our main concern is to preserve our forces. On the contrary, if such a policy did not suit the particular situation it would lead our forces to disaster. A great many generals have failed through this mistaken assumption.

The one certain effect a preponderantly negative policy will have is to retard the decision: in other words, action is transposed into waiting for the decisive moment. This usually means that *action is postponed* in time and space to the extent that space is relevant and circumstances permit. If the time arrives when further waiting would bring excessive disadvantages, then the benefit of the negative policy has been exhausted. The destruction of the enemy—an aim that has until then been postponed but not displaced by another consideration—now reemerges.

Our discussion has shown that while in war many different roads can lead to the goal, to the attainment of the political object, fighting is the only possible means. Everything is governed by a supreme law, the *decision by force of arms.* If the opponent does seek battle, this recourse can never be denied him. A commander who prefers another strategy must first be sure that his opponent either will not appeal to that supreme tribunal—force—or that he will lose the verdict if he does. To sum up: of all the possible aims in war, the destruction of the enemy's armed forces always appears as the highest.

At a later stage and by degrees we shall see what other kinds of strategies can achieve in war. All we need to do for the moment is to admit the general *possibility of their existence,* the possibility of deviating from the basic concept of war under the pressure of special circumstances. But even at this point we must not fail to emphasize that the *violent resolution of the crisis,* the wish to annihilate the enemy's forces, is the first-born son of war. If the political aims are small, the motives slight and tensions low, a prudent general may look for any way to avoid major crises and decisive actions, exploit any weaknesses in the opponent's military and political strategy, and finally reach a peaceful settlement. If his assumptions are sound and promise success we are not entitled to criticize him. But he must never forget that he is moving on devious paths where the god of war may catch him unawares. He must always keep an eye on his opponent so that he does not, if the latter has taken up a sharp sword, approach him armed only with an ornamental rapier.

These conclusions concerning the nature of war and the function of its purposes and means; the manner in which war in practice deviates in varying degrees from its basic, rigorous concept, taking this form or that, but always remaining subject to that basic concept, as to a supreme law; all these points must be kept in mind in our subsequent analyses if we are to perceive the real connections between all aspects of war, and the true significance of each; and if we wish to avoid constantly falling into the wildest inconsistencies with reality and even with our own arguments.

CHAPTER THREE

On Military Genius

Any complex activity, if it is to be carried on with any degree of virtuosity, calls for appropriate gifts of intellect and temperament. If they are outstanding and reveal themselves in exceptional achievements, their possessor is called a "genius."

We are aware that this word is used in many senses, differing both in degree and in kind. We also know that some of these meanings make it difficult to establish the essence of genius. But since we claim no special expertise in philosophy or grammar, we may be allowed to use the word in its ordinary meaning, in which "genius" refers to a very highly developed mental aptitude for a particular occupation.

Let us discuss this faculty, this distinction of mind for a moment, setting out its claims in greater detail, so as to gain a better understanding of the concept. But we cannot restrict our discussion to *genius* proper, as a superlative degree of talent, for this concept lacks measurable limits. What we must do is to survey all those gifts of mind and temperament that in combination bear on military activity. These, taken together, constitute *the essence of military genius*. We have said *in combination*, since it is precisely the essence of military genius that it does not consist in a single appropriate gift—courage, for example—while other qualities of mind or temperament are wanting or are not suited to war. Genius consists *in a harmonious combination of elements*, in which one or the other ability may predominate, but none may be in conflict with the rest.

If every soldier needed some degree of military genius our armies would be very weak, for the term refers to a special cast of mental or moral powers which can rarely occur in an army when a society has to employ its abilities in many different areas. The smaller the range of activities of a nation and the more the military factor dominates, the greater will be the incidence of military genius. This, however, is true only of its distribution, not of its quality. The latter depends on the *general intellectual development* of a given society. In any primitive, warlike race, the warrior spirit is far more common than among civilized peoples. It is possessed by almost every warrior: but in civilized societies only necessity will stimulate it in the people as a whole, since they lack the natural disposition for it. On the other hand, we will never find a savage who is a truly great commander, and very rarely one who would be considered a military genius, since this requires a degree of intellectual powers beyond anything that a primitive people can develop. Civilized societies, too, can obviously possess a warlike character to greater or lesser degree, and the more they develop it, the greater will be the number of men with military spirit in their armies. Possession of military genius coincides with the higher degrees of civilization: the most highly developed societies produce the most brilliant soldiers, as the Romans and the French have shown us. With them, as with every people renowned in war, the greatest names do not appear before a high level of civilization has been reached.

We can already guess how great a role intellectual powers play in the higher forms of military genius. Let us now examine the matter more closely.

War is the realm of danger; therefore *courage* is the soldier's first requirement.

Courage is of two kinds: courage in the face of personal danger, and courage to accept responsibility, either before the tribunal of some outside power or before the court of one's own conscience. Only the first kind will be discussed here.

Courage in face of personal danger is also of two kinds. It may be indifference to danger, which could be due to the individual's constitution, or to his holding life cheap, or to habit. In any case, it must be regarded as a permanent *condition*. Alternatively, courage may result from such positive motives as ambition, patriotism, or enthusiasm of any kind. In that case courage is a feeling, an emotion, not a permanent state.

These two kinds of courage act in different ways. The first is the more dependable; having become second nature, it will never fail. The other will often achieve more. There is more reliability in the first kind, more boldness in the second. The first leaves the mind calmer; the second tends to stimulate, but it can also blind. *The highest kind of courage is a compound of both.*

War is the realm of physical exertion and suffering. These will destroy us unless we can make ourselves indifferent to them, and for this birth or training must provide us with a certain strength of body and soul. If we do possess those qualities, then even if we have nothing but common sense to guide them we shall be well equipped for war: it is exactly these qualities that primitive and semicivilized peoples usually possess.

If we pursue the demands that war makes on those who practice it, we come to the region dominated by the *powers of intellect*. War is the realm of uncertainty; three quarters of the factors on which action in war is based are wrapped in a fog of greater or lesser uncertainty. A sensitive and discriminating judgment is called for; a skilled intelligence to scent out the truth.

Average intelligence may recognize the truth occasionally, and exceptional courage may now and then retrieve a blunder; but usually intellectual inadequacy will be shown up by indifferent achievement.

War is the realm of chance. No other human activity gives it greater scope: no other has such incessant and varied dealings with this intruder. Chance makes everything more uncertain and interferes with the whole course of events.

Since all information and assumptions are open to doubt, and with chance at work everywhere, the commander continually finds that things are not as he expected. This is bound to influence his plans, or at least the assumptions underlying them. If this influence is sufficiently powerful to cause a change in his plans, he must usually work out new ones; but for these the necessary information may not be immediately available. During an operation decisions have usually to be made at once: there may be no time to review the situation or even to think it through. Usually, of course, new information and reevaluation are not enough to make us give up our intentions: they only call them in question. We now know more, but this makes us more, not less uncertain. The latest reports do not arrive all at once: they merely trickle in. They continually impinge on our decisions, and our mind must be permanently armed, so to speak, to deal with them.

If the mind is to emerge unscathed from this relentless struggle with the unforeseen, two qualities are indispensable: *first, an intellect that, even in the darkest hour, retains some glimmerings of the inner light which leads to truth; and second, the courage to follow this faint light wherever it may lead.* The first of these qualities is described by the French term, *coup d'oeil*; the second is *determination.*

The aspect of war that has always attracted the greatest attention is the engagement. Because time and space are important elements of the engagement, and were particularly significant in the days when the cavalry attack was the decisive factor, the *idea of a rapid and accurate decision* was first based on an evaluation of time and space, and consequently received a name which refers to visual estimates only. Many theorists of war have employed the term in that limited sense. But soon it was also used of any sound decision taken in the midst of action—such as recognizing the right point to attack, etc. *Coup d'oeil* therefore refers not alone to the physical but, more commonly, to the inward eye. The expression, like the quality itself, has certainly always been more applicable to tactics, but it must also have its place in strategy, since here as well quick decisions are often needed. Stripped of metaphor and of the restrictions imposed on it by the phrase, the concept merely refers to the quick recognition of a truth that the mind would ordinarily miss or would perceive only after long study and reflection.

Determination in a single instance is an expression of courage; if it becomes characteristic, a mental habit. But here we are referring not to physical courage but to the courage to accept responsibility, courage in the face of a moral danger. This has often been called *courage d'esprit*, because it is created by the intellect. That, however, does not make it an act of the intellect: it is an act of temperament. Intelligence alone is not courage; we often see that the most intelligent people are irresolute. Since in the rush of events a man is governed by feelings rather than by thought, the intellect needs to arouse the quality of courage, which then supports and sustains it in action.

Looked at in this way, the role of determination is to limit the agonies of doubt and the perils of hesitation when the motives for action are inadequate. Colloquially, to be sure, the term "determination" also applies to a propensity for daring, pugnacity, boldness, or temerity. But when a man has adequate grounds for action—whether subjective or objective, valid or false—he cannot properly be called "determined." This would amount to putting oneself in his position and weighting the scale with a doubt that he never felt. In such a case it is only a question of strength or weakness. I am not such a pedant as to quarrel with common usage over a slight misuse of a word; the only purpose of these remarks is to preclude misunderstandings.

Determination, which dispells doubt, is a quality that can be aroused only by the intellect, and by a specific cast of mind at that. More is required to create determination than a mere conjunction of superior insight with the appropriate emotions. Some may bring the keenest brains to the most formidable problems, and may possess the courage to accept serious responsibilities; but when faced with a difficult situation they still find themselves unable to reach a decision. Their courage and their intellect work in separate compartments, not together; determination, therefore, does not result. It is engendered only by a *mental act*; the mind tells man that boldness is required, and thus gives direction to his will. This particular cast of mind, which employs the fear of *wavering* and *hesitating* to suppress all other fears, is the force that makes strong men determined. Men of low intelligence, therefore, cannot possess determination in the sense in which we use the word. They may act without hesitation in a crisis, but if they do, they act *without reflection*; and a man who acts without reflection cannot, of course, be torn by doubt. From time to time action of this type may even be appropriate; but, as I have said before, it is the *average result* that indicates the existence of military genius. The statement may surprise the reader who knows some determined cavalry officers who are little given to deep thought: but he must remember that we are talking about a special kind of intelligence, not about great powers of meditation.

In short, we believe that determination proceeds from a special type of mind, from a strong rather than a brilliant one. We can give further proof of this interpretation by pointing to the many examples of men who show great determination as junior officers, but lose it as they rise in rank. Conscious of the need to be decisive, they also recognize the risks entailed by a *wrong* decision; since they are unfamiliar with the problems now facing them, their mind loses its former incisiveness. The more used they had been to instant action, the more their timidity increases as they realize the dangers of the vacillation that ensnares them.

Having discussed *coup d'oeil* and determination it is natural to pass to a related subject: *presence of mind*. This must play a great role in war, the domain of the unexpected, since it is nothing but an increased capacity of dealing with the unexpected. We admire presence of mind in an apt repartee, as we admire quick thinking in the face of danger. Neither needs to be exceptional, so long as it meets the situation. A reaction following long and deep reflection may seem quite commonplace; as an immediate response, it may give keen pleasure. The expression "presence of mind" precisely conveys the speed and immediacy of the help provided by the intellect.

Whether this splendid quality is due to a special cast of mind or to steady nerves depends on the nature of the incident, but neither can ever be entirely lacking. A quick retort shows wit; resourcefulness in sudden danger calls, above all, for steady nerve.

Four elements make up the climate of war: danger, exertion, uncertainty, and chance. If we consider them together, it becomes evident how much fortitude of mind and character are needed to make progress in these impeding elements with safety and success. According to circumstance, reporters and historians of war use such terms as *energy, firmness, staunchness, emotional balance*, and *strength of character*. These products of a heroic nature could almost be treated as one and the same force—strength of will—which adjusts itself to circumstances: but though closely linked, they are not identical. A closer study of the interplay of psychological forces at work here may be worth while.

To begin with, clear thought demands that we keep one point in mind: of the weight, the burden, the resistance—call it what you like—that challenges the psychological strength of the soldier, only a small part is the *direct result of the enemy's activity, his resistance, or his operations*. The direct and primary impact of enemy activity falls, initially, on the soldier's person without affecting him in his capacity as commander. If, for example, the enemy resists four hours instead of two, the commander is in danger twice as long; but the higher an officer's rank, the less significant this factor becomes, and to the commander-in-chief it means nothing at all.

A second way in which the enemy's resistance *directly* affects the commander is the loss that is caused by prolonged resistance and the influence this exerts on his sense of responsibility. The deep anxiety which he must experience works on his strength of will and puts it to the test. Yet we believe that this is not by any means the heaviest burden he must bear, for he is answerable to himself alone. All other effects of enemy action, however, are felt by the men under his command, and *through them react on him*.

So long as a unit fights cheerfully, with spirit and elan, great strength of will is rarely needed; but once conditions become difficult, as they must when much is at stake, things no longer run like a well-oiled machine. The machine itself begins to resist, and the commander needs tremendous willpower to overcome this resistance. The machine's *resistance* need not consist of disobedience and argument, though this occurs often enough in individual soldiers. It is the impact of the ebbing of moral and physical strength, of the heart-rending spectacle of the dead and wounded, that the commander has to withstand—first in himself, and then in all those who, directly or indirectly, have entrusted him with their thoughts and feelings, hopes and fears. As each man's strength gives out, as it no longer responds to his will, the inertia of the whole gradually comes to rest on the commander's will alone. The ardor of his spirit must rekindle the flame of purpose in all others;

his inward fire must revive their hope. Only to the extent that he can do this will he retain his hold on his men and keep control. Once that hold is lost, once his own courage can no longer revive the courage of his men, the mass will drag him down to the brutish world where danger is shirked and shame is unknown. Such are the burdens in battle that the commander's courage and strength of will must overcome if he hopes to achieve outstanding success. The burdens increase with the number of men in his command, and therefore the higher his position, the greater the strength of character he needs to bear the mounting load.

Energy in action varies in proportion to the strength of its motive, whether the motive be the result of intellectual conviction or of emotion. Great strength, however, is not easily produced where there is no emotion.

Of all the passions that inspire man in battle, none, we have to admit, is so powerful and so constant as the longing for honor and renown. The German language unjustly tarnishes this by associating it with two ignoble meanings in the terms "greed for honor" (*Ehrgeiz*) and "hankering after glory" (*Ruhmsucht*). The abuse of these noble ambitions has certainly inflicted the most disgusting outrages on the human race; nevertheless their origins entitle them to be ranked among the most elevated in human nature. In war they act as the essential breath of life that animates the inert mass. Other emotions may be more common and more venerated—patriotism, idealism, vengeance, enthusiasm of every kind—but they are no substitute for a thirst for fame and honor. They may, indeed, rouse the mass to action and inspire it, but they cannot give the commander the ambition to strive higher than the rest, as he must if he is to distinguish himself. They cannot give him, as can ambition, a personal, almost proprietary interest in every aspect of fighting, so that he turns each opportunity to best advantage—plowing with vigor, sowing with care, in the hope of reaping with abundance. It is primarily this spirit of endeavor on the part of commanders at all levels, this inventiveness, energy, and competitive enthusiasm, which vitalizes an army and makes it victorious. And so far as the commander-in-chief is concerned, we may well ask whether history has ever known a great general who was not ambitious; whether, indeed, such a figure is conceivable.

Staunchness indicates the will's resistance to a single blow; *endurance* refers to prolonged resistance.

Though the two terms are similar and are often used interchangeably, the difference between them is significant and unmistakable. Staunchness in face of a single blow may result from strong emotion, whereas intelligence helps sustain endurance. The longer an action lasts, the more deliberate endurance becomes, and this is one of its sources of strength.

We now turn to *strength of mind*, or of *character*, and must first ask what we mean by these terms.

Not, obviously, vehement display of feeling, or passionate temperament: that would strain the meaning of the phrase. We mean the ability to keep one's head at times of exceptional stress and violent emotion. Could strength of intellect alone account for such a faculty? We doubt it. Of course the opposite does not flow from the fact that some men of outstanding intellect do lose their self-control; it could be argued that a powerful rather than a capacious mind is what is needed. But it might be closer to the truth to assume that the faculty known as *self-control*—the gift of keeping calm even under the greatest stress—is rooted in temperament. It is itself an emotion which serves to balance the passionate feelings in strong characters without destroying them, and it is this balance alone that assures the dominance of the intellect. The counterweight we mean is simply the sense of human dignity, the noblest pride and deepest need of all: the urge *to act rationally at all times.* Therefore we would argue that a strong character is one *that will not be unbalanced by the most powerful emotions.*

If we consider how men differ in their emotional reactions, we first find a group with small capacity for being roused, usually known as "stolid" or "phlegmatic."

Second, there are men who are extremely active, but whose feelings never rise above a certain level, men whom we know to be sensitive but calm.

Third, there are men whose passions are easily inflamed, in whom excitement flares up suddenly but soon burns out, like gunpowder. And finally we come to those who do not react to minor matters, who will be moved only very gradually, not suddenly, but whose emotions attain great strength and durability. These are the men whose passions are strong, deep, and concealed.

These variants are probably related to the *physical forces* operating in the human being—they are part of that dual organism we call the nervous system, one side of which is physical, the other psychological. With our slight scientific knowledge we have no business to go farther into that obscure field; it is important nonetheless to note the ways in which these various psychological combinations can affect military activity, and to find out how far one can look for great strength of character among them.

Stolid men are hard to throw off balance, but total lack of vigor cannot really be interpreted as strength of character. It cannot be denied, however, that the imperturbability of such men gives them a certain narrow usefulness in war. They are seldom strongly motivated, lack initiative and consequently are not particularly active; on the other hand they seldom make a serious mistake.

The salient point about the second group is that trifles can suddenly stir them to act, whereas great issues are likely to overwhelm them. This kind of man will gladly help an individual in need, but the misfortune of an entire people will only sadden him; they will not stimulate him to action.

In war such men show no lack of energy or balance, but they are unlikely to achieve anything significant unless they possess a *very powerful intellect* to provide the needed stimulus. But it is rare to find this type of temperament combined with a strong and independent mind.

Inflammable emotions, feelings that are easily roused, are in general of little value in practical life, and therefore of little value in war. Their impulses are strong but brief. If the energy of such men is joined to courage and ambition they will often prove most useful at a modest level of command, simply because the action controlled by junior officers is of short duration. Often a single brave decision, a burst of emotional force, will be enough. A daring assault is the work of a few minutes, while a hard-fought battle may last a day, and a campaign an entire year.

Their volatile emotions make it doubly hard for such men to preserve their balance; they often lose their heads, and nothing is worse on active service. All the same, it would be untrue to say that highly excitable minds could never be strong—that is, could never keep their balance even under the greatest strain. Why should they not have a sense of their own dignity, since as a rule they are among the finer natures? In fact, they usually have such a sense, but there is not time for it to take effect. Once the crisis is past, they tend to be ashamed of their behavior. If training, self-awareness, and experience sooner or later teaches them how to be on guard against themselves, then in times of great excitement an internal counterweight will assert itself so that they too can draw on great strength of character.

Lastly, we come to men who are difficult to move but have strong feelings—men who are to the previous type like heat to a shower of sparks. These are the men who are best able to summon the titanic strength it takes to clear away the enormous burdens that obstruct activity in war. Their emotions move as great masses do—slowly but irresistibly.

These men are not swept away by their emotions so often as is the third group, but experience shows that they too can lose their balance and be overcome by blind passion. This can happen whenever they lack the noble pride of self-control, or whenever it is inadequate. We find this condition mostly among great men in primitive societies, where passion tends to rule for lack of intellectual discipline. Yet even among educated peoples and civilized societies men are often swept away by passion, just as in the Middle Ages poachers chained to stags were carried off into the forest.

We repeat again: strength of character does not consist solely in having powerful feelings, but in maintaining one's balance in spite of them. Even with the violence of emotion, judgment and principle must still function like a ship's compass, which records the slightest variations however rough the sea.

We say a man has strength of character, or simply has character, if he sticks to his convictions, whether these derive from his own opinions or someone else's, whether they represent principles, attitudes, sudden insights, or any other mental force. Such *firmness* cannot show itself, of course, if a man keeps changing his mind. This need not be the consequence of external influence; the cause may be the workings of his own intelligence, but this would suggest a peculiarly insecure mind. Obviously a man whose opinions are constantly changing, even though this is in response to his own reflections, would not be called a *man of character*. The term is applied only to men whose views are *stable and constant*. This may be because they are well thought-out, clear, and scarcely open to revision; or, in the case of indolent men, because such people are not in the habit of mental effort and therefore have no reason for altering their views; and finally, because a firm decision, based on fundamental principle derived from reflection, is relatively immune to changes of opinion.

With its mass of vivid impressions and the doubts which characterize all information and opinion, there is no activity like war to rob men of confidence in themselves and in others, and to divert them from their original course of action.

In the dreadful presence of suffering and danger, emotion can easily overwhelm intellectual conviction, and in this psychological fog it is so hard to form clear and complete insights that changes of view become more understandable and excusable. Action can never be based on anything firmer than instinct, a sensing of the truth. Nowhere, in consequence, are differences of opinion so acute as in war, and fresh opinions never cease to batter at one's convictions. No degree of calm can provide enough protection: new impressions are too powerful, too vivid, and always assault the emotions as well as the intellect.

Only those general principles and attitudes that result from clear and deep understanding can provide a *comprehensive* guide to action. It is to these that opinions on specific problems should be anchored. The difficulty is to hold fast to these results of contemplation in the torrent of events and new opinions. Often there is a gap between principles and actual events that cannot always be bridged by a succession of logical deductions. Then a measure of self-confidence is needed, and a degree of skepticism is also salutary. Frequently nothing short of an imperative principle will suffice, which is not part of the immediate thought-process, but dominates it: that principle is in all doubtful cases *to stick to one's first opinion and to refuse to change unless forced to do so by a clear conviction.* A strong faith in the overriding truth of tested principles is needed; the *vividness* of transient impressions must not make us forget that such truth as they contain is of a lesser stamp. By giving precedence, in case of doubt, to our earlier convictions, by holding to them stubbornly, our actions acquire that quality of steadiness and consistency which is termed strength of character.

It is evident how greatly strength of character depends on balanced temperament; most men of emotional strength and stability are therefore men of powerful character as well.

Strength of character can degenerate into *obstinacy.* The line between them is often hard to draw in a specific case; but surely it is easy to distinguish them in theory.

Obstinacy *is not an intellectual defect*; it comes from reluctance to admit that one is wrong. To impute this to the mind would be illogical, for the mind is the seat of judgment. Obstinacy *is a fault of temperament.* Stubbornness and intolerance of contradiction result from a special kind of *egotism,* which elevates above everything else *the pleasure of its autonomous intellect, to which others must bow.* It might also be called vanity, if it were not something superior: vanity is content with the appearance alone; obstinacy demands the material reality.

We would therefore argue that strength of character turns to obstinacy as soon as a man resists another point of view not from superior insight or attachment to some higher principle, but because he *objects instinctively*. Admittedly, this definition may not be of much practical use; but it will nevertheless help us avoid the interpretation that obstinacy is simply a more intense form of strong character. There is a basic difference between the two. They are closely related, but one is so far from being *a higher degree* of the other that we can even find extremely obstinate men who are too dense to have much strength of character.

So far our survey of the attributes that a great commander needs in war has been concerned with qualities in which mind and temperament work together. Now we must address ourselves to a special feature of military activity—possibly the most striking even though it is not the most important—which is not related to temperament, and involves merely the intellect. I mean the relationship between warfare and terrain.

This relationship, to begin with, is *a permanent factor*—so much so that one cannot conceive of a regular army operating except in a definite space. Second, its importance is *decisive in the highest degree*, for it affects the operations of all forces, and at times entirely alters them. Third, its influence may be felt in the *very smallest feature of the ground*, but it can also dominate *enormous areas*.

In these ways the relationship between warfare and terrain determines the peculiar character of military action. If we consider other activities connected with the soil—gardening, for example, farming, building, hydraulic engineering, mining, game-keeping, or forestry—none extends to more than a very limited area, and a working knowledge of that area is soon acquired. But a commander must submit his work to a partner, space, which he can never completely reconnoiter, and which because of the constant movement and change to which he is subject he can never really come to know. To be sure, the enemy is generally no better off; but the handicap, though shared, is still a handicap, and the man with enough talent and experience to overcome it will have a real advantage. Moreover it is only in a general sense that the difficulty is the same for both sides; in any particular case the defender usually knows the area far better than his opponent.

This problem is unique. To master it a special gift is needed, which is given the too restricted name of *a sense of locality*. It is the faculty of *quickly and accurately grasping the topography of any area* which enables a man to find his way about at any time. Obviously this is an act of the imagination. Things are perceived, of course, partly by the naked eye and partly by the mind, which fills the gaps with guesswork based on learning and experience, and thus constructs a whole out of the fragments that the eye can see; but if the whole is to be vividly present to the mind, imprinted like a picture, like a map, upon the brain, without fading or blurring in detail, *it can only be achieved by the mental gift that we call imagination*. A poet or painter may be shocked to find that his Muse dominates these activities as well: to him it might seem odd to say that a young gamekeeper needs an unusually powerful imagination in order to be competent. If so, we gladly admit that this is to apply the concept narrowly and to a modest task. But however remote the connection, his skill must still derive from this natural gift, for if imagination is entirely lacking it would be difficult to combine details into a clear, coherent image. We also admit that a good memory can be a great help; but are we then to think of memory as a separate gift of the mind, or does imagination, after all, imprint those pictures in the memory more clearly? The question must be left unanswered, especially since it seems difficult even to conceive of these two forces as operating separately.

That practice and a trained mind have much to do with it is undeniable. Puységur, the celebrated quarter-master-general of Marshal Luxembourg, writes that at the beginning of his career he had little faith in his sense of locality; when he had to ride any distance at all to get the password, he invariably lost his way.

Scope for this talent naturally grows with increased authority. A hussar or scout leading a patrol must find his way easily among the roads and tracks. All he needs are a few landmarks and some modest powers of observation and imagination. A commander-in-chief, on the other hand, must aim at acquiring an overall knowledge of the configuration of a province, of an entire country. His mind must hold a vivid picture of the road-network, the river-lines and the mountain ranges, without ever losing a sense of his immediate surroundings. Of course he can draw general information from reports of all kinds, from maps, books, and memoirs. Details will be furnished by his staff. Nevertheless it is true that with a quick, unerring sense of locality his dispositions will be more rapid and assured; he will run less risk of a certain awkwardness in his concepts, and be less dependent on others.

We attribute this ability to the imagination; but that is about the only service that war can demand from this frivolous goddess, who in most military affairs is liable to do more harm than good.

With this, we believe, we have reached the end of our review of the intellectual and moral powers that human nature needs to draw upon in war. The vital contribution of intelligence is clear throughout. No wonder then, that war, though it may appear to be uncomplicated, cannot be waged with distinction except by men of outstanding intellect.

Once this view is adopted, there is no longer any need to think that it takes a great intellectual effort to outflank an enemy position (an obvious move, performed innumerable times) or to carry out a multitude of similar operations.

It is true that we normally regard the plain, efficient soldier as the very opposite of the contemplative scholar, or of the inventive intellectual with his dazzling range of knowledge. This antithesis is not entirely unrealistic; but it does not prove that courage alone will make an efficient soldier, or that having brains and using them is not a necessary part of being a good fighting man. Once again we must insist: no case is more common than that of the officer whose energy declines as he rises in rank and fills positions that are beyond his abilities. But we must also remind the reader that outstanding effort, the kind that gives men a distinguished name, is what we have in mind. Every level of command has its own intellectual standards, its own prerequisites for fame and honor.

A major gulf exists between a commander-in-chief—a general who leads the army as a whole or commands in a theater of operations—and the senior generals immediately subordinate to him. The reason is simple: the second level is subjected to much closer control and supervision, and thus gives far less scope for independent thought. People therefore often think outstanding intellectual ability is called for only at the top, and that for all other duties common intelligence will suffice. A general of lesser responsibility, an officer grown gray in the service, his mind well-blinkered by long years of routine, may often be considered to have developed a certain stodginess; his gallantry is respected, but his simplemindedness makes us smile. We do not intend to champion and promote these good men; it would contribute nothing to their efficiency, and little to their happiness. We only wish to show things as they are, so that the reader should not think that a brave but brainless fighter can do anything of outstanding significance in war.

Since in our view even junior positions of command require outstanding intellectual qualities for outstanding achievement, and since the standard rises with every step, it follows that we recognize the abilities that are needed if the second positions in an army are to be filled with distinction. Such officers may appear to be rather simple compared to the polymath scholar, the far-ranging business executive, the statesman; but we should not dismiss the value of their practical intelligence. It sometimes happens of course that someone who made his reputation in one rank carries it with him when he is promoted, without really deserving to. If not much is demanded of him, and he can avoid exposing his incompetence, it is difficult to decide what reputation he really deserves. Such cases often cause one to hold in low estimate soldiers who in less responsible positions might do excellent work.

Appropriate talent is needed at all levels if distinguished service is to be performed. But history and posterity reserve the name of "genius" for those who have excelled in the highest positions—as commanders-in-chief—since here the demands for intellectual and moral powers are vastly greater.

To bring a war, or one of its campaigns, to a successful close requires a thorough grasp of national policy. On that level strategy and policy coalesce: the commander-in-chief is simultaneously a statesman.

Charles XII of Sweden is not thought of as a great genius, for he could never subordinate his military gifts to superior insights and wisdom, and could never achieve a great goal with them. Nor do we think of Henry IV of France in this manner: he was killed before his skill in war could affect the relations between states. Death denied him the chance to prove his talents in this higher sphere, where noble feelings and a generous disposition, which effectively appeased internal dissension, would have had to face a more intractable opponent.

The great range of business that a supreme commander must swiftly absorb and accurately evaluate has been indicated in the first chapter. We argue that a commander-in-chief must also be a statesman, but he must not cease to be a general. On the one hand, he is aware of the entire political situation; on the other, he knows exactly how much he can achieve with the means at his disposal.

Circumstances vary so enormously in war, and are so indefinable, that a vast array of factors has to be appreciated—mostly in the light of probabilities alone. The man responsible for evaluating the whole must bring to his task the quality of intuition that perceives the truth at every point. Otherwise a chaos of opinions and considerations would arise, and fatally entangle judgment. Bonaparte rightly said in this connection that many of the decisions faced by the commander-in-chief resemble mathematical problems worthy of the gifts of a *Newton* or an *Euler*.

What this task requires in the way of higher intellectual gifts is a sense of unity and a power of judgment raised to a marvelous pitch of vision, which easily grasps and dismisses a thousand remote possibilities which an ordinary mind would labor to identify and wear itself out in so doing. Yet even that superb display of divination, the sovereign eye of genius itself, would still fall short of historical significance without the qualities of character and temperament we have described.

Truth in itself is rarely sufficient to make men act. Hence the step is always long from cognition to volition, from knowledge to ability. The most powerful springs of action in men lie in his emotions. He derives his most vigorous support, if we may use the term, from that blend of brains and temperament which we have learned to recognize in the qualities of determination, firmness, staunchness, and strength of character.

Naturally enough, if the commander's superior intellect and strength of character did not express themselves in the final success of his work, and were only taken on trust, they would rarely achieve historical importance.

What the layman gets to know of the course of military events is usually nondescript. One action resembles another, and from a mere recital of events it would be impossible to guess what obstacles were faced and overcome. Only now and then, in the memoirs of generals or of their confidants, or as the result of close historical study, are some of the countless threads of the tapestry revealed. Most of the arguments and clashes of opinion that precede a major operation are deliberately concealed because they touch political interests, or they are simply forgotten, being considered as scaffolding to be demolished when the building is complete.

Finally, and without wishing to risk a closer definition of the higher reaches of the spirit, let us assert that the human mind (in the normal meaning of the term) is far from uniform. If we then ask what sort of mind is likeliest to display the qualities of military genius, experience and observation will both tell us that it is the inquiring rather than the creative mind, the comprehensive rather than the specialized approach, the calm rather than the excitable head to which in war we would choose to entrust the fate of our brothers and children, and the safety and honor of our country.

CHAPTER FOUR

On Danger in War

To someone who has never experienced danger, the idea is attractive rather than alarming. You charge the enemy, ignoring bullets and casualties, in a surge of excitement. Blindly you hurl yourself toward icy death, not knowing whether you or anyone else will escape him. Before you lies that golden prize, victory, the fruit that quenches the thirst of ambition. Can that be so difficult? No, and it will seem even less difficult than it is. But such moments are rare; and even they are not, as is commonly thought, brief like a heartbeat, but come rather like a medicine, in recurring doses, the taste diluted by time.

Let us accompany a novice to the battlefield. As we approach the rumble of guns grows louder and alternates with the whir of cannonballs, which begin to attract his attention. Shots begin to strike close around us. We hurry up the slope where the commanding general is stationed with his large staff. Here cannonballs and bursting shells are frequent, and life begins to seem more serious than the young man had imagined. Suddenly someone you know is wounded; then a shell falls among the staff. You notice that some of the officers act a little oddly; you yourself are not as steady and collected as you were: even the bravest can become slightly distracted. Now we enter the battle raging before us, still almost like a spectacle, and join the nearest divisional commander. Shot is falling like hail, and the thunder of our own guns adds to the din. Forward to the brigadier, a soldier of acknowledged bravery, but he is careful to take cover behind a rise, a house or a clump of trees. A noise is heard that is a certain indication of increasing danger—the rattling of grapeshot on roofs and on the ground. Cannonballs tear past, whizzing in all directions, and musketballs begin to whistle around us. A little further we reach the firing line, where the infantry endures the hammering for hours with incredible steadfastness. The air is filled with hissing bullets that sound like a sharp crack if they pass close to one's head. For a final shock, the sight of men being killed and mutilated moves our pounding hearts to awe and pity.

The novice cannot pass through these layers of increasing intensity of danger without sensing that here ideas are governed by other factors, that the light of reason is refracted in a manner quite different from that which is normal in academic speculation. It is an exceptional man who keeps his powers of quick decision intact if he has never been through this experience before. It is true that (with habit) as we become accustomed to it the impression soon wears off, and in half-an-hour we hardly notice our surroundings any more; yet the ordinary man can never achieve a state of perfect unconcern in which his mind can work with normal flexibility. Here again we recognize that ordinary qualities are not enough; and the greater the area of responsibility, the truer this assertion becomes. Headlong, dogged, or innate courage, overmastering ambition, or long familiarity with danger—all must be present to a considerable degree if action in this debilitating element is not to fall short of achievements that in the study would appear as nothing out of the ordinary.

Danger is part of the friction of war. Without an accurate conception of danger we cannot understand war. That is why I have dealt with it here.

CHAPTER FIVE

On Physical Effort in War

If no one had the right to give his views on military operations except when he is frozen, or faint from heat and thirst, or depressed from privation and fatigue, objective and accurate views would be even rarer than they are. But they would at least be subjectively valid, for the speaker's experience would precisely determine his judgment. This is clear enough when we observe in what a deprecatory, even mean and petty way men talk about the failure of some operation that they have witnessed, and even more if they actually took part. We consider that this indicates how much influence physical effort exerts, and shows how much allowance has to be made for it in all our assessments.

Among the many factors in war that cannot be measured, physical effort is the most important. Unless it is wasted, physical effort is a coefficient of all forces, and its exact limit cannot be determined. But it is significant that, just as it takes a powerful archer to bend the bow beyond the average, so it takes a powerful mind to drive his army to the limit. It is one thing for an army that has been badly defeated, is beset by danger on all sides, and is disintegrating like crumbling masonry, to seek its safety in utmost physical effort. It is altogether different when a victorious army, buoyed up by its own exhilaration, remains a willing instrument in the hands of its commander. The same effort, which in the former case can at most arouse sympathy, must be admired in the other, where it is much harder to maintain.

The inexperienced observer now comes to recognize one of the elements that seem to chain the spirit and secretly wear away men's energies.

Although we are dealing only with the efforts that a general can demand of his troops, a commander of his subordinates, in other words although we are concerned with the courage it takes to make the demand and the skill to keep up the response, we must not forget the physical exertion required of the commander himself. Since we have pursued our analysis of war conscientiously to this point, we must deal with this residue as well.

Our reason for dealing with physical effort here is that like danger it is one of the great sources of friction in war. Because its limits are uncertain, it resembles one of those substances whose elasticity makes the degree of its friction exceedingly hard to gauge.

To prevent these reflections, this assessment of the impeding conditions of war, from being misused, we have a natural guide in our sensibilities. No one can count on sympathy if he accepts an insult or mistreatment because he claims to be physically handicapped. But if he manages to defend or revenge himself, a reference to his handicap will be to his advantage. In the same way, a general and an army cannot remove the stain of defeat by explaining the dangers, hardships, and exertions that were endured; but to depict them adds immensely to the credit of a victory. We are prevented from making an apparently justified statement by *our feelings*, which themselves act as a higher judgment.

CHAPTER SIX

Intelligence in War

By "intelligence" we mean every sort of information about the enemy and his country—the basis, in short, of our own plans and operations. If we consider the actual basis of this information, how unreliable and transient it is, we soon realize that war is a flimsy structure that can easily collapse and bury us in its ruins. The textbooks agree, of course, that we should only believe reliable intelligence, and should never cease to be suspicious, but what is the use of such feeble maxims? They belong to that wisdom which for want of anything better scribblers of systems and compendia resort to when they run out of ideas.

Many intelligence reports in war are contradictory; even more are false, and most are uncertain. What one can reasonably ask of an officer is that he should possess a standard of judgment, which he can gain only from knowledge of men and affairs and from common sense. He should be guided by the laws of probability. These are difficult enough to apply when plans are drafted in an office, far from the sphere of action; the task becomes infinitely harder in the thick of fighting itself, with reports streaming in. At such times one is lucky if their contradictions cancel each other out, and leave a kind of balance to be critically assessed. It is much worse for the novice if chance does not help him in that way, and on the contrary one report tallies with another, confirms it, magnifies it, lends it color, till he has to make a quick decision—which is soon recognized to be mistaken, just as the reports turn out to be lies, exaggerations, errors, and so on. In short, most intelligence is false, and the effect of fear is to multiply lies and inaccuracies. As a rule most men would rather believe bad news than good, and rather tend to exaggerate the bad news. The dangers that are reported may soon, like waves, subside; but like waves they keep recurring, without apparent reason. The commander must trust his judgment and stand like a rock on which the waves break in vain. It is not an easy thing to do. If he does not have a buoyant disposition, if experience of war has not trained him and matured his judgment, he had better make it a rule to suppress his personal convictions, and give his hopes and not his fears the benefit of the doubt. Only thus can he preserve a proper balance.

This difficulty of *accurate recognition* constitutes one of the most serious sources of friction in war, by making things appear entirely different from what one had expected. The senses make a more vivid impression on the mind than systematic thought—so much so that I doubt if a commander ever launched an operation of any magnitude without being forced to repress new misgivings from the start. Ordinary men, who normally follow the initiative of others, tend to lose self-confidence when they reach the scene of action: things are not what they expected, the more so as they still let others influence them. But even the man who planned the operation and now sees it being carried out may well lose confidence in his earlier judgment; whereas self-reliance is his best defense against the pressures of the moment. War has a way of masking the stage with scenery crudely daubed with fearsome apparitions. Once this is cleared away, and the horizon becomes unobstructed, developments will confirm his earlier convictions—this is one of the great chasms between *planning and execution*.

CHAPTER SEVEN

Friction in War

If one has never personally experienced war, one cannot understand in what the difficulties constantly mentioned really consist, nor why a commander should need any brilliance and exceptional ability. Everything looks simple; the knowledge required does not look remarkable, the strategic options are so obvious that by comparison the simplest problem of higher mathematics has an impressive scientific dignity. Once war has actually been seen the difficulties become clear; but it is still extremely hard to describe the unseen, all-pervading element that brings about this change of perspective.

Everything in war is very simple, but the simplest thing is difficult. The difficulties accumulate and end by producing a kind of friction that is inconceivable unless one has experienced war. Imagine a traveler who late in the day decides to cover two more stages before nightfall. Only four or five hours more, on a paved highway with relays of horses: it should be an easy trip. But at the next station he finds no fresh horses, or only poor ones; the country grows hilly, the road bad, night falls, and finally after many difficulties he is only too glad to reach a resting place with any kind of primitive accommodation. It is much the same in war. Countless minor incidents—the kind you can never really foresee—combine to lower the general level of performance, so that one always falls far short of the intended goal. Iron will-power can overcome this friction; it pulverizes every obstacle, but of course it wears down the machine as well. We shall often return to this point. The proud spirit's firm will dominates the art of war as an obelisk dominates the town square on which all roads converge.

Friction is the only concept that more or less corresponds to the factors that distinguish real war from war on paper. The military machine—the army and everything related to it—is basically very simple and therefore seems easy to manage. But we should bear in mind that none of its components is of one piece: each part is composed of individuals, every one of whom retains his potential of friction. In theory it sounds reasonable enough: a battalion commander's duty is to carry out his orders; discipline welds the battalion together, its commander must be a man of tested capacity, and so the great beam turns on its iron pivot with a minimum of friction. In fact, it is different, and every fault and exaggeration of the theory is instantly exposed in war. A battalion is made up of individuals, the least important of whom may chance to delay things or somehow make them go wrong. The dangers inseparable from war and the physical exertions war demands can aggravate the problem to such an extent that they must be ranked among its principal causes.

This tremendous friction, which cannot, as in mechanics, be reduced to a few points, is everywhere in contact with chance, and brings about effects that cannot be measured, just because they are largely due to chance. One, for example, is the weather. Fog can prevent the enemy from being seen in time, a gun from firing when it should, a report from reaching the commanding officer. Rain can prevent a battalion from arriving, make another late by keeping it not three but eight hours on the march, ruin a cavalry charge by bogging the horses down in mud, etc.

We give these examples simply for illustration, to help the reader follow the argument. It would take volumes to cover all difficulties. We could exhaust the reader with illustrations alone if we really tried to deal with the whole range of minor troubles that must be faced in war. The few we have given will be excused by those readers who have long since understood what we are after.

Action in war is like movement in a resistant element. Just as the simplest and most natural of movements, walking, cannot easily be performed in water, so in war it is difficult for normal efforts to achieve even moderate results. A genuine theorist is like a swimming teacher, who makes his pupils practice motions on land that are meant to be performed in water. To those who are not thinking of swimming the motions will appear grotesque and exaggerated. By the same token, theorists who have never swum, or who have not learned to generalize from experience, are impractical and even ridiculous: they teach only what is already common knowledge: how to walk.

Moreover, every war is rich in unique episodes. Each is an uncharted sea, full of reefs. The commander may suspect the reefs' existence without ever having seen them; now he has to steer past them in the dark. If a contrary wind springs up, if some major mischance appears, he will need the greatest skill and personal exertion, and the utmost presence of mind, though from a distance everything may seem to be proceeding automatically. An understanding of friction is a large part of that much-admired sense of warfare which a good general is supposed to possess. To be sure, the best general is not the one who is most familiar with the idea of friction, and who takes it most to heart (he belongs to the anxious type so common among experienced commanders). The good general must know friction in order to overcome it whenever possible, and in order not to expect a standard of achievement in his operations which this very friction makes impossible. Incidentally, it is a force that theory can never quite define. Even if it could, the development of instinct and tact would still be needed, a form of judgment much more necessary in an area littered by endless minor obstacles than in great, momentous questions, which are settled in solitary deliberation or in discussion with others. As with a man of the world instinct becomes almost habit so that he always acts, speaks, and moves appropriately, so only the experienced officer will make the right decision in major and minor matters—at every pulsebeat of war. Practice and experience dictate the answer: "this is possible, that is not." So he rarely makes a serious mistake, such as can, in war, shatter confidence and become extremely dangerous if it occurs often.

Friction, as we choose to call it, is the force that makes the apparently easy so difficult. We shall frequently revert to this subject, and it will become evident that an eminent commander needs more than experience and a strong will. He must have other exceptional abilities as well.

CHAPTER EIGHT

Concluding Observations on Book One

We have identified danger, physical exertion, intelligence, and friction as the elements that coalesce to form the atmosphere of war, and turn it into a medium that impedes activity. In their restrictive effects they can be grouped into a single concept of general friction. Is there any lubricant that will reduce this abrasion? Only one, and a commander and his army will not always have it readily available: combat experience.

Habit hardens the body for great exertions, strengthens the heart in great peril, and fortifies judgment against first impressions. Habit breeds that priceless quality, calm, which, passing from hussar and rifleman up to the general himself, will lighten the commander's task.

In war the experienced soldier reacts rather in the same way as the human eye does in the dark: the pupil expands to admit what little light there is, discerning objects by degrees, and finally seeing them distinctly. By contrast, the novice is plunged into the deepest night.

No general can accustom an army to war. Peacetime maneuvers are a feeble substitute for the real thing; but even they can give an army an advantage over others whose training is confined to routine, mechanical drill. To plan maneuvers so that some of the elements of friction are involved, which will train officers' judgment, common sense, and resolution is far more worthwhile than inexperienced people might think. It is immensely important that no soldier, whatever his rank, should wait for war to expose him to those aspects of active service that amaze and confuse him when he first comes across them. If he has met them even once before, they will begin to be familiar to him. This is true even of physical effort. Exertions must be practiced, and the mind must be made even more familiar with them than the body. When exceptional efforts are required of him in war, the recruit is apt to think that they result from mistakes, miscalculations, and confusion at the top. In consequence, his morale is doubly depressed. If maneuvers prepare him for exertions, this will not occur.

Another very useful, though more limited, way of gaining familiarity with war in peacetime is to attract foreign officers who have seen active service. Peace does not often reign everywhere in Europe, and never throughout the whole world. A state that has been at peace for many years should try to attract some experienced officers—only those, of course, who have distinguished themselves. Alternatively, some of its own officers should be sent to observe operations, and learn what war is like.

However few such officers may be in proportion to an army, their influence can be very real. Their experience, their insights, and the maturity of their character will affect their subordinates and brother officers. Even when they cannot be given high command they should be considered as guides who know the country and can be consulted in specific eventualities.

BOOK TWO

On the Theory of War

CHAPTER ONE

Classifications of the Art of War

Essentially war is fighting, for fighting is the only effective principle in the manifold activities generally designated as war. Fighting, in turn, is a trial of moral and physical forces through the medium of the latter. Naturally moral strength must not be excluded, for psychological forces exert a decisive influence on the elements involved in war.

The need to fight quickly led man to invent appropriate devices to gain advantages in combat, and these brought about great changes in the forms of fighting. Still, no matter how it is constituted, the concept of fighting remains unchanged. That is what we mean by war.

The first inventions consisted of weapons and equipment for the individual warrior. They have to be produced and tested before war begins; they suit the nature of the fighting, which in turn determines their design. Obviously, however, this activity must be distinguished from fighting proper; it is only the preparation for it, not its conduct. It is clear that weapons and equipment are not essential to the concept of fighting, since even wrestling is fighting of a kind.

Fighting has determined the nature of the weapons employed. These in turn influence the combat; thus an interaction exists between the two. But fighting itself still remains a distinct activity; the more so as it operates in a peculiar element—that of danger.

Thus, if there was ever a need to distinguish between two activities, we find it here. In order to indicate the practical importance of this idea, we would suggest how often it is that the ablest man in one area is shown up as the most useless pedant in another.

In fact, it is not at all difficult to consider these two activities separately if one accepts the idea of an armed and equipped fighting force as *given*: a means about which one does not need to know anything except its chief effects in order to use it properly.

Essentially, then, the art of war is the art of using the given means in combat; there is no better term for it than the *conduct of war*. To be sure in its wider sense the art of war includes all activities that exist for the sake of war, such as the creation of the fighting forces, their raising, armament, equipment, and training.

It is essential to the validity of a theory to distinguish between these two activities. It is easy to see that if the art of war were always to start with raising armed forces and adapting them to the requirements of the particular case, it would be applicable only to those few instances where the forces available exactly matched the need. If, on the other hand, one wants a theory that is valid for the great majority of cases and not completely unsuitable for any, it must be based on the most prevalent means and their most significant effects.

The conduct of war, then, consists in the planning and conduct of fighting. If fighting consisted of a single act, no further subdivision would be needed. However, it consists of a greater or lesser number of single *acts, each complete in itself,* which, as we pointed out in Chapter 1 of Book I,[1] are called "engagements" and which form new entities. This gives rise to the completely different activity of *planning and executing these engagements themselves,* and of *coordinating* each of them with the others in order to further the object of the war. One has been called *tactics,* and the other *strategy.*

The distinction between tactics and strategy is now almost universal, and everyone knows fairly well where each particular factor belongs without clearly understanding why. Whenever such categories are blindly used, there must be a deep-seated reason for it. We have tried to discover the distinction, and have to say that it was just this common usage that led to it. We reject, on the other hand, the artificial definitions of certain writers, since they find no reflection in general usage.

According to our classification, then, tactics teaches *the use of armed forces in the engagement;* strategy, *the use of engagements for the object of the war.*

The concept of a single or a self-contained engagement and the conditions on which its unity depends can be more accurately defined only when we examine it more closely. For the moment, it is enough to say that in terms of space (that is, of simultaneous engagements) its unity is bounded by the range of *personal command.* In terms of time, however (that is, of a close succession of engagements) it lasts until the turning point, which is characteristic of all engagements, has been passed.

There may be doubtful cases—those, for instance, in which a number of engagements could perhaps also be regarded as a single one. But that will not spoil our basis for classification, since the point is common to all practical systems of classification where distinctions gradually merge on a descending scale. Thus there may be individual acts which, without a shift in point of view, may belong either to strategy or to tactics; for instance, very extended positions that are little more than a chain of posts, or arrangements for certain river-crossings.

Our classification applies to and exhausts *only the utilization of the fighting forces.* But war is served by many activities that are quite different from it; some closely related, others far removed. All these activities concern the *maintenance of the fighting forces.* While their creation and training precedes their use, maintenance is concurrent with and a necessary condition

[1] Clausewitz means Chapter Two. Eds.

for it. Strictly speaking, however, all these should be considered as activities preparatory to battle, of the type that are so closely related to the action that they are part of military operations and alternate with actual *utilization*. So one is justified in excluding these as well as all other preparatory activities from the narrower meaning of the art of war—the actual conduct of war. Indeed, it is necessary to do this if theory is to serve its principal purpose of *discriminating between dissimilar elements*. One would not want to consider the whole business of maintenance and administration as part of the *actual conduct of war*. While it may be in constant interaction with the utilization of the troops, the two are essentially very different.

In the third chapter of Book I we pointed out that, if combat or the engagement *is defined* as the only directly effective activity, the threads of all other activities will be included because they all lead to combat. The statement meant that all these activities are thus provided with a purpose, which they will have to pursue in accordance with their individual laws. Let us elaborate further on this subject.

Activities that exist in addition to the engagement differ widely.

Some of these are in one respect part of combat proper and identical with it, while in another respect they serve to maintain the fighting forces. Others are related to maintenance alone; which has an effect on combat only because of its interaction with the outcome of the fighting.

The matters that in one respect are still part of the combat are *marches, camps, and billets*: each concerns a separate phase of existence of the troops, and when one thinks of troops, the idea of the engagement must always be present.

The rest, concerned with maintenance alone, consists of *supply, medical services, and maintenance of arms and equipment.*

Marches are completely identical with the utilization of troops. *Marching in the course of an engagement* (usually known as "deployment")[2] while not entailing the actual use of weapons, is so closely and inescapably linked with it as to be an integral part of what is considered an engagement. A march that is not undertaken in the course of an engagement is simply the execution of a strategic plan. The latter determines *when, where* and *with what forces* an engagement is to be fought. The march is only the means of carrying out this plan.

A march that is not part of an engagement is thus a tool of strategy, but it is not a matter of strategy exclusively. Since the forces undertaking it may at any time become involved in an engagement, the execution of the march is subject to the laws of both tactics and strategy. If a column is ordered to take a route on the near side of a river or a range of hills, that is a strategic measure: it implies that if an engagement has to be fought in the course of the march, one prefers to offer it on the near rather than the far side.

If on the other hand a column takes a route along a ridge instead of fol-

[2] In German: *Evolution*. This signifies the deployment of troops within battle as opposed to general operational maneuvers. Eds.

lowing the road through a valley, or breaks up into several smaller columns for the sake of convenience, these are tactical measures: they concern the *manner* in which the forces are to be used in the event of an engagement.

The internal order of march bears a constant relationship to readiness for combat and is therefore of a tactical nature: it is nothing more than the first preliminary disposition for a possible engagement.

The march is the tool by which strategy deploys its effective elements, the engagements. But these often become apparent only in their effect, and not in their actual course. Inevitably, therefore, in discussion the tool has often been confused with the effective element. One speaks of decisive skillful marches, and really means the combinations of engagements to which they lead. This substitution of concept is too natural, and the brevity of expression too desirable, to call for change. But it is only a telescoped chain of ideas, and one must keep the proper meaning in mind to avoid errors.

One such error occurs when strategic combinations are believed to have a value irrespective of their tactical results. One works out marches and maneuvers, achieves one's objective without fighting an engagement, and then deduces that it is possible to defeat the enemy without fighting. Only at a later stage shall we be able to show the immense implications of this mistake.

Although marching can be seen as an integral part of combat, it has certain aspects that do not belong here, and that therefore are neither tactical nor strategic. These include all measures taken solely for the convenience of the troops, such as building roads and bridges, and so forth. These are merely preconditions; under certain circumstances they may be closely linked with the use of troops and be virtually identical with them—for instance, when a bridge is built in full view of the enemy. But essentially these activities are alien to the conduct of war, and the theory of the latter does not cover them.

The term "camp" is a term for any concentration of troops in readiness for action, as distinct from "billets." Camps are places for rest and recuperation, but they also imply strategic willingness to fight wherever they may be. But their siting does determine the engagement's basic lines—a precondition of all defensive engagements. So they are essential parts both of strategy and of tactics.

Camps are replaced by billets whenever troops are thought to need more extensive recuperation. Like camps, they are therefore strategic in location and extent, and tactical in their internal organization which is geared to readiness for action.

As a rule, of course, camps and billets serve a purpose besides that of resting the troops; they may, for instance, serve to protect a certain area or maintain a position. But their purpose may simply be to rest the troops. We have to remember that strategy may pursue a wide variety of objectives: anything that seems to offer an advantage can be the purpose of an engagement, and the maintenance of the instrument of war will often itself become the object of a particular strategic combination.

So in a case where strategy merely aims at preserving the troops, we need not have strayed far afield: the use of troops is still the main concern, since that is the point of their disposition anywhere in the theater of war.

On the other hand, the maintenance of troops in camps or billets may call for activities that do not constitute a use of the fighting forces, such as the building of shelters, the pitching of tents, and supply and sanitary services. These are neither tactical nor strategic in nature.

Even entrenchments, where site and preparation are obviously part of the order of battle and therefore tactical, are not part of the conduct of war so far as *their actual construction* is concerned. On the contrary, troops must be taught the necessary skills and knowledge as part of their training, and the theory of combat takes all that for granted.

Of the items wholly unconnected with engagements, serving only to maintain the forces, supply is the one which most directly affects the fighting. It takes place almost every day and affects every individual. Thus it thoroughly permeates the strategic aspects of all military action. The reason why we mention the strategic aspect is that in the course of a given engagement supply will rarely tend to cause an alteration of plans—though such a change remains perfectly possible. Interaction therefore will be most frequent between strategy and matters of supply, and nothing is more common than to find considerations of supply affecting the strategic lines of a campaign and a war. Still, no matter how frequent and decisive these considerations may be, the business of supplying the troops remains an activity essentially separate from their use; its influence shows in its results alone.

The other administrative functions we have mentioned are even further removed from the use of troops. Medical services, though they are vital to an army's welfare, affect it only through a small portion of its men, and therefore exert only a weak and indirect influence on the utilization of the rest. Maintenance of equipment, other than as a constant function of the fighting forces, takes place only periodically, and will therefore rarely be taken into account in strategic calculations.

At this point we must guard against a misunderstanding. In any individual case these things may indeed be of decisive importance. The distance of hospitals and supply depots may easily figure as the sole reason for very important strategic decisions—a fact we do not want to deny or minimize. However, we are not concerned with the actual circumstances of any individual case, but with pure theory. Our contention therefore is that this type of influence occurs so rarely that we should not give the theory of medical services and replacement of munitions any serious weight in the theory of the conduct of war. Unlike the supplying of the troops, therefore, it would not seem worth while to incorporate the various ways and systems those theories might suggest, and their results, into the theory of the conduct of war.

To sum up: we clearly see that the activities characteristic of war may be split into two main categories: those *that are merely preparations for war*, and *war proper*. The same distinction must be made in theory as well.

The knowledge and skills involved in the preparations will be concerned with the creation, training and maintenance of the fighting forces. It is immaterial what label we give them, but they obviously must include such matters as artillery, fortification, so-called elementary tactics, as well as all the organization and administration of the fighting forces and the like. The theory of war proper, on the other hand, is concerned with the use of these means, once they have been developed, for the purposes of the war. All that it requires from the first group is the end product, an understanding of their main characteristics. That is what we call "the art of war" in a narrower sense, or "the theory of the conduct of war," or "the theory of the use of the fighting forces." For our purposes, they all mean the same thing.

That narrower theory, then, deals with the engagement, with fighting itself, and treats such matters as marches, camps, and billets as conditions that may be more or less identical with it. It does not comprise questions of supply, but will take these into account on the same basis *as other given factors.*

The art of war in the narrower sense must now in its turn be broken down into tactics and strategy. The first is concerned with the form of the individual engagement, the second with its use. Both affect the conduct of marches, camps, and billets only through the engagement; they become tactical or strategic questions insofar as they concern either the engagement's form or its significance.

Many readers no doubt will consider it superfluous to make such a careful distinction between two things so closely related as tactics and strategy, because they do not directly affect the conduct of operations. Admittedly only the rankest pedant would expect theoretical distinctions to show direct results on the battlefield.

The primary purpose of any theory is to clarify concepts and ideas that have become, as it were, confused and entangled. Not until terms and concepts have been defined can one hope to make any progress in examining the question clearly and simply and expect the reader to share one's views. Tactics and strategy are two activities that permeate one another in time and space but are nevertheless essentially different. Their inherent laws and mutual relationship cannot be understood without a total comprehension of both.

Anyone for whom all this is meaningless either will admit no theoretical analysis at all, or his intelligence has never been insulted by the confused and confusing welter of ideas that one so often hears and reads on the subject of the conduct of war. These have no fixed point of view; they lead to no satisfactory conclusion; they appear sometimes banal, sometimes absurd, sometimes simply adrift in a sea of vague generalization; and all because this subject has seldom been examined in a spirit of scientific investigation.

CHAPTER TWO

On the Theory of War

Originally the Term "Art of War" Only Designated the Preparation of the Forces

Formerly, the terms "art of war" or "science of war" were used to designate only the total body of knowledge and skill that was concerned with material factors. The design, production, and use of weapons, the construction of fortifications and entrenchments, the internal organization of the army, and the mechanism of its movements constituted the substance of this knowledge and skill. All contributed to the establishment of an effective fighting force. It was a case of handling a material substance, a unilateral activity, and was basically nothing but a gradual rise from a craft to a refined mechanical art. It was about as relevant to combat as the craft of the swordsmith to the art of fencing. It did not yet include the use of force under conditions of danger, subject to constant interaction with an adversary, nor the efforts of spirit and courage to achieve a desired end.

True War First Appears in Siege Warfare

Siege warfare gave the first glimpse of the conduct of operations, of intellectual effort; but this usually revealed itself only in such new techniques as approaches, trenches, counterapproaches, batteries and so forth, and marked each step by some such product. It was only the thread needed to link these material inventions. Since in siege warfare that is almost the only way in which the intellect can manifest itself, the matter usually rested there.

Next the Subject Was Touched on by Tactics

Later, tactics attempted to convert the structure of its component parts into a general system, based on the peculiar properties of its instrument.[1] This certainly led to the battlefield, but not yet to creative intellectual activity. The result was rather armies which had been transformed by their formations and orders of battle into automata, designed to discharge their activity like pieces of clockwork set off by a mere word of command.

[1] The armed forces. Eds.

The Actual Conduct of War Occurred Only Incidentally and Incognito

The actual conduct of war—the free use of the given means, appropriate to each individual occasion—was not considered a suitable subject for theory, but one that had to be left to natural preference. Gradually, war progressed from medieval hand-to-hand fighting toward a more orderly and complex form. Then, admittedly, the human mind was forced to give some thought to this matter; but as a rule its reflections appear only incidentally and, so to speak, incognito, in memoirs and histories.

Reflections on the Events of War Led to the Need for a Theory

As these reflections grew more numerous and history more sophisticated, an urgent need arose for principles and rules whereby the controversies that are so normal in military history—the debate between conflicting opinions—could be brought to some sort of resolution. This maelstrom of opinions, lacking in basic principles and clear laws round which they could be crystallized, was bound to be intellectually repugnant.

Efforts To Formulate a Positive Theory

Efforts were therefore made to equip the conduct of war with principles, rules, or even systems. This did present a positive goal, but people failed to take adequate account of the endless complexities involved. As we have seen, the conduct of war branches out in almost all directions and has no definite limits; while any system, any model, has the finite nature of a synthesis. An irreconcilable conflict exists between this type of theory and actual practice.

Limitation to Material Factors

Theorists soon found out how difficult the subject was, and felt justified in evading the problem by again directing their principles and systems only to physical matters and unilateral activity. As in the science concerning *preparation for war,* they wanted to reach a set of sure and positive conclusions, and for that reason considered only factors that could be mathematically calculated.

Numerical Superiority

Numerical superiority was a material factor. It was chosen from all elements that make up victory because, by using combinations of time and space, it could be fitted into a mathematical system of laws. It was thought that all other factors could be ignored if they were assumed to be equal on both sides and thus cancelled one another out. That might have been acceptable as a temporary device for the study of the characteristics of this single factor; but to make the device permanent, to accept superiority of numbers as the one and only rule, and to reduce the whole secret of the art of war to the formula of numerical superiority *at a certain time in a certain place* was an oversimplification that would not have stood up for a moment against the realities of life.

Supply

Another theoretical treatment sought to reduce a different material factor to a system: supply. Based on the assumption that an army was organized in a certain manner, its supply was set up as a final arbiter for the conduct of war.

That approach also produced some concrete figures, but these rested on a mass of arbitrary assumptions. They were therefore not able to stand the test of practical experience.

Base

One ingenious mind sought to condense a whole array of factors, some of which did indeed stand in intellectual relation to one another, into a single concept, that of the *base*. This included *feeding the army, replacing its losses in men and equipment, assuring its communications with home, and even the safety of its retreat* in case that should become necessary. He started by substituting this concept for all these individual factors; next substituting the area or extent of this base for the concept itself, and ended up by substituting for this area the angle which the fighting forces created with their base line.[2] All this led to a purely geometrical result, which is completely useless. This uselessness is actually inevitable in view of the fact that none of these substitutions could be made without doing violence to the facts and without dropping part of the content of the original idea. The concept of a base is a necessary tool in strategy and the author deserves credit for having discovered it; but it is completely inadmissible to use it in the manner described. It was bound to lead to one-sided conclusions which propelled that theorist into the rather contradictory direction of believing in the superior effectiveness of enveloping positions.

Interior Lines

As a reaction to that fallacy, another geometrical principle was then exalted: that of so-called interior lines. Even though this tenet rests on solid ground—

[2] The reference is to Clausewitz's contemporary, H. D. v. Bülow. See P. Paret, "The Genesis of *On War*," p. 10 above. Eds.

on the fact that the engagement is the only effective means in war—its purely geometrical character, still makes it another lopsided principle that could never govern a real situation.[3]

All These Attempts Are Objectionable

It is only analytically that these attempts at theory can be called advances in the realm of truth; synthetically, in the rules and regulations they offer, they are absolutely useless.

They aim at fixed values; but in war everything is uncertain, and calculations have to be made with variable quantities.

They direct the inquiry exclusively toward physical quantities, whereas all military action is intertwined with psychological forces and effects.

They consider only unilateral action, whereas war consists of a continuous interaction of opposites.

They Exclude Genius from the Rule

Anything that could not be reached by the meager wisdom of such one-sided points of view was held to be beyond scientific control: it lay in the realm of genius, *which rises above all rules.*

Pity the soldier who is supposed to crawl among these scraps of rules, not good enough for genius, which genius can ignore, or laugh at. No; what genius does is the best rule, and theory can do no better than show how and why this should be the case.

Pity the theory that conflicts with reason! No amount of humility can gloss over this contradiction; indeed, the greater the humility, the sooner it will be driven off the field of real life by ridicule and contempt.

Problems Facing Theory When Moral Factors Are Involved

Theory becomes infinitely more difficult as soon as it touches the realm of moral values. Architects and painters know precisely what they are about as long as they deal with material phenomena. Mechanical and optical structures are not subject to dispute. But when they come to the aesthetics of their work, when they aim at a particular effect on the mind or on the senses, the rules dissolve into nothing but vague ideas.

Medicine is usually concerned only with physical phenomena. It deals with the animal organism, which, however, is subject to constant change, and thus is never exactly the same from one moment to the next. This renders the task of medicine very difficult, and makes the physician's judgment count for more than his knowledge. But how greatly is the difficulty

[3] The reference is to A. H. Jomini. See P. Paret, "The Genesis of *On War*," pp. 10–11 above. Eds.

increased when a mental factor is added, and how much more highly do we value the psychiatrist!

Moral Values Cannot Be Ignored in War

Military activity is never directed against material force alone; it is always aimed simultaneously at the moral forces which give it life, and the two cannot be separated.

But moral values can only be perceived by the inner eye, which differs in each person, and is often different in the same person at different times.

Since danger is the common element in which everything moves in war, courage, the sense of one's own strength, is the principal factor that influences judgment. It is the lens, so to speak, through which impressions pass to the brain.

And yet there can be no doubt that experience will by itself provide a degree of objectivity to these impressions.

Everyone knows the moral effects of an ambush or an attack in flank or rear. Everyone rates the enemy's bravery lower once his back is turned, and takes much greater risks in pursuit than while being pursued. Everyone gauges his opponent in the light of his reputed talents, his age, and his experience, and acts accordingly. Everyone tries to assess the spirit and temper of his own troops and of the enemy's. All these and similar effects in the sphere of mind and spirit have been proved by experience: they recur constantly, and are therefore entitled to receive their due as objective factors. What indeed would become of a theory that ignored them?

Of course these truths must be rooted in experience. No theorist, and no commander, should bother himself with psychological and philosophical sophistries.

Principal Problems in Formulating a Theory of the Conduct of War

In order to get a clear idea of the difficulties involved in formulating a theory of the conduct of war and so be able to deduce its character, we must look more closely at the major characteristics of military activity.

First Property: Moral Forces and Effects

HOSTILE FEELINGS

The first of these attributes consists of moral forces and the effects they produce.

Essentially combat is an expression of *hostile feelings*. But in the large-scale combat that we call war hostile feelings often have become merely hostile *intentions*. At any rate there are usually no hostile feelings between individuals. Yet such emotions can never be completely absent from war.

Modern wars are seldom fought without hatred between nations; this serves more or less as a substitute for hatred between individuals. Even where there is no national hatred and no animosity to start with, the fighting itself will stir up hostile feelings: violence committed on superior orders will stir up the desire for revenge and retaliation against the perpetrator rather than against the powers that ordered the action. That is only human (or animal, if you like), but it is a fact. Theorists are apt to look on fighting in the abstract as a trial of strength without emotion entering into it. This is one of a thousand errors which they quite consciously commit because they have no idea of the implications.

Apart from emotions stimulated by the nature of combat, there are others that are not so intimately linked with fighting; but because of a certain affinity, they are easily associated with fighting: ambition, love of power, enthusiasms of all kinds, and so forth.

The Effects of Danger

COURAGE

Combat gives rise to the element of danger in which all military activity must move and be maintained like birds in air and fish in water. The effects of danger, however, produce an emotional reaction, either as a matter of immediate instinct, or consciously. The former results in an effort to avoid the danger, or, where that is not possible, in fear and anxiety. Where these effects do not arise, it is because instinct has been outweighed by *courage*. But courage is by no means a conscious act; like fear, it is an emotion. Fear is concerned with physical and courage with moral survival. Courage is the nobler instinct, and as such cannot be treated as an inanimate instrument that functions simply as prescribed. So courage is not simply a counterweight to danger, to be used for neutralizing its effects: it is a quality on its own.

Extent of the Influence Exercised by Danger

In order properly to appreciate the influence which danger exerts in war, one should not limit its sphere to the physical hazards of the moment. Danger dominates the commander not merely by threatening him personally, but by threatening all those entrusted to him; not only at the moment where it is actually present, but also, through the imagination, at all other times when it is relevant; not just directly but also indirectly through the sense of responsibility that lays a tenfold burden on the commander's mind. He could hardly recommend or decide on a major battle without a certain feeling of strain and distress at the thought of the danger and responsibility such a major decision implies. One can make the point that action in war, insofar as it is true action and not mere existence, is never completely free from danger.

CHAPTER TWO

Other Emotional Factors

In considering emotions that have been aroused by hostility and danger as being peculiar to war, we do not mean to exclude all others that accompany man throughout his life. There is a place for them in war as well. It may be true that many a petty play of emotions is silenced by the serious duties of war; but that holds only for men in the lower ranks who, rushed from one set of exertions and dangers to the next, lose sight of the other things in life, forego duplicity because death will not respect it, and thus arrive at the soldierly simplicity of character that has always represented the military at its best. In the higher ranks it is different. The higher a man is placed, the broader his point of view. Different interests and a wide variety of passions, good and bad, will arise on all sides. Envy and generosity, pride and humility, wrath and compassion—all may appear as effective forces in this great drama.

Intellectual Qualities

In addition to his emotional qualities, the intellectual qualities of the commander are of major importance. One will expect a visionary, high-flown and immature mind to function differently from a cool and powerful one.

The Diversity of Intellectual Quality Results in a Diversity of Roads to the Goal

The influence of the great diversity of intellectual qualities is felt chiefly in the higher ranks, and increases as one goes up the ladder. It is the primary cause for the diversity of roads to the goal—already discussed in Book I—and for the disproportionate part assigned to the play of probability and chance in determining the course of events.

Second Property: Positive Reaction

The second attribute of military action is that it must expect positive reactions, and the process of interaction that results. Here we are not concerned with the problem of calculating such reactions—that is really part of the already mentioned problem of calculating psychological forces—but rather with the fact that the very nature of interaction is bound to make it unpredictable. The effect that any measure will have on the enemy is the most singular factor among all the particulars of action. All theories, however, must stick to categories of phenomena and can never take account of a truly unique case; this must be left to judgment and talent. Thus it is natural that military activity, whose plans, based on general circumstances, are so frequently disrupted by unexpected particular events; should remain largely a matter of talent, and that theoretical *directives* tend to be less useful here than in any other sphere.

Third Property: Uncertainty of All Information

Finally, the general unreliability of all information presents a special problem in war: all action takes place, so to speak, in a kind of twilight, which, like fog or moonlight, often tends to make things seem grotesque and larger than they really are.

Whatever is hidden from full view in this feeble light has to be guessed at by talent, or simply left to chance. So once again for lack of objective knowledge one has to trust to talent or to luck.

A Positive Doctrine Is Unattainable

Given the nature of the subject, we must remind ourselves that it is simply not possible to construct a model for the art of war that can serve as a scaffolding on which the commander can rely for support at any time. Whenever he has to fall back on his innate talent, he will find himself outside the model and in conflict with it; no matter how versatile the code, the situation will always lead to the consequences we have already alluded to: *talent and genius operate outside the rules, and theory conflicts with practice.*

Alternatives Which Make a Theory Possible

THE DIFFICULTIES VARY IN MAGNITUDE

There are two ways out of this dilemma.

In the first place, our comments on the nature of military activity in general should not be taken as applying equally to action at all levels. What is most needed in the lower ranks is courage and self-sacrifice, but there are far fewer problems to be solved by intelligence and judgment. The field of action is more limited, means and ends are fewer in number, and the data more concrete: usually they are limited to what is actually visible. But the higher the rank, the more the problems multiply, reaching their highest point in the supreme commander. At this level, almost all solutions must be left to imaginative intellect.

Even if we break down war into its various *activities,* we will find that the difficulties are not uniform throughout. The more physical the activity, the less the difficulties will be. The more the activity becomes intellectual and turns into motives which exercise a determining influence on the commander's will, the more the difficulties will increase. Thus it is easier to use theory to organize, plan, and conduct an engagement than it is to use it in determining the engagement's purpose. Combat is conducted with physical weapons, and although the intellect does play a part, material factors will dominate. But when one comes to the *effect* of the engagement, where material successes turn into motives for further action, the intellect alone is decisive. In brief, *tactics* will present far fewer difficulties to the theorist than will *strategy*.

Theory Should Be Study, Not Doctrine

The second way out of this difficulty is to argue that a theory need not be a positive doctrine, a sort of *manual* for action. Whenever an activity deals primarily with the same things again and again—with the same ends and the same means, even though there may be minor variations and an infinite diversity of combinations—these things are susceptible of rational study. It is precisely that inquiry which is the most essential part of any *theory*, and which may quite appropriately claim that title. It is an analytical investigation leading to a close *acquaintance* with the subject; applied to experience—in our case, to military history—it leads to thorough *familiarity* with it. The closer it comes to that goal, the more it proceeds from the objective form of a science to the subjective form of a skill, the more effective it will prove in areas where the nature of the case admits no arbiter but talent. It will, in fact, become an active ingredient of talent. Theory will have fulfilled its main task when it is used to analyze the constituent elements of war, to distinguish precisely what at first sight seems fused, to explain in full the properties of the means employed and to show their probable effects, to define clearly the nature of the ends in view, and to illuminate all phases of warfare in a thorough critical inquiry. Theory then becomes a guide to anyone who wants to learn about war from books; it will light his way, ease his progress, train his judgment, and help him to avoid pitfalls.

A specialist who has spent half his life trying to master every aspect of some obscure subject is surely more likely to make headway than a man who is trying to master it in a short time. Theory exists so that one need not start afresh each time sorting out the material and plowing through it, but will find it ready to hand and in good order. It is meant to educate the mind of the future commander, or, more accurately, to guide him in his self-education, not to accompany him to the battlefield; just as a wise teacher guides and stimulates a young man's intellectual development, but is careful not to lead him by the hand for the rest of his life.

If the theorist's studies automatically result in principles and rules, and if truth spontaneously crystallizes into these forms, theory will not resist this natural tendency of the mind. On the contrary, where the arch of truth culminates in such a keystone, this tendency will be underlined. But this is simply in accordance with the scientific law of reason, to indicate the point at which all lines converge, but never to construct an algebraic formula for use on the battlefield. Even these principles and rules are intended to provide a thinking man with a frame of reference for the movements he has been trained to carry out, rather than to serve as a guide which at the moment of action lays down precisely the path he must take.

This Point of View Makes Theory Possible and Eliminates Its Conflict with Reality

This point of view will admit the feasibility of a satisfactory theory of war—one that will be of real service and will never conflict with reality. It only needs intelligent treatment to make it conform to action, and to end the absurd difference between theory and practice that unreasonable theories have so often evoked. That difference, which defies common sense, has often been used as a pretext by limited and ignorant minds to justify their congenital incompetence.

Theory Thus Studies the Nature of Ends and Means

ENDS AND MEANS IN TACTICS

It is the task of theory, then, to study the nature of ends and means.

In tactics the means are the fighting forces trained for combat; the end is victory. A more precise definition of this concept will be offered later on, in the context of "the engagement." Here, it is enough to say that the enemy's withdrawal from the battlefield is the sign of victory. Strategy thereby gains the end it had ascribed to the engagement, the end that constitutes its real *significance*. This significance admittedly will exert a certain influence on the kind of victory achieved. A victory aimed at weakening the enemy's fighting forces is different from one that is only meant to seize a certain position. The significance of an engagement may therefore have a noticeable influence on its planning and conduct, and is therefore to be studied in connection with tactics.

Factors That Always Accompany the Application of the Means

There are certain constant factors in any engagement that will affect it to some extent; we must allow for them in our use of armed forces.

These factors are the locality or terrain, the time of day, and the weather.

Terrain

Terrain, which can be resolved into a combination of the geographical surroundings and the nature of the ground, could, strictly speaking, be of no influence at all on an engagement fought over a flat, uncultivated plain.

This does actually occur in the steppes, but in the cultivated parts of Europe it requires an effort of the imagination to conceive it. Among civilized nations combat uninfluenced by its surroundings and the nature of the ground is hardly conceivable.

CHAPTER TWO

Time of Day

The time of day affects an engagement by the difference between day and night. By implication, of course, these precise limits may be exceeded: every engagement takes a certain time, and major ones may last many hours. When a major battle is being planned, it makes a decisive difference whether it is to start in the morning or in the afternoon. On the other hand there are many engagements where the time of day is a neutral factor; in the general run of cases it is of minor importance.

Weather

It is rarer still for weather to be a decisive factor. As a rule only fog makes any difference.

Ends and Means in Strategy

The original means of strategy is victory—that is, tactical success; its ends, in the final analysis, are those objects which will lead directly to peace. The application of these means for these ends will also be attended by factors that will influence it to a greater or lesser degree.

Factors That Affect the Application of the Means

These factors are the geographical surroundings and nature of the terrain (the former extended to include the country and people of the entire theater of war); the time of day (including the time of year); and the weather (particularly unusual occurrences such as severe frost, and so forth).

These Factors Form New Means

Strategy, in connecting these factors with the outcome of an engagement, confers a special significance on that outcome and thereby on the engagement: *it assigns a particular aim to it*. Yet insofar as that aim is not the one that will lead directly to peace, it remains subsidiary and is also to be thought of as a means. Successful engagements or victories in all stages of importance may therefore be considered as strategic means. The capture of a position is a successful engagement in terms of terrain. Not only individual engagements with particular aims are to be classified as means: any greater unity formed in a combination of engagements by being directed toward a common aim can also be considered as *a means*. A winter campaign is such a combination in terms of the time of year.

What remains in the way of ends, then, are only those objects that lead *directly* to peace. All these ends and means must be examined by the theorist in accordance with their effects and their relationships to one another.

Strategy Derives the Means and Ends To Be Examined Exclusively from Experience

The first question is, how an exhaustive list of these objects is arrived at. If a scientific examination were meant to produce this result, it would become involved in all those difficulties which logical necessity has excluded both from the conduct and from the theory of war. We therefore turn to experience and study those sequences of events related in military history. The result will, of course, be a limited theory, based only on facts recorded by military historians. But that is inevitable, since theoretical results must have been derived from military history or at least checked against it. Such a limitation is in any case more theoretical than real.

A great advantage offered by this method is that theory will have to remain realistic. It cannot allow itself to get lost in futile speculation, hairsplitting, and flights of fancy.

How Far Should an Analysis of the Means be Carried?

A second question is, how far theory should carry its analysis of the means. Obviously only so far as the separate attributes will have significance in practice. The range and effectiveness of different firearms is tactically most important; but their construction, though it governs their performance, is irrelevant. The conduct of war has nothing to do with making guns and powder out of coal, sulphur, saltpeter, copper and tin; its given quantities are weapons that are ready for use and their effectiveness. Strategy uses maps without worrying about trigonometric surveys; it does not inquire how a country should be organized and a people trained and ruled in order to produce the best military results. It takes these matters as it finds them in the European community of nations, and calls attention only to unusual circumstances that exert a marked influence on war.

Substantial Simplification of Knowledge

Clearly, then, the range of subjects a theory must cover may be greatly simplified and the knowledge required for the conduct of war can be greatly reduced. Military activity in general is served by an enormous amount of expertise and skills, all of which are needed to place a well-equipped force in the field. They coalesce into a few great results before they attain their final purpose in war, like streams combining to form rivers before they flow into the sea. The man who wishes to control them must familiarize himself only with those activities that empty themselves into the great ocean of war.

CHAPTER TWO

This Simplification Explains the Rapid Development of Great Commanders, and Why Commanders Are Not Scholars

In fact, this result of our investigation is so inescapable that if it were any different its validity would be in doubt. Only this explains why in war men have so often successfully emerged in the higher ranks, and even as supreme commanders, whose former field of endeavor was entirely different; the fact, indeed, that distinguished commanders have never emerged from the ranks of the most erudite or scholarly officers, but have been for the most part men whose station in life could not have brought them a high degree of education. That is why anyone who thought it necessary or even useful to begin the education of a future general with a knowledge of all the details has always been scoffed at as a ridiculous pedant. Indeed, that method can easily be proved to be harmful: for the mind is formed by the knowledge and the direction of ideas it receives and the guidance it is given. Great things alone can make a great mind, and petty things will make a petty mind unless a man rejects them as completely alien.

EARLIER CONTRADICTIONS

The simplicity of the knowledge required in war has been ignored: or rather, that knowledge has always been lumped together with the whole array of ancillary information and skills. This led to an obvious contradiction with reality, which could only be resolved by ascribing everything to genius that needs no theory and for which no theory ought to be formulated.

ACCORDINGLY, THE USEFULNESS OF ALL KNOWLEDGE WAS DENIED, AND EVERYTHING WAS ASCRIBED TO NATURAL APTITUDE

Everyone with a grain of common sense realized the vast distance between a genius of the highest order and a learned pedant. Men arrived at a type of free thinking that rejected all belief in theory and postulated that the conduct of war was a natural function of man which he performed as well as his aptitude permitted. It cannot be denied that this view is closer to the truth than the emphasis on irrelevant expertise; still, on closer examination it will be found to be an overstatement. No activity of the human mind is possible without a certain stock of ideas; for the most part these are not innate but acquired, and constitute a man's knowledge. The only question therefore is what type of ideas they should be. We believe that we have answered this by saying that they should refer only to those things with which he will be immediately concerned as a soldier.

Knowledge Will Be Determined by Responsibility

Within this field of military activity, ideas will differ in accordance with the commander's area of responsibility. In the lower ranks they will be focused upon minor and more limited objectives; in the more senior, upon wider and more comprehensive ones. There are commanders-in-chief who could not have led a cavalry regiment with distinction, and cavalry commanders who could not have led armies.

The Knowledge Required in War Is Very Simple, but at the Same Time It Is Not Easy to Apply

Knowledge in war *is very simple,* being concerned with so few subjects, and only with their final results at that. But this does not make its application easy. The obstacles to action in general have already been discussed in Book I. Leaving aside those that can be overcome only by courage, we argue that genuine intellectual activity is simple and easy only in the lower ranks. The difficulty increases with every step up the ladder; and at the top—the position of commander-in-chief—it becomes among the most extreme to which the mind can be subjected.

The Nature of Such Knowledge

A commander-in-chief need not be a learned historian nor a pundit, but he must be familiar with the higher affairs of state and its innate policies; he must know current issues, questions under consideration, the leading personalities, and be able to form sound judgments. He need not be an acute observer of mankind or a subtle analyst of human character; but he must know the character, the habits of thought and action, and the special virtues and defects of the men whom he is to command. He need not know how to manage a wagon or harness a battery horse, but he must be able to gauge how long a column will take to march a given distance under various conditions. This type of knowledge cannot be forcibly produced by an apparatus of scientific formulas and mechanics; it can only be gained through a talent for judgment, and by the application of accurate judgment to the observation of man and matter.

The knowledge needed by a senior commander is distinguished by the fact that it can only be attained by a special talent, through the medium of reflection, study and thought: an intellectual instinct which extracts the essence from the phenomena of life, as a bee sucks honey from a flower. In addition to study and reflection, life itself serves as a source. Experience, with its wealth of lessons, will never produce a *Newton* or an *Euler,* but it may well bring forth the higher calculations of a *Condé* or a *Frederick.*

To save the intellectual repute of military activity there is no need to resort to lies and simple-minded pedantry. No great commander was ever a man of limited intellect. But there are numerous cases of men who served with the greatest distinction in the lower ranks and turned out barely mediocre in the highest commands, because their intellectual powers were inadequate. Even among commanders-in-chief a distinction must of course be made according to the extent of their authority.

Knowledge Must Become Capability

One more requisite remains to be considered—a factor more vital to military knowledge than to any other. Knowledge must be so absorbed into the mind that it almost ceases to exist in a separate, objective way. In almost any other art or profession a man can work with truths he has learned from musty books, but which have no life or meaning for him. Even truths that are in constant use and are always to hand may still be externals. When an architect sits down with pen and paper to determine the strength of an abutment by a complicated calculation, the truth of the answer at which he arrives is not an expression of his own personality. First he selects the data with care, then he submits them to a mental process not of his own invention, of whose logic he is not at the moment fully conscious, but which he applies for the most part mechanically. It is never like that in war. Continual change and the need to respond to it compels the commander to carry the whole intellectual apparatus of his knowledge within him. He must always be ready to bring forth the appropriate decision. By total assimilation with his mind and life, the commander's knowledge must be transformed into a genuine capability. That is why it all seems to come so easily to men who have distinguished themselves in war, and why it is all ascribed to natural talent. We say *natural talent* in order to distinguish it from the talent that has been trained and educated by reflection and study.

These observations have, we believe, clarified the problems that confront any theory of warfare, and suggested an approach to its solution.

We have divided the conduct of war into the two fields of tactics and strategy. The theory of the latter, as we have already stated, will unquestionably encounter the greater problems since the former is virtually limited to material factors, whereas for strategic theory, dealing as it does with ends which bear directly on the restoration of peace, the range of possibilities is unlimited. As these ends will have to be considered primarily by the *commander-in-chief*, the problems mainly arise in those fields that lie within his competence.

In the field of strategy, therefore, even more than in tactics, theory will be content with the simple consideration of material and psychological factors, especially where it embraces the highest of achievements. It will be sufficient if it helps the commander acquire those insights that, once absorbed into his way of thinking, will smooth and protect his progress, and will never force him to abandon his convictions for the sake of any objective fact.

CHAPTER THREE

Art of War or Science of War

Usage Is Still Unsettled

ABILITY AND KNOWLEDGE. THE OBJECT OF SCIENCE IS KNOWLEDGE; THE OBJECT OF ART IS CREATIVE ABILITY

The use of these terms seems still to be unsettled, and simple though the matter may be, we apparently still do not know on what basis we should choose between them. We have already argued that *knowledge* and *ability* are different things—so different that there should be no cause for confusion. A book cannot really teach us how to do anything, and therefore "art" should have no place in its title. But we have become used to summarizing the knowledge required for the practice of art (individual branches of which may be complete sciences in themselves) by the term "theory of art," or simply "art." It is therefore consistent to keep this basis of distinction and call everything "art" whose object is creative ability, as, for instance, architecture. The term "science" should be kept for disciplines such as mathematics or astronomy, whose object is pure knowledge. That every theory of art may contain discrete sciences goes without saying, and need not worry us. But it is also to be noted that no science can exist without some element of art: in mathematics, for instance, the use of arithmetic and algebra is an art. But art may go still further. The reason is that, no matter how obvious and palpable the difference between knowledge and ability may be in the totality of human achievement, it is still extremely difficult to separate them entirely in the individual.

The Difficulty of Separating Perception from Judgment

ART OF WAR

Of course all thought is art. The point where the logician draws the line, where the premises resulting from perceptions end and where judgment starts, is the point where art begins. But further: perception by the mind is already a judgment and therefore an art; so too, in the last analysis, is perception by the senses. In brief, if it is impossible to imagine a human being capable of perception but not of judgment or vice versa, it is likewise impossible to separate art and knowledge altogether. The more these delicate motes of light are personified in *external forms* of being, the more will their realms separate. To repeat, creation and production lie in the realm of art; science will dominate where the object is inquiry and knowledge. It follows that the term "art of war" is more suitable than "science of war."

We have discussed this at length because these concepts are indispensable. But we must go on to say that strictly speaking war is neither an art nor a science. To take these concepts as a point of departure is misleading in that it has unintentionally caused war to be put on a par with other arts or sciences, resulting in a mass of incorrect analogies.

This difficulty was already recognized in the past, and it was therefore suggested that war was a craft. That, however, proved more of a loss than a gain, because a craft is simply an *inferior* form of art and as such subject to stricter and more rigorous laws. Actually, there was a time—the age of the *condottieri*—when the art of war was akin to a craft. But this tendency had no *internal,* only an *external* basis. Military history shows how unnatural and unsatisfactory it turned out to be.

War Is an Act of Human Intercourse

We therefore conclude that war does not belong in the realm of arts and sciences; rather it is part of man's social existence. War is a clash between major interests, which is resolved by bloodshed—that is the only way in which it differs from other conflicts. Rather than comparing it to art we could more accurately compare it to commerce, which is also a conflict of human interests and activities; and it is *still* closer to politics, which in turn may be considered as a kind of commerce on a larger scale. Politics, moreover, is the womb in which war develops—where its outlines already exist in their hidden rudimentary form, like the characteristics of living creatures in their embryos.

Difference

The essential difference is that war is not an exercise of the will directed at inanimate matter, as is the case with the mechanical arts, or at matter which is animate but passive and yielding, as is the case with the human mind and emotions in the fine arts. In war, the will is directed at an animate object that *reacts.* It must be obvious that the intellectual codification used in the arts and sciences is inappropriate to such an activity. At the same time it is clear that continual striving after laws analogous to those appropriate to the realm of inanimate matter was bound to lead to one mistake after another. Yet it was precisely the mechanical arts that the art of war was supposed to imitate. The fine arts were impossible to imitate, since they themselves do not yet have sufficient laws and rules of their own. So far all attempts at formulating any have been found too limited and one-sided and have constantly been undermined and swept away by the currents of opinion, emotion and custom.

Part of the object of this book is to examine whether a conflict of living forces as it develops and is resolved in war remains subject to general laws, and whether these can provide a useful guide to action. This much is clear: this subject, like any other that does not surpass man's intellectual capacity, can be elucidated by an inquiring mind, and its internal structure can to some degree be revealed. That alone is enough to turn the concept of theory into reality.

CHAPTER FOUR

Method and Routine[1]

In order to explain succinctly the concepts of method and routine, which play such an important role in war, we must glance briefly at the logical hierarchy that governs the world of action like a duly constituted authority.

Law is the broadest concept applicable to both perception and action. In its literal sense the term obviously contains a subjective, arbitrary element, and yet it expresses the very thing on which man and his environment essentially depend. Viewed as a matter of cognition, law is the relationship between things and their effects. Viewed as a matter of the will, law is a determinant of action; at that point, it is synonymous with *decree* and *prohibition*.

Principle is also a law for action, but not in its *formal, definitive meaning*; it represents only the spirit and the sense of the law: in cases where the diversity of the real world cannot be contained within the rigid form of law, the application of principle allows for a greater latitude of judgment. Cases to which principle cannot be applied must be settled by judgment; principle thus becomes essentially a support, or lodestar, to the man responsible for the action.

A principle is *objective* if it rests on objective truth and is therefore equally valid for all; it is *subjective* and is generally called a maxim if subjective considerations enter into it. In that case it has value only for the person who adopts it.

Rule is a term often used in the sense of law; it then becomes synonymous with principle. The proverb goes "there is an exception to every rule" and not "to every law," which shows that in the case of a rule one reserves the right to a more liberal interpretation.

In another sense, the term "rule" is used for "means": to recognize an underlying truth through a single obviously relevant feature enables us to derive a general law of action from this feature. Rules in games are like this, and so are the short cuts used in mathematics, and so on.

Regulations and *directions* are directives dealing with a mass of minor, more detailed circumstances, too numerous and too trivial for general laws.

"*Method*," finally, or "mode of procedure," is a constantly recurring procedure that has been selected from several possibilities. It becomes routine when action is prescribed by method rather than by general principles or individual regulation. It must necessarily be assumed that all cases to which such a routine is applied will be essentially alike. Since this will not be

[1] The German word *Methodismus* has no precise English equivalent. Eds.

entirely so, it is important that it be true of at least *as many as possible*. In other words, methodical procedure should be designed to meet the most probable cases. Routine is not based on definite individual premises, but rather on the *average probability* of analogous cases. Its aim is to postulate an average truth, which, when applied evenly and constantly, will soon acquire some of the nature of a mechanical skill, which eventually does the right thing almost automatically.

In the conduct of war, perception cannot be governed by laws: the complex phenomena of war are not so uniform, nor the uniform phenomena so complex, as to make laws more useful than the simple truth. Where a simple point of view and plain language are sufficient, it would be pedantic and affected to make them *complex* and *involved*. Nor can the theory of war apply the concept of law to action, since no prescriptive formulation universal enough to deserve the name of law can be applied to the constant change and diversity of the phenomena of war.

Principles, rules, regulations, and methods are, however, indispensable concepts to or for that part of the theory of war that leads to positive doctrines; for in these doctrines the truth can express itself only in such compressed forms.

Those concepts will appear most frequently in tactics, which is that part of war in which theory can develop most fully into a positive doctrine. Some examples of tactical principles are: except in emergencies cavalry is not to be used against unbroken infantry; firearms should not be used until the enemy is within effective range; in combat, as many troops as possible should be preserved for the final phase. None of these concepts can be dogmatically applied to every situation, but a commander must always bear them in mind so as not to lose the benefit of the truth they contain in cases where they do apply.

Cooking in the enemy camp at unusual times suggests that he is about to move. The intentional exposure of troops in combat indicates a feint. This manner of inferring the truth may be called a rule because one deduces the enemy's intentions from a single visible fact connected with them.

If the rule enjoins that one should resume attacking the enemy as soon as he starts to withdraw his artillery, then a whole course of action is determined by this single phenomenon which has revealed his entire condition: the fact that he is ready to give up the fight. While he is doing so, he cannot offer serious resistance or even avoid action as he could once he is fully on the move.

To the extent that *regulations* and *methods* have been drilled into troops as active principles, theoretical preparation for war is part of its actual conduct. All standing instructions on formations, drill, and field-service are regulations and methods. Drill instructions are mainly regulations; field manuals, mainly methods. The actual conduct of war is based on these things; they are accepted as given procedures and as such must have their place in the theory of the conduct of war.

In the employment of these forces, some activities remain a matter of choice. Regulations, or prescriptive directions, do not apply to them, precisely because regulations preclude freedom of choice. Routines, on the other hand, represent a general way of executing tasks as they arise based, as we have said, on average probability. They represent the dominance of principles and rules, carried through to actual application. As such they may well have a place in the theory of the conduct of war, provided they are not falsely represented as absolute, binding frameworks for action (systems); rather they are the best of the general forms, short cuts, and options that may be substituted for individual decisions.

The frequent application of routine in war will also appear essential and inevitable when we consider how often action is based on pure conjecture or takes place in complete ignorance, either because the enemy prevents us from knowing all the circumstances that might affect our dispositions, or because there is not enough time. Even if we did know all the circumstances, their implications and complexities would not permit us to take the necessary steps to deal with them. Therefore our measures must always be determined by a limited number of possibilities. We have to remember the countless minor factors implicit in every case. The only possible way of dealing with them is to treat each case as implying all the others, and base our dispositions on the general and the probable. Finally we have to remember that as the number of officers increases steadily in the lower ranks, the less the trust that can be placed on their true insight and mature judgment. Officers whom one should not expect to have any greater understanding than regulations and experience can give them have to be helped along by routine methods tantamount to rules. These will steady their judgment, and also guard them against eccentric and mistaken schemes, which are the greatest menace in a field where experience is so dearly bought.

Routine, apart from its sheer inevitability, also contains one positive advantage. Constant practice leads to *brisk, precise,* and *reliable* leadership, reducing natural friction and easing the working of the machine.

In short, routine will be more frequent and indispensable, the lower the level of action. As the level rises, its use will decrease to the point where, at the summit, it disappears completely. Consequently, it is more appropriate to tactics than to strategy.

War, in its highest forms, is not *an infinite mass of minor events,* analogous despite their diversities, which can be controlled with greater or lesser effectiveness depending on the methods applied. War consists rather of *single, great decisive actions,* each of which needs to be handled individually. War is not like a field of wheat, which, without regard to the individual stalk, may be mown more or less efficiently depending on the quality of the scythe; it is like a stand of mature trees in which the axe has to be used judiciously according to the characteristics and development of each individual trunk.

The highest level that routine may reach in military action is of course determined not by rank but by the nature of each situation. The highest ranks are least affected by it simply because the scope of their operations is the most comprehensive. A standard order of battle or system of advance guards and outposts are methods by which a general may be fettering not only his subordinates, but, in certain cases, also himself. Of course these methods may be his own inventions, and adapted to particular conditions; to the extent that they are based on the general properties of troops and weapons, they can also be a subject of theory. But any method by which strategic plans are turned out ready-made, as if from some machine, must be totally rejected.

So long as no acceptable theory, no intelligent analysis of the conduct of war exists, routine methods will tend to take over even at the highest levels. Some of the men in command have not had the opportunities of self-improvement afforded by education and contact with the higher levels of society and government. They cannot cope with the impractical and contradictory arguments of theorists and critics even though their common sense rejects them. Their only insights are those that have been gained by experience. For this reason, they prefer to use the means with which their experience has equipped them, even in cases that could and should be handled freely and individually. They will copy their supreme commander's favorite device—thus automatically creating a new routine. When we find generals under Frederick the Great using the so-called oblique order of battle; generals of the French Revolution using turning movements with a much extended front; and commanders under Bonaparte attacking with a brutal rush of concentric masses, then we recognize in these repetitions a ready-made method, and see that even the highest ranks are not above the influence of routine. Once an improved theory helps the study of the conduct of war, and educates the mind and judgment of the senior commanders, routine methods will no longer reach so high. Those types of routine that must be considered indispensable will then at least be based on a theory rather than consist in sheer imitation. No matter how superbly a great commander operates, there is always a subjective element in his work. If he displays a certain style, it will in large part reflect his own personality; but that will not always blend with the personality of the man who copies that style.

Yet it would be neither possible nor correct to eliminate subjective routine or personal style entirely from the conduct of war. They should be seen, rather, as manifestations of the influence exerted on individual phenomena by the total character of the war—an influence which, if it has not been foreseen and allowed for by accepted theory, may find no other means of adequate expression. What could be more natural than the fact that the War of the French Revolution had its characteristic style, and what theory could have been expected to accommodate it? The danger is that this kind of style, developed out of a single case, can easily outlive the situation that gave rise to it; for conditions change imperceptibly. That danger is the very thing a theory should prevent by lucid, rational criticism. When in 1806 the Prussian generals, Prince Louis at Saalfeld, Tauentzien on the Dornberg near Jena, Grawert on one side of Kapellendorf and Rüchel on the other, plunged into the open jaws of disaster by using Frederick the Great's oblique order of battle, it was not just a case of a style that had outlived its usefulness but the most extreme poverty of the imagination to which routine has ever led. The result was that the Prussian army under Hohenlohe was ruined more completely than any army has ever been ruined on the battlefield.

CHAPTER FIVE

Critical Analysis[1]

The influence of theoretical truths on practical life is always exerted more through critical analysis than through doctrine. Critical analysis being the application of theoretical truths to actual events, it not only reduces the gap between the two but also accustoms the mind to these truths through their repeated application. We have established a criterion for theory, and must now establish one for critical analysis as well.

We distinguish between the *critical approach* and the plain narrative of a historical event, which merely arranges facts one after another, and at most touches on their immediate causal links.

Three different intellectual activities may be contained in the critical approach.

First, the discovery and interpretation of equivocal facts. This is historical research proper, and has nothing in common with theory.

Second, the tracing of effects back to their causes. This is *critical analysis proper*. It is essential for theory; for whatever in theory is to be defined, supported, or simply described by reference to experience can only be dealt with in this manner.

Third, the investigation and evaluation of means employed. This last is criticism proper, involving praise and censure. Here theory serves history, or rather the lessons to be drawn from history.

In the last two activities which are the truly critical parts of historical inquiry, it is vital to analyze everything down to its basic elements, to incontrovertible truth. One must not stop half-way, as is so often done, at some arbitrary assumption or hypothesis.

The deduction of effect from cause is often blocked by some insuperable extrinsic obstacle: the true causes may be quite unknown. Nowhere in life is this so common as in war, where the facts are seldom fully known and the underlying motives even less so. They may be intentionally concealed by those in command, or, if they happen to be transitory and accidental, history may not have recorded them at all. That is why critical narrative must usually go hand in hand with historical research. Even so, the disparity between cause and effect may be such that the critic is not justified in considering the effects as inevitable results of known causes. This is bound to produce gaps—historical results that yield no useful lesson. All a theory demands is that investigation should be resolutely carried on till such a gap

[1] The German term *Kritik* here means "critique, critical analysis, evaluation, and interpretation" rather than "criticism." Eds.

is reached. At that point, judgment has to be suspended. Serious trouble arises only when known facts are forcibly stretched to explain effects; for this confers on these facts a spurious importance.

Apart from that problem, critical research is faced with a serious intrinsic one: effects in war seldom result from a single cause; there are usually several concurrent causes. It is therefore not enough to trace, however honestly and objectively, a sequence of events back to their origin: each identifiable cause still has to be correctly assessed. This leads to a closer analysis of the nature of these causes, and in this way critical investigation gets us into theory proper.

A critical *inquiry*—the examination of the means—poses the question as to what are the peculiar effects of the means employed, and whether these effects conform to the intention with which they were used.

The particular effects of the means leads us to an investigation of their nature—in other words, into the realm of theory again.

We have seen that in criticism it is vital to reach the point of incontrovertible truth; we must never stop at an arbitrary assumption that others may not accept, lest different propositions, equally valid perhaps, be advanced against them; leading to an unending argument, reaching no conclusions, and resulting in no lesson.

We have also seen that both investigation of the causes and examination of the means leads to the realm of theory—that is, to the field of universal truth that cannot be inferred merely from the individual instance under study. If a usable theory does indeed exist, the inquiry can refer to its conclusions and at that point end the investigation. However, where such theoretical criteria do not exist, analysis must be pressed until the basic elements are reached. If this happens often, it will lead the writer into a labyrinth of detail: he will have his hands full and find it almost impossible to give each point the attention it demands. As a result, in order to set a limit to his inquiries, he will have to stop short of arbitrary assumptions after all. Even if they would not seem arbitrary to him, they would to others, because they are neither self-evident nor have they been proved.

In short a working theory is an essential basis for criticism. Without such a theory it is generally impossible for criticism to reach that point at which it becomes truly instructive—when its arguments are convincing and cannot be refuted.

But it would be wishful thinking to imagine that any theory could cover every abstract truth, so that all the critic had to do would be to classify the case studied under the appropriate heading. Equally, it would be ridiculous to expect criticism to reverse course whenever it came up against the limits of a sacrosanct theory. The same spirit of analytical investigation which creates a theory should also guide the work of the critic who both may and should often cross into the realm of theory in order to elucidate any points of special importance. The function of criticism would be missed entirely if criticism were to degenerate into a mechanical application of theory. All the positive results of theoretical investigation—all the principles, rules, and methods—will increasingly lack universality and absolute truth the closer they come to being positive doctrine. They are there to be used when needed, and their suitability in any given case must always be a matter of judgment. A critic should never use the results of theory as laws and standards, but only—as the soldier does—as *aids to judgment*. If, in tactics, it is generally agreed that in the standard line of battle cavalry should be posted not in line with but behind the infantry, it would nevertheless be foolish to condemn every different deployment simply because it is different. The critic should analyze the reasons for the exception. He has no right to appeal to theoretical principles unless these reasons are inadequate. Again, if theory lays it down that an attack with divided forces reduces the probability of success, it would be equally unreasonable, without further analysis, to attribute failure to the separation of forces whenever both occur together; or when an attack with divided forces is successful to conclude that the original theoretical assertion was incorrect. The inquiring nature of criticism can permit neither. In short, criticism largely depends on the results of the theorist's analytic studies. What theory has already established the critic need not go over again, and it is the theorist's function to provide the critic with these findings.

The critic's task of investigating the relation of cause and effect and the appropriateness of means to ends will be easy when cause and effect, means and ends, are closely linked.

When a surprise attack renders an army incapable of employing its powers in an orderly and rational manner, then the effect of the surprise cannot be questioned. When theory has established that an enveloping attack leads to greater, if less certain, success, we have to ask whether the general who used this envelopment was primarily concerned with the magnitude of success. If so, he chose the right way to go about it. But if he used it in order to make *more certain* of success, basing his action not so much on individual circumstances as on the general nature of enveloping attacks, as has happened innumerable times, then he misunderstood the nature of the means he chose and committed an error.

The business of critical analysis and proof is not very difficult in cases of this kind; it is bound to be easy if one restricts oneself to the most immediate aims and effects. This may be done quite arbitrarily if one isolates the matter from its setting and studies it only under those conditions.

But in war, as in life generally, all parts of a whole are interconnected and thus the effects produced, however small their cause, must influence all subsequent military operations and modify their final outcome to some degree, however slight. In the same way, every means must influence even the ultimate purpose.

One can go on tracing the effects that a cause produces so long as it seems worth while. In the same way, a means may be evaluated, not merely with respect to its immediate end: that end itself should be appraised as a means for the next and highest one; and thus we can follow a chain of sequential objectives until we reach one that requires no justification, because its necessity is self-evident. In many cases, particularly those involving great and decisive actions, the analysis must extend to the *ultimate objective*, which is to bring about peace.

Every stage in this progression obviously implies a new basis for judgment. That which seems correct when looked at from one level may, when viewed from a higher one, appear objectionable.

In a critical analysis of the action, the search for the causes of phenomena and the testing of means in relation to ends always go hand in hand, for only the search for a cause will reveal the questions that need to be studied.

The pursuit of this chain, upward and downward, presents considerable problems. The greater the distance between the event and the cause that we are seeking, the larger the number of other causes that have to be considered at the same time. Their possible influence on events has to be established and allowed for, since the greater the magnitude of any event, the wider the range of forces and circumstances that affect it. When the causes for the loss of a battle have been ascertained, we shall admittedly also know some of the causes of the effects that this lost battle had upon the whole—but only some, since the final outcome will have been affected by other causes as well.

In the analysis of the means, we encounter the same multiplicity as our viewpoint becomes more comprehensive. The higher the ends, the greater the number of means by which they may be reached. The final aim of the war is pursued by all armies simultaneously, and we therefore have to consider the full extent of everything that has happened, or might have happened.

We can see that this may sometimes lead to a broad and complex field of inquiry in which we may easily get lost. A great many assumptions have to be made about things that did not actually happen but seemed possible, and that, therefore, cannot be left out of account.

When in March 1797 Bonaparte and the Army of Italy advanced from the Tagliamento to meet the Archduke Charles, their object was to force a decision on the Austrians before the arrival of their reinforcements from the Rhine. If we consider only the immediate objective, the means were well-chosen, as the result showed. The Archduke's forces were still so weak that he made only an attempt at resistance on the Tagliamento. On seeing the strength and resolution of his enemy, he abandoned the area and the approaches to the Norican Alps. How could Bonaparte make use of this success? Should he press on into the heart of the Austrian Empire, ease the advance of the two armies of the Rhine under Moreau and Hoche, and work in close conjunction with them? That was how Bonaparte saw it, and from his point of view he was right. But the critic may take a wider view—that of the French Directory; whose members could see, and must have realized, that the campaign on the Rhine would not begin for another six weeks. From that standpoint, then, Bonaparte's advance through the Norican Alps could only be considered an unjustifiable risk. If the Austrians had moved sizable reserves from the Rhine to Styria with which the Archduke Charles could have attacked the Army of Italy, not only would that Army have been destroyed, but the entire campaign would have been lost. Bonaparte realized this by the time he reached Villach, and this persuaded him to sign the Armistice of Leoben with alacrity.

If the critic takes a still wider view, he can see that the Austrians had no reserves between the Archduke's army and Vienna, and that the advance of the Army of Italy was a threat to the capital itself.

Let us assume that Bonaparte knew the capital to be vulnerable and his own superiority over the Archduke even in Styria to be decisive. His rapid advance into the heart of Austria would then no longer have been pointless. The value of the attack would now depend merely on the value the Austrians set on the retention of Vienna. If, rather than lose the capital, they would accept whatever conditions for peace Bonaparte offered them, the threat to Vienna could be considered as his final aim. If Bonaparte had somehow known of this, the critic would have no more to say. But if the issue was still uncertain, the critic must take a more comprehensive point of view, and ask what would have happened if the Austrians had abandoned Vienna, and withdrawn into the vast expanse of territory they still controlled. That, however, is obviously a question which cannot possibly be answered without reference to the probable encounter between the two armies on the Rhine. There the French were so decisively superior in numbers—130,000 against 80,000—that the issue would not have been much in doubt. But then the question would again have arisen, what use would the French Directory have made of the victory? Would the French have pursued their advantage to the far frontiers of the Austrian monarchy, breaking Austrian power and shattering the Empire, or would they have been satisfied with the conquest of a sizable part of it as a surety for peace? We have to ascertain the probable consequences of both possibilities before determining the probable choice of the Directory. Let us assume that this consideration led to the answer that the French forces were far too weak to bring about the total collapse of Austria, so that the mere attempt to do so would have reversed the situation and even the conquest and occupation of a significant segment of Austrian territory would have placed the French in a strategic situation with which their forces could hardly have coped. This argument would have colored their view of the situation in which the Army of Italy found itself, and reduced its likely prospects. No doubt this is what persuaded Bonaparte, although he realized the Archduke's hopeless situation, to sign the peace of Campo Formio, on conditions that imposed on the Austrians no greater sacrifices than the loss of some provinces which even the most successful campaign could not have recovered. But the French could not have counted even on the moderate gains of Campo Formio, and therefore could not have made them the objectives of their offensive, had it not been for two considerations. The first was the value the Austrians placed on the two possible outcomes. Though both of them made eventual success appear probable, would the Austrians have thought it worth the sacrifices they entailed—the continuation of the war—when that price could have been avoided by concluding a peace on not too unfavorable terms? The second consideration consists in the question whether the Austrian government would even pursue its reflections and thoroughly evaluate the potential limits of French success, rather than be disheartened by the impression of current reverses?

The first of these considerations is not simply idle speculation. On the contrary, it is of such decisive practical importance that it always arises whenever one aims at total victory. It is this which usually prevents such plans from being carried out.

The second consideration is just as essential, for war is not waged against an abstract enemy, but against a real one who must always be kept in mind. Certainly a man as bold as Bonaparte was conscious of this, confident as he was in the terror inspired by his approach. The same confidence led him to Moscow in 1812, but there it left him. In the course of the gigantic battles, the terror had already been somewhat blunted. But in 1797 it was still fresh, and the secret of the effectiveness of resisting to the last had not yet been discovered. Still, even in 1797 his boldness would have had a negative result if he had not, as we have seen, sensed the risk involved and chosen the moderate peace of Campo Formio as an alternative.

We must now break off this discussion. It will suffice to show the comprehensive, intricate and difficult character which a critical analysis may assume if it extends to ultimate objectives—in other words, if it deals with the great and decisive measures which must necessarily lead up to them. It follows that in addition to theoretical insight into the subject, natural talent will greatly enhance the value of critical analysis: for it will primarily depend on such talent to illuminate the connections which link things together and to determine which among the countless concatenations of events are the essential ones.

But talent will be needed in another way as well. Critical analysis is not just an evaluation of the means actually employed, but of *all possible means*—which first have to be formulated, that is, invented. One can, after all, not condemn a method without being able to suggest a better alternative. No matter how small the range of possible combinations may be in most cases, it cannot be denied that listing those that have not been used is not a mere analysis of existing things but an achievement that cannot be performed to order since it depends on the creativity of the intellect.

We are far from suggesting that the realm of true genius is to be found in cases where a handful of simple, practical schemes account for everything. In our view it is quite absurd, though it is often done, to treat the turning of a position as an invention of great genius. And yet such individual creative evaluations are necessary, and they significantly influence the value of critical analysis.

When on 30 July 1796, Bonaparte decided to raise the siege of Mantua in order to meet Wurmser's advance, and fell with his entire strength on each of the latter's columns separately while they were divided by Lake

Garda and the Mincio, he did so because this seemed the surest way to decisive victories. These victories in fact did occur, and were repeated even more decisively in the same way against later attempts to relieve Mantua. There is only one opinion about this: unbounded admiration.

And yet, Bonaparte could not choose this course on 30 July without renouncing all hope of taking the city; for it was impossible to save the siege train, and it could not be replaced during the current campaign. In point of fact, the siege turned into a mere blockade and the city, which would have fallen within a week if the siege had been maintained, held out for six more months despite all Bonaparte's victories in the field.

Critics, unable to recommend a better way of resistance, have considered this an unavoidable misfortune. Resisting a relieving army behind lines of circumvallation had fallen into such disrepute and contempt that it occurred to no one. And yet in the days of Louis XIV it had so often been successfully employed that one can only call it a whim of fashion that a hundred years later it never occurred to anyone *at least to weigh* its merits. If that possibility had been admitted, closer scrutiny of the situation would have shown that 40,000 of the finest infantrymen in the world whom Bonaparte could have placed behind a line of circumvallation at Mantua, would, if they were well entrenched, have had so little cause to fear the 50,000 Austrians whom Wurmser was bringing to relieve the town, that the lines were in little danger even of being attacked. This is not the place to labor the point; we believe we have said enough to show that the possibility deserved notice. We cannot tell whether Bonaparte himself ever considered the plan. There is no trace of it in his memoirs and the rest of the published sources; none of the later critics touched upon it, because they were no longer in the habit of considering this scheme. There is no great merit in recalling its existence; one only has to shed the tyranny of fashion in order to think of it. One does, however, have to think of it in order to consider it and to compare it with the means which Bonaparte in fact employed. Whatever the result of this comparison the critic should not fail to make it.

The world was filled with admiration when Bonaparte, in February 1814, turned from Blücher after beating him at Etoges, Champ-Aubert, Montmirail, and elsewhere, to fall on Schwarzenberg, and beat him at Montereau and Mormant. By rapidly moving his main force back and forth, Bonaparte brilliantly exploited the allies' mistake of advancing with divided forces. If, people thought, these superb strokes in all directions failed to save him, at least it was not his fault. No one has yet asked what would have happened if, instead of turning away from Blücher, and back to Schwarzenberg, he had gone on hammering Blücher and had pursued him back to the Rhine. We are convinced that the complexion of the whole campaign would have been changed and that, instead of marching on Paris, the allied armies would have withdrawn across the Rhine. We do not require others to share our view, but no expert can doubt that the critic is bound to consider that alternative once it has been raised.

The option is much more obvious in this case than in the previous one.

Nevertheless it has been overlooked, because people are biased and blindly follow a single line of thought.

The need for suggesting a better method than the one that is condemned has created the type of criticism which is used almost exclusively: the critic thinks he must only indicate the method which he considers to be better, without having to furnish proof. In consequence not everyone is convinced; others follow the same procedure, and a controversy starts without any basis for discussion. The whole literature on war is full of this kind of thing.

The proof that we demand is needed whenever the advantage of the means suggested is not plain enough to rule out all doubts; it consists in taking each of the means and assessing and comparing the particular merits of each in relation to the objective. Once the matter has thus been reduced to simple truths, the controversy must either stop, or at least lead to new results. By the other method, the pros and cons simply cancel out.

Suppose, for instance, that in the case of the last example, we had not been satisfied, and wanted to prove that the relentless pursuit of Blücher would have served Napoleon better than turning against Schwarzenberg. We would rely on the following simple truths:

1. Generally speaking, it is better to go on striking in the same direction than to move one's forces this way and that, because shifting troops back and forth involves losing time. Moreover, it is easier to achieve further successes where the enemy's morale has already been shaken by substantial losses; in this way, none of the superiority that has been attained will go unexploited.

2. Even though Blücher was weaker than Schwarzenberg, his enterprising spirit made him more important. The center of gravity lay with him, and he pulled the other forces in his direction.

3. The losses Blücher suffered were on the scale of a serious defeat. Bonaparte had thus gained so great a superiority over him as to leave no doubt that he would have to retreat as far as the Rhine, for no reserves of any consequence were stationed on that route.

4. No other possible success could have caused so much alarm or so impressed the allies' mind. With a staff which was known to be as timid and irresolute as Schwarzenberg's, this was bound to be an important consideration. The losses incurred by the Crown Prince of Württemberg at Montereau and by Count Wittgenstein at Mormant were sure to be fairly well known to Prince Schwarzenberg; on the other hand, news of the misfortunes that Blücher met with along his distant and discontinuous line between the Marne and the Rhine could have reached him only as an avalanche of rumors. Bonaparte's desperate thrust toward Vitry at the end of March was an attempt to test the effect that the threat of a strategic envelopment would have on the allies. It was obviously based on the principle of terror, but in wholly different circumstances now that Bonaparte had been defeated at Laon and Arcis, and Blücher had joined Schwarzenberg with 100,000 men.

Some people, of course, will not be convinced by these arguments, but at least they will not be able to reply that "as Bonaparte, in his thrust towards the Rhine, was threatening Schwarzenberg's base, so Schwarzenberg was threatening Paris, which was Bonaparte's." The reasons we have cited above should make it clear that it would not have occurred to Schwarzenberg to advance on Paris.

In the instance from 1796 which we have touched on above we would say that Bonaparte considered the plan that he adopted as the one best guaranteed to beat the Austrians. Even if this had been true, the outcome would have been an empty triumph which could hardly have significantly affected the fall of Mantua. Our own proposal would have been much more likely to prevent Mantua from being relieved; but even if we put ourselves in Bonaparte's place and take the opposite view—that it offered a smaller prospect of success—the choice would have been based on balancing a likelier but almost useless, and therefore minor, victory against a less likely but far greater one. If the matter is looked at in that light, boldness would surely have opted for the second course: but looked at superficially, the opposite was what occurred. Bonaparte certainly held to the bolder intention, so there can be no doubt that he did not think the matter through to the point where he could assess the consequences as fully as we can in the light of experience.

In the study of means, the critic must naturally frequently refer to military history, for in the art of war experience counts more than any amount of abstract truths. Historical proof is subject to conditions of its own, which will be dealt with in a separate chapter; but unfortunately these conditions are so seldom met with that historical references usually only confuse matters more.

Another important point must now be considered: how far is the critic free, or even duty-bound, to assess a single case in the light of his greater knowledge, including as it does a knowledge of the outcome? Or when and where should he ignore these things in order to place himself exactly in the situation of the man in command?

If the critic wishes to distribute praise or blame, he must certainly try to put himself exactly in the position of the commander; in other words, he must assemble everything the commander knew and all the motives that affected his decision, and ignore all that he could not or did not know, especially the outcome. However, this is only an ideal to be aimed at, if never fully achieved: a situation giving rise to an event can never look the same to the analyst as it did to the participant. A mass of minor circumstances that may have influenced his decision are now lost to us, and many subjective motives may never have been exposed at all. These can only be discovered from the memoirs of the commanders, or from people very close to them. Memoirs often treat such matters pretty broadly, or, perhaps deliberately, with something less than candor. In short, the critic will always lack much that was present in the mind of the commander.

But it is even more difficult for the critic to shut off his superfluous knowledge. That is possible only with regard to accidental factors that impinge on the situation without being basic to it; in all really essential matters, however, it is very difficult and never fully attainable.

Let us first consider the outcome. Unless this was the result of chance, it is almost impossible to prevent the knowledge of it from coloring one's judgment of the circumstances from which it arose: we see these things in the light of their result, and to some extent come to know and appreciate them fully only because of it. Military history in all its aspects is itself a *source of instruction* for the critic, and it is only natural that he should look at all particular events in the light of the whole. Therefore, even if in some cases he did try to disregard results altogether, he could never entirely succeed.

But this is true not only of the outcome (that is, with what happens subsequently) but also of facts that were present from the beginning—the factors that determine the action. The critic will, as a rule, have more information than the participant. One would think he could easily ignore it, but he cannot. This is because knowledge of previous and simultaneous circumstances does not rest on specific information alone but on numerous conjectures and assumptions. Completely accidental matters apart, very little information does come to hand which has not been preceded by assumptions or conjectures. If specifics do not materialize, these assumptions and conjectures will take their place. Now we can understand why later critics who know all the previous and attendant circumstances must not be influenced by their knowledge when they ask which among the unknown facts they themselves would have considered probable at the time of the action. We maintain that complete insulation is as impossible here as it is when we consider the final outcome, and for the same reasons.

Therefore, if a critic wishes to praise or blame any specific action, he will only partly be able to put himself in the situation of the participant. In many cases he can do this well enough to suit practical purposes, but we must not forget that sometimes it is completely impossible.

It is, however, neither necessary nor desirable for the critic to identify himself completely with the commander. In war, as in all skills, a trained natural aptitude is called for. This virtuosity may be great or small. If it is great, it may easily be superior to that of the critic: what student would lay claim to the talent of a Frederick or a Bonaparte? Hence, unless we are to hold our peace in deference to outstanding talent, we must be allowed to profit from the wider horizons available to us. A critic should therefore not check a great commander's solution to a problem as if it were a sum in arithmetic. Rather, he must recognize with admiration the commander's success, the smooth unfolding of events, the higher workings of his genius. The essential interconnections that genius had divined, the critic has to reduce to factual knowledge.

To judge even the slightest act of talent, it is necessary for the critic to take a more comprehensive point of view, so that he, in possession of any number of objective reasons, reduces subjectivity to the minimum, and so avoids judging by his own, possibly limited, standards.

This elevated position of criticism, dispensing praise or blame with a full knowledge of all the circumstances, will not insult our feelings. The critic will do this only if he pushes himself into the limelight and implies that all the wisdom that is in fact derived from his complete knowledge of the case is due to his own abilities. No matter how crass that fraud, vanity may very easily lead to it, and it will naturally give offense. More often the critic does not mean to be arrogant; but, unless he makes a point of denying it, a hasty reader will suspect him of it, and this will at once give rise to a charge of lack of critical judgment.

If the critic points out that a Frederick or a Bonaparte made mistakes, it does not mean that he would not have made them too. He may even admit that in the situation of these generals he might have made far greater errors. What it does mean is that he can recognize these mistakes from the pattern of events and feels that the commander's sagacity should have seen them as well.

This is a judgment based on the pattern of events and therefore also *on their outcome*. But, in addition, the outcome may have a completely different effect on judgment—when the outcome is simply used as proof that an action was either correct or incorrect. This may be called a judgment *by results*. At first sight such a judgment would seem entirely inadmissible, but that is not the case.

When in 1812 Bonaparte advanced on Moscow the crucial question was whether the capture of the capital, together with everything else that had already happened, would induce Czar Alexander to make peace. That had happened in 1807 after the battle of Friedland, and it had also worked in 1805 and 1809 with the Emperor Francis after the battles of Austerlitz and Wagram. If, however, peace was not made at Moscow, Bonaparte would have no choice but to turn back, which would have meant a strategic defeat. Let us leave aside the steps by which he advanced on Moscow, and the question whether, in the process, he missed a number of opportunities that might have made the Czar decide on peace. Let us also leave aside the terrible circumstances of the retreat, which may have had their root in the conduct of the entire campaign. The crucial question remains the same: no matter how much more successful the advance on Moscow might have been, it would still have been uncertain whether it could have frightened the Czar into suing for peace. And even if the retreat had not led to the annihilation of the army, it could never have been anything but a major strategic defeat. If the Czar had concluded a disadvantageous peace, the campaign of 1812 would have ranked with those of Austerlitz, Friedland, and Wagram. But if these campaigns had not resulted in peace, they would probably have led to similar catastrophes. Regardless of the power, skill, and wisdom shown by the conqueror of the world, the final fatal question remained everywhere the same. Should we then ignore the actual results of the campaigns of 1805, 1807, and 1809, and, by the test of 1812 alone, proclaim them to be products of imprudence, and their success to be a breach of natural law? Should we maintain that in 1812 strategic justice finally overcame blind chance? That would be a very forced conclusion, an arbitrary judgment where half the evidence is missing, because the human eye cannot trace the interconnection of events back to the decisions of the vanquished monarchs.

Still less can it be said that the campaign of 1812 ought to have succeeded like the others, and that its failure was due to something extraneous: there was nothing extraneous about Alexander's steadfastness.

What can be more natural than to say that in 1805, 1807, and 1809 Bonaparte had gauged his enemy correctly, while in 1812 he did not? In the earlier instances he was right, in the latter he was wrong, and we can say that *because the outcome proves it.*

In war, as we have already pointed out, all action is aimed at probable rather than at certain success. The degree of certainty that is lacking must in every case be left to fate, chance, or whatever you like to call it. One may of course ask that this dependence should be as slight as possible, but only in reference to a particular case—in other words, it should be as *small as possible in that individual case.* But we should not habitually prefer the course that involves the least uncertainty. That would be an enormous mistake, as our theoretical arguments will show. There are times when the utmost daring is the height of wisdom.

It would seem that a commander's personal merits, and thus also his responsibility, become irrelevant to all questions that have to be left to chance. Nevertheless, we cannot deny an inner satisfaction whenever things turn out right; when they do not, we feel a certain intellectual discomfort. *That is all the meaning that should be attached to a judgment of right and wrong that we deduce from success, or rather that we find in success.*

But it is obvious that the intellectual pleasure at success and the intellectual discomfort at failure arise from an obscure sense of some delicate link, invisible to the mind's eye, between success and the commander's genius. It is a gratifying assumption. The truth of this is shown by the fact that our sympathy increases and grows keener as success and failure are repeated by the same man. That is why luck in war is of higher quality than luck in gambling. So long as a successful general has not done us any harm, we follow his career with pleasure.

The critic, then, having analyzed everything within the range of human calculation and belief, will let the outcome speak for that part whose deep, mysterious operation is never visible. The critic must protect this unspoken result of the workings of higher laws against the stream of uninformed opinion on the one hand, and against the gross abuses to which it may be subjected on the other.

Success enables us to understand much that the workings of human intelligence alone would not be able to discover. That means that it will be useful mainly in revealing intellectual and psychological forces and effects, because these are least subject to reliable evaluation, and also because they are so closely involved with the will that they may easily control it. Wherever decisions are based on fear or courage, they can no longer be judged objectively; consequently, intelligence and calculation can no longer be expected to determine the probable outcome.

We must now be allowed to make a few remarks about the instruments critics use—their idiom; for in a sense it accompanies action in war. Critical analysis, after all, is nothing but thinking that should precede the action. We therefore consider it essential that the language of criticism should have the same character as thinking must have in wars; otherwise it loses its practical value and criticism would lose contact with its subject.

In our reflections on the theory of the conduct of war, we said that it ought to train a commander's mind, or rather, guide his education; theory is not meant to provide him with positive doctrines and systems to be used as intellectual tools. Moreover, if it is never necessary or even permissible to use scientific guidelines in order to judge a given problem in war, if the truth never appears in systematic form, if it is not acquired deductively but always *directly* through the natural perception of the mind, then that is the way it must also be in critical analysis.

We must admit that wherever it would be too laborious to determine the facts of the situation, we must have recourse to the relevant principles established by theory. But in the same way as in war these truths are better served by a commander who has absorbed their meaning in his mind rather than one who treats them as rigid external rules, so the critic should not apply them like an external law or an algebraic formula whose relevance need not be established each time it is used. These truths should always be allowed to become self-evident, while only the more precise and complex proofs are left to theory. We will thus avoid using an arcane and obscure language, and express ourselves in plain speech, with a sequence of clear, lucid concepts.

Granted that while this cannot always be completely achieved, it must remain the aim of critical analysis. The complex forms of cognition should be used as little as possible, and one should never use elaborate scientific guidelines as if they were a kind of truth machine. Everything should be done through the natural workings of the mind.

However, this pious aspiration, if we may call it that, has rarely prevailed in critical studies; on the contrary, a kind of vanity has impelled most of them to an ostentatious exhibition of ideas.

The first common error is an awkward and quite impermissible use of certain narrow systems as formal bodies of laws. It is never difficult to demonstrate the one-sidedness of such systems; and nothing more is needed to discredit their authority once and for all. We are dealing here with a limited problem, and since the number of possible systems is after all finite, this error is the lesser of two evils that concern us.

A far more serious menace is the retinue of *jargon, technicalities, and metaphors* that attends these systems. They swarm everywhere—a lawless rabble of camp followers. Any critic who has not seen fit to adopt a system—

either because he has not found one that he likes or because he has not yet got that far—will still apply an occasional scrap of one as if it were a ruler, to show the crookedness of a commander's course. Few of them can proceed without the occasional support of such scraps of scientific military theory. The most insignificant of them—mere technical expressions and metaphors—are sometimes nothing more than ornamental flourishes of the critical narrative. But it is inevitable that all the terminology and technical expressions of a given system will lose what meaning they have, if any, once they are torn from their context and used as general axioms or nuggets of truth that are supposed to be more potent than a simple statement.

Thus it has come about that our theoretical and critical literature, instead of giving plain, straightforward arguments in which the author at least always knows what he is saying and the reader what he is reading, is crammed with jargon, ending at obscure crossroads where the author loses his readers. Sometimes these books are even worse: they are just hollow shells. The author himself no longer knows just what he is thinking and soothes himself with obscure ideas which would not satisfy him if expressed in plain speech.

Critics have yet a third failing: showing off their erudition, and the misuse of historical examples. We have already stated what the history of the art of war is, and our views on historical examples and military history in general will be developed in later chapters. A fact that is cited in passing may be used to support *the most contradictory views*; and three or four examples from distant times and places, dragged in and piled up from the widest range of circumstances, tend to distract and confuse one's judgment without proving anything. The light of day usually reveals them to be mere trash, with which the author intends to show off his learning.

What is the practical value of these obscure, partially false, confused and arbitrary notions? Very little—so little that they have made theory, from its beginnings, the very opposite of practice, and not infrequently the laughing stock of men whose military competence is beyond dispute.

This could never have happened if by means of simple terms and straightforward observation of the conduct of war theory had sought to determine all that was determinable; if, without spurious claims, with no unseemly display of scientific formulae and historical compendia, it had stuck to the point and never parted company with those who have to manage things in battle by the light of their native wit.

CHAPTER SIX

On Historical Examples

Historical examples clarify everything and also provide the best kind of proof in the empirical sciences. This is particularly true of the art of war. General Scharnhorst, whose manual is the best that has ever been written about actual war, considers historical examples to be of prime importance to the subject, and he makes admirable use of them. If he had survived the wars of 1813-1815, the fourth part of his revised work on artillery would have demonstrated even better the powers of observation and instruction with which he treated his experiences.

Historical examples are, however, seldom used to such good effect. On the contrary, the use made of them by theorists normally not only leaves the reader dissatisfied but even irritates his intelligence. We therefore consider it important to focus attention on the proper and improper uses of examples.

Undoubtedly, the knowledge basic to the art of war is empirical. While, for the most part, it is derived from the nature of things, this very nature is usually revealed to us only by experience. Its application, moreover, is modified by so many conditions that its effects can never be completely established merely from the nature of the means.

The effects of gunpowder—that major agent of military activity—could only be demonstrated by experience. Experiments are still being conducted to study them more closely.

It is, of course, obvious that an iron cannonball, impelled by powder to a speed of 1,000 feet per second, will smash any living creature in its path. One needs no experience to believe that. But there are hundreds of relevant details determining this effect, some of which can only be revealed empirically. Nor is the physical effect the only thing that matters: the psychological effect is what concerns us, and experience is the only means by which it can be established and appreciated. In the Middle Ages firearms were a new invention, so crude that their physical effect was much less important than today; but their psychological impact was considerably greater. One has to have seen the steadfastness of one of the forces trained and led by Bonaparte in the course of his conquests—seen them under fierce and unrelenting fire—to get some sense of what can be accomplished by troops steeled by long experience of danger, in whom a proud record of victories has instilled the noble principle of placing the highest demands on themselves. As an idea alone it is unbelievable. On the other hand, there are European armies that still have troops such as Tartars, Cossacks, and Croats whose ranks can easily be scattered by a few rounds of artillery.

Still, the empirical sciences, the theory of the art of war included, cannot always back their conclusions with historical proofs. The sheer range to be covered would often rule this out; and, apart from that, it might be difficult to point to actual experience on every detail. If, in warfare, a certain means turns out to be highly effective, it will be used again; it will be copied by others and become fashionable; and so, backed by experience, it passes into general use and is included in theory. Theory is content to refer to experience in general to indicate the origin of the method, but not to prove it.

It is a different matter when experience is cited in order to displace a method in current usage, confirm a dubious, or introduce a new one. In those cases, individual instances from history must be produced as evidence.

A closer look at the use of historical examples will enable us to distinguish four points of view.

First, a historical example may simply be used as an *explanation* of an idea. Abstract discussion, after all, is very easily misunderstood, or not understood at all. When an author fears that this might happen, he may use a historical example to throw the necessary light on his idea and to ensure that the reader and the writer will remain in touch.

Second, it may serve to show the *application* of an idea. An example gives one the opportunity of demonstrating the operation of all those minor circumstances which could not be included in a general formulation of the idea. Indeed, this is the difference between theory and experience. Both the foregoing cases concerned true examples; the two that follow concern historical proof.

Third, one can appeal to historical fact to support a statement. This will suffice wherever one merely wants to prove the *possibility* of some phenomenon or effect.

Fourth and last, the detailed presentation of a historical event, and the combination of several events, make it possible to deduce a doctrine: the proof is in the evidence itself.

The use of the first type generally calls only for a brief mention of the case, for only one aspect of it matters. Historical truth is not even essential here: an imaginary case would do as well. Still, historical examples always have the advantage of being more realistic and of bringing the idea they are illustrating to life.

The second type of usage demands a more detailed presentation of events; but authenticity, once again, is not essential. In this respect, we repeat what we said about the first case.

The third purpose is sufficiently met, as a rule, by the simple statement of an undisputed fact. If one is trying to show that an entrenched position can under certain circumstances prove effective, a mention of the Bunzelwitz position will support the statement.

If, however, some historical event is being presented in order to demonstrate a general truth, care must be taken that every aspect bearing on the truth at issue is fully and circumstantially developed—carefully assembled, so to speak, before the reader's eyes. To the extent that this cannot be done, the proof is weakened, and the more necessary it will be to use a number of cases to supply the evidence missing in that one. It is fair to assume that where we cannot cite more precise details, the average effect will be decided by a greater number of examples.

Suppose one wants to prove from experience that cavalry should be placed in the rear of infantry rather than in line with it; or that, without definite numerical superiority, it is extremely dangerous to use widely separated columns in attempting to envelop the enemy, both on the battlefield and in the theater of operations—tactically or strategically, in other words. As to the first instance, it is not enough to cite a few defeats where the cavalry was on the flanks, and a few victories where it was behind the infantry; in the second case, it will not be enough to refer to the battles of Rivoli or Wagram, and the Austrian attacks on the Italian theater, or those of the French on the German theater of war, in 1796. Instead one must accurately trace all the circumstances and individual events, to show the way in which those types of position and attack definitely contributed to the defeat. The result will show *to what degree* these types are objectionable—a point that must be settled in any case, because a general condemnation would conflict with the truth.

We have already agreed that where a detailed factual account cannot be given, any lack of evidence may be made up by the number of examples; but this is clearly a dangerous expedient, and is frequently misused. Instead of presenting a fully detailed case, critics are content merely to *touch on* three or four, which give the *semblance* of strong proof. But there are occasions where nothing will be proved by a dozen examples—if, for instance, they frequently recur and one could just as easily cite a dozen cases that had opposite results. If anyone lists a dozen defeats in which the losing side attacked with divided columns, I can list a dozen victories in which that very tactic was employed. Obviously this is no way to reach a conclusion.

Reflection upon these diverse circumstances will show how easily examples may be misused.

An event that is lightly touched upon, instead of being carefully detailed, is like an object seen at a great distance: it is impossible to distinguish any detail, and it looks the same from every angle. Such examples have actually been used to support the most conflicting views. Daun's campaigns are, to some, models of wisdom and foresight; to others, of timidity and vacillation. Bonaparte's thrust across the Norican Alps in 1797 strikes some as a splendid piece of daring; others will call it completely reckless. His strategic defeat in 1812 may be put down to an excess of energy, but also to a lack of it. All these views have been expressed, and one can easily see why: the pattern of events was interpreted in different ways. Nevertheless, these conflicting opinions cannot coexist; one or the other must be wrong.

Feuquières, that excellent man, deserves our thanks for the wealth of examples that adorn his memoirs. He not only records a number of events that would otherwise have been forgotten; he was the first to make really useful comparisons between abstract theoretical ideas and real life insofar as the cases cited can be considered as explanations and closer definitions of his theoretical assertions. Still, to an impartial modern reader, he has hardly achieved the aim he usually set himself, that of proving theoretical principles by historical examples. Though he occasionally records events in some detail, he still falls short of proving that the conclusions he has drawn are the inevitable consequences of their inherent patterns.

Another disadvantage of merely touching on historical events lies in the fact that some readers do not know enough about them, or do not remember them well enough to grasp what the author has in mind. Such readers have no choice but to be impressed by the argument, or to remain untouched by it altogether.

It is hard of course to recount a historical event or reconstruct it for the reader in the way required if it is to be used as evidence. The writer rarely has the means, the space, or the time for that. We maintain, however, that where a new or debatable point of view is concerned, a single thoroughly detailed event is more instructive than ten that are only touched on. The main objection to this superficial treatment is not that the writer pretends he is trying to prove something but that he himself has never mastered the events he cites, and that such superficial, irresponsible handling of history leads to hundreds of wrong ideas and bogus theorizing. None of this would come about if the writer's duty were to show that the new ideas he is presenting as guaranteed by history are indisputably derived from the precise pattern of events.

Once one accepts the difficulties of using historical examples, one will come to the most obvious conclusion that examples should be drawn from modern military history, insofar as it is properly known and evaluated.

Not only were conditions different in more distant times, with different ways of waging war, so that earlier wars have fewer practical lessons for us; but military history, like any other kind, is bound with the passage of time to lose a mass of minor elements and details that were once clear. It loses some element of life and color, like a picture that gradually fades and darkens. What remains in the end, more or less at random, are large masses and isolated features, which are thereby given undue weight.

If we examine the conditions of modern warfare, we shall find that the wars that bear a considerable resemblance to those of the present day, especially with respect to armaments, are primarily campaigns beginning with the War of the Austrian Succession. Even though many major and minor circumstances have changed considerably, these are close enough to modern warfare to be instructive. The situation is different with the War of the Spanish Succession; the use of firearms was much less advanced, and cavalry was still the most important arm. The further back one goes, the less useful military history becomes, growing poorer and barer at the same time. The history of antiquity is without doubt the most useless and the barest of all.

This uselessness is of course not absolute; it refers only to matters that depend on a precise knowledge of the actual circumstances, or on details in which warfare has changed. Little as we may know about the battles the

Swiss fought against the Austrians, the Burgundians, and the French, it is they that afford the first and strongest demonstration of the superiority of good infantry against the best cavalry. A general glance at the age of the *Condottieri* is enough to show that the conduct of war depends entirely on the instrument employed; at no other time were the forces used so specialized in character or so completely divorced from the rest of political and civil life. The peculiar way in which Rome fought Carthage in the Second Punic War—by attacking Spain and Africa while Hannibal was still victorious in Italy—can provide a most instructive lesson: we still know enough about the general situation of the states and armies that enabled such a roundabout method of resistance to succeed.

But the further one progresses from broad generalities to details, the less one is able to select examples and experiences from remote times. We are in no position to evaluate the relevant events correctly, nor to apply them to the wholly different means we use today.

Unfortunately, writers have always had a pronounced tendency to refer to events in ancient history. How much of this is due to vanity and quackery can remain unanswered; but one rarely finds any honesty of purpose, any earnest attempt to instruct or convince. Such allusions must therefore be looked upon as sheer decoration, designed to cover gaps and blemishes.

To teach the art of war entirely by historical examples, which is what Feuquières tried to do, would be an achievement of the utmost value; but it would be more than the work of a lifetime: anyone who set out to do it would first have to equip himself with a thorough personal experience of war.

Anyone who feels the urge to undertake such a task must dedicate himself for his labors as he would prepare for a pilgrimage to distant lands. He must spare no time or effort, fear no earthly power or rank, and rise above his own vanity or false modesty in order to tell, in accordance with the expression of the *Code Napoleón, the truth, the whole truth, and nothing but the truth*.

BOOK THREE

On Strategy in General

CHAPTER ONE

Strategy

The general concept of strategy was defined in the second chapter of Book Two.[1] It is the use of an engagement for the purpose of the war. Though strategy in itself is concerned only with engagements, the theory of strategy must also consider its chief means of execution, the fighting forces. It must consider these in their own right and in their relation to other factors, for they shape the engagement and it is in turn on them that the effect of the engagement first makes itself felt. Strategic theory must therefore study the engagement in terms of its possible results and of the moral and psychological forces that largely determine its course.

Strategy is the use of the engagement for the purpose of the war. The strategist must therefore define an aim for the entire operational side of the war that will be in accordance with its purpose. In other words, he will draft the plan of the war, and the aim will determine the series of actions intended to achieve it: he will, in fact, shape the individual campaigns and, within these, decide on the individual engagements. Since most of these matters have to be based on assumptions that may not prove correct, while other, more detailed orders cannot be determined in advance at all, it follows that the strategist must go on campaign himself. Detailed orders can then be given on the spot, allowing the general plan to be adjusted to the modifications that are continuously required. The strategist, in short, must maintain control throughout.

This has not always been the accepted view, at least so far as the general principle is concerned. It used to be the custom to settle strategy in the capital, and not in the field—a practice that is acceptable only if the government stays so close to the army as to function as general headquarters.

Strategic theory, therefore, deals with planning; or rather, it attempts to shed light on the components of war and their interrelationships, stressing those few principles or rules that can be demonstrated.

The reader who recalls from the first chapter of Book I how many vitally important matters are involved in war will understand what unusual mental gifts are needed to keep the whole picture steadily in mind.

A prince or a general can best demonstrate his genius by managing a campaign exactly to suit his objectives and his resources, doing neither too much nor too little. But the effects of genius show not so much in novel forms of action as in the ultimate success of the whole. What we should admire is

[1] The definition is in fact first stated in Book Two, Chapter One, see p. 128 above. Eds.

the accurate fulfillment of the unspoken assumptions, the smooth harmony of the whole activity, which only become evident in final success.

The student who cannot discover this harmony in actions that lead up to a final success may be tempted to look for genius in places where it does not and cannot exist.

In fact, the means and forms that the strategist employs are so very simple, so familiar from constant repetition, that it seems ridiculous in the light of common sense when critics discuss them, as they do so often, with ponderous solemnity. Thus, such a commonplace maneuver as turning an opponent's flank may be hailed by critics as a stroke of genius, of deepest insight, or even of all-inclusive knowledge. Can one imagine anything more absurd?

It is even more ridiculous when we consider that these very critics usually exclude all moral qualities from strategic theory, and only examine material factors. They reduce everything to a few mathematical formulas of equilibrium and superiority, of time and space, limited by a few angles and lines. If that were really all, it would hardly provide a scientific problem for a schoolboy.

But we should admit that scientific formulas and problems are not under discussion. The relations between material factors are all very simple; what is more difficult to grasp are the intellectual factors involved. Even so, it is only in the highest realms of strategy that intellectual complications and extreme diversity of factors and relationships occur. At that level there is little or no difference between strategy, policy and statesmanship, and there, as we have already said, their influence is greater in questions of quantity and scale than in forms of execution. Where execution is dominant, as it is in the individual events of a war whether great or small, then intellectual factors are reduced to a minimum.

Everything in strategy is very simple, but that does not mean that everything is very easy. Once it has been determined, from the political conditions, what a war is meant to achieve and what it can achieve, it is easy to chart the course. But great strength of character, as well as great lucidity and firmness of mind, is required in order to follow through steadily, to carry out the plan, and not to be thrown off course by thousands of diversions. Take any number of outstanding men, some noted for intellect, others for their acumen, still others for boldness or tenacity of will: not one may possess the combination of qualities needed to make him a greater than average commander.

It sounds odd, but everyone who is familiar with this aspect of warfare will agree that it takes more strength of will to make an important decision in strategy than in tactics. In the latter, one is carried away by the pressures of the moment, caught up in a maelstrom where resistance would be fatal, and, suppressing incipient scruples, one presses boldly on. In strategy, the pace is much slower. There is ample room for apprehensions, one's own and those of others; for objections and remonstrations and, in consequence, for premature regrets. In a tactical situation one is able to see at least half the problem with the naked eye, whereas in strategy everything has to be guessed at and presumed. Conviction is therefore weaker. Consequently most generals, when they ought to act, are paralyzed by unnecessary doubts.

Now a glance at history. Let us consider the campaign that Frederick the Great fought in 1760, famous for its dazzling marches and maneuvers, praised by critics as a work of art—indeed a masterpiece. Are we to be beside ourselves with admiration at the fact that the King wanted first to turn Daun's right flank, then his left, then his right again, and so forth? Are we to consider this profound wisdom? Certainly not, if we are to judge without affectation. What is really admirable is the King's wisdom: pursuing a major objective with limited resources, he did not try to undertake anything beyond his strength, but always *just enough* to get him what he wanted. This campaign was not the only one in which he demonstrated his judgment as a general. It is evident in all the three wars fought by the great King.

His object was to bring Silesia into the safe harbor of a fully guaranteed peace.

As head of a small state resembling other states in most respects, and distinguished from them only by the efficiency of some branches of its administration, Frederick could not be an Alexander. Had he acted like Charles XII, he too would have ended in disaster. His whole conduct of war, therefore, shows an element of restrained strength, which was always in balance, never lacking in vigor, rising to remarkable heights in moments of crisis, but immediately afterward reverting to a state of calm oscillation, always ready to adjust to the smallest shift in the political situation. Neither vanity, ambition, nor vindictiveness could move him from this course; and it was this course alone that brought him success.

How little these few words can do to appreciate that characteristic of the great general! One only has to examine carefully the causes and the miraculous outcome of this struggle to realize that it was only the King's acute intelligence that led him safely through all hazards.

This is the characteristic we admire in all his campaigns, but especially in the campaign of 1760. At no other time was he able to hold off such a superior enemy at so little cost.

The other aspect to be admired concerns the difficulties of execution. Maneuvers designed to turn a flank are easily planned. It is equally easy to conceive a plan for keeping a small force concentrated so that it can meet a scattered enemy on equal terms at any point, and to multiply its strength by rapid movement. There is nothing admirable about the ideas themselves. Faced with such simple concepts, we have to admit that they are simple.

But let a general try to imitate Frederick! After many years eye-witnesses still wrote about the risk, indeed the imprudence, of the King's positions; and there can be no doubt that the danger appeared three times as threatening at the time as afterward.

It was the same with the marches undertaken under the eyes, frequently under the very guns, of the enemy. Frederick chose these positions and made these marches, confident in the knowledge that Daun's methods, his dispositions, his sense of responsibility and his character would make such maneuvers risky but not reckless. But it required the King's boldness, resolution, and strength of will to see things in this way, and not to be confused and intimidated by the danger that was still being talked and written about thirty years later. Few generals in such a situation would have believed such simple means of strategy to be feasible.

Another difficulty of execution lay in the fact that throughout this campaign the King's army was constantly on the move. Twice, in early July and early August, it followed Daun while itself pursued by Lacy, from the Elbe into Silesia over wretched country roads. The army had to be ready for battle at any time, and its marches had to be organized with a degree of ingenuity that required a proportionate amount of exertion. Though the army was accompanied, and delayed, by thousands of wagons, it was always short of supplies. For a week before the battle of Liegnitz in Silesia, the troops marched day and night, alternatively deploying and withdrawing along the enemy's front. This cost enormous exertions and great hardship.

Could all this be done without subjecting the military machine to serious friction? Is a general, by sheer force of intellect, able to produce such mobility with the ease of a surveyor manipulating an astrolabe? Are the generals and the supreme commander not moved by the sight of the misery suffered by their pitiful, hungry, and thirsty comrades in arms? Are complaints and misgivings about such conditions not reported to the high command? Would an ordinary man dare to ask for such sacrifices, and would these not automatically lower the morale of the troops, corrupt their discipline, in short undermine their fighting spirit unless an overwhelming belief in the greatness and infallibility of their commander outweighed all other considerations? It is this which commands our respect; it is these miracles of execution that we have to admire. But to appreciate all this in full measure one has to have had a taste of it through actual experience. Those who know war only from books or the parade-ground cannot recognize the existence of these impediments to action, and so we must ask them to accept on faith what they lack in experience.

We have used the example of Frederick to bring our train of thought into focus. In conclusion, we would point out that in our exposition of strategy we shall describe those material and intellectual factors that seem to us to be the most significant. We shall proceed from the simple to the complex, and conclude with the unifying structure of the entire military activity—that is, with the plan of campaign.

An earlier manuscript of Book Two contains the following passages, marked by the author: "To be used in the first chapter of Book Three." The projected revision of this chapter was never made, and these passages are therefore inserted here in full.

In itself, the deployment of forces at a certain point merely makes an engagement possible; it does not necessarily take place. Should one treat this possibility as a reality, as an actual occurrence? Certainly. It becomes real because of its consequences, and *consequences of some kind will always follow.*

Possible Engagements Are To Be Regarded As Real Ones Because of Their Consequences

If troops are sent to cut off a retreating enemy and he thereupon surrenders without further fight, his decision is caused solely by the threat of a fight posed by those troops.

If part of our army occupies an undefended enemy province and thus denies the enemy substantial increments to his strength, the factor making it possible for our force to hold the province is the engagement that the enemy must expect to fight if he endeavors to retake it.

In both cases results have been produced by the mere possibility of an engagement; the possibility has acquired reality. But let us suppose that in each case the enemy had brought superior forces against our troops, causing them to abandon their goal without fighting. This would mean that we had fallen short of our objective; but still the engagement that we offered the enemy was not without effect—it did draw off his forces. Even if the whole enterprise leaves us worse off than before, we cannot say that no effects resulted from using troops in this way, by producing the *possibility of an engagement*; the effects were similar to those of a lost engagement.

This shows that the destruction of the enemy's forces and the overthrow of the enemy's power can be accomplished only as the result of an engagement, no matter whether it really took place or was merely offered but not accepted.

The Twofold Object of the Engagement

These results, moreover, are of two kinds: direct and indirect. They are indirect if other things intrude and become the object of the engagement—things which cannot in themselves be considered to involve the destruction of the enemy's forces, but which lead up to it. They may do so by a circuitous route, but are all the more powerful for that. The possession of provinces, cities, fortresses, roads, bridges, munitions dumps, etc., may be the *immediate* object of an engagement, but can never be the final one. Such acquisitions should always be regarded merely as means of gaining greater superiority, so that in the end we are able to offer an engagement to the enemy when he is in no position to accept it. These actions should be considered as intermediate links, as steps leading to the operative principle, never as the operative principle itself.

EXAMPLES

With the occupation of Bonaparte's capital in 1814, the objective of the war had been achieved. The political cleavages rooted in Paris came to the surface, and that enormous split caused the Emperor's power to collapse. Still, all this should be considered in the light of the military implications. The occupation caused a substantial diminution in Bonaparte's military strength and his capacity to resist, and a corresponding increase in the superiority of the allies. Further resistance became impossible, and it was this which led to peace with France. Suppose the allied strength had suddenly been similarly reduced by some external cause: their superiority would have vanished, and with it the whole effect and significance of their occupation of Paris.

We have pursued this argument to show that this is the natural and only sound view to take, and this is what makes it important. We are constantly brought back to the question: what, at any given stage of the war or campaign, will be the likely outcome of all the major and minor engagements that the two sides can offer one another? In the planning of a campaign or a war, this alone will decide the measures that have to be taken from the outset.

If This View Is Not Adopted, Other Matters Will Be Inaccurately Assessed

If we do not learn to regard a war, and the separate campaigns of which it is composed, as a chain of linked engagements each leading to the next, but instead succumb to the idea that the capture of certain geographical points or the seizure of undefended provinces are *of value in themselves,* we are liable to regard them as windfall profits. In so doing, and in ignoring the fact that they are links in a continuous chain of events, we also ignore the possibility that their possession may later lead to definite disadvantages. This mistake is illustrated again and again in military history. One could almost put the matter this way: just as a businessman cannot take the profit from a single transaction and put it into a separate account, so an isolated advantage gained in war cannot be assessed separately from the overall result. A businessman must work on the basis of his total assests, and in war the advantages and disadvantages of a single action could only be determined by the final balance.

By looking on each engagement as part of a series, at least insofar as events are predictable, the commander is always on the high road to his goal. The forces gather momentum, and intentions and actions develop with a vigor that is commensurate with the occasion, and impervious to outside influences.

CHAPTER TWO

Elements of Strategy

The strategic elements that affect the use of engagements may be classified into various types: moral, physical, mathematical, geographical, and statistical.

The first type covers everything that is created by intellectual and psychological qualities and influences; the second consists of the size of the armed forces, their composition, armament and so forth; the third includes the angle of lines of operation, the convergent and divergent movements wherever geometry enters into their calculation; the fourth comprises the influence of terrain, such as commanding positions, mountains, rivers, woods, and roads; and, finally, the fifth covers support and maintenance. A brief consideration of each of these various types will clarify our ideas and, in passing, assess the relative value of each. Indeed if they are studied separately some will automatically be stripped of any undue importance. For instance, it immediately becomes clear that the value of the base of operations, even if we take this in its simplest form as meaning a *base-line*, depends less on its geometric forms than on the nature of the roads and terrain through which they run.

It would however be disastrous to try to develop our understanding of strategy by analyzing these factors in isolation, since they are usually interconnected in each military action in manifold and intricate ways. A dreary analytical labyrinth would result, a nightmare in which one tried in vain to bridge the gulf between this abstract basis and the facts of life. Heaven protect the theorist from such an undertaking! For our part, we shall continue to examine the picture as a whole, and take our analysis no further than is necessary in each case to elucidate the idea we wish to convey, which will always have its origins in the impressions made by the sum total of the phenomena of war, rather than in speculative study.

CHAPTER THREE

Moral Factors

We must return once more to this subject, already touched upon in Chapter Three of Book Two[1] since the moral elements are among the most important in war. They constitute the spirit that permeates war as a whole, and at an early stage they establish a close affinity with the will that moves and leads the whole mass of force, practically merging with it, since the will is itself a moral quantity. Unfortunately they will not yield to academic wisdom. They cannot be classified or counted. They have to be seen or felt.

The spirit and other moral qualities of an army, a general or a government, the temper of the population of the theater of war, the moral effects of victory or defeat—all these vary greatly. They can moreover influence our objective and situation in very different ways.

Consequently, though next to nothing can be said about these things in books, they can no more be omitted from the theory of the art of war than can any of the other components of war. To repeat, it is paltry philosophy if in the old-fashioned way one lays down rules and principles in total disregard of moral values. As soon as these appear one regards them as exceptions, which gives them a certain scientific status, and thus makes them into rules. Or again one may appeal to genius, which is above all rules; which amounts to admitting that rules are not only made for idiots, but are idiotic in themselves.

If the theory of war did no more than remind us of these elements, demonstrating the need to reckon with and give full value to moral qualities, it would expand its horizon, and simply by establishing this point of view would condemn in advance anyone who sought to base an analysis on material factors alone.

Another reason for not placing moral factors beyond the scope of theory is their relation to all other so-called rules. The effects of physical and psychological factors form an organic whole which, unlike a metal alloy, is inseparable by chemical processes. In formulating any rule concerning physical factors, the theorist must bear in mind the part that moral factors may play in it; otherwise he may be misled into making categorical statements that will be too timid and restricted, or else too sweeping and dogmatic. Even the most uninspired theories have involuntarily had to stray into the area of intangibles; for instance, one cannot explain the effects of a victory without taking psychological reactions into account. Hence most of the matters dealt with in this book are composed in equal parts of physical and of moral

[1] Book One is meant. Eds.

causes and effects. One might say that the physical seem little more than the wooden hilt, while the moral factors are the precious metal, the real weapon, the finely-honed blade.

History provides the strongest proof of the importance of moral factors and their often incredible effect: this is the noblest and most solid nourishment that the mind of a general may draw from a study of the past. Parenthetically, it should be noted that the seeds of wisdom that are to bear fruit in the intellect are sown less by critical studies and learned monographs than by insights, broad impressions, and flashes of intuition.

We might list the most important moral phenomena in war and, like a diligent professor, try to evaluate them one by one. This method, however, all too easily leads to platitudes, while the genuine spirit of inquiry soon evaporates, and unwittingly we find ourselves proclaiming what everybody already knows. For this reason we prefer, here even more than elsewhere, to treat the subject in an incomplete and impressionistic manner, content to have pointed out its general importance and to have indicated the spirit in which the argument of this book are conceived.

CHAPTER FOUR

The Principal Moral Elements

They are: *the skill of the commander, the experience and courage of the troops, and their patriotic spirit.* The relative value of each cannot be universally established; it is hard enough to discuss their potential, and even more difficult to weigh them against each other. The wisest course is not to underrate any of them—a temptation to which human judgment, being fickle, often succumbs. It is far preferable to muster historical evidence of the unmistakable effectiveness of all three.

Nevertheless it is true that at this time the armies of practically all European states have reached a common level of discipline and training. To use a philosophic expression: the conduct of war has developed in accordance with its natural laws. It has evolved methods that are common to most armies, and that no longer even allow the commander scope to employ special artifices (in the sense, for example, of Frederick the Great's oblique order of battle). It cannot be denied, therefore, that as things stand at present proportionately greater scope is given to the troops' patriotic spirit and combat experience. A long period of peace may change this again.

The troops' national feeling (enthusiasm, fanatical zeal, faith, and general temper) is most apparent in mountain warfare where every man, down to the individual soldier, is on his own. For this reason alone mountainous areas constitute the terrain best suited for action by an armed populace.

Efficiency, skill, and the tempered courage that welds the body of troops into a single mold will have their greatest scope in operations in open country.

The commander's talents are given greatest scope in rough hilly country. Mountains allow him too little real command over his scattered units and he is unable to control them all; in open country, control is a simple matter and does not test his ability to the fullest.

These obvious affinities should guide our planning.

CHAPTER FIVE

Military Virtues of the Army

Military virtues should not be confused with simple bravery, and still less with enthusiasm for a cause. Bravery is obviously a necessary component. But just as bravery, which is part of the natural make-up of a man's character, can be developed in a soldier—a member of an organization—it must develop differently in him than in other men. In the soldier the natural tendency for unbridled action and outbursts of violence must be subordinated to demands of a higher kind: obedience, order, rule, and method. An army's efficiency gains life and spirit from enthusiasm for the cause for which it fights, but such enthusiasm is not indispensable.

War is a special activity, different and separate from any other pursued by man. This would still be true no matter how wide its scope, and though every able-bodied man in the nation were under arms. An army's military qualities are based on the individual who is steeped in the spirit and essence of this activity; who trains the capacities it demands, rouses them, and makes them his own; who applies his intelligence to every detail; who gains ease and confidence through practice, and who completely immerses his personality in the appointed task.

No matter how clearly we see the citizen and the soldier in the same man, how strongly we conceive of war as the business of the entire nation, opposed diametrically to the pattern set by the *condottieri* of former times, the business of war will always remain individual and distinct. Consequently for as long as they practice this activity, soldiers will think of themselves as members of a kind of guild, in whose regulations, laws, and customs the spirit of war is given pride of place. And that does seem to be the case. No matter how much one may be inclined to take the most sophisticated view of war, it would be a serious mistake to underrate professional pride (*esprit de corps*) as something that may and must be present in an army to greater or lesser degree. Professional pride is the bond between the various natural forces that activate the military virtues; in the context of this professional pride they crystallize more readily.

An army that maintains its cohesion under the most murderous fire; that cannot be shaken by imaginary fears and resists well-founded ones with all its might; that, proud of its victories, will not lose the strength to obey orders and its respect and trust for its officers even in defeat; whose physical power, like the muscles of an athlete, has been steeled by training in privation and effort; a force that regards such efforts as a means to victory rather than a curse on its cause; that is mindful of all these duties and qualities by virtue of the single powerful idea of the honor of its arms—such an army is imbued with the true military spirit.

It is possible to fight superbly, like the men of the Vendée, and to achieve great results, like the Swiss, the Americans, and the Spaniards without developing the kind of virtues discussed here; it is even possible to be the victorious commander of a regular army, like Prince Eugene and Marlborough, without drawing substantially on their help. No one can maintain that it is impossible to fight a successful war without these qualities. We stress this to clarify the concept, and not lose sight of the idea in a fog of generalities and give the impression that military spirit is all that counts in the end. That is not the case. The spirit of an army may be envisaged as a definite moral factor that can be mentally subtracted, whose influence may therefore be estimated—in other words, it is a tool whose power is measurable.

Having thus characterized it, we shall attempt to describe its influence and the various ways of developing it.

Military spirit always stands in the same relation to the parts of an army as does a general's ability to the whole. The general can command only the overall situation and not the separate parts. At the point where the separate parts need guidance, the military spirit must take command. Generals are chosen for their outstanding qualities, and other high-ranking officers are carefully tested; but the testing process becomes less thorough the further we descend on the scale of command, and we must be prepared for a proportionate diminution of personal talent. What is missing here must be made up by military virtues. The same role is played by the natural qualities of a people mobilized for war: *bravery, adaptability, stamina, and enthusiasm.* These, then, are the qualities that can act as substitutes for the military spirit and vice-versa, leading us to the following conclusions:

1. Military virtues are found only in regular armies, and they are the ones that need them most. In national uprisings and peoples' wars their place is taken by natural warlike qualities, which develop faster under such conditions.

2. A regular army fighting another regular army can get along without military virtues more easily than when it is opposed by a people in arms; for in the latter case, the forces have to be split up, and the separate units will more frequently have to fend for themselves. Where the troops can remain concentrated, however, the talents of the commander are given greater scope, and can make up for any lack of spirit among the troops. Generally speaking, the need for military virtues becomes greater the more the theater of operations and other factors tend to complicate the war and disperse the forces.

If there is a lesson to be drawn from these facts, it is that when an army lacks military virtues, every effort should be made to keep operations as simple as possible, or else twice as much attention should be paid to other aspects of the military system. The mere fact that soldiers belong to a "regular army" does not automatically mean they are equal to their tasks.

Military spirit, then, is one of the most important moral elements in war. Where this element is absent, it must either be replaced by one of the others, such as the commander's superior ability or popular enthusiasm, or else the results will fall short of the efforts expended. How much has been accomplished by this spirit, this sterling quality, this refinement of base ore into precious metal, is demonstrated by the Macedonians under Alexander, the Roman legions under Caesar, the Spanish infantry under Alexander Farnese, the Swedes under Gustavus Adolphus and Charles XII, the Prussians under Frederick the Great, and the French under Bonaparte. One would have to be blind to all the evidence of history if one refused to admit that the outstanding successes of these commanders and their greatness in adversity were feasible only with the aid of an army possessing these virtues.

There are only two sources for this spirit, and they must interact in order to create it. The first is a series of victorious wars; the second, frequent exertions of the army to the utmost limits of its strength. Nothing else will show a soldier the full extent of his capacities. The more a general is accustomed to place heavy demands on his soldiers, the more he can depend on their response. A soldier is just as proud of the hardships he has overcome as of the dangers he has faced. In short, the seed will grow only in the soil of constant activity and exertion, warmed by the sun of victory. Once it has grown into a strong tree, it will survive the wildest storms of misfortune and defeat, and even the indolent inertia of peace, at least for a while. Thus, this spirit can be *created* only in war and by great generals, though admittedly it may endure, for several generations at least, even under generals of average ability and through long periods of peace.

One should be careful not to compare this expanded and refined solidarity of a brotherhood of tempered, battle-scarred veterans with the self-esteem and vanity of regular armies which are patched together only by service-regulations and drill. Grim severity and iron discipline may be able to preserve the military virtues of a unit, but it cannot create them. These factors are valuable, but they should not be overrated. Discipline, skill, goodwill, a certain pride, and high morale, are the attributes of an army trained in times of peace. They command respect, but they have no strength of their own. They stand or fall together. One crack, and the whole thing goes, like a glass too quickly cooled. Even the highest morale in the world can, at the first upset, change all too easily into despondency, an almost boastful fear; the French would call it *sauve qui peut*. An army like this will be able to prevail only by virtue of its commander, never on its own. It must be led with more than normal caution until, after a series of victories and exertions, its inner strength will grow to fill its external panoply. We should take care never to confuse the real spirit of an army with its mood.

CHAPTER SIX

Boldness

In the chapter dealing with the certainty of success, we discussed the place that boldness occupies in the dynamic system of forces, and the part it plays when opposed to prudence and discretion. We tried to show that the theorist has no right to restrict boldness on doctrinal grounds.

But this noble capacity to rise above the most menacing dangers should also be considered as a principle in itself, separate and active. Indeed, in what field of human activity is boldness more at home than in war?

A soldier, whether drummer boy or general, can possess no nobler quality; it is the very metal that gives edge and luster to the sword.

Let us admit that boldness in war even has its own prerogatives. It must be granted a certain power over and above successful calculations involving space, time, and magnitude of forces, for wherever it is superior, it will take advantage of its opponent's weakness. In other words, it is a genuinely creative force. This fact is not difficult to prove even scientifically. Whenever boldness encounters timidity, it is likely to be the winner, because timidity in itself implies a loss of equilibrium. Boldness will be at a disadvantage only in an encounter with deliberate caution, which may be considered bold in its own right, and is certainly just as powerful and effective; but such cases are rare. Timidity is the root of prudence in the majority of men.

In most soldiers, the development of boldness can never be detrimental to other qualities, because the rank and file is bound by duty and the conditions of the service to a higher authority, and thus is led by external intelligence. With them boldness acts like a coiled spring, ready at any time to be released.

The higher up the chain of command, the greater is the need for boldness to be supported by a reflective mind, so that boldness does not degenerate into purposeless bursts of blind passion. Command becomes progressively less a matter of personal sacrifice and increasingly concerned for the safety of others and for the common purpose. The quality that in most soldiers is disciplined by service regulations that have become second nature to them, must in the commanding officer be disciplined by reflection. In a commander a bold act may prove to be a blunder. Nevertheless, it is a laudable error, not to be regarded on the same footing as others. Happy the army where ill-timed boldness occurs frequently; it is a luxuriant weed, but indicates the richness of the soil. Even foolhardiness—that is, boldness without any object—is not to be despised: basically it stems from daring, which in this case has erupted with a passion unrestrained by thought. Only when boldness rebels against obedience, when it defiantly ignores an expressed command, must it be treated as a dangerous offense; then it must be prevented, not for its innate qualities, but because an order has been disobeyed, and in war obedience is of cardinal importance.

Given the same amount of intelligence, timidity will do a thousand times more damage in war than audacity. The truth of this observation will be self-evident to our readers.

In fact, the supervention of a rational purpose ought to make it easier to be bold, and therefore less meritorious. Yet the opposite is true.

The power of the various emotions is sharply reduced by the intervention of lucid thought and, more, by self control. Consequently, boldness grows *less common in the higher ranks*. Even if the growth of an officer's perception and intelligence does not keep pace with his rise in rank, the realities of war will impose their conditions and concerns on him. Indeed their influence on him will be greater the less he really understands them. In war, this is the main basis for the experience expressed in the French proverb, "Tel brille au second qui s'éclipse au premier."[1] Nearly every general known to us from history as mediocre, even vacillating, was noted for dash and determination as a junior officer.

A distinction should be made among acts of boldness that result from sheer necessity. Necessity comes in varying degrees. If it is pressing, a man in pursuit of his aim may be driven to incur one set of risks in order to avoid others just as serious. In that event one can admire only his powers of resolution, which, however, are also of value. The young man who leaps across a deep chasm to show off his horsemanship displays boldness; if he takes the same leap to escape a band of savage janissaries all he shows is resolution. The greater the distance between necessity and action, the more numerous the possibilities that have to be identified and analyzed before action is taken, the less is the factor of boldness reduced. When Frederick the Great perceived in 1756 that war was unavoidable and that he was lost unless he could forestall his enemies, it became a necessity for him to initiate hostilities; but at the same time it was an act of boldness, because few men in his position would have dared to act in this way.

While strategy is exclusively the province of generals and other senior officers, boldness in the rest of the army is as important a factor in planning as any other military virtue. More can be achieved with an army drawn from people known for their boldness, an army in which a daring spirit has always been nurtured, than with an army that lacks this quality. For that reason, boldness in general has been mentioned here, even though our actual subject is the boldness of the commander. After having given a broad description of this military virtue, however, there is not much left to say. The higher the military rank, the greater is the degree to which activity is governed by the mind, by the intellect, by insight. Consequently boldness,

[1] The same man who shines at the second level is eclipsed at the top. Eds.

which is a quality of temperament, will tend to be held in check. This explains why it is so rare in the higher ranks, and why it is all the more admirable when found there. Boldness governed by superior intellect is the mark of a hero. This kind of boldness does not consist in defying the natural order of things and in crudely offending the laws of probability; it is rather a matter of energetically supporting that higher form of analysis by which genius arrives at a decision: rapid, only partly conscious weighing of the possibilities. Boldness can lend wings to intellect and insight; the stronger the wings then, the greater the heights, the wider the view, and the better the results; though a greater prize, of course, involves greater risks. The average man, not to speak of a hesitant or weak one, may in an imaginary situation, in the peace of his room far removed from danger and responsibility, arrive at the right answer—that is, insofar as this is possible without exposure to reality. But beset on every side with danger and responsibility he will lose perspective. Even if this is provided by others, he will lose his powers of decision, for here no one else can help him.

In other words a distinguished commander without boldness is unthinkable. No man who is not born bold can play such a role, and therefore we consider this quality the first prerequisite of the great military leader. How much of this quality remains by the time he reaches senior rank, after training and experience have affected and modified it, is another question. The greater the extent to which it is retained, the greater the range of his genius. The magnitude of the risks increases, but so does that of the goal. To the critical student there is not much difference between actions governed by some compelling long-range aim and those that are dictated by pure ambition—between the policies of a Frederick and an Alexander. The actions of the latter may fascinate the imagination because of their supreme boldness, while those of the former may be more satisfying to the intellect because they are dictated by an inner necessity.

We must mention one more factor of importance.

An army may be imbued with boldness for two reasons: it may come naturally to the people from which the troops are recruited, or it may be the result of a victorious war fought under bold leadership. If the latter is the case, boldness will at the outset be lacking.

Today practically no means other than war will educate a people in this spirit of boldness; and it has to be a war waged under daring leadership. Nothing else will counteract the softness and the desire for ease which debase the people in times of growing prosperity and increasing trade.

A people and nation can hope for a strong position in the world only if national character and familiarity with war fortify each other by continual interaction.

CHAPTER SEVEN

Perseverance

The reader expects to hear of strategic theory, of lines and angles, and instead of these denizens of the scientific world he finds himself encountering only creatures of everyday life. But the author cannot bring himself to be in the slightest degree more scientific than he considers his subject to warrant—strange as this attitude may appear.

In war more than anywhere else things do not turn out as we expect. Nearby they do not appear as they did from a distance. With what assurance an architect watches the progress of his work and sees his plans gradually take shape! A doctor, though much more exposed to chance and to inexplicable results, knows his medicines and the effects they produce. By contrast, a general in time of war is constantly bombarded by reports both true and false; by errors arising from fear or negligence or hastiness; by disobedience born of right or wrong interpretations, of ill will, of a proper or mistaken sense of duty, of laziness, or of exhaustion; and by accidents that nobody could have foreseen. In short, he is exposed to countless impressions, most of them disturbing, few of them encouraging. Long experience of war creates a knack of rapidly assessing these phenomena; courage and strength of character are as impervious to them as a rock to the rippling waves. If a man were to yield to these pressures, he would never complete an operation. *Perseverance* in the chosen course is the essential counterweight, provided that no compelling reasons intervene to the contrary. Moreover, there is hardly a worthwhile enterprise in war whose execution does not call for infinite effort, trouble, and privation; and as man under pressure tends to give in to physical and intellectual weakness, only great strength of will can lead to the objective. It is steadfastness that will earn the admiration of the world and of posterity.

CHAPTER EIGHT

Superiority of Numbers

In tactics, as in strategy, superiority of numbers is the most common element in victory. Let us first consider this general characteristic, which calls for the following exposition.

Strategy decides the time when, the place where, and the forces with which the engagement is to be fought, and through this threefold activity exerts considerable influence on its outcome. Once the tactical encounter has taken place and the result—be it victory or defeat—is assured, strategy will use it to serve the object of the war. This object of course is usually remote, and only rarely lies very near at hand. A series of secondary objectives may serve as means to the attainment of the ultimate goal; these intermediate ends, which are means to higher ends, may in practice be of various types. Even the ultimate object, the purpose of the entire war, differs in almost every case. We shall become better acquainted with these matters as we go into the various details that they affect. We do not propose here to enumerate them completely, even if this were possible. For the time being, therefore, we will not discuss the use of the engagement.

Nor are the factors by which strategy influences the outcome of the engagement simple enough to be dealt with in a single statement. In determining the time and place of the engagement, and the forces to be used, strategy poses numerous possibilities, each of which will have a different effect on the outcome of the engagement. Here again we shall become acquainted with the subject gradually as we study the various factors that bear on it.

If we thus strip the engagement of all the variables arising from its purpose and circumstances, and disregard the fighting value of the troops involved (which is a given quantity), we are left with the bare concept of the engagement, a shapeless battle in which the only distinguishing factor is the number of troops on either side.

These numbers, therefore, will determine victory. It is, of course, evident from the mass of abstractions I have made to reach this point that superiority of numbers in a given engagement is only one of the factors that determines victory. Superior numbers, far from contributing everything, or even a substantial part, to victory, may actually be contributing very little, depending on the circumstances.

But superiority varies in degree. It can be two to one, or three or four to one, and so on; it can obviously reach the point where it is overwhelming.

In this sense superiority of numbers admittedly is the most important factor in the outcome of an engagement, so long as it is great enough to counterbalance all other contributing circumstances. It thus follows that as many troops as possible should be brought into the engagement at the decisive point.

Whether these forces prove adequate or not, we will at least have done everything in our power. This is the first principle of strategy. In the general terms in which it is expressed here it would hold true for Greeks and Persians, for Englishmen and Mahrattas, for Frenchmen and Germans. But in order to be more concrete, let us examine the military conditions in Europe.

European armies are comparable in equipment, organization, and training. Such differences as may exist are to be found in the spirit of the troops and the ability of the commander. If we look at recent European history, we shall not find another Marathon.

At Leuthen Frederick the Great, with about 30,000 men, defeated 80,000 Austrians; at Rossbach he defeated 50,000 allies with 25,000 men. These however are the only examples of victories won over an opponent two or even nearly three times as strong. Charles XII at the battle of Narva is not in the same category. The Russians at that time could hardly be considered as Europeans; moreover, we know too little about the main features of that battle. Bonaparte commanded 120,000 men at Dresden against 220,000—not quite half. At Kolin, Frederick the Great's 30,000 men could not defeat 50,000 Austrians; similarly, victory eluded Bonaparte at the desperate battle of Leipzig, though with his 160,000 men against 280,000, his opponent was far from being twice as strong.

These examples may show that in modern Europe even the most talented general will find it very difficult to defeat an opponent twice his strength. When we observe that the skill of the greatest commanders may be counterbalanced by a two-to-one ratio in the fighting forces, we cannot doubt that in ordinary cases, whether the engagement be great or small, a significant superiority in numbers (it does not have to be more than double) will suffice to assure victory, however adverse the other circumstances. It is possible, of course, to imagine a mountain pass where even a tenfold superiority would not be sufficient, but in such a situation we cannot really speak of an engagement.

We believe then that in our circumstances and all similar ones, a main factor is the possession of strength at the really vital point. Usually it is actually the most important factor. To achieve strength at the decisive point depends on the strength of the army and on the skill with which this strength is employed.

The first rule, therefore, should be: put the largest possible army into the field. This may sound a platitude, but in reality it is not.

To show for how long the strength of armies was not considered to be of major significance, we need only point out that most of the military histories of the eighteenth century—even the most extensive ones—either do not mention the size of armies, or do so only in a very casual way; certainly they never emphasize it. Tempelhoff, in his history of the Seven Years War, is the first author to give figures regularly, though they are only approximations.

Even Massenbach's account of the Prussian campaign in the Vosges in 1793–1794, with its frequently critical observations, has much to say about hills and valleys, roads and tracks, but not a syllable about the strength of the opposing forces.

Further proof is found in the strange ideas that haunted some authors: that there is a certain optimum size for an army, an ideal norm, and that any troops in excess of it are more trouble than they are worth.[1]

Finally there are many cases in which not all forces available were actually used in a battle or war, because numerical superiority was not given its due importance.

If one is genuinely convinced that a great deal can be achieved by significant superiority, this conviction is bound to influence the preparations for war. The aim will then be to take the field in the greatest possible strength, either in order to get the upper hand, or at least in order to make sure that the enemy does not. So much for the overall strength that should be used in waging war.

In practice, the size will be decided by the government. This decision marks the start of military activity—it is indeed a vital part of strategy—and the general who is to command the army in the field usually has to accept the size of his forces as a given factor. Either he was not consulted in the matter, or circumstances may have prevented the raising of a sufficiently large force.

Consequently, the forces available must be employed with such skill that even in the absence of absolute superiority, relative superiority is attained at the decisive point.

To achieve this, the calculation of space and time appears as the most essential factor, and this has given rise to the belief that in strategy space and time cover practically everything concerning the use of the forces. Indeed, some have gone so far as to ascribe to great generals a special organ to deal with strategy and tactics.

But although the equation of time and space does underlie everything else and is, so to speak, the daily bread of strategy, it is neither the most difficult nor the decisive factor.

If we consider past wars with an open mind, we will find that at least on the strategic plane there have been very few cases where errors in such a calculation led to serious defeat. Moreover, if the concept of a skillful correlation of time and space is to explain every instance in which a determined and enterprising general was able, by means of rapid marches, to beat several forces with a single army (Frederick the Great, Bonaparte), we would con-

[1] We immediately think of Tempelhoff and Montalembert: the former in a passage on page 148 of the first part of his work, the latter in a reference in his "Correspondance" to the Russian plan of Operations for 1759. Cl. Clausewitz refers to Tempelhoff's German version of Henry Lloyd's *History of the Late War in Germany*: *Geschichte des Siebenjährigen Krieges in Deutschland*, 1783–1801; and to the *Correspondance de Mr. le marquis de Montalembert*, 1777. Eds.

fuse ourselves unnecessarily by conventional jargon. If concepts are to be clear and fruitful, things must be called by their right names.

The true reasons for such victories were the correct appraisal of the opposing generals (Daun, Schwarzenberg), willingness to risk facing them for a time with inferior forces, energy for rapid movement, boldness for quick attacks, and the increased activity which danger generates in great men. What does this have to do with the ability to calculate the relationship of two such simple elements as time and space?

Even the ricochet effect of forces to which great generals have frequently entrusted their defense, by which the victories of Rossbach and Montmirail gave impetus to those of Leuthen and Montereau, is, if we wish to be clear and accurate, rare in history.

Relative superiority, that is, the skillful concentration of superior strength at the decisive point, is much more frequently based on the correct appraisal of this decisive point, on suitable planning from the start; which leads to appropriate disposition of the forces, and on the resolution needed to sacrifice nonessentials for the sake of essentials—that is, the courage to retain the major part of one's forces united. This is particularly characteristic of Frederick the Great and Bonaparte.

With this discussion, we believe we have shown how significant superiority of numbers really is. It must be regarded as fundamental—to be achieved in every case and to the fullest possible extent.

But it would be seriously misunderstanding our argument, to consider numerical superiority as indispensable to victory; we merely wished to stress the relative importance. The principle is served if we use the largest possible force; the question whether to avoid a fight for lack of strength can be decided only in the light of all other circumstances.

CHAPTER NINE

Surprise

The subject of the previous chapter—the universal desire for relative numerical superiority—leads to another desire, which is consequently no less universal: that *to take the enemy by surprise*. This desire is more or less basic to all operations, for without it superiority at the decisive point is hardly conceivable.

Surprise therefore becomes the means to gain superiority, but because of its psychological effect it should also be considered as an independent element. Whenever it is achieved on a grand scale, it confuses the enemy and lowers his morale; many examples, great and small, show how this in turn multiplies the results. We are not speaking here of a surprise assault, which falls under the general category of "attack," but of the desire to surprise the enemy by our plans and dispositions, especially those concerning the distribution of forces. This is just as feasible in defense, and indeed it is a major weapon of the tactical defense.

We suggest that surprise lies at the root of all operations without exception, though in widely varying degrees depending on the nature and circumstances of the operation.

These variations may already originate in the characteristics of the army, of the general, or even of the government.

The two factors that produce surprise are secrecy and speed. Both presuppose a high degree of energy on the part of the government and the commander; on the part of the army, they require great efficiency. Surprise will never be achieved under lax conditions and conduct. But while the wish to achieve surprise is common and, indeed, indispensable, and while it is true that it will never be completely ineffective, it is equally true that by its very nature surprise can rarely be *outstandingly* successful. It would be a mistake, therefore, to regard surprise as a key element of success in war. The principle is highly attractive in theory, but in practice it is often held up by the friction of the whole machine.

Basically surprise is a tactical device, simply because in tactics time and space are limited in scale. Therefore in strategy surprise becomes more feasible the closer it occurs to the tactical realm, and more difficult, the more it approaches the higher levels of policy.

Preparations for war usually take months. Concentrating troops at their main assembly points generally requires the installation of supply dumps and depots, as well as considerable troop movements, whose purpose can be guessed soon enough.

It is very rare therefore that one state surprises another, either by an attack or by preparations for war. In the seventeenth and eighteenth centuries, when war often turned on sieges, a frequent and important aim was to invest a fortress by surprise; but this too rarely proved successful.

On the other hand, surprise is more easily carried out in operations requiring little time. It is often relatively simple to steal a march on the enemy and in this way occupy a position, a topographical feature, or a road. It is obvious, however, that the greater the ease with which surprise is achieved, the smaller is its effectiveness, and vice versa. In the abstract, we may believe that small surprises often lead to greater things, such as a victorious battle or the capture of an important depot, but history does not bear this out. Cases in which such surprises led to major results are very rare. From this we may conclude how considerable are the inherent difficulties.

Of course anyone who consults history must not allow historians to divert him with their favorite theories, or with maxims and a smug parade of technicalities. He must look at the facts. Take, for example, a certain day in the Silesian campaign of 1761, which has achieved a kind of notoriety in this connection. On 22 July Frederick the Great stole a march on Laudon, moved to Nossen near Neisse, and thereby it is claimed prevented the Austrian and Russian armies in upper Silesia from joining forces, thus gaining a breathing spell of four weeks. If we study this event in the works of the principal authorities,[1] and consider the facts with an open mind, we will find no such significance in this march, but rather inconsistencies in the entire argument, fashionable as it has become, and much that is unaccountable in Laudon's movements during these famous maneuvers. No one looking for truth and understanding could be satisfied with such a historical example.

When we expect great results from the element of surprise in the course of a campaign, we think of strenuous activity, quick decisions, and forced marches. But even in instances where these are present to a high degree, they may not always produce the intended results, as is shown by two commanders who can be considered supreme in these matters: Frederick the Great and Bonaparte. In July 1760 the former suddenly pounced on Lacy from Bautzen and then turned toward Dresden. But the interlude accomplished little; indeed, it left Frederick considerably worse off than before, for in the meantime Glatz had fallen.

In 1813 Bonaparte twice turned suddenly from Dresden against Blücher, not to mention his descent from upper Lusatia on Bohemia, but he was unable to achieve his goal. Both actions were thrusts into thin air, which cost him time and casualties and might have seriously endangered his position at Dresden.

[1] Tempelhoff, *Der Veteran*, Frederick the Great. Cl. *Der Veteran* refers either to a collection of articles with that title, or more likely to the memoirs of an Austrian officer who served in the Seven Years War, Jacob de Cogniazo, *Geständnisse eines östereichischen Veteranen*, Breslau, 1781-91. Eds.

Major success in a surprise action therefore does not depend on the energy, forcefulness, and resolution of the commander: it must be favored by other circumstances. We do not wish to deny the possibility of success, but merely want to establish the fact that it does require favorable conditions, which are not often present, and can rarely be created by the general.

Both the commanders whom we have just cited provide striking examples of this: first Bonaparte, in 1814, in his famous operation against Blücher's forces, which were moving along the Marne, separated from the main allied army. We can hardly imagine a greater result from an unexpected advance carried out in two days. Blücher's troops, strung out over a distance of three days' marches, were beaten separately, and suffered casualties on the scale of a major battle. This was entirely due to surprise, for Blücher's order of march would have been different if he had known that an attack by Bonaparte might be imminent. The French success depended on Blücher's mistake. Bonaparte, to be sure, did not know how Blücher saw the situation; he benefited from a fortunate coincidence.

The battle of Liegnitz in 1760 is another case in point. Frederick the Great won this battle because during the night he moved from a position that he had only just occupied. Laudon was taken completely by surprise and lost 70 cannon and 10,000 men. At that time Frederick was acting on the principle of moving frequently in order to avoid battle, or at least in order to frustrate the enemy's plans; but this had not been his intention when he changed his position on the night of 14–15 June. He moved, as he says himself, because he was dissatisfied with the position he had occupied that day. Here too chance played a large part, and the outcome would have been different had it not been for the difficult, hilly terrain, and the coincidence of Frederick's nocturnal shift of position with the preliminary phases of Laudon's attack.

Even the higher, and highest, realms of strategy provide some examples of momentous surprises. It will suffice to recall the brilliant campaigns of the Great Elector against Sweden, sweeping from Franconia to Pomerania, and from the Mark Brandenburg to the river Pregel. The campaign of 1757 and Bonaparte's famous crossing of the Alps in 1800 are other examples. In the latter case, the Austrian army surrendered its entire theater of operations, and in 1757 another army came very close to surrendering not merely its operational theater but itself as well. Finally Frederick's invasion of Silesia may be cited as an example of a totally unexpected war. In all these cases the results were massive and far-reaching. Yet history has few such events to report—unless, of course, we confuse them with instances of states being ill-prepared for war because of sheer inactivity and lack of energy, such as Saxony in 1756 and Russia in 1812.

One more observation needs to be made, which goes to the very heart of the matter. Only the commander who imposes his will can take the enemy by surprise; and in order to impose his will, he must act correctly. If we surprise the enemy with faulty measures, we may not benefit at all, but instead suffer sharp reverses. Our surprise, in that case, will cause the enemy little worry; by exploiting our mistakes, he will find ways of warding off any ill-effects. Since the offensive offers much more scope for positive action than the defensive, the element of surprise is more often related to the attack—but far from exclusively so, as we shall see later on. Mutual surprises by the offensive and defensive may collide, in which case the side will be justified and succeed that has hit the nail most squarely on the head.

That, at any rate, is how it ought to be. But for a simple reason it does not always happen in real life. For the side that can benefit from the psychological effects of surprise, the worse the situation is, the better it may turn out, while the enemy finds himself incapable of making coherent decisions. This holds true not only for senior commanders, but for everyone involved; for one peculiar feature of surprise is that it loosens the bonds of cohesion, and individual action can easily become significant.

Much depends on the relationship established between the two sides. If general moral superiority enables one opponent to intimidate and outdistance the other, he can use surprise to greater effect, and may even reap the fruits of victory where ordinarily he might expect to fail.

CHAPTER TEN

Cunning

The term "cunning" implies secret purpose. It contrasts with the straightforward, simple, direct approach much as wit contrasts with direct proof. Consequently, it has nothing in common with methods of persuasion, of self-interest, or of force, but a great deal with deceit, which also conceals its purpose. It is itself a form of deceit, when it is completed; yet not deceit in the ordinary sense of the word, since no outright breach of faith is involved. The use of a trick or stratagem permits the intended victim to make his own mistakes, which, combined in a single result, suddenly change the nature of the situation before his very eyes. It might be said, that, as wit juggles with ideas and beliefs, so cunning juggles with actions.

At first glance, it seems not unjust that the term "strategy" should be derived from "cunning" and that, for all the real and apparent changes that war has undergone since the days of ancient Greece, this term still indicates its essential nature.

If we leave the actual execution of force, the engagement, to tactics, and consider strategy as the art of skillfully exploiting force for a larger purpose, and if we disregard for the moment such characteristics as fierce ambition which acts like a compressed spring, great will-power which yields only reluctantly, etc., no human characteristic appears so suited to the task of directing and inspiring strategy as the gift of cunning. The universal urge to surprise, discussed in the previous chapter, already points to this conclusion: since each surprise action is rooted in at least some degree of cunning.

Yet however much one longs to see opposing generals vie with one another in craft, cleverness, and cunning, the fact remains that these qualities do not figure prominently in the history of war. Rarely do they stand out amid the welter of events and circumstances.

The reason for this is obvious, and is closely related to the substance of the previous chapter.

Strategy is exclusively concerned with engagements and with the directions relating to them. Unlike other areas of life it is not concerned with actions that consist only of words, such as statements, declarations, and so forth. But words, being cheap, are the most common means of creating false impressions.

Analogous things in war—plans and orders issued for appearances only, false reports designed to confuse the enemy, etc.—have as a rule so little strategic value that they are used only if a ready-made opportunity presents itself. They should not be considered as a significant independent field of action at the disposal of the commander.

To prepare a sham action with sufficient thoroughness to impress an enemy requires a considerable expenditure of time and effort, and the costs increase with scale of the deception. Normally they call for more than can be spared, and consequently so-called strategic feints rarely have the desired effect. It is dangerous, in fact, to use substantial forces over any length of time merely to create an illusion; there is always the risk that nothing will be gained and that the troops deployed will not be available when they are really needed.

In war generals are always mindful of this sobering truth, and thus tend to lose the urge to play with sly mobility. Stern necessity usually permeates direct action to such an extent that no room is left for such a game. In brief, the strategist's chessmen do not have the kind of mobility that is essential for stratagem and cunning.

We conclude that an accurate and penetrating understanding is a more useful and essential asset for the commander than any gift for cunning—though the latter will do no harm so long as it is not employed, as it all too often is, at the expense of more essential qualities of character.

However, the weaker the forces that are at the disposal of the supreme commander, the more appealing the use of cunning becomes. In a state of weakness and insignificance, when prudence, judgment, and ability no longer suffice, cunning may well appear the only hope. The bleaker the situation, with everything concentrating on a single desperate attempt, the more readily cunning is joined to daring. Released from all future considerations, and liberated from thoughts of later retribution, boldness and cunning will be free to augment each other to the point of concentrating a faint glimmer of hope into a single beam of light which may yet kindle a flame.

CHAPTER ELEVEN

Concentration of Forces in Space

The best strategy is always *to be very strong*; first in general, and then at the decisive point. Apart from the effort needed to create military strength, which does not always emanate from the general, there is no higher and simpler law of strategy than that of *keeping one's forces concentrated.* No force should ever be detached from the main body unless the need is definite and *urgent.* We hold fast to this principle, and regard it as a reliable guide. In the course of our analysis, we shall learn in what circumstances dividing one's forces may be justified. We shall also learn that the principle of concentration will not have the same results in every war, but that those will change in accordance with means and ends.

Incredible though it sounds, it is a fact that armies have been divided and separated countless times, without the commander having any clear reason for it, simply because he vaguely felt that this was the way things ought to be done.

This folly can be avoided completely, and a great many unsound reasons for dividing one's forces never be proposed, as soon as concentration of force is recognized as the norm, and every separation and split as an exception that has to be justified.

CHAPTER TWELVE

Unification of Forces in Time

We have come to a concept that is likely to be misleading when applied to real life. A clear definition and development seems necessary, and we hope we may be forgiven another short analysis.

War is the impact of opposing forces. It follows that the stronger force not only destroys the weaker, but that its impetus carries the weaker force along with it. This would seem not to allow a protracted, consecutive, employment of force: instead, the simultaneous use of all means intended for a given action appears as an elementary law of war.

In practice this is true, but only when war really resembles a mechanical thrust. When it consists of a lengthy interaction of mutually destructive forces, the successive employment of force certainly becomes feasible. That is the case in tactics, primarily because tactics are chiefly based on fire power; and there are other reasons as well. If in a fire-fight a thousand men face five hundred, the sum of their losses may be calculated from the total forces involved on both sides. A thousand men fire twice as many rounds as five hundred, but of the thousand, more will be hit than of the five hundred, for it must be assumed that the thousand will be deployed more closely. If we suppose that they suffer twice as many hits, the losses on each side would be equal. The five hundred, for example, would suffer two hundred casualties, as would the thousand. Now, if the force of five hundred had kept an equal number of men in reserve, out of range, eight hundred able-bodied men would be available to each opponent. But on one side five hundred men would be fresh and fully supplied with ammunition, while all of the eight hundred facing them would be to some extent disorganized, tired and short of ammunition. To be sure, it is not correct to assume that because of their greater number the thousand would lose twice as many men as the five hundred would have lost in their place. The greater loss sustained by the side that held half of its strength in reserve must be counted as a disadvantage. It must also be admitted that as a general rule the thousand may initially have an opportunity of driving the enemy from his position and forcing him to withdraw. Whether these two advantages balance the disadvantage of opposing, with eight hundred somewhat battle-weary men, an enemy not appreciably weaker and who has five hundred completely fresh men, cannot be decided by further analysis. We must rely on experience; and few officers who have seen action would not grant superiority to the side with fresh troops.

It becomes clear why the deployment of too great a force may be detrimental: no matter how great the advantage which superiority offers in the first moment of the engagement, we may have to pay for it in the next.

The danger, however, applies only to the *phase of confusion, the condition of disarray and weakness*—in brief, the crisis that occurs in every engagement, even on the victorious side. In the context of such a weakened condition, the appearance of relatively fresh troops will be decisive.

On the other hand, as soon as the disorganizing effect of victory ceases, and all that remains is the moral superiority caused by every victory, fresh troops alone can no longer save the situation—they too will be swept away. A beaten army cannot make a comeback the following day, merely by being reinforced with strong reserves. *Here we arrive at the source of a vital difference between strategy and tactics.*

Tactical successes, those attained *in the course* of the engagement, *usually occur during the phase of disarray and weakness.* On the other hand, the strategic success, the overall effect of the engagement, the completed victory, whether great or insignificant, *already lies beyond that phase.* The strategic outcome takes shape only when the fragmented results have combined into a single, independent whole. But at that point the crisis is over, the forces regain their original cohesion, weakened only by the casualties they have actually suffered.

The consequence of this difference is that in the tactical realm force can be used successively, while strategy knows only the simultaneous use of force.

If in a tactical situation initial success does not lead to a conclusive victory, we have reason to fear the immediate future. It follows that for the first phase we should use only the amount of force that seems absolutely necessary. The rest should be kept out of range of fire and out of hand-to-hand fighting, so that we can oppose the enemy's reserves with fresh troops of our own or defeat his weakened forces with them. In a strategic situation this does not hold true. For one thing, as has been shown, once a strategic success is achieved, a reaction is less likely to set in, because the crisis has passed; for another, not all strategic forces have necessarily been *weakened.* The only troops that have suffered losses are those that have been *tactically* engaged—those, in other words, that have fought. Provided they have not been wasted, only the irreducible minimum will have been in action, far from the total that has been strategically committed. Units that were scarcely or not at all involved in the fighting, because of the army's superiority, and which contributed to success by their mere presence, are the same after victory as they were before, and are as ready for further tasks as if they had been completely inactive. It is obvious how greatly the margin of strength provided by these units can contribute to the successful outcome; it is equally understandable that their presence can substantially reduce the losses sustained by the troops that are actually tactically engaged.

Since in strategy casualties do not increase with the size of the forces used, and may even be reduced, and since obviously greater force is more likely to lead to success, it naturally follows that we can never use too great a force, and further, that all available force must be used *simultaneously*.

The truth of this proposition, however, needs to be established in another area as well. So far we have only discussed combat itself. It is the essential activity of war, but we must also consider men, time, and space, which are the components of this activity. We must consider the effects of their influence.

Fatigue, exertion, and privation constitute a separate destructive factor in war—a factor not essentially belonging to combat, but more or less intricately involved in it, and pertaining especially to the realm of strategy. This factor is also present in tactical situations, and possibly in its most intense form; but since tactical actions are of shorter duration, the effects of exertion and privation will be limited. On the strategic plane, however, where the dimensions of time and space are enlarged, the effects are always perceptible, and often decisive. It is not unusual that a victorious army suffers greater losses from sickness than from battle.

If we consider this category of destruction in the strategic realm as we consider that of artillery fire and hand-to-hand fighting in the tactical realm, we may well conclude that by the end of the campaign or some other strategic period, everything exposed to the factor of destruction will be in a weakened state, and that the appearance of new forces would be *decisive*. In a strategic as well as in a tactical situation, therefore, we might be tempted to seek initial success with a minimum of troops, in order to retain strong reserves for the final struggle.

Plausibility is lent to this argument by many actual cases. In order to evaluate it correctly, we must scrutinize the separate ideas involved. First of all, the notion of reinforcements must not be confused with the idea of fresh and unused troops. Few campaigns end without the addition of new troops seeming most desirable, indeed decisive, both to the winner and loser; but that is not here an issue: reinforcements would not be needed at all if a large enough force had been used in the first place. The notion that a fresh army taking the field would have higher morale than troops already in action (comparable to a tactical reserve, which does indeed rate higher than men who have already suffered much) is completely discounted by experience. Just as an unsuccessful campaign diminishes the courage and morale of the troops, a successful one heightens these values. By and large then, these factors tend to cancel each other out; the gain in experience is left as clear profit. In any case, here we should study successful rather than unsuccessful campaigns, since whenever failure can be predicted with any degree of certainty, adequate force is lacking in the first place, and holding any part in reserve for later use would be unthinkable.

That point being settled, the question arises whether the losses sustained by a unit from exertion and privation will increase in proportion to its size, as would be the case in an engagement. The answer to that question must be in the negative.

Exertion is mainly caused by the dangers which in varying degree are inseparable from military operations. To counter these dangers everywhere, to proceed on our course with confidence, is the object of much of the activity comprising the tactical and strategic duties of the army. The weaker an army, the more arduous this duty becomes; while the greater the army's superiority, the easier it becomes. Who can doubt this? A campaign against a substantially weaker opponent will call for less exertion than a campaign against an equally strong force, not to speak of one that is superior to our own.

So much for physical exertion. Privation is a somewhat different matter. It consists mainly of lack of food and lack of shelter for the troops, either in quarters or in comfortable camps. The problems of food and shelter admittedly increase with the size of a force concentrated in one place. On the other hand, does not this very superiority provide the best method of spreading out over a greater area and thus finding more means of supply and shelter?

During the advance in Russia in 1812, Bonaparte kept his forces massed along a single road in an unheard-of manner, causing equally unheard-of shortages. This may be attributed to his principle that one can never be too strong at the decisive point. Whether or not he pushed this principle too far in this instance cannot be discussed here, but it is certain that, if he had wanted to avoid these shortages, all he had to do was to advance on a broader front. There was room enough in Russia; indeed, there would almost always be enough space. Therefore this example provides no basis for the claim that the simultaneous use of greatly superior forces will produce greater suffering. Supposing, however, that wind and weather and the unavoidable exertions of war had indeed weakened—in spite of the relief afforded to the whole—even that part of the army which, as a surplus force, could have been held in reserve for later use: it becomes even more essential to look at the situation as a whole and ask whether this loss would have equaled the gains that superior strength might have achieved in one way or another.

We must consider another very important point. In a minor engagement it is not too difficult to judge approximately how much force is needed to achieve substantial success, and what would be superfluous. In strategy this is practically impossible, because strategic success cannot be defined and delineated with the same precision. What may be regarded as surplus strength in a tactical situation must be considered in strategy as a means of exploiting success if the opportunity arises. Since the margin of profit increases with the scale of victory, superiority of force can quickly reach a level which the most careful calculation of strength could never have determined.

By means of his vastly superior strength Bonaparte was able to penetrate to Moscow in 1812 and occupy the city. If his superior strength had also enabled him to crush the Russian army, he would probably have concluded a peace in Moscow that would have been less readily attainable by other means. We cite this example simply as an illustration; proof would require a detailed exposition, which would here be out of place.

All these reflections deal only with the successive use of force. They are not concerned with the idea of a reserve as such, though the two touch in many points. That subject has additional ramifications, as the following chapter will show.

What we are trying to establish is that while tactically the mere *duration* of an engagement weakens the forces so that time becomes a factor in the outcome, this is not the case in strategy. To the extent that in strategy time does exert destructive effects on the forces involved, these are partly mitigated by the size of the forces, and partly offset in other ways. Consequently it cannot be the intent of the strategist to make an ally of time *for its own sake,* by committing force gradually, step by step.

We say *for its own sake* because time can possess significance as a result of factors that derive from but are not identical with it. Indeed it must be significant for one opponent or the other. This is quite a different matter, by no means trivial or unimportant, and will form the subject of later study.

The rule, then, that we have tried to develop is this: all forces intended and available for a strategic purpose should be applied *simultaneously*; their employment will be the more effective the more everything can be concentrated a single action at a single moment.

That does not mean that successive effort and sustained effect have no place in strategy. They cannot be ignored, the less so since they form one of the principal means toward a final success: the continuous deployment of new forces. This too will be the subject of another chapter. We mention it here only to avoid misunderstanding.

We now turn to a subject closely linked to our previous discussion, which will clarify the whole matter—we mean, the *strategic reserve.*

CHAPTER THIRTEEN

The Strategic Reserve

A reserve has two distinct purposes. One is to prolong and renew the action; the second, to counter unforeseen threats. The first purpose presupposes the value of the successive use of force, and therefore does not belong to strategy. The case of a unit being sent to a point that is about to be overrun is clearly an instance of the second category, since the amount of resistance necessary at that point had obviously not been foreseen. A unit that is intended merely to prolong the fighting in a particular engagement and for that purpose is kept in reserve, will be available and subordinate to the commanding officer, though posted out of the reach of fire. Thus it will be a tactical rather than a strategic reserve.

But the need to hold a force in readiness for emergencies may also arise in strategy. Hence there can be such a thing as a strategic reserve, but only when emergencies are conceivable. In a tactical situation, where we frequently do not even know the enemy's measures until we see them, where they may be hidden by every wood and every fold of undulating terrain, we must always be more or less prepared for unforeseen developments, so that positions that turn out to be weak can be reinforced, and so that we can in general adjust our dispositions to the enemy's actions.

Such cases also occur in strategy, since strategy is directly linked to tactical action. In strategy too decisions must often be based on direct observation, on uncertain reports arriving hour by hour and day by day, and finally on the actual outcome of battles. It is thus an essential condition of strategic leadership that forces should be held in reserve according to the degree of strategic uncertainty.

In the defensive generally, particularly in the defense of certain natural features such as rivers, mountain ranges, and so forth, we know this is constantly required.

But uncertainty decreases the greater the distance between strategy and tactics; and it practically disappears in that area of strategy that borders on the political.

The movement of the enemy's columns into battle can be ascertained only by actual observation—the point at which he plans to cross a river by the few preparations he makes, which become apparent a short time in advance; but the direction from which he threatens our country will usually be announced in the press before a single shot is fired. The greater the scale of preparations, the smaller the chance of achieving a surprise. Time and space involved are vast, the circumstances that have set events in motion so well known and so little subject to change, that his decisions will either be apparent early enough, or can be discovered with certainty.

Moreover even if a strategic reserve should exist, in this area of strategy its value will decrease the less specific its intended employment.

We have seen that the outcome of a skirmish or single engagement is in itself of no significance; all such partial actions await resolution in the outcome of the battle as a whole.

In turn, the outcome of the battle as a whole has only relative significance, which varies in numerous gradations according to the size and overall importance of the defeated force. The defeat of a corps may be made up for by the victory of an army, and even the defeat of one army may be balanced or even turned into a victory by the successes of a larger army, as happened in the two days' fighting at Kulm in 1813. No one can doubt this; but it is equally clear that the impact of every victory, the successful outcome of every battle, gains in absolute significance with the importance of the defeated force, and consequently the possibility of recouping such losses at a later encounter also becomes less likely. This point will be examined more closely later on; for the present, it is enough to call attention to the existence of this progression.

Let us add a third observation. While the successive use of force in a tactical situation always postpones the main decision to the end of the action, in strategy the law of the simultaneous use of forces nearly always advances the main decision, which need not necessarily be the ultimate one, to the beginning. These three conclusions, therefore, justify the view that a strategic reserve becomes less essential, less useful, and more dangerous to use, the more *inclusive* and general its intended purpose.

The point at which the concept of a strategic reserve begins to be self-contradictory is not difficult to determine: it comes when the *decisive stage* of the battle has been reached. All forces must be used to achieve it, and any idea of reserves, of *available combat units* that are not meant to be used until after this decision, is an absurdity.

Thus, while a tactical reserve is a means not only of meeting any unforeseen maneuver by the enemy but also of reversing the unpredictable outcome of combat when this becomes necessary, strategy must renounce this means, at least so far as the overall decision is concerned. Setbacks in one area can, as a rule, be offset only by achieving gains elsewhere, and in a few cases by transferring troops from one area to another. Never must it occur to a strategist to deal with such a setback by holding forces in reserve.

We have called it an absurdity to maintain a strategic reserve that is not meant to contribute to the overall decision. The point is so obvious that we should not have devoted two chapters to it if it were not for the fact that the idea can look somewhat more plausible when veiled in other concepts, as indeed it frequently is. One man thinks of a strategic reserve as the peak of wise and cautious planning, another rejects the whole idea, including that of a tactical reserve. This kind of confused thinking does actually affect reality. For a striking example, we should recall that in 1806 Prussia billeted a reserve of 20,000 men under Prince Eugene of Württemberg in Brandenburg and could not get them to the Saale River in time, while another 25,000 men were kept in East and south Prussia *to be mobilized at some later stage, to act as a reserve.*

These examples will, we hope, spare us the reproach of tilting at windmills.

CHAPTER FOURTEEN

Economy of Force

As we have already said, principles and opinions can seldom reduce the path of reason to a simple line. As in all practical matters, a certain latitude always remains. Beauty cannot be defined by abscissas and ordinates; neither are circles and ellipses created by their algebraic formulas. The man of action must at times trust in the sensitive instinct of judgment, derived from his native intelligence and developed through reflection, which almost unconsciously hits on the right course. At other times he must simplify understanding to its dominant features, which will serve as rules; and sometimes he must support himself with the crutch of established routine.

One of these simplified features, or aids to analysis, is always to make sure that all forces are involved—always to ensure that no part of the whole force is idle. If a segment of one's force is located where it is not sufficiently busy with the enemy, or if troops are on the march—that is, idle—while the enemy is fighting, then these forces are being managed uneconomically. In this sense they are being wasted, which is even worse than using them inappropriately. When the time for action comes, the first requirement should be that all parts must act: even the least appropriate task will occupy some of the enemy's forces and reduce his overall strength, while completely inactive troops are neutralized for the time being. Obviously this view is a corollary of the principles developed in the last three chapters. It is the same truth, restated from a somewhat broader point of view, and reduced to a single concept.

CHAPTER FIFTEEN

The Geometrical Factor

The extent to which geometry, or form and pattern in the deployment of forces in war, can become a dominant principle is shown in the art of fortification, in which geometry applies to almost everything, large or small. In tactics too it plays a major part. Geometry forms the basis of tactics in the narrower sense—the theory of moving troops. In field fortification and in the theory of entrenched positions and their attack, the lines and angles of geometry rule like judges who will decide the contest. In the past, a good deal of this was misapplied, and some of it was no more than playing at soldiers; yet in today's tactics, where the outflanking of the enemy is the aim of every engagement, the geometrical factor has again achieved great significance. Although its application is simple, it recurs constantly. Nevertheless geometry cannot govern tactics as it governs siege warfare: when troops face one another everything is more mobile, and psychological forces, individual differences, and chance play a more influential part. In strategy the influence of geometry is even less significant. Though here too types of troop formations and configurations of countries and states are significant, the principle of geometry is not *decisive* as in the art of fortification, and not nearly so important as in tactics. The manner of its influence will gradually be demonstrated wherever it is relevant and needs to be considered. For the present we shall call attention to the difference between tactics and strategy in this respect.

In tactics time and space are rapidly reduced to their absolute minimum. A unit that is attacked in flank and rear will soon reach the stage where its chance of retreat has vanished: in such a situation it comes close to being completely unable to continue the fight, and the commander must either try to extricate himself from this predicament or prevent it from occurring at all. For this reason all tactical arrangements aimed at envelopment are highly effective, and their effectiveness consists largely in the concern they induce about their consequences. That is why the geometric factor in the disposition of forces is so important.

These considerations are reflected only faintly in strategy, with its concern for great spans of time and space. Armies do not burst from one theater of war into another; rather, a projected strategic envelopment may easily take weeks and months to carry out. Besides, distances are so great that the chances of even the best measures finally achieving the desired result remain slight.

In strategy therefore, the effect of such combinations, that is of the geometric pattern, is much smaller; on the other hand the effect of an advantage gained at one point is much greater. This advantage can be exploited to the full, before countermeasures interfere or even cancel it out. In consequence we do not hesitate to consider it an established truth that in strategy the number and scale of the engagements won are more meaningful than the pattern of the major lines connecting them.

The very opposite view was a favorite of recent theorists, who believed that in this way they would increase the importance of strategy. Strategy, they thought, expressed the higher functions of the intellect; they thought that war would be ennobled by its study, and, according to a modern substitution of concepts, be made *more scientific*. We believe that it is one of the chief functions of a comprehensive theory of war to expose such vagaries, and it is because the geometrical element usually provides the point of departure for these fantasies that we have drawn special attention to it.

CHAPTER SIXTEEN

The Suspension of Action in War

If we regard war as an act of mutual destruction, we are bound to think of both sides as usually being in action and advancing. But as soon as we consider each moment separately, we are almost equally bound to think of only one side as advancing while the other is expectantly waiting; for conditions will never be exactly identical on both sides, nor will their mutual relationship remain the same. In time changes will occur, and it follows that any given moment will favor one side more than the other. If we assume that both generals are completely cognizant of their own and their opponent's conditions, one of them will be motivated to act, which becomes in turn to the other a reason for waiting. Both cannot simultaneously want to advance, or on the other hand to wait. This mutual exclusion of identical aims does not, in the present context, derive from the principle of polarity, and therefore it does not contradict the assertion made in Chapter Five of Book Two.[1] Rather, its basis lies in the fact that the determinant is really the same for both commanders: the probability of improvement, or deterioration, of the situation in the future.

Even if we suppose that circumstances could be completely balanced, or if we assume that insufficient knowledge of their mutual circumstances gives the commanders the impression that such equality exists, the differences in their political purpose will still rule out the possibility of a standstill. Politically, only one can be the aggressor: there can be no war if both parties seek to defend themselves. The aggressor has a positive aim, while the defender's aim is merely negative. Positive action is therefore proper to the former, since it is the only means by which he can achieve his ends. Consequently when conditions are equal for both parties the attacker ought to act, since his is the positive aim.

Seen in this light, suspension of action in war is a contradiction in terms. Like two incompatible elements, armies must continually destroy one another. Like fire and water they never find themselves in a state of equilibrium, but must keep on interacting until one of them has completely disappeared. Imagine a pair of wrestlers deadlocked and inert for hours on end! In other words, military action ought to run its course steadily like a wound-up clock. But no matter how savage the nature of war, it is fettered by human weaknesses; and no one will be surprised at the contradiction that man seeks and creates the very danger that he fears.

[1] Sic. This point is discussed not in Chapter Five of Book Two but in Chapter One of Book One, pp. 81–85 above. Eds.

The history of warfare so often shows us the very opposite of unceasing progress toward the goal, that it becomes apparent that *immobility* and *inactivity* are the normal *state* of armies in war, and *action is the exception.* This might almost make us doubt the accuracy of our argument. But if this is the burden of much of military history, the most recent series of wars does substantiate the argument. Its validity was demonstrated and its necessity was proved only too plainly by the revolutionary wars. In these wars, and even more in the campaigns of Bonaparte, warfare attained the unlimited degree of energy that we consider to be its elementary law. We see it is possible to reach this degree of energy; and if it is possible, it is necessary.

How, in fact, could we reasonably defend the exertion of so much effort in war, unless action is intended! A baker fires his oven only when he is ready to bake bread; horses are harnessed to a carriage only when we intend to drive; why should we make the enormous exertions inherent in war if our only object is to produce a similar effort on the part of the enemy?

So much in justification of the general principle. Now for its modifications, insofar as they arise from the nature of the subject and do not depend on individual circumstances.

Let us note three determinants that function as inherent counterweights and prevent the clockwork from running down rapidly or without interruption.

The first of these, which creates a permanent tendency toward delay and thus becomes a retarding influence, is the fear and indecision native to the human mind. It is a sort of moral force of gravity, which, however, works by repulsion rather than attraction: namely, aversion to danger and responsibility.

In the fiery climate of war, ordinary natures tend to move more ponderously; stronger and more frequent stimuli are therefore needed to ensure that momentum is maintained. To understand why the war is being fought is seldom sufficient in itself to overcome this ponderousness. Unless an enterprising martial spirit is in command, a man who is as much at home in war as a fish is in water, or unless great responsibilities exert a pressure, inactivity will be the rule, and progress the exception.

The second cause is the imperfection of human perception and judgment, which is more pronounced in war than anywhere else. We hardly know accurately our own situation at any particular moment, while the enemy's, which is concealed from us, must be deduced from very little evidence. Consequently it often happens that both sides see an advantage in the same objective, even though in fact it is more in the interest of only one of them. Each may therefore think it wiser to await a better moment, as I have already explained in Chapter Five of Book Two.[2]

The third determinant, which acts like a rachet-wheel, occasionally stopping the works completely, is the greater strength of the defensive. A may

[2] Sic. See previous note, p. 216 above. Eds.

not feel strong enough to attack B, which does not, however, mean that B is strong enough to attack A. The additional strength of the defensive is not only lost when the offensive is assumed but is transferred to the opponent. Expressed in algebraic terms, the difference between A + B and A − B equals 2 B. It therefore happens that both sides at the same time not only feel too weak for an offensive, but that they really are too weak.

Thus, in the midst of the conflict itself, concern, prudence, and fear of excessive risks find reason to assert themselves and to tame the elemental fury of war.

But these determinants are hardly adequate explanations for the long periods of inactivity that occurred in earlier wars, in which no vital issues were at stake, and in which nine-tenths of the time that the troops spent under arms was occupied by idleness. As stated in the chapter on the Purpose and Means in War, this phenomenon is mainly due to the influence that the demands of the one belligerent, and the condition and state of mind of the other, exert on the conduct of the war.

These factors can become so influential that they reduce war to something tame and half-hearted. War often is nothing more than armed neutrality, a threatening attitude meant to support negotiations, a mild attempt to gain some small advantage before sitting back and letting matters take their course, or a disagreeable obligation imposed by an alliance, to be discharged with as little effort as possible.

In all such cases, where the impetus of interest is slight and where there is little hostile spirit, where we neither want to do much harm to the enemy nor have much to fear from him, in short where no great motive presses and promotes action, governments will not want to risk much. This explains the tame conduct of such conflicts, in which the hostile spirit of true war is held in check.

The more these factors turn war into something half-hearted, the less solid are the bases that are available to theory: essentials become rarer, and accidents multiply.

Nevertheless, even this type of conflict gives scope to intelligence; possibly even wider and more varied scope. Gambling for high stakes seems to have turned into haggling for small change. In this type of war, where military action is reduced to insignificant, time-killing flourishes, to skirmishes that are half in earnest and half in jest; to lengthy orders that add up to nothing; to positions and marches that in retrospect are described as scientific, simply because their minute original motive has been forgotten and common sense cannot make anything of them—in this type of conflict many theorists see the real, authentic art of war. In these feints, parries, and short lunges of earlier wars they find the true end of all theory and the triumph of mind over matter. More recent wars appear to them as crude brawls that can teach nothing and that are to be considered as relapses into barbarism. This view is as petty as its subject. In the absence of great forces and passions it is indeed simpler for ingenuity to function; but is not guiding great forces, navigation through storms and surging waves, a higher exercise of the intellect? That other, formalized type of swordsmanship is surely included and implicit in the more energetic mode of conducting war. It has the same relation to it as the movements on a ship have to the motion of the ship. It can only be carried on so long as it is tacitly understood that the opponent follows suit. But is it possible to tell how long this condition will be observed? The French Revolution surprised us in the false security of our ancient skills, and drove us from Châlons to Moscow. With equal suddenness, Frederick the Great surprised the Austrians in the quiet of their antiquated ways of war, and shook their monarchy to its foundations. Woe to the government, which, relying on half-hearted politics and a shackled military policy, meets a foe who, like the untamed elements, knows no law other than his own power! Any defect of action and effort will turn to the advantage of the enemy, and it will not be easy to change from a fencer's position to that of a wrestler. A slight blow may then often be enough to cause a total collapse.

All of these reasons explain why action in war is not continuous but spasmodic. Violent clashes are interrupted by periods of observation, during which both sides are on the defensive. But usually one side is more strongly motivated, which tends to affect its behavior: the offensive element will dominate, and usually maintain its continuity of action.

CHAPTER SEVENTEEN

The Character of Contemporary Warfare

All planning, particularly strategic planning, must pay attention to the character of contemporary warfare.

Bonaparte's audacity and luck have cast the old accepted practices to the winds. Major powers were shattered with virtually a single blow. The stubborn resistance of the Spaniards, marred as it was by weakness and inadequacy in particulars, showed what can be accomplished by arming a people and by insurrection. The Russian campaign of 1812 demonstrated in the first place that a country of such size could not be conquered (which might well have been foreseen),[1] and in the second that the prospect of eventual success does not always decrease in proportion to lost battles, captured capitals, and occupied provinces, which is something that diplomats used to regard as dogma, and made them always ready to conclude a peace however bad. On the contrary, the Russians showed us that one often attains one's greatest strength in the heart of one's own country, when the enemy's offensive power is exhausted, and the defensive can then switch with enormous energy to the offensive. Prussia taught us in 1813 that rapid efforts can increase an army's strength six times if we make use of a militia, and, what is more, that the militia can fight as well in foreign countries as at home. All these cases have shown what an enormous contribution the heart and temper of a nation can make to the sum total of its politics, war potential, and fighting strength. Now that governments have become conscious of these resources, we cannot expect them to remain unused in the future, whether the war is fought in self-defense or in order to satisfy intense ambition.

Obviously, wars waged by both sides to the full extent of their national strength must be conducted on different principles from wars in which policy was based on the comparative size of the regular armies. In those days, regular armies resembled navies, and were like them in their relation to the country and to its institutions. Fighting on land therefore had something in common with naval tactics, a quality which has now completely disappeared.

[1] Clausewitz himself had predicted in 1804 that if Napoleon ever invaded Russia he would be defeated. P. Paret, *Clausewitz and the State* (New York, 1976), p. 224. Eds.

CHAPTER EIGHTEEN

Tension and Rest

The Dynamic Law in War

The sixteenth chapter of this book stated that in most campaigns periods of inaction and repose have been much longer than periods of action. Even though modern war is of a totally different character, as has been pointed out in the preceding chapter, it remains true that periods of active warfare will always be interspersed with greater or smaller periods of rest. We must now take a closer look at the nature of these two phases of warfare.

When fighting is interrupted, in other words when neither side has a positive aim, a state of rest and equilibrium results; equilibrium, naturally, in its widest sense, covering not only physical and psychological forces, but all circumstances and motives. As soon as one side adopts a new and positive aim and begins to pursue it however tentatively, as soon as the opponent resists tension of forces builds up. This tension lasts until the immediate issue has been decided: either one side renounces its goal, or the other side concedes it.

This decision, which is always derived from the results of the combinations of actions that are developed by both sides, is followed by movement in one direction or the other.

When this movement has been exhausted, either through the difficulties it has met, such as the frictions that are inherent in any action, or through new opposing forces, inactivity returns, or a new cycle of tension and decision begins, followed by further movement—usually in the opposite direction.

This theoretical distinction between balance, tension, and movement has a greater practical application than may at first appear.

A state of rest and equilibrium can accommodate a good deal of activity; that is to say, the kind of activity arising from incidental causes, and not designed to lead to major changes. Significant engagements, even major battles, may take place; but these actions are still of a different nature and therefore usually have different results.

In a state of tension a decision will always have greater effect; partly because greater will-power and greater pressure of circumstances are involved, and partly because everything is already prepared for major action. In this situation the effect resembles the explosion of a carefully sealed mine, while an event of similar proportions which takes place during a period of rest is more like a flare-up of gunpowder in the open air.

The state of tension is obviously a matter of degree. Numerous gradations are possible as it approaches the state of rest, the last stage being so close as to make it difficult to distinguish one condition from the other.

The most significant lesson drawn from these observations is that any move made in a state of tension will be more important, and will have more results, than it would have if made in a state of equilibrium. In times of maximum tension this importance will rise to an infinite degree.

The cannonade of Valmy decided more than the battle of Hochkirch.

If the enemy abandons territory because he is unable to defend it, we can use this territory in a different manner than if the enemy's retreat were undertaken with the intention of fighting under more favorable circumstances. A badly chosen position or a single miscalculated move on our part during the course of a strategic attack by the enemy may have fatal results, while in a state of equilibrium such blunders would have to be glaring indeed to arouse the enemy's reaction at all.

As we have mentioned, most former wars were waged largely in this state of equilibrium, or at least expressed tensions that were so limited, so infrequent, and feeble, that the fighting that did occur during these periods was seldom followed by important results. Instead a battle might be fought to celebrate the birthday of a monarch (Hochkirch), to satisfy military honor (Kunersdorf), or to assuage a commander's vanity (Freiberg).

In our opinion it is essential that a commander should recognize these circumstances and act in concert with their spirit. The experience of the campaign of 1806 has shown the extent to which this capacity may at times be missing. During that period of enormous tension, events were pressing toward a major decision which, with all its consequences, ought to have absorbed the full attention of the commander; yet at that very time, plans were proposed and even partly carried out, such as the reconnaissance in Franconia, that in a state of equilibrium could at most have caused a slight tremor. But these confusing schemes and time-consuming ideas dissipated activity and energy that should have gone into the really urgent measures that alone could have saved the day.

The distinction we have made is also necessary to the further development of our theory. Everything we shall have to say about the relation between attack and defense and the way in which this polarity develops refers to the state of crisis in which the forces find themselves during periods of tension and movement. By contrast, all activity that occurs during a state of equilibrium will be regarded and treated as a mere corollary. The state of crisis is the real war; the equilibrium is nothing but its reflex.

BOOK FOUR

The Engagement

CHAPTER ONE

Introduction

In the last book we examined the factors that may be called the operative elements in war. We now turn to the essential military activity, fighting, which by its material and psychological effects comprises in simple or compound form the object of the war. The operative elements must therefore be contained in this activity and in its effects.

The framework of the engagement is tactical; a broad survey will familiarize us with its general appearence. Every engagement has a specific purpose that gives it its peculiar characteristics, and these special purposes will be examined later. Compared with the general characteristics of fighting, the peculiarities tend to be relatively unimportant, with the result that most engagements are very much alike. Rather than have to refer repeatedly to these common features, we intend to deal with them at once, before discussing their special applications.

In the following chapter we shall first give a short description of the tactical course of battle today, since it is the basis of our concept of fighting.

CHAPTER TWO

The Nature of Battle Today

Our assumptions about tactics and strategy being what they are, it will be self-evident that a change in the nature of tactics will automatically react on strategy. If tactical phenomena differ completely from one case to another, strategic ones must also differ, if they are to remain consistent and rational. It is important therefore to describe a major battle in its contemporary form before discussing its strategic use.

What usually happens in a major battle today? The troops move calmly into position in great masses deployed in line and depth. Only a relatively small proportion is involved, and is left to conduct a firefight for several hours, interrupted now and then by minor blows—charges, bayonet assaults and cavalry attacks—which cause the fighting to sway to some extent to and fro. Gradually, the units engaged are burned out, and when nothing is left but cinders, they are withdrawn and others take their place.

So the battle slowly smolders away, like damp gunpowder. Darkness brings it to a halt: no one can see, and no one cares to trust himself to chance. The time has come to reckon up how much in the way of serviceable troops is left on either side—troops, that is, which are not yet burned out like dead volcanoes. One estimates how much ground has been won or lost, as well as the degree of security in one's rear. The results, along with personal impressions of the bravery and cowardice, intelligence and stupidity that one thinks one has observed in one's own troops and the enemy's, are then combined in an overall impression on which a decision is based: either to quit the field or to renew the fight in the morning.

This description does not claim to be a full picture of a modern battle—it is merely meant to give a broad impression. It applies equally to attacker and defender. Specific features, such as the particular objective or the nature of the terrain, may be added without changing the general impression.

But it is no accident that contemporary battles should be like this. Contemporary armies have developed almost identically in military organization and methods; the element of war itself, stirred up by great national interests, has become dominant and is pursuing its natural ccurse. Battles will not change their character so long as both these conditions hold good.

This general picture of modern battle will be useful later on when we have to determine the value of various coefficients, such as strength, terrain, and so forth. The description is valid only for general, great and decisive engagements and those that approximate to them; lesser ones have changed in the same manner, but not to the same extent. Proof of this will be found in tactics, but we shall have further opportunities for adding details that will make this clearer.

CHAPTER THREE

The Engagement in General

Fighting is the central military act; all other activities merely support it. Its nature consequently needs close examination.

Engagements mean fighting. The object of fighting is the destruction or defeat of the enemy. The enemy in the individual engagement is simply the opposing fighting force.

This is the simple concept, and we shall return to it. But first we must introduce a number of other considerations.

If a state with its fighting forces is thought of a single unit, a war will naturally tend to be seen in terms of a single great engagement. Under the primitive conditions of savage peoples this generally holds true. But our wars today consist of a large number of engagements, great and small, simultaneous or consecutive, and this fragmentation of activity into so many separate actions is the result of the great variety of situations out of which wars can nowadays arise.

Even the ultimate aim of contemporary warfare, the political object, cannot always be seen as a single issue. Even if it were, action is subject to such a multitude of conditions and considerations that the aim can no longer be achieved by a single tremendous act of war. Rather it must be reached by a large number of more or less important actions, all combined into one whole. Each of these separate actions has a specific purpose relating to the whole.

We have already said that the concept of the engagement lies at the root of all strategic action, since strategy is the use of force, the heart of which, in turn, is the engagement. So in the field of strategy we can reduce all military activity to the unitary concept of the single engagement, and concern ourselves exclusively with its purposes. We will come to identify these purposes as we discuss the circumstances that give rise to them in the engagement. Here it is enough to say that every engagement, large or small, has its own particular purpose which is subordinate to the general one. That being so, the destruction and subjugation of the enemy must be regarded simply as a means toward the general end, which it obviously is.

But this conclusion is true only in a formal sense, and is significant only because of the connection between these various concepts. We have brought up this connection only in order to get it out of the way.

What do we mean by the defeat of the enemy? Simply the destruction of his forces, whether by death, injury, or any other means—either completely or enough to make him stop fighting. Leaving aside all specific purposes of any particular engagement, the complete or partial destruction of the enemy must be regarded as the sole object of all engagements.

We maintain that in the majority of cases, and especially in major actions, the particular purpose that both distinguishes the action and links it to the war as a whole is only a slight modification of the general purpose of the war or a subsidiary purpose connected with it. It is important enough to give the action its particular character, but of little weight compared to that overall purpose. If the subsidiary purpose alone is fulfilled, only an unimportant part of the objective has been achieved. If we are right, then the notion according to which the destruction of the enemy forces is only the means, while the ends are bound to be quite different, is only generally true. We would reach the wrong conclusion unless we bear in mind that this very destruction of the enemy's forces is also part of the final purpose. That purpose itself is only a slight modification of the destructive aim.

Ignoring this point led to completely wrong ideas before the recent wars, creating fashions and fragmentary systems through which theory became elevated far above everyday practice; the more so, since theory attached less importance to the use of the real instrument—the destruction of the enemy's forces.

No such system could, of course, have been conceived without other erroneous assumptions, and without replacing the concept of destruction of the enemy's forces by different ideas wrongly assumed to be effective. We shall expose these as opportunity arises, but we cannot deal with the engagement without reasserting its importance and true value, and suggesting the errors to which a strictly formal view might lead.

How are we to prove that usually, and in all the most important cases, the destruction of the enemy's forces must be the main objective? How are we to counter the highly sophisticated theory that supposes it possible for a particularly ingenious method of inflicting minor direct damage on the enemy's forces to lead to major indirect destruction; or that claims to produce, by means of limited but skillfully applied blows, such paralysis of the enemy's forces and control of his will-power as to a constitute a significant shortcut to victory? Admittedly, an engagement at one point may be worth more than at another. Admittedly, there is a skillful ordering of priority of engagements in strategy; indeed that is what strategy is all about, and we do not wish to deny it. We do claim, however, that direct annihilation of the enemy's forces must always be the *dominant consideration.* We simply want to establish this dominance of the destructive principle.

But we must repeat that the subject that here concerns us is strategy, not tactics. We are therefore not discussing the tactical means used to destroy a maximum of enemy force with a minimum of effort. By direct destruction we mean tactical success. We maintain therefore that only great tactical successes can lead to great strategic ones; or as we have already said more specifically, *tactical* successes are of *paramount importance* in war.

The proof of our assertion is fairly simple. It can be found in the time absorbed by complex operations. The question whether a simple attack or a more complex one will be the more effective will certainly be answered in favor of the latter if one assumes the enemy to be passive. But every complex operation takes time; and this time must be available without a counter-attack on one of its parts interfering with the development of the whole. If the enemy decides on a simpler attack, one that can be carried out quickly, he will gain the advantage and wreck the grand design. So, in the evaluation of a complex attack, every risk that may be run during its preparatory stages must be weighed. The scheme should only be adopted if there is no danger that the enemy can wreck it by more rapid action. Wherever this is possible we ourselves must choose the shorter path. We must further simplify it to whatever extent the character and situation of the enemy and any other circumstances make necessary. If we abandon the weak impressions of abstract concepts for reality, we will find that an active, courageous, and resolute adversary will not leave us time for long-range intricate schemes; but that is the very enemy against whom we need these skills most. It seems to us that this is proof enough of the *superiority* of the simple and direct over the complex.

This does not mean that the simple attack is best. It means rather that one should not swing wider than latitude allows. The probability of direct confrontation increases with the aggressiveness of the enemy. So, rather than try to outbid the enemy with complicated schemes, one should, on the contrary, try to outdo him in simplicity.

The foundation-stones of these opposites are for the one, intelligence, for the other, courage. One may be tempted to believe that moderate courage coupled with great intelligence will be more effective than moderate intelligence and great courage. But, unless these factors are assumed to be unreasonably disproportionate, one is not entitled to consider intelligence as superior to courage in a field whose very name is danger—one which must be regarded as courage's proper realm.

After this abstract reasoning, we wish to add that experience, far from leading to a different conclusion, is the very source of our conviction, and lies at the root of this train of thought.

If we read history with an open mind, we cannot fail to conclude that, among all the military virtues, *the energetic conduct of war* has always contributed most to glory and success.

Later, we will show how we shall apply the principle that the destruction of enemy forces must be regarded as the main objective; not just in the war generally, but in each individual engagement and within all the different conditions necessitated by the circumstances out of which the war has arisen. For the moment we have simply been concerned to establish its general importance, and can now return to the engagement.

CHAPTER FOUR

The Engagement in General—Continued

In the previous chapter we defined the purpose of the engagement as being the destruction of the enemy. We have tried to prove this to be true in the majority of cases and in major actions, since the destruction of the enemy's forces must always be the dominant consideration in war. Other objectives that may be added and that to some degree may even dominate, will be examined in the following chapter, and we shall gradually learn more about them. For the present, we shall ignore them and treat the destruction of the enemy by itself as a completely adequate purpose for the individual engagement.

What do we mean by "destruction of the enemy's forces?" A reduction of strength relatively larger than our own. Equal absolute losses will, of course, mean smaller relative losses to the side with numerical superiority, and can therefore be considered an advantage. But having stripped the engagement of all other objects, we must also exclude that of using it to effect indirectly a greater destruction of the enemy forces. Consequently, only the direct profit gained in the process of mutual destruction may be considered as having been the object. This profit is absolute: it remains fixed throughout the entire balance sheet of the campaign and in the end will always prove pure gain. Any other type of victory over the enemy would either have its basis in other objectives, which we are not discussing here; or would yield only a temporary and relative profit. An example will clarify this.

If by skillful deployment one can place the enemy at such a disadvantage that he cannot continue fighting without risk, and if after some resistance he retreats, we can say that at this point we have beaten him. But, if we have lost proportionately as many men in the process as he did, no trace of this so-called victory will show up in the final balance-sheet of the campaign. Getting the better of an enemy—that is, placing him in a position where he has to break off the engagement—cannot in itself be considered as an objective, and for this reason cannot be included in the definition of the objective. Nothing remains, therefore, but the direct profit gained in the process of destruction. This gain includes not merely the casualties inflicted during the action, but also those which occur as a direct result of his retreat.

It is a familiar experience that the winner's casualties in the course of an engagement show little difference from the loser's. Frequently there is no difference at all, and sometimes even an inverse one. The really crippling losses, those the vanquished does not share with the victor, only start with his retreat. The feeble remnants of badly shaken battalions are cut down by cavalry; exhausted men fall by the wayside; damaged guns and caissons are abandoned, while others are unable to get away quickly enough on poor roads and are taken by the enemy's cavalry; small detachments get lost in the night and fall defenseless into the enemy's hands. Thus a victory usually only starts to gather weight after the issue has already been decided. This would be a paradox, if it were not resolved as follows.

Physical casualties are not the only losses incurred by both sides in the course of the engagement: their moral strength is also shaken, broken and ruined. In deciding whether or not to continue the engagement it is not enough to consider the loss of men, horses and guns; one also has to weigh the loss of order, courage, confidence, cohesion, and plan. The decision rests chiefly on the state of morale, which, in cases where the victor has lost as much as the vanquished, has always been the single decisive factor.

The ratio of physical loss on either side is in any case hard to gauge in the course of an engagement; but this does not apply to loss of morale. There are two main indicators of this. One is loss of the ground on which one has fought; the other is the preponderance of enemy reserves. The faster one's own reserves have shrunk in relation to the enemy's, the more it has cost to maintain the balance. That alone is palpable proof of the enemy's superior morale, and it seldom fails to cause some bitterness in a general—a certain loss of respect for the forces he commands. But the main point is that soldiers, after fighting for some time, are apt to be like burned-out cinders. They have shot off their ammunition, their numbers have been diminished, their strength and their morale are drained, and possibly their courage has vanished as well. As an organic whole, quite apart from their loss in numbers, they are far from being what they were before the action; and thus the amount of reserves spent is an accurate measure on the loss of morale.

As a rule, then, loss of ground and lack of fresh reserves are the two main reasons for retreat. There may, however, be others, which we do not wish to exclude or minimize, having to do with the interdependence of the parts or with the overall plan.

Every engagement is a bloody and destructive test of physical and moral strength. Whoever has the greater sum of both left at the end is the victor.

In the engagement, the loss of morale has proved the major decisive factor. Once the outcome has been determined, the loss continues to increase, and reaches its peak only at the end of the action. This becomes the means of achieving the margin of profit in the destruction of the enemy's physical forces which is the real purpose of the engagement. Loss of order and cohesion often makes even the resistance of individual units fatal for them. The spirit of the whole is broken; nothing is left of the original obsession with triumph or disaster that made men ignore all risks; for most of them danger is no longer a challenge to their courage, but harsh punishment to be endured. Thus the tool is weakened and blunted at the first impact of the enemy's victory, and is no longer suitable for countering danger with danger.

This is the time for the victor to consolidate his gains by physical destruction—the only advantage that will be permanently his. The enemy's morale will gradually recover, order will be restored, his courage will return; and in most cases only a very small portion, if any, of the hard-earned superiority will remain. In some, admittedly rare, instances a thirst for revenge and an increased surge of animosity may even produce the opposite effect. But the advantages gained by inflicting casualties, in dead, wounded, prisoners, and captured material, can never disappear from the ledger.

Losses incurred during the battle consist mostly of dead and wounded; after the battle, they are usually greater in terms of captured guns and prisoners. While the former are shared more or less evenly by winner and loser, the latter are not. For that reason they are usually only found on one side, or at any rate in significant numbers on one side.

That is why guns and prisoners have always counted as the real trophies of victory: they are also its measure, for they are tangible evidence of its scale. They are a better index to the degree of superior morale than any other factors, even when one relates them to the casualty figures. And through this, the factor of morale makes itself felt in yet another way.

We have pointed out that morale destroyed during the engagement and its immediate aftermath is gradually restored and often shows no trace of its disruption. While this is the case with smaller parts of the whole, it also happens with larger segments. It can even be the case with the entire army; but rarely, if ever, can it be true of the state and the government that the army serves. At that level, things are seen more objectively and from a higher standpoint. The degree of one's own weakness and inadequacy shows up only too clearly in the amount of trophies taken by the enemy and their ratio to the casualties sustained.

All in all, loss of moral equilibrium must not be underestimated merely because it has no absolute value and does not always show up in the final balance. It can attain such massive proportions that it overpowers everything by its irresistible force. For this reason it may in itself become a main objective of the action, which we shall discuss elsewhere. Here we have still to consider some of its other basic features.

The psychological effect of a victory does not merely grow in proportion to the amount of the military forces involved, but does so at an accelerating rate. This is because the increase is one not merely of size but of intensity. In a defeated division, order is easily restored. As the body's warmth restores the circulation in a limb numb with cold, a division's spirit is quickly revived by the spirit of the army when the two join up again. Thus even if the effects of a minor victory do not disappear altogether, they are partially lost to the enemy. But this is not the case where the army itself has suffered a disastrous defeat: everything collapses together. One large fire is a great deal hotter than several small ones.

Another factor to be considered in determining the psychological value of a victory is the ratio of the opposing forces. If a small force beats a larger one, its gain is not only doubled but it shows a greater margin of general superiority, which the loser knows he may have to face again and again. Actually, however, this effect is *hardly noticeable* in such a case. At the moment of battle, information about the strength of the enemy is usually uncertain, and the estimate of one's own is usually unrealistic. The stronger party either simply refuses to admit the disproportion, or at least will underrate it, and is therefore to a large extent protected from the psychological disadvantage to which it would give rise. The actual facts, which have been suppressed by ignorance, vanity, or even deliberate prudence, will emerge only much later, when history is written. History by that time will probably glorify the army and its commander, but its contribution to morale will no longer be of any help in a situation that has long since passed.

If prisoners and captured guns are the objects by which victory is mainly personified, its true crystallization then, the engagement, will most likely be planned so as to obtain them. In this, destruction of the enemy by killing and wounding appears only as a means.

The influence that this choice exerts on tactical deployment is not the concern of strategy. However, it does affect the engagement when it comes to threatening the enemy's rear and protecting one's own. It is upon this that the number of prisoners and captured guns will mainly depend, and tactical measures will not suffice on their own when strategic circumstances are unfavorable.

The risk of having to fight on two fronts, and the even greater risk of finding one's retreat cut off, tend to paralyze movement and the ability to resist, and so affect the balance between victory and defeat. What is more, in the case of defeat, they increase the losses and can raise them to their very limit—to annihilation. A threat to the rear can, therefore, make a defeat *more probable*, as well as *more decisive*.

Out of this then arises an instinctive determination in the conduct of war and particularly in engagements, large and small, to protect one's own rear and to gain control of the enemy's. The instinct is derived from the concept of victory itself, which as we have shown, is more than mere killing.

This determination constitutes an *immediate purpose* of battle and a universal one. No engagement is conceivable in which it would not accompany, in one or both its forms, the naked application of force. Not even the smallest unit will attack the enemy without thinking of its line of retreat, and usually it will seek out the enemy's as well.

It would go too far to discuss how easily a complex situation can deflect this instinct from its natural course, how often it has to yield to weightier considerations. We shall content ourselves by stating it as a general law of the nature of engagements. It must be considered universally valid; its natural pressure is ubiquitous and thus it becomes the point around which almost all tactical and strategic moves revolve.

If in conclusion we consider the total concept of a victory, we find that it consists of three elements:

1. The enemy's greater loss of material strength
2. His loss of morale
3. His open admission of the above by giving up his intentions.

Casualty reports on either side are never accurate, seldom truthful, and in most cases deliberately falsified. Even the number of trophies is usually unreliably reported; thus, where they are not considerable, they may leave the victory in doubt as well. Trophies apart, there is no accurate measure of loss of morale; hence in many cases the abandonment of the fight remains the only authentic proof of victory. In lowering one's colors one acknowledges that one has been at fault and concedes in this instance that both might and right lie with the opponent. This shame and humiliation, which must be distinguished from all other psychological consequences of the transformation of the balance, is an essential part of victory. It is the only element that affects public opinion outside the army; that impresses the people and the governments of the two belligerents and of their allies. Granted, abandoning an intention is not the same as abandoning the battlefield, even after prolonged and stubborn fighting. An outpost may retire after a stubborn resistance without being accused of abandoning its task. Even in engagements intended to destroy the enemy, withdrawal from the battlefield does not always imply that the aim has been abandoned, as for instance in planned retreats in which it is intended to dispute every foot of ground. All this will be discussed later under the heading of the particular purposes of engagements. For the present, we only wish to draw attention to the fact that in the majority of cases it is difficult to distinguish between the abandonment of intentions and the abandonment of the battlefield; the impression produced by the former, both in military and civilian circles, should not be underrated.

For generals and armies without an established reputation, this is a difficult aspect of otherwise sound operations. A series of engagements followed by retreats may appear to be a series of reverses. This may be quite untrue, but it can make a very bad impression. It is not possible for a general in retreat to forestall this moral effect by making his true intentions known. To do so effectively, he would have to disclose his overall plan of action, and that would be contrary to his main interests.

To demonstrate the exceptional importance of this concept of victory, we would recall the battle of Soor, in which the trophies taken were insignificant (a few thousand prisoners and twenty guns). Frederick the Great proclaimed his victory by remaining on the battlefield for five more days, although his retreat toward Silesia already had been decided on and was demanded by the general situation. As he said himself, he counted on the psychological impact of the victory to bring him nearer peace. Even though a few more victories (such as the engagement of Katholisch-Hennersdorf in Lusatia and the battle of Kesselsdorf) were needed to establish this peace, one cannot say that the battle of Soor had no moral effect.

If a victory has primarily shaken the opponent's confidence, and thus increased the number of trophies to an unusual degree, then the lost engagement turns into defeat on a scale not produced by every victory. Since in this type of débacle the morale of the defeated is affected to a far higher degree, a total inability to offer resistance is frequently the result, and action will now consist in evasion—that is to say, flight.

Jena and Belle-Alliance[1] were defeats on this scale; Borodino was not. Since the difference is merely a matter of degree, it would be pedantic to draw an arbitrary line. Still, for the clear distinction of theoretical ideas, it is essential to maintain certain concepts as focal points. It is in fact a weakness in our terminology that in the case of a major defeat we do not have a *single* word to designate the victory that corresponds to it, or in the case of a less far-reaching victory a word to designate the corresponding degree of defeat.

[1] Waterloo. Eds.

CHAPTER FIVE

The Significance of the Engagement

In the previous chapter we discussed the engagement in its absolute form, as though it were a microcosm of war as a whole. We now turn to the relationship that it bears as one part to the other parts of a greater whole. We begin by inquiring into the precise significance that an engagement may possess.

Since war is nothing but mutual destruction, it would seem most natural to conceive, and it is possibly also most natural in fact, that all the forces on each side should unite in one great mass, and all successes should consist of one great thrust of these forces. There is much to be said for this idea and on the whole it would be salutary to adhere to it and, to begin with, to consider the smaller engagements as necessary by-products, like wood-shavings. However, the matter is never disposed of so easily as this.

The multiplication of engagements obviously results from the splitting up of forces, and we shall therefore deal with the specific purposes of individual engagements in that context. These purposes, and with them the whole range of engagements, can be classified; and a study of these classifications will help to elucidate our discussion.

The destruction of the enemy's forces is admittedly the purpose of all engagements. But, other purposes may well be linked to this and may even predominate. A distinction must therefore be made between a case in which destruction of enemy forces is the main consideration and one in which it is more of a means. Apart from the destruction of the enemy's forces, the conquest of a locality or of a physical object may also be a general motive, either by itself or in conjunction with other motives, in which case one motive will usually predominate. The two main forms of war, attack and defense, which we shall discuss shortly while not affecting the first of these objectives, do affect the other two. A chart would show the following:

Offensive Engagement	*Defensive Engagement*
1. Destruction of the enemy's forces	1. Destruction of the enemy's forces
2. Conquest of a locality	2. Defense of a locality
3. Conquest of an object	3. Defense of an object

But, these objectives do not cover the whole ground if we think of reconnaissance and demonstrations, in which obviously none of the above fits the purpose of the action. Indeed, a fourth category ought to be allowed. Strictly speaking, reconnaissances aimed at making the enemy show himself, feints designed to wear him out, demonstrations meant to pin him down at one place or to draw him off toward another are aims that can be achieved only indirectly and *on the pretext of one of the three objectives named above* (usually the second); for the enemy who wants to reconnoiter must act as if he were about to attack and defeat or dislodge our forces, and so forth. But such pretexts are not the real aim, which is what concerns us here. Therefore, to the three objectives of the attacker we must add a fourth: that of misleading the enemy—in other words, putting up a sham fight. The very nature of the matter makes it obvious that this object is conceivable only in the context of attack.

On the other hand, we must observe that the defense of a locality may be of two kinds: either absolute, if the locality is not to be given up at all, or relative, if it must be held only for a certain time. The latter form constantly recurs in engagements fought by advance posts and rear guards.

There is probably no need to stress that the differing purposes of an engagement affect the preparations that are made for it. We make one plan for dislodging an enemy post and another for annihilating it; one plan for holding a locality at all costs and another for merely delaying the enemy. In the former case, there is little need to worry about the retreat; in the latter, the retreat is of paramount importance; and so on.

These reflections come under the heading of tactics, and are here cited only as examples. How the various aims of an engagement look from the strategic angle will be studied in the chapters that deal with them. For the present, we will confine ourselves to a few general observations.

To begin with, the relative importance of aims diminishes roughly in the order of the foregoing table. The first of them should always predominate in a major battle. Finally, the last two in the defensive engagement are of a kind that do not really bring results: they are completely negative, and their value can only be the indirect one of making some positive purpose elsewhere easier to achieve. *If engagements of this type become too frequent, it obviously indicates an unfavorable strategic situation.*

CHAPTER SIX

Duration of the Engagement

If we turn from a discussion of the engagement itself to its relation to other factors in war, its duration acquires special importance.

In a sense, the duration of an engagement can be interpreted as a separate, secondary success. The decision can never be reached too soon to suit the winner or delayed long enough to suit the loser. A victory is greater for having been gained quickly; defeat is compensated for by having been long postponed.

This is true in general. It assumes practical importance in engagements whose object is a delaying action.[1]

In such a case, the whole success often consists in nothing but the time the action takes. That is why we include duration in the spectrum of strategic elements.

The duration of an engagement and the broad conditions under which it is fought are necessarily connected. These conditions are the size of the force, its relation in men and material to the opponent, and the character of the terrain. Twenty thousand men will not wear each other down so rapidly as will two thousand. An enemy with a two- or threefold superiority cannot be resisted for as long as one of equal strength. A cavalry engagement is decided faster than an infantry engagement, and this in turn is decided faster than an action in which artillery is involved. One cannot make progress so rapidly in mountains and forests as on a plain. All this is obvious enough.

It follows that the strength, composition, and deployment of the two sides must all be taken into account if the objective of the engagement lies in its duration. But it is less important to state this rule than to show its connection with its principal results, which we know about from experience.

The resistance of a normal division of eight to ten thousand men of all arms, even against a significantly superior enemy and on not very favorable terrain, lasts for several hours; if the enemy is only slightly superior, if at all, it may last half a day. A corps of three or four divisions can hold out for twice that time, and an army of eighty to one hundred thousand men for three or four times as long. For that length of time, then, these forces may be left to their own resources. And if within this time additional forces are brought to bear, no second engagement takes place, but their effectiveness

[1] *Relative Vertheidigung*, literally "relative defense," obviously means "delaying action" here. Eds.

quickly blends with the success of the original engagement into a single whole.

We have taken these figures from actual experience. We must now define more closely the moment of decision and, consequently, termination of the engagement.

CHAPTER SEVEN

Decision of the Engagement

No engagement is decided in a single moment, although in each there are crucial moments which are primarily responsible for the outcome. Losing an engagement is therefore like the gradual sinking of a scale. But each engagement reaches a point when it may be regarded as decided, so that to reopen it would constitute a new engagement rather than the continuation of the old one. The accurate perception of that point is very important in order to decide whether reinforcements would be profitably employed in renewing the action.

New troops are often vainly sacrificed in an engagement that is past retrieving; and the chance of reversing a decision is often missed while it could still be done. Let us cite two examples that can hardly be more striking.

At Jena in 1806 Prince Hohenlohe, with 35,000 men, accepted battle against Bonaparte's 60 to 70,000. He lost so badly that his whole force was virtually annihilated. At that point, General Rüchel, with about 12,000 men, decided to reopen the action, with the result that his force too was demolished on the spot.

The same day at Auerstädt saw 25,000 men fight until noon against Davout's 28,000. Admittedly the force was not successful, but neither was it in a state of dissolution, and suffered no greater losses than the enemy, who had no cavalry. General Kalckreuth's reserve, 18,000 men strong, was not used to turn the tide of battle; if it had been, a Prussian defeat would have been impossible.

Every engagement is a whole, made up of subsidiary engagements that add up to the overall result. The decision of the engagement consists in this total result. The result need not be the type of victory described in Chapter Six. Often no preparations have been made for one, or no opportunity arises because the enemy gives way too soon. Even after stubborn resistance, the decision is normally reached sooner than is the kind of success generally associated with the idea of victory.

We may, therefore, ask what normally constitutes this moment of decision, this point of no return at which fresh (though of course not disproportionate) forces will be too late to save the day?

Excluding feints, which by their very nature do not lead to a decision, we arrive at the following answers:

1. Where the purpose of the engagement is the possession of some mobile object, the decisive moment is reached when this object is lost.

2. Where the purpose of the engagement is the possession of a certain locality, the decisive moment is usually though not invariably reached, when this locality is lost. This holds true only if the locality is of great defensive strength; terrain that is easily overrun, no matter how great its importance in other respects, can be recaptured without great difficulty.

3. In all cases in which the above conditions have not already led to a decision, and in particular where the main objective is the destruction of the enemy's forces, the moment of decision comes when the victor ceases to be in a state of disarray and thus to some extent ineffective; in other words, when the successive application of force as discussed in Chapter 12 of Book III, is no longer advantageous. That is why we designate this point as central to the strategic unity of the engagement.

Thus, an engagement cannot be retrieved if the attacking force has lost little if any of its cohesion and effectiveness, or if it has recovered from a temporary loss of effectiveness, while the defender has become more or less disorganized.

The smaller the proportion of troops in actual combat, and the larger the proportion that contributed to victory merely by being present as a reserve, the less is a new enemy force likely to deprive us of victory. The commander and the army who have come closest to conducting an engagement with the utmost economy of force and the maximum psychological effect of strong reserves are on the surest road to victory. In modern times the French must be credited with great mastery in this respect, particularly under the leadership of Bonaparte.

Further the smaller the total force, the sooner will the victor master the crisis and recover his former effectiveness. A cavalry picket chasing the enemy posthaste can regain its proper order in a few minutes, and that is the extent of the whole crisis. A whole cavalry regiment will need more time. Even more time is required by infantry after it has been deployed in skirmishing order. Divisions of all arms, which have become dispersed, need more time still; the engagement has caused a disruption of order aggravated by the fact that no part knows exactly where any other is. Thus the larger the total effort involved, the longer is the moment delayed when the victor is able to retrieve, repair, and rearrange his used tools from their state of untidy confusion, and restore order in the workshop of combat.

An additional delaying factor will arise if night falls while the victor is still in the critical phase; another, if the terrain is rough and wooded. On the other hand, it must be remembered in connection with these factors that night is a great *source of protection*. Conditions rarely provide a night attack with much promise of success. An excellent case in point is York's attack on Marmont at Laon on 10 March 1814. Like night, rough and wooded country can shield an army from counterattack while victory is still in the critical phase. Both of these factors—night, and rough and wooded country—therefore make it harder rather than easier to renew the fight.

So far we have treated the rapid reinforcement of the losing side as a simple addition to its strength, with support coming up from the rear, which is what normally happens. But an entirely different situation arises when the reinforcements attack the enemy's flank or rear.

The effectiveness of flank or rear attacks is a subject we shall deal with later from the point of view of strategy. The kind of attack we are discussing now, designed to turn the tide of battle, is in the main a tactical affair; we shall discuss it at this point only because we are now concerned with tactical results, and the subject therefore does overlap into the field of tactics.

The impact of a force may be substantially heightened if it is directed at the enemy's flank or rear. But this is not invariably the case: the impact may just as easily be weakened. The conditions under which the engagement is being fought will dictate its plan in this as in all other respects, but this is not the place to go into detail. For our present purpose, there are two important considerations.

First, flank and rear attacks as a rule affect the consequences of the outcome more favorably than they do the decision itself. When an engagement has to be retrieved, obviously the prime consideration is its favorable conclusion rather than the magnitude of the victory. In this respect, one would think, therefore, that reinforcements brought up to save the situation would be less effective if they attacked the opponent's flank and rear, operating separately, than if they united with us. There is certainly no lack of examples for this; but we maintain that because of a second consideration the opposite will usually be true.

This second point is the psychological impact of surprise, which as a rule accompanies the appearance of reinforcements sent to restore a situation.

The effect of surprise is always heightened if it takes place in flank or rear; in the critical phase of victory an army is strung out and dispersed, and less capable of dealing with it. At the beginning, while the troops are still concentrated and always prepared for such an eventuality, an attack in flank or rear would carry relatively little weight; during the last moments of an engagement it will mean a great deal more.

Therefore we do not hesitate to state that in most cases reinforcements are much more effective when approaching the enemy from flank and rear, just as a longer handle gives greater leverage. In that way it is possible to restore an engagement with a force that would have been insufficient if used against the front. Here, in operations on the flank or rear, where effectiveness almost defies precise calculation because moral effects become dominant, boldness and daring are given fullest scope.

In marginal situations all these factors must be considered, and their combined effect assessed, if we wish to decide whether anything can still be done to retrieve an engagement that is going badly.

If the original engagement is not considered ended, then a new one, opened by the arrival of reinforcements, will merge with it and lead to a combined result. The initial loss would then be entirely erased. The situation is different if the original engagement has already been decided: in that case there will be two distinct results. When the reinforcements are only of moderate strength—in other words not up to that of the enemy—prospects for the second round are hardly bright. But if they are strong enough to fight the second engagement irrespective of the first, a favorable outcome may possibly compensate or outweigh the initial defeat, though it will never be able to cancel its effects completely.

At the battle of Kunersdorf, Frederick the Great overran the Russians' left wing at the first assault and captured seventy guns. By the time the battle ended, both gains had been lost again, and the whole result of the first attack had vanished from the record. Had the King been able to stand on his first success and delay the second round until the following day, the first day's gains would still have been a compensation even if he had lost the second part of the battle.

If a losing battle can be caught before its conclusion and turned into a success, the initial loss not only disappears from the record, but becomes the basis for a greater victory. For on closely examining the tactical progress of an engagement it becomes obvious that, up to its very end, the results of each of the subsidiary engagements are only suspended verdicts, which not only may be revoked by the final outcome, but may be turned into their very opposites. The more our own forces have suffered, the more exhausted the enemy will be. His own crisis is likely to be that much greater and the superiority of our fresh forces will weigh all the more. If the final result turns out to be in our favor, if we manage to recapture battlefield and trophies from the enemy, then all the forces that these have cost him will turn out to the credit of our account; our earlier defeat becomes a stepping-stone to greater triumph. The most brilliant military exploits, which in victory would have meant so much to the enemy that he could have ignored their cost, now leave him with nothing but remorse over the strength thus sacrificed. The magic of victory and the curse of defeat can change the specific gravity of the elements of battle.

Even in a situation where one is decisively stronger than the enemy, and could easily requite his victory with a greater one, it is better to retrieve a losing battle (provided it is sufficiently important) before its close, rather than fight a second engagement later on.

At Liegnitz in 1760, Field Marshal Daun tried to go to General Laudon's help while the latter was still in action; but after the battle had been lost he did not attempt to attack Frederick, though that was well within his powers.

Fierce advance guard engagements preceding a battle, therefore, should be considered merely as a necessary evil, and avoided wherever they are not essential.

There is yet another deduction to be examined.

The outcome of a lost battle must not be taken as an argument for deciding on a new one; rather, any such decision must be based on the rest of the circumstances. That precept, however, is counteracted by a psychological factor that must be reckoned with: the instinct for retaliation and revenge. It is a universal instinct, shared by the supreme commander and the youngest drummer boy; the morale of troops is never higher than when it comes to repaying that kind of debt. All this of course presupposes that the defeated troops do not constitute too large a portion of the whole; if they do, such a feeling would be overpowered by that of impotence.

There is thus a natural propensity to exploit this psychological factor in order to recapture what has been lost by seeking a new engagement, particularly if the rest of the situation warrants it. The very nature of such a second engagement decrees that in most cases it should be an attack.

The history of minor engagements will show numerous examples of this kind of retribution; major battles, on the other hand, usually spring from too many other causes to be based on such a relatively trivial motive.

Undoubtedly, however, it was the desire for revenge which led the noble Blücher, on 14 February 1814, to return to the field with a third force after two of his corps had been beaten at Montmirail three days earlier. Had he known it would bring him face to face with Bonaparte himself, he would surely have felt justified in saving his vengeance for another day; but he expected to avenge himself on Marmont. Far from garnering the fruits of a noble desire for retaliation, he had to pay the price for his miscalculation.

The distance between units that have to *coordinate* their action depends on the duration of the engagement and the moment of its decision. If a single engagement is intended, their deployment will be a matter of tactics; but it can be considered as such only where the forces are so close as to exclude the possibility of two separate engagements—in other words, where the area within which all of the action takes place may be strategically considered as a single point. It does frequently happen in war, however, that forces meant to fight in concert have to be placed so far apart that, while their conjunction in battle remains the primary intention, the possibility of seperate action has also to be considered. Such a deployment is therefore strategic.

Dispositions of this type include marches by separate columns and divisions, advance guards and flanking corps, reserves intended to support more than one strategic point, the concentration of individual corps from widely spread billets, and so forth. It is obvious that this is a constantly recurring type of operation—the small change, so to speak, of the strategic budget, while important battles and other operations comparable in scale may be considered its gold and silver.

CHAPTER EIGHT

Mutual Agreement to Fight

There can be no engagement unless both sides are willing. That concept, which is fundamental to the duel, is responsible for a terminology used by military historians that often results in vague and misleading notions.

The discussions of these writers frequently turn on the point that one commander has offered battle and that the other has refused to accept it.

An engagement is however a very peculiar form of duel. Its basis does not consist only in mutual desire or willingness to fight, but in the purposes involved; and those always belong to a larger whole—the more so because the war itself, considered as a single conflict, is governed by political aims and conditions that themselves belong to a larger whole. As a result, the mutual desire for victory assumes a minor role; rather, it ceases to be independent, and has to be regarded as no more than the nerve which enables the higher political will to act.

In ancient times, and again more recently when standing armies first came into being, the expression "vainly offering battle to the enemy" had more meaning than it has today. In the ancient world a battle meant a trial of strength on open ground, free of all obstacles. The art of war consisted wholly in organization and formation—in other words, in the order of battle.

In those days armies were generally so well entrenched in their camps that these positions were thought to be impregnable. A battle became possible only after the enemy left camp and entered the lists, so to speak, on accessible terrain.

Hence, when we read that Hannibal vainly offered battle to Fabius, all we learn about Fabius is that a battle did not figure in his plan. This does not prove either the material or moral superiority of Hannibal. Yet the expression is correct so far as the latter is concerned: Hannibal genuinely wanted to fight.

In the early days of modern armies, conditions governing major engagements and battles were similar. Thus large masses of troops were deployed in action and directed in accordance with a set order of battle. This was a large, unwieldy arrangement requiring fairly even ground: in rough or wooded, not to speak of mountainous areas the system was unsuitable for attack or even for defense. To some degree, then, the defending side was able to find ways of avoiding battle. These conditions lasted, though steadily to a lesser extent, until the first Silesian Wars. Only in the Seven Years War did an attack in *difficult* terrain become feasible and customary. While terrain was still an asset to those who chose to utilize it, it had ceased to be a magic circle that was regarded as out of bounds to the natural forces of war.

In the course of the last three decades, war has evolved a great deal further in this respect. Today there is nothing to prevent a commander bent on a decisive battle from seeking out his enemy and attacking him. If he does not, he cannot be considered to have wanted the engagement; today if he says that he has offered battle but that the enemy refused it, it merely means that he did not think conditions favorable for an engagement. It is an admission on his part to which that expression does not apply; he uses it only as an excuse.

It is true that while the defender cannot nowadays decline an engagement, he can avoid it by abandoning his position, and thereby his object in holding it. But this kind of success already constitutes the better part of victory for the attacker—the recognition of his provisional superiority.

It is therefore no longer possible to talk of "a challenge refused" (which implies a tacit agreement between opponents) in order to justify inertia on the side whose move it is—that of the attacker. The defender, on the other hand, must be considered as wanting battle so long as he does not retreat. For his part, he may claim to have offered battle if he is not attacked; but that will be considered self-evident.

The commander who wishes to *retreat* and is able to do so can hardly be forced into battle by his opponent. Frequently, however, the attacker is not content with the advantages provided by such a retreat and feels the need for an actual victory. In such a case remarkable skill is often used to find and apply the few available means of *forcing* even an evasive opponent to stand and fight.

There are two principal ways of accomplishing this: first, to *surround* the enemy and cut off his retreat, or make it so difficult that battle seems to him the lesser evil; and second, *to take him by surprise*. The advantages of the latter were in the past based on the difficulty of all movement, but surprise has lost its usefulness today. Modern armies are so flexible and mobile that a general will not shrink from retreat even in full view of the enemy. Real difficulties will arise only for forces operating in exceptionally disadvantageous terrain.

A case in point would be the battle of Neresheim. On 11 August 1796 Archduke Charles fought Moreau in the Rauhe Alp for no other reason than to facilitate his own retreat. We must confess, however, that in this instance we have never entirely understood the reasoning of the famous general and writer.

The battle of Rossbach is another example, if indeed the commander of the allied armies never really intended to attack Frederick the Great.

As for the battle of Soor, Frederick himself said he accepted battle only because a retreat in full view of the enemy seemed a risky operation. But the King has also cited other reasons for the battle.

Except for actual night attacks, such cases will, generally speaking, always be very rare; and those in which the enemy is forced to fight because he has been surrounded will usually involve only isolated corps, such as Fink's at the battle of Maxen.

CHAPTER NINE

The Battle: Its Decision

What is the battle? It is a struggle by the main force—but not just an insignificant action fought for secondary objectives, not simply an attempt to be abandoned if one realizes early enough that its object is difficult to attain: it is a struggle for real victory, waged with all available strength.

Secondary objectives may combine with the principal one even in a battle, and the battle itself will be colored by the circumstances that gave rise to it. Even a battle is connected with a still larger entity of which it is only a part. But since the essence of war is fighting, and since the battle is the fight of the main force, the battle must always be considered as the true center of gravity of the war. All in all, therefore, its distinguishing feature is that, more than any other type of action, battle exists for its own sake alone.

This has a bearing on the *manner of its decision*, and on the *effect of the victory that is gained*, and determines *the value that theory must ascribe to the battle as a means to the end*. That is why, at this point, we are making it a subject of special study. Later we will discuss the particular ends that may also be involved, but which leave its character—assuming it deserves the name of "Battle"—essentially unchanged.

If a battle is primarily an end in itself, the elements of its decision must be contained in it. In other words, victory must be pursued so long as it lies within the realm of the possible; battle must never be abandoned because of particular circumstances, but only when the strength available has quite clearly become inadequate.

How can this stage be more accurately determined?

Where some degree of complex integration and deployment of the army is the principal condition under which the courage of the troops can gain a victory (as was the case for a considerable time in modern warfare), then the *destruction of this line of battle* is itself the decision. A wing overrun and driven out of line decides the fate of the flank that has held fast. If, as in an earlier period, the essence of defense lies in the close integration of troops with the terrain and its obstacles, so that the army and the position are one, then the capture of an *essential point* of the position brings about the decision. We say that the key to the position is lost; it can no longer be defended, the battle cannot be continued. In both cases the defeated armies may be compared to the broken strings of an instrument that no longer function.

The geometrical as well as the topographical principles discussed here tended to keep the armies in a state of rigid tension which prevented them from using their strength down to the last man. These principles have now lost so much of their influence that they are no longer dominant. An army still goes into battle in a certain order, but this order is no longer decisive. Defense is still improved by exploiting the accidents of terrain, but it no longer relies on these alone.

In the second chapter of this book we attempted to give a general idea of the character of a modern battle. This description pictures the order of battle simply as a disposition of troops designed to facilitate their use, and the course of battle as a slow process of mutual attrition that will reveal which side can first exhaust its opponent.

In a major battle more than any other type of engagement, the decision to give up the fight depends on the relative strength of unused reserves still available. They are the forces whose morale is still intact; mauled and battered battalions—dying embers left by the furnace of destruction—cannot be compared to them. Lost ground, as we have pointed out, is also an index of impaired morale. It must also be taken into account, though more as an indicator of losses suffered than as a loss in itself. The main concern of both commanders will always be the number of reserves available on both sides.

Usually a battle takes shape from the start, though not in any obvious manner. Often this shape has already been decisively determined by the preliminary dispositions made for the battle, and then it shows lack of insight in the commander who opens the engagement under these unfavorable conditions without being aware of them. Even if the course of battle is not predetermined, it is in the nature of things that it consists in a slowly shifting balance, which starts early, but, as we have said, is not easily detectable. As time goes on, it gathers momentum and becomes more obvious. It is less a matter of oscillating to and fro, as fanciful accounts of combat have misled many people into thinking.

But whether the equilibrium remains undisturbed for some time, or whether it swings to one side, rights itself, and then swings to the other, it is certain that a commander usually knows that he is losing the battle long before he orders retreat. Battles in which one unexpected factor has a major effect on the course of the whole usually exist only in the stories told by people who want to explain away their defeats.

On this subject we can only appeal to impartial and experienced soldiers who we are sure will confirm our argument for those of our readers who have no personal experience of war. A thorough analysis of the process would take us too far into the realm of tactics, where it really belongs. What matters at this point is only the outcome.

While we believe that the defeated commander is usually aware of the likelihood of defeat long before he decides to concede the battle, we also admit that there are contrary cases; otherwise we should be stating a self-contradictory tenet. If a battle were to be considered lost each time it took a definite turn, no additional forces would be committed in the hope of saving it. It follows that such a definite turn could not precede the moment of retreat by any appreciable amount of time. There are certainly cases in which a battle, after taking a definite turn in favor of one side, ended up in favor of the other, but such cases are *not common*; in fact they are unusual. But it is just this exceptional case that every general hopes for when his luck is out; he has to hope for it so long as there is any chance of a turn for the better. He hopes that by dint of greater efforts, by raising whatever morale is left in the troops, by surpassing himself or by sheer good fortune, he will be able to reverse his fortunes just once more, and he will keep at it for as long as his courage and his judgment allow. We shall have more to say about this later, but first we wish to enumerate the signs that indicate a change in equilibrium.

The outcome of the battle as a whole is made up of the results of its constituent engagements; these, in turn, may be recognized by three distinct signs.

The first is the psychological effect exerted by the commanding officer's moral stamina. If a divisional commander sees his battalions being worsted, it will show in his attitude and his reports, and these in turn will affect the commander-in-chief's decisions. Even local setbacks that appear to have been retrieved will therefore count for something in the end. The impressions made on the mind of the commander-in-chief will easily accumulate even against his better judgment.

The second is a wasting away of one's own troops at a rate faster than that of the enemy's. This may be estimated quite accurately, since the tempo of our battles is deliberate and seldom very tumultuous.

The third is the amount of ground lost.

All these indicators serve as a kind of compass by which a commander can tell the direction in which his battle is going. The loss of entire batteries while none are captured from the enemy; the crushing of his battalions by the enemy's cavalry while the enemy's own battalions remain impenetrable; the involuntary retreat of his firing line from point to point; futile efforts to capture certain positions, which end in the scattering of the assault troops by well-aimed grape and case-shot; a weakening of the rate of fire of his guns as opposed to the enemy's; an abnormally rapid thinning out of his battalions under fire caused by groups of able-bodied men accompanying the wounded to the rear; units cut off and captured because the battle line is disrupted; evidence of the line of retreat being imperiled: all this will indicate to a commander where he and his battle are heading. The longer they are headed in that direction, the more definite the movement becomes, the more difficult it will be to effect a change, and the closer comes the time when the battle has to be conceded. That is the moment we shall now consider.

We have explicitly stated more than once that as a rule the final outcome turns on the ratio of unused reserves still available. A commander who recognizes his enemy's distinct superiority in reserves will decide to retreat. It is a peculiarity of modern battles that all the mishaps and losses sustained in its course can be retrieved by fresh troops. The reason lies in the modern order of battle and the way in which troops are brought into action, permitting the use of reserves almost everywhere and in any situation. Therefore so long as a commander has more reserves than his enemy, he will not give up even though the battle shows signs of going badly. But once his reserves start to become weaker than the enemy's, the end is a foregone conclusion. His remaining moves depend partly on the circumstances, and partly on the degree of the commander's personal courage and endurance, which may well deteriorate into unwise obstinacy. Just how a general arrives at a correct estimate of the ratio of reserves on each side is a matter of skill and experience, and does not concern us here. What does concern us is the result as it emerges from his thinking. Even this is not yet the real moment of decision: an answer that emerges only gradually is not the proper catalyst for that; it can not do more than broadly influence the ultimate decision which will in turn be triggered off by immediate considerations. Of these there are two that recur constantly: a threat to the line of retreat and the approach of night.

If every new turn in the course of the battle implies a growing threat to the line of retreat, and if reserves have been reduced to the point at which they can no longer relieve the pressure, there is no other solution than to submit to fate, and to save, by means of an orderly retreat, all that would be lost by further delay and scattered through flight and defeat.

Night is a different matter in that it normally puts a stop to all engagements, for special conditions must obtain to justify night operations. Since for a retreat darkness is more advantageous than daylight, a commander who feels that a retreat is unavoidable, or at least very likely, will prefer to use night for this purpose.

It goes without saying that, apart from these principal and most common factors, there may be many others which are less important, more individual, and less predictable. The more the balance of a battle threatens to be upset, the more sensitive it will be to anything that happens to any of its constituent parts. The loss of a single battery, or dispersal by a cavalry charge, may serve to confirm a general's already partially formed decision to retreat.

A final word on this subject must deal with the point at which a general's courage and his better judgment come, so to speak, into conflict with one another.

On the one hand, there is the domineering pride of a victorious conqueror, the inflexible determination that goes with innate obstinacy, and the desperate resistance of noble enthusiasm, all of which refuse to abandon a field of battle where honor is involved. On the other, there is the voice of reason counseling against spending all one has, against gambling away one's last resources, and in favor of retaining whatever is necessary for an orderly retreat. No matter how highly rated the qualities of courage and steadfastness may be in war, no matter how small the chance of victory may be for the leader who hesitates to go for it with all the power at his disposal, there is a point beyond which persistence becomes desperate folly, and can therefore never be condoned. In that most famous of all battles, Belle-Alliance, Bonaparte staked his last remaining strength on an effort to retrieve a battle that was beyond retrieving; he spent every last penny, and then fled like a beggar from the battlefield and the Empire.

CHAPTER TEN

The Battle—Continued: The Effects of Victory

Depending on one's point of view, one may marvel just as much at the remarkable results of some victories as at the lack of results of others. Let us take a moment to consider the nature of the effect that a major victory may have.

Three things are easily distinguished here: the effect upon the instruments themselves—the generals and their armies; the effect on the belligerent states; and the actual influence that those effects can have on the future course of the war.

If one considers only the insignificant difference between the winning and the losing side that exists on the battlefield in terms of killed and wounded, prisoners and captured arms, the consequences resulting from such an unimportant feature often seem quite inconceivable. Yet as a rule the course of events is all too natural.

As we have already mentioned in Chapter Seven, the scale of a victory does not increase simply at a rate commensurate with the increase in size of the defeated armies, but progressively. The outcome of a major battle has a greater psychological effect on the loser than on the winner. This, in turn, gives rise to additional loss of material strength, which is echoed in loss of morale; the two become mutually interactive as each enhances and intensifies the other. So one must place a special emphasis on the moral effect, which works in opposite directions on each side: while sapping the strength of the loser, it raises the vigor and energy of the winner. But the defeated side is the one most affected by it, since it becomes the direct cause of additional loss. Moreover it is closely related to the dangers, exertions, and hardships—in brief, to all the wear and tear inseparable from war. It merges with these conditions and is nurtured by them.

On the victor's side, however, all these factors only serve to increase the scope of his courage. So what happens is that the loser's scale falls much further below the original line of equilibrium than the winner's scale rises above it. That is why, in considering the effects of a victory, we are particularly interested in those that manifest themselves on the losing side.

These effects are greater after a large-scale action than after a minor one, and greater still after a major battle than after an ancillary engagement. A major battle exists for its own sake, for the sake of the victory it is to bring and which it seeks by means of maximum exertion. A defeat of the enemy in this place and at this time is the intention on which all strands of the war plan converge, uniting all remote hopes and vague concepts concerning the future. Here fate presents us with the answer to our bold question. This is what causes the tension that weighs not only on the commander, but on the whole army, down to the last wagon driver; in diminishing degrees, to be sure, but in diminishing importance as well. A major battle, in all ages and under whatever conditions, has never been fought as an extemporaneous, unexpected, or meaningless discharge of military duty. It is a grandiose event, well above the run of daily life, partly on its own merits and partly because the commander has so planned it in order to raise the general psychological tension. The higher the degree of suspense concerning the outcome of the issue, the greater its effect will be.

The moral effect of victory in battle is even greater today than it was in the earlier wars of the modern period. If modern battle is, as we have depicted it, a fight to the finish, the outcome is decided more by the sum of all strengths, physical as well as moral, than by individual dispositions or mere chance.

It is possible to avoid repeating a mistake, and one can always hope that another day will bring a better deal from luck or chance, but the sum total of physical and moral strength is not so susceptible to rapid change. Therefore the judgment pronounced by a victory seems to be of greater importance for the future. While only a few of those involved in a battle, inside the army or out, will be conscious of that difference, the very course of the battle itself will impress the result on the minds of all who actually took part in it. Public accounts of the battle, even if they are embellished by a few added details, will make it fairly evident to the rest of the world as well that the causes were general rather than particular.

Those who have never been through a serious defeat will naturally find it hard to form a vivid and thus altogether true picture of it: abstract concepts of this or that minor loss will never match the reality of a major defeat. The matter is worth closer examination.

When one is losing, the first thing that strikes one's imagination, and indeed one's intellect, is the melting away of numbers. This is followed by a loss of ground, which almost always happens, and can even happen to the attacker if he is out of luck. Next comes the break-up of the original line of battle, the confusion of units, and the dangers inherent in the retreat, which, with rare exceptions, are always present to some degree. Then comes the retreat itself, usually begun in darkness, or at any rate continued through the night. Once that begins, you have to leave stragglers and a mass of exhausted men behind; among them generally the bravest—those who have ventured out farthest or held out longest. The feeling of having been defeated, which on the field of battle had struck only the senior officers, now runs through the ranks down to the very privates. It is aggravated by the horrible necessity of having to abandon to the enemy so many worthy comrades, whom one had come to appreciate especially in the heat of battle. Worse still is the growing loss of confidence in the high command, which is held more or less responsible by every subordinate for his own wasted efforts. What is worse, the sense of being beaten is not a mere nightmare that may pass; it has become a palpable fact that the enemy is stronger. It is a fact for which the reasons may have lain too deep to be predictable at the outset, but it emerges clearly and convincingly in the end. One may have been aware of it all along, but for the lack of more solid alternatives this awareness was countered by one's trust in chance, good luck, Providence, and in one's own audacity and courage. All this has now turned out to have been insufficient, and one is harshly and inexorably confronted by the terrible truth.

All these impressions are still far removed from panic. An army with spirit will never panic in the face of defeat; even others panic in the wake of a lost battle only in exceptional cases. The impressions themselves are unavoidable in the best of armies. Here and there they may be tempered by long familiarity with war and victory, by solid trust in the high command, but they are never entirely absent at the outset. Moreover, they are not a mere consequence of the loss of trophies; these are normally lost at a later stage, and the fact does not become common knowledge right away. No matter how slowly and gradually the shift in balance occurs, emotions of this kind are certain to appear; they produce an effect on which one may infallibly depend. We have already mentioned that the amount of trophies lost will add to this effect.

The effectiveness of an army in such a state is considerably impaired. In this weakened condition (which, to repeat, is aggravated by all the routine difficulties of war) it can hardly be expected to retrieve its losses by renewed exertions. Before the battle, the two sides were in balance, real or assumed; this balance has now been upset, and an outside cause is needed to restore it. Without such external support, any further exertions will only result in further loss.

It follows then that even a modest victory by the main force is enough to start a steady sinking of the opponent's scale, until a change in external factors produces a new turn of events. Failing that, if the winner presses on in search of greater prizes and greater glory, only an outstanding commander and an army filled with military spirit, steeled and tempered in numerous campaigns, will be able to keep the swollen torrent of power within bounds and to slow its tide by making small but frequent stands until the force of victory has run its course and spent itself.

The effect of all of this outside the army—on the people and on the government—is a sudden collapse of the most anxious expectations, and a complete crushing of self-confidence. This leaves a vacuum that is filled by a corrosively expanding fear which completes the paralysis. It is as if the electric charge of the main battle had sparked a shock to the whole nervous system of one of the contestants. This effect may differ from case to case, but it always exists to some degree. In place of an immediate and determined effort by everyone to hold off further misfortune, there is a general fear that any effort will be useless. Men will hesitate where they should act, or will even dejectedly resign themselves and leave everything to fate.

The consequences of these effects of victory on the future course of the wai depend partly on the character and talent of the victorious commander; but even more on the conditions that gave rise to the victory and on those conditions that victory creates in turn. Unless a commander is bold and enterprising, no great results can be expected from even the most brilliant victory; but it can be rendered ineffectual even more quickly by major adverse circumstances. A Frederick the Great, had he been in Daun's place, would have made a completely different use of the victory at Kolin; and by how much more could France have exploited the battle of Leuthen, if she had been in Prussia's place!

The conditions that entitle one to look for great results from a major victory will be explored in their proper context. At that point we shall be able to explain the disparity which may, at first glance, seem to exist between the dimensions of a victory and those of its consequences. All too often it is blamed on lack of energy on the part of the victor. What concerns us here is only the battle itself. Our argument is that the effects of victory that we have described will always be present; that they increase in proportion to the scale of the victory; and that they increase the more the battle is a major one—that is, the more the army's full strength is committed, the more this strength represents the total military force, and the more the latter represents the whole state.

In that case, is the theorist justified in assuming that these effects of victory must necessarily be accepted as given? Should he not rather try to find an effective way of dealing with them? It seems natural to answer in the affirmative; but Heaven protect us from being misled into this blind alley which has trapped so many theorists, and where the argument becomes self-annihilating.

These effects are indeed quite inevitable, being based on the nature of the case. They must follow even if we find ways of counteracting them, as the movement of a cannonball must continue in the direction of the earth's rotation, even if it loses some of its speed by being fired in the opposite direction—that is, from east to west.

All war presupposes human weakness, and seeks to exploit it.

While, then, at a later stage and in another context, we may discuss what should be done after a major defeat; while we shall review the assets that still remain in an almost hopeless case: and while we shall assume it possible, even in such a situation, that everything can still be put to rights, we certainly do not suggest that the effects of a major defeat can gradually be wiped out altogether. The forces and the means employed to restore the situation could have been used for a positive purpose; and this applies to moral as well as to physical forces.

It is another question whether defeat in a major battle may be instrumental in arousing forces that would otherwise have remained dormant. That is not impossible; it has actually occurred in many countries. But to evoke such an intensified reaction lies outside the limits of the art of war;

only where there is reason to expect it can the strategist take it into consideration.

If there are cases, then, in which the consequences of a victory may actually appear to be injurious because of the reaction aroused—cases that are very rare exceptions indeed—we must be the more ready to recognize the possibility of differences in the consequences of a given victory—here dependent on the character of the people or state defeated.

CHAPTER ELEVEN

The Battle—Continued: The Use of the Battle

No matter how a particular war is conducted and what aspects of its conduct we subsequently recognize as being essential, the very concept of war will permit us to make the following unequivocal statements:

1. Destruction of the enemy forces is the overriding principle of war, and, so far as positive action is concerned, the principal way to achieve our object.
2. Such destruction of forces can *usually* be accomplished only by fighting.
3. Only major engagements involving all forces lead to major success.
4. The greatest successes are obtained where all engagements coalesce into one great battle.
5. Only in a great battle does the commander-in-chief control operations in person; it is only natural that he should prefer to entrust the direction of the battle to himself.

These facts lead to a dual law whose principles support each other: destruction of the enemy's forces is generally accomplished by means of great battles and their results; and, the primary object of great battles must be the destruction of the enemy's forces.

No doubt the principle of destruction is also present to greater or lesser extent in other types of action. Certainly there have been minor engagements (such as Maxen) in which favorable circumstances have resulted in the destruction of a disproportionate number of enemy forces. And on the other hand, the capture or defense of a single position may be so crucial as to dominate a great battle. But in general it remains true that great battles are fought only to destroy the enemy's forces, and that the destruction of these forces can be accomplished only by a major battle.

The major battle is therefore to be regarded as concentrated war, as the center of gravity of the entire conflict or campaign. Just as the focal point of a concave mirror causes the sun's rays to converge into a perfect image and heats them to maximum intensity, so all forces and circumstances of war are united and compressed to maximum effectiveness in the major battle.

The massing of troops into a single whole, which happens to some degree in every campaign, indicates a belligerent's intention to use this mass in a major blow, either on his own initiative (as the attacker) or at the instigation of the other side (as the defender). If major action fails to develop, it is because outside factors have appeared, modifying and restraining the original animosity, and weakening, altering, or halting any movement. Even under general conditions of inactivity—so characteristic of many wars—the possibility of a battle always remains a focus for both sides, a distant aim toward which their courses of action can be directed. The more earnestly a war is waged, the more it is charged with hatred and animosity, and the more it becomes a struggle for mastery on both sides, the more all activity will tend to erupt into bitter fighting, and the greater the importance that will then attach to a great battle.

Wherever a great and positive goal exists, one that will seriously affect the enemy, a great battle is not only the most natural but also the best means of attaining it, as we shall later show in detail. As a rule, shrinking from a major decision by evading such a battle carries its own punishment.

The attacking side is the one that has a positive purpose and is therefore likely to regard the great battle as its own preferred means of action. Without intending at this point to define the concepts of attack and defense in any detail, we must add that even for the defender a battle is the only effective means of sooner or later coming to grips with his situation and solving his problem.

Battle is the bloodiest solution. While it should not simply be considered as mutual murder—its effect, as we shall see in the next chapter, is rather a killing of the enemy's spirit than of his men—it is always true that the character of battle, like its name, is slaughter [*Schlacht*], and its price is blood. As a human being the commander will recoil from it.

But the human spirit recoils even more from the idea of a decision brought about by a single blow. Here all action is compressed into *a single point* in time and space. Under these conditions a man may dimly feel that his powers cannot be developed and brought to bear in so short a period, that much would be gained if he could have more time even if there is no reason to suppose that this would work in his favor. All this is sheer illusion, yet not to be dismissed on that account. The very weakness that assails anyone who has to make an important decision may affect even more strongly a military commander who is called upon to decide a matter of such far-reaching consequences by a single blow.

That is why governments and commanders have always tried to find ways of avoiding a decisive battle and of reaching their goal by other means or of quietly abandoning it. Historians and theorists have taken great pains, when describing such campaigns and conflicts, to point out that other means not only served the purpose as well as a battle that was never fought, but were indeed evidence of higher skill. This line of thought had brought us almost to the point of regarding, in the economy of war, battle as a kind of evil brought about by mistake—a morbid manifestation to which an orthodox, correctly managed war should never have to resort. Laurels were to be reserved for those generals who knew how to conduct a war without bloodshed; and it was to be the specific purpose of the theory of war to teach this kind of warfare.

Recent history has scattered such nonsense to the winds. Still, one cannot be certain that it will not recur here or there for shorter or longer periods, betray those responsible into mistakes which, because they cater to weakness, cater to human nature. It is quite possible that at some time in the future, Bonaparte's campaigns and battles will be considered brutalities, almost blunders, while the old-fashioned dress sword of antiquated and desiccated manners and institutions will be relied upon and praised. If the theorist can point out the dangers of this attitude, he will have provided an essential service to those who care to listen. We hope we may be able to do this for those in our beloved country who occupy positions of influence; serving them as guides and calling on them to subject these matters to profound study.

Our conviction that only a great battle can produce a major decision is founded not on an abstract concept of war alone, but also on experience. Since time began, only great victories have paved the way for great results; certainly for the attacking side, and to some degree also for the defense. The surrender at Ulm was a unique event, which would not have happened even to Bonaparte if he had not been willing to shed blood. It must in fact be looked upon as the aftermath of the victories that he had won in earlier campaigns. All fortunate generals, and not only the bold, the daring, and the stubborn, seek to crown their achievements by risking everything in decisive battles. Their answer to this transcendental question, then, should be enough for us.

We are not interested in generals who win victories without bloodshed. The fact that slaughter is a horrifying spectacle must make us take war more seriously, but not provide an excuse for gradually blunting our swords in the name of humanity. Sooner or later someone will come along with a sharp sword and hack off our arms.

We regard a great battle as a decisive factor in the outcome of a war or campaign, but not necessarily as the only one. Campaigns whose outcome have been determined by a single battle have become fairly common only in recent times, and those cases in which they have settled an entire war are very rare exceptions.

The decision that is brought about by a great battle does not of course depend entirely on the battle itself—that is, the scale of the forces engaged or the intensity of the victory. It depends also on countless other factors that affect the war potential of each side and the belligerent states. But by committing the major part of their available strength to this gigantic duel, both sides initiate a major decision. In some respects its scope can be predicted, but not in all. It may not be the only decision, but it is the *first*, and as such will affect all those that follow. Therefore, the purpose of a great battle is to act—more or less according to circumstances, but always to some extent—as the provisional center of gravity of the entire campaign. A commander who enters each battle with the true military spirit—the faith, the feeling, in short the conviction that he must and will defeat his enemy—will most likely try to tip the scales of the first battle with everything he has, hoping and striving to win everything. We doubt whether Bonaparte in any of his campaigns ever took the field without the idea of crushing the enemy in the very first encounter. Frederick the Great, in more limited circumstances and with less scope, had the same idea whenever at the head of his small force, he tried to fend off the Russians or the Imperial armies.

To repeat: the decision that is brought about by the battle partly depends on the battle itself—its scale, and the size of the forces involved—and partly on the magnitude of the success.

What a commander can do to heighten the significance of the battle in the first respect is quite obvious; we only wish to point out that, as the scale of the battle grows, so does the number of additional circumstances that are decided by it. Therefore, commanders with enough self-confidence to go for great decisions have always managed to deploy the great bulk of their forces in a great battle without seriously neglecting other areas.

The success, or, more properly, the degree of a victory depends mainly on four factors:

1. On the tactical pattern according to which the battle is fought
2. On the terrain
3. On the composition of the forces
4. On the relative strength of the opposing armies.

A battle fought with parallel fronts and without an enveloping action is not so likely to bring great results as one in which the defeated army has been turned, or made to change its front to greater or lesser degree. In rough or hilly country the impact is weakened, and therefore the results will also be less.

If the loser's cavalry is equal to or stronger than that of the victor, the effects of pursuit are lost, and with them some important results of the victory.

Finally, it must be obvious that the effect of a victory will be greater in cases where the victor is numerically superior and has used his superiority to turn the enemy's flank or make him change his front than in those in which the victor was the weaker party. To be sure, the battle of Leuthen might cause some doubt about the practical soundness of this principle, but we hope we may for once use a phrase we normally avoid: *there is an exception to every rule.*

A general can use all these means to make a battle decisive. Of course, they carry their own risks; but all his actions are subject to this dynamic law of the moral world.

There is then no factor in war that rivals the battle in importance; and the greatest strategic skill will be displayed in creating the right conditions for it, choosing the right place, time and line of advance, and making the fullest use of its results.

The fact that these matters are important does not mean that they are complex and obscure. Far from it: everything is quite simple, and needs only moderate skill in planning. The great requirements are the gifts of quickly sizing up a situation, of vigor, persistency, and a youthful, enterprising spirit—all of them heroic qualities to which we shall have to refer again. Clearly, most of these are not qualities that can be acquired through book learning. If they can be taught at all, a general will have to receive his instruction from sources other than the printed word.

The impulse to fight a great battle, the unhampered instinctive movement toward it, must emanate from a sense of one's own powers and the absolute conviction of necessity—in other words, from innate courage and perception, sharpened by experience of responsibility.

Apt examples are the best teachers, but one must never let a cloud of preconceived ideas get in the way; for even the rays of the sun are refracted and diffused by clouds. It is the theorist's most urgent task to dissipate such preconceptions which at times form and infiltrate like a miasma. The errors intellect creates, intellect can again destroy.

CHAPTER TWELVE

Strategic Means of Exploiting Victory

The preparations leading up to victory are a most difficult task, and one for which the strategist seldom receives due credit. His hour of glory and praise comes when he exploits his victory.

A number of questions arise to which we shall address ourselves in time: what the actual purpose of a battle may be, how it fits into the general pattern of the war, to what extent conditions may allow a victory to run its course, and at what point it reaches its culmination. Meanwhile, what remains true under all imaginable conditions is that no victory will be effective without pursuit; and no matter how brief the exploitation of victory, it must always go further than an immediate follow-up. Rather than repeat that fact at every opportunity, we will spend a moment on it now.

Pursuit of a beaten enemy begins the moment he concedes the fight and abandons his position. Previous movement either way has nothing to do with this—that is part of the development of the battle itself. At this juncture victory, while assured, is usually still limited and modest in its dimensions. Little positive advantage would be gained in the normal course of events unless victory were consummated by pursuit on the first day. It is usually only then, as we have said, that the trophies tend to be taken which will embody the victory. We shall discuss this phase first.

Normally both sides are already physically tired when they go into battle, since the movements directly preceding an engagement are usually of a very strenuous kind. A prolonged struggle on the battlefield calls for exertions that complete the exhaustion. Moreover the winning side is in almost as much disorder and confusion as the losers, and will, therefore, have to pause so that order can be restored, stragglers collected, and ammunition distributed. For the victor, these conditions create the critical phase that has already been mentioned. If the defeated troops are only a minor part of the enemy's forces and have other units to fall back on or can look to strong reinforcements to arrive, the victor may easily run the risk of losing his gains at any moment. This consideration will cut short the pursuit or at least keep it within very narrow bounds. But even without any risk of the enemy being reinforced, the circumstances already described will counterbalance the victor's elasticity in pursuit. Though the victory itself is not in danger, reverses are possible that may reduce the advantage. At this point, too, a general's freedom of action bears a heavy handicap—the whole weight of human needs and weaknesses. Each of the thousands under his command needs food and rest, and longs for nothing so much as a few hours free of danger and fatigue. Very few men—and they are the exceptions—are able to follow and feel beyond the present moment. Only these few, having accomplished the urgent task at hand, are left with enough mental energy to think of making further gains—gains which at such a time may seem trifling embellishments of victory, indeed an extravagance. The voice of the other thousands, however, is what is heard in the general's council; it is conducted up a channel of senior officers who urge these human needs on the general's sympathy. The general's own energies have been sapped by mental and physical exertion, and so it happens that for purely human reasons less is achieved than was possible. What does get accomplished is due to the supreme commander's *ambition, energy,* and quite possibly his *callousness.* Only thus can we explain the timorous way in which so many generals exploit a victory that has given them the upper hand. The "immediate pursuit" after a victory is a term we would as a rule only apply to pursuit on the very same day, including at most the following night. Beyond that point, the pursuer's own need for rest will in any case call a halt.

There are various degrees of immediate pursuit, according to its nature.

The first is pursuit by cavalry alone. This usually amounts more to keeping the enemy under surveillance and in a state of alarm than to the application of serious pressure, since it can easily be disrupted by the slightest natural obstacle. Cavalry may be effective against isolated units of demoralized and weak troops, but if faced with the enemy's main force, it can only act as an auxiliary arm. The enemy's fresh reserves can cover his retreat, and at the first insignificant natural obstacle can unite all arms and make an effective stand. The only exception is an army actually in flight and on the way to total dissolution.

The second degree of pursuit is undertaken by a strong vanguard of all arms, including of course the bulk of the cavalry. This kind of pursuit presses the enemy until he reaches a place where his rear guard can take a strong stand, or until his whole army can take up a new position. Since there is not likely to be an immediate opportunity for either, the pursuit reaches farther. It will, however, not last longer than an hour or a few hours at most; otherwise the vanguard may tend to lose contact with its support.

The third and highest degree of pursuit keeps the whole victorious army advancing so long as its strength holds out. In that event the merest threat of an attack or of a flank being turned will cause the beaten force to abandon most positions that the terrain might afford. Also its rear guard will be less likely to get involved in stiff delaying actions.

In all three cases nightfall ends the action, even where it cannot be considered closed. The rare exceptions in which pursuit is carried on throughout the night must rank as unusually intensified.

Considering the fact that in night operations more or less everything is left to chance, and that regular formation and routine have vanished anyway by the time the fighting ends, one can well understand that both commanders will shrink from the idea of continuing operations in the dark.

Unless success is assured by the loser's complete disarray, or by the victorious army's exceptional military virtues, everything will be left very much to fate, and no commander, not even the most audacious, will find this to his liking. As a rule, then, night puts an end to pursuit, even if the decision has taken place only a short time before it grew dark. Night affords the loser either the immediate opportunity to rest and reassemble, or a headstart if he chooses to continue the retreat under cover of darkness.

After such a break, the loser will no doubt find his situation to have improved considerably. A substantial part of the confusion has been sorted out, fresh ammunition has been issued, and the force as a whole has been reorganized. Any further encounters with the victor will constitute a new engagement rather than a continuation of the old; and while it does not by any means hold out promise of absolute success, at least it is a fresh start and not just a mopping-up operation on the part of the victor.

Thus whenever the victor can keep up the pursuit throughout the night—even if only with a strong advance guard of all arms—the effects of the victory will be on a vastly greater scale. The battles of Leuthen and Belle-Alliance provide examples.

This kind of operation is basically tactical, and we mention it only in order to clarify in our own minds the difference it can make in the effect of a victory.

This immediate pursuit to the next halt is the victor's prerogative, and hardly bears any relation to his further plans and situation. These may considerably diminish the success of a major victory, but they cannot prevent this immediate exploitation. Even if one can conceive of such cases, they would be so rare that they could not significantly influence theory.

This is one of the points where recent military experience has opened up a whole new field of energy. In earlier wars, smaller in scope and more narrowly circumscribed, conventions had developed that unnecessarily restricted many aspects of operations and this one in particular. *The very idea, the honor* of victory, appeared to be the whole point so far as the commanders were concerned. Actual destruction of enemy forces was to them only one of many means of war—certainly not the main, even less the only one. They were only too ready to sheathe their swords as soon as the enemy lowered his. Once a decision had been reached, one stopped fighting as a matter of course: further bloodshed was considered unnecessarily brutal.

This spurious philosophy was not the complete basis for a decision. It did, however, express an attitude that assured a ready hearing and attached much weight to the plea of general exhaustion and the physical impossibility of continuing the battle. Admittedly, it is only natural to spare one's victorious troops where no others are available, the more so if one expects that the tasks ahead will pretty soon be more than they can cope with, as they often are when an offensive is maintained. But this kind of reasoning was incorrect: clearly, any further losses caused by a continued pursuit would be proportionately a great deal smaller than those the enemy would suffer.

Only where the fighting forces were not considered to be the vital factor could the former view possibly originate. In earlier wars, accordingly, one finds that only the greatest heroes—Charles XII, Marlborough, Prince Eugene, Frederick the Great—would drive home a victory already decisive enough by vigorous pursuit. Other generals as a rule were content to remain in possession of the field. Contemporary war, which is waged with increased vigor in response to the increased scope of circumstances, has broken these conventional bounds: pursuit is now one of the victor's main concerns, and the trophies are thus substantially increased. Even if there are instances among more recent battles where this has not happened, these were exceptions, and unusual factors were always at work.

At the battles of Gross-Görschen and Bautzen, a total rout was prevented only by the superior allied cavalry; at Grossbeeren and Dennewitz, by pique on the part of the Crown Prince of Sweden; and at Laon, by Blücher's age and poor state of health.

Borodino also offers a relevant example, and we cannot refrain from saying more about it—partly because we do not think the matter can be settled just by blaming Bonaparte, and partly because it might appear that this case, along with a considerable number of similar ones, may possibly be placed in that category which we have considered as extremely rare, when from the very start of the battle the commander was bound hand and foot by his general situation.

Bonaparte has been severely censured, particularly by French historians and great admirers of his (Vaudoncourt, Chambray, Ségur) for failing to drive the Russian army off the field or to use his last remaining strength to crush it. They argue that what was merely a lost battle could have been an absolute rout. We should be led too far afield if we gave a detailed picture of the relative situations of both armies. But so much is clear: that when Bonaparte crossed the Niemen he had 300,000 men in those corps that would participate in the battle at Borodino; now only 120,000 were left, and he may well have wondered if he had enough to march on Moscow—and it was on Moscow that everything appeared to hinge. The victory he had just won made him reasonably confident of taking the capital; it seemed exceedingly unlikely that the Russians could fight another battle within a week; and it was in Moscow that he hoped to make peace. Admittedly, he could have been more sure of making peace if the Russian army had been completely destroyed; but his first priority was still to get to Moscow, and to get there in enough strength to be in a position to enforce his will on the capital, and thereby on the government and the Russian Empire.

As it turned out, the force that eventually reached Moscow was inadequate to this task. But it would have been even less adequate if, in the course of smashing the Russian army, Napoleon had also destroyed his own. He was thoroughly conscious of this fact, and, to our mind, was entirely justified. In spite of this, we must not count it among those cases in which the general situation obliged the commander to refrain from following up his victory by an immediate pursuit. In fact there was never any question of pursuit as such. The battle was decided about 4 P.M.; but the Russians still held most of the battlefield, and had as yet no intention of withdrawing. What is more, they would have met a renewed attack with stubborn resistance; and this, while leading them to certain disaster, would also have inflicted heavy additional losses on the French. The battle of Borodino, like that of Bautzen, is therefore among those that were *never completely fought out*. At Bautzen, the defeated party chose to leave the battlefield early; at Borodino, the victor chose to content himself with only a partial victory—not because he thought the issue was still in doubt, but because a total victory would have cost him more than he was able to pay.

In returning to our subject, observations concerning immediate pursuit lead us to the following conclusion: the importance of the victory is chiefly determined by the vigor with which the immediate pursuit is carried out. In other words, pursuit makes up the second act of the victory and in many cases is more important than the first. Strategy at this point draws near to tactics in order to receive the completed assignment from it; and its first exercise of authority is to demand that the victory should really be complete.

Yet the repercussions of a victory seldom cease at the end of the first pursuit: it is only the beginning of the actual course of events for which victory has provided the momentum. As we have stated previously, this course of events will be influenced by other factors which are not yet under discussion. But we will now go on to examine the more general aspects of pursuit in order to avoid repetition later.

In the continuation of the pursuit we can again distinguish three gradations. The first consists in merely following the enemy; the second in exerting pressure on him; and the third in a parallel march to cut him off.

If we merely follow the enemy, he will keep retreating until he feels ready for another engagement. In other words, this kind of pursuit would suffice to exhaust the effect of the superiority gained in the battle. In addition, the victor will capture everything the loser cannot take with him: sick and wounded, stragglers, baggage, and wagons of all kinds. In itself, however, merely following the enemy will not accelerate the break-up of his forces, while pressure and parallel marches will.

In employing the next highest degree of further pursuit, we are not satisfied with merely following the enemy to his previous position and occupying as much of the terrain as he is willing to cede. Instead, we arrange at each stage to demand something more: our leading units are geared to attack his rear guard every time it tries to take up a position, thereby speeding up the enemy's retreat and promoting his disintegration. This is so mainly because his retreat has to take the form of continuous, uninterrupted flight. Nothing is more repugnant to a soldier than hearing the enemy's guns yet again just as he is settling down to rest after a strenuous march. This sensation, repeated day after day, can lead to absolute panic. As a constant recognition of the fact that the enemy has the upper hand and that resistance is beyond one's capability, it is bound to be extremely damaging to the troops' morale. The pressure is worst when it forces the retreating army into night marches. If at sunset the victor drives his enemy from the campsite he has chosen either for his whole army or for his rear guard, the defeated army will have to march by night or at least shift its position and retreat still further—it comes to much the same. Meanwhile, the victorious army passes an undisturbed night.

Dispositions for marches and the choice of positions are dependent, here as elsewhere, on a great variety of factors: supplies particularly, outstanding features of the terrain, large towns, and so forth. It would therefore be arrant pedantry to demonstrate by geometrical analysis how the pursuer, having the upper hand, can keep the retreating army marching night after night while his own men are resting. But it remains a fact, and a useful one, that pursuit so planned may have such a tendency, and will thereby become immensely more effective. The reason it is seldom put into practice is that the pursuing army itself would find such a procedure more difficult than keeping regular hours and stopping at the usual time of day. It is far easier to break camp early in the morning, occupy the next at noon, spend the rest of the day in resupplying, and sleep at night than to base your moves exactly on the enemy's, take all decisions at short notice, break camp at dawn one day and dusk the next; always facing the enemy for hours at a time, trading shot for shot with him, skirmishing, planning to turn his flank; in short, deploying every artifice of tactics which the situation demands. This obviously places a heavy burden on the pursuer, and war brings so many burdens that it is only human to eliminate any that appear at all avoidable. These considerations are valid whether they apply to the whole army or, as is more usual, to a strong vanguard. The above reasons account for the relative rarity of this type of pursuit—the application of constant pressure on the defeated army. Even Bonaparte made little use of it in his Russian campaign of 1812. The reason is obvious: the difficulties and hardships of this campaign were already enough to threaten his army with complete disaster before he had reached his goal. But in other campaigns the French have distinguished themselves by their energy in this respect as well.

Finally, the third and most effective degree of pursuit takes the form of marching parallel with the enemy toward the immediate goal of his retreat.

Every defeated army has a first point of retreat—more or less distant—which it is most anxious to reach. Possibly it poses a threat to its further retreat, such as a defile; or it may be essential to reach this objective before the enemy because it is a major city, a supply base, or something of the kind; or, finally, it may be a point where the army expects to gain new powers of resistance, as in the case of a strong position, a junction with additional forces, and so forth.

If the pursuer, by using a secondary road, aims for this objective, he can obviously make the enemy increase the speed of his retreat at heavy cost; he can make it a scramble and finally a stampede. Only three ways of meeting this are open to the beaten force. The first is to turn upon the enemy, and try, by means of a surprise attack, to create a turn for the better which the situation does not really afford. Obviously this requires a bold and enterprising commander and first-rate troops—beaten, perhaps, but far from being utterly defeated. A retreating army is therefore seldom able to make use of this stratagem.

The second way is to speed up the retreat. But that is just what the victor wants. It is likely to exhaust the troops and result in stragglers by the score, guns and transport rendered unserviceable, and consequently will cause enormous losses.

The third way is to take a detour, thereby avoiding the nearest point of interception, on the theory that traveling at a greater distance from the enemy will involve less exertion, and will prevent haste from causing further loss. This is the worst way of all. It may be compared to additional debts incurred by a bankrupt, and usually leads to still greater embarrassment. There may, indeed, be cases where this course is advisable, and others where it is the only way out. There may even be cases where it worked. Generally, however, it is certainly chosen not so much from the conviction that it is the best way to reach one's objective safely, but from another, less creditable motive: fear of closing with the enemy. Woe to the general who succumbs to such fear! No matter how low the morale of the troops, how justified the apprehension of being at a disadvantage in an encounter with the enemy, the situation will only be aggravated by timidly shirking every opportunity of contact. Bonaparte would never have been able to return across the Rhine in 1813 with even the thirty to forty thousand men still left to him after the battle of Hanau if he had refused a battle there and tried to cross at Mannheim or Coblenz. Such minor engagements, carefully prepared and carried out, in which the defeated army, being on the defensive, is able to reap the benefit of the terrain, *are the very means of starting a recovery in the morale of the troops.*

Even a minor success can do wonders. Still, most generals have to overcome great reluctance to make the attempt. At first sight evasion seems so much easier that it is usually preferred. This very evasion often furthers the purpose of the victor more than anything, and frequently ends in the total destruction of the defeated force. Of course we are speaking of an army as a whole, not of a part that has been cut off and is trying to rejoin the rest by a detour. The latter constitutes a different situation, and success in such a case is not unlikely. But there is a condition attached to this race for a given objective: part of the victor's force must follow the retreating army on the road it has taken, mopping up everything it leaves behind, and to impress the pursued force with its presence. Blücher, whose pursuit after Belle Alliance was a model in all other respects, missed this point.

Marches of this type do indeed weaken the pursuing force. They are not advisable when the enemy can fall back on another (sizeable) force, where a first-rate general is in command of the enemy forces, and their destruction is not already well under way. However, where one can afford to use this means, it works like an efficient machine. The beaten army's loss from sickness and fatigue is grossly disproportionate, and its whole morale is weakened and depressed by constant fear of imminent disaster, until, in the end, organized resistance is inconceivable. Thousands of prisoners are taken every day without a blow being struck. At such a time of good fortune, the victor must not be afraid to divide his forces in order to envelop everything within reach of his army, isolate outlying units, take fortresses that are caught off guard, occupy large towns, and so forth. He may do whatever he wants until the situation changes; the more liberties he takes, the later that moment will come.

There are many examples of such brilliant effects of major victories and first-rate pursuits in the Napoleonic Wars. It should be enough to recall the battles of Jena, Regensburg, Leipzig, and Belle-Alliance.

CHAPTER THIRTEEN

Retreat after a Lost Battle

When a battle is lost, the strength of the army is broken—its moral even more than its physical strength. A second battle without the help of new and favorable factors would mean outright defeat, perhaps even absolute destruction. That is a military axiom. It is in the nature of things that a retreat should be continued until the balance of power is reestablished—whether by means of reinforcements or the cover of strong fortresses or major natural obstacles or the overextension of the enemy. The magnitude of the losses, the extent of the defeat, and, what is even more important, the nature of the enemy, will determine how soon the moment of equilibrium will return. There are indeed many instances of a beaten force being able to rally only a short distance away without its situation having changed at all since the battle. The explanation lies either in the low morale of the victor or in the fact that the superiority won in battle was not enough to drive its impact home.

In order to utilize any weakness or mistake on the part of the enemy, not giving an inch more ground than the force of circumstances requires, and especially in order to keep morale as high as possible, it is absolutely necessary to make a slow fighting retreat, boldly confronting the pursuer whenever he tries to make too much of his advantage. The retreats of great commanders and experienced armies are always like the retreat of a wounded lion, and this unquestionably is theoretically preferable as well.

When a dangerous position has to be abandoned, time is often wasted on trivial formalities, thereby compounding the danger. In such a case, everything depends on getting away as quickly as possible. Experienced commanders consider this to be very important. But it should not be confused with a general retreat. Anyone who then believes that a few forced marches will give him a good start and help him make a stand is dangerously wrong. The first movements have to be almost imperceptibly short, and it must be a general principle not to let the enemy impose his will. This principle cannot be put into practice without fighting fierce engagements with the pursuing enemy, but it is a principle worth the cost. Otherwise the pace is bound to increase till withdrawal turns into rout. More men will be lost as stragglers than would have been lost in rear guard actions. And the last vestiges of courage will have disappeared.

The means of putting the above-mentioned principle into practice consist of a number of factors: a strong rear guard, made up of the best troops, led by the most courageous general, and supported at crucial moments by the rest of the army; skillful use of the terrain; strong ambushes wherever the daring of the enemy's vanguard and the terrain permit. In short, it consists of planning and initiating regular small-scale engagements.

The degree of difficulty involved in a retreat depends, of course, on whether the battle was fought on favorable terms and on the severity of the fighting. Jena and Belle-Alliance show that any sort of regular retreat becomes impossible if one fights to the last against a superior foe.

Here and there it has been suggested (by Lloyd and Bülow, for instance) that the retreating forces should be divided; they should withdraw in separate columns, or even by divergent routes. It must be clear that we are not here discussing a separation simply for convenience of march, in which the option and the intention of fighting in concert are retained. Any other kind of separation is extremely dangerous; it goes against the grain, and would be a great mistake. A lost battle always tends to have an enfeebling, disintegrating effect; the immediate need is to reassemble, and to recover order, courage, and confidence in the concentration of troops. It is absurd to think that an enemy, at the moment he is following up his victory, can be harassed on both his flanks by a divided force. This may possibly impress an opposing general who is nothing but a timid pedant, and in such a case one can try it. But unless one can be sure of this kind of weakness, it would be better not to try. Where the strategic situation after the battle requires that one's flanks be covered by separate detachments, it will have to be done to the most limited extent possible. But such a separation must always be regarded as a drawback, and one is usually in no position to effect it on the day after the battle.

Frederick the Great, after the battle of Kolin and the raising of the siege of Prague, did withdraw in three columns. But he did not do so of his own free will, but because the position of his forces and the need to cover Saxony left him no alternative. Bonaparte, after the battle of Brienne, ordered Marmont back to the Aube while he crossed the Seine and turned toward Troyes. This did not end in disaster only because the allies, instead of following him, also divided their forces: one part under Blücher headed for the Marne while the other under Schwarzenberg, fearing itself to be too weak, advanced extremely slowly.

CHAPTER FOURTEEN

Night Operations

The conduct and the peculiar aspects of night operations are matters of tactics. Here we shall only consider them to the extent that they constitute a distinct form of war.

Basically, a night attack is only an intensified raid. At first glance it looks highly effective: supposedly the defender is taken unawares, while the attacker, of course, is well prepared for what is about to happen. What an uneven contest! One imagines complete confusion on one side, and on the other an attacker concerned merely to profit by it. This image explains the many schemes for night attacks put forward by those who have neither to lead them nor accept responsibility for them. In practice they are very rare.

All such ideas assume that the attacker knows the complete layout of the defense, which, having been previously planned and carried out, could not escape his reconnaissance and intelligence. On the other hand, the attacker's dispositions, made only at the moment of execution, must remain unknown to the other side. But even the latter does not always happen, and the former is even less common. Unless the enemy is so close as to be in full view (as Frederick the Great was to the Austrians before the battle of Hochkirch) knowledge of his position will be incomplete. It will be acquired from reconnaissance, patrols, prisoners' statements and spies, and it can never really be reliable for the simple reason that all such reports are always a little out of date, and the enemy may in the meantime have changed his position. Moreover with the old system of tactics and encampment it was far easier to find out the position of the enemy than it is today. A line of tents is more easily distinguished than a collection of huts, much less a bivouac; and a line of battle deployed in linear formation can be distinguished more easily than one consisting of divisions in column, as is now the rule. It is possible to have a perfect view of the area in which a division is encamped in such a manner, and still not be able to form a clear picture of its layout.

And the layout of the defense is not all one needs to know; it is just as important to know the dispositions that he will take in the course of the action. After all, he will not just shoot his guns off blindly. Since these tactical decisions have come to be more important than the initial positions, night raids have become more difficult to carry out in war today than they were formerly. Nowadays a defender can station his troops with more flexibility, and contemporary warfare thus enables him to surprise the enemy with unexpected blows.

In a night operation, then, the attacker seldom if ever knows enough about the defense to make up for his lack of visual observation.

The defender has another slight advantage: the ground he occupies is better known to him than it is to the attacker, in the same way as a man can find his way around his own room in the dark more easily than can a stranger. He can find and round up all the component parts of his forces more quickly than can his assailant.

It follows from all this that the attacker needs his eyes in night operations just as much as does the defense. Therefore special reasons are needed to justify a night attack.

Generally, these reasons affect subordinate parts of the army, rarely the army as a whole. It follows that night raids will as a rule occur in minor engagements, and seldom in major battles.

A subordinate part of the enemy's force can easily be attacked and surrounded in superior strength. One can either capture it intact, or inflict severe loss on it by forcing it to fight against heavy odds—provided all other conditions are favorable. But this type of plan can only be carried out as a complete surprise. No part of the enemy's force would ever be willing to fight on such unequal terms; it would more likely withdraw. The element of surprise in turn can, with the rare exceptions of thickly wooded areas, only be present with the help of darkness. So if one wants to take this type of advantage of an enemy unit's weak position, one must make use of night, at least for the completion of one's preparations, even if the actual engagement will not take place much before dawn. This accounts for all the minor night operations against outposts and other minor units. Essentially they are meant, by superiority and envelopment, to involve the unsuspecting enemy in an engagement on such unequal terms that he cannot escape without heavy loss.

The larger the force under attack, the more difficult the operation. A larger force has great internal resources and can keep on fighting until help arrives.

That is why under ordinary circumstances this kind of attack cannot be used against the whole of the enemy's force. While an entire army cannot expect outside aid, its own resources are sufficient to ward off an attack from several sides, especially nowadays when such attacks are so common that everyone is trained to deal with them. The success of an enveloping attack usually depends on factors that have nothing to do with surprise. We need not go into them here; it is enough to say that while envelopment can offer great rewards, it can also involve great risks. So except in special circumstances, it can be justified only by great superiority, such as can, of course, be concentrated on a subordinate part of an army.

Enveloping and surrounding a lesser enemy unit, especially under cover of night, is a more feasible operation for another reason. The troops detailed for the task, no matter how superior, will in all probability constitute only a fraction of the total strength; and it is safer to gamble only a part on a risky operation than the whole of the army. And the risks themselves are reduced still further by the fact that the assault force will be covered and supported by a larger force, or even by the whole army.

Night operations are not merely risky; they are also difficult to execute. This too limits their scale. Their essence being surprise, it must be a prime consideration to approach without being seen. This, too, is easier for small detachments than for large ones, and rarely feasible for the columns of an entire army. Hence operations of this kind are generally aimed at single outposts. They can be used against a larger corps only if its outposts are inadequate, as was the case of Frederick the Great at Hochkirch. Such conditions are in turn less likely to obtain in the case of a whole army than in that of its subsidiary parts.

War has recently been waged with so much greater energy and speed that armies have at times had to camp very near each other, and without a strong system of outposts. This always coincides with the critical stages that usually precede a decision. At that stage, however, both armies are in a greater state of readiness for action. In earlier wars, on the other hand, the practice was often for armies to camp in full view of each other, even when all they aimed at was to hold each other in check—and this could last for some time. For weeks on end Frederick the Great would camp so near the Austrians that both sides could have exchanged cannon shots.

While such encampments certainly favored night raids, they have been abandoned in more recent wars. Armies nowadays are no longer independent organisms, self-sufficient in matters of supply and encampment: as a rule they think it wise to leave a full day's march between the enemy and themselves.

If we now examine the question of night raids on a whole army, it will be evident that adequate reasons for this type of operation are unusual. They may be traced to the following causes:

1. Exceptional carelessness or provocation on the part of the enemy. This is infrequent. Where it does occur it is usually balanced by decidedly superior morale.

2. A wave of panic in the enemy army, or, in general, a case in which the attacker's morale is so superior as to be able to supply the place of guidance in action.

3. Fighting one's way through a superior enemy force by which one is surrounded. In this case, the element of surprise is crucial; the single aim—escape—permits a far greater concentration of forces.

4. Finally, desperate situations in which one's troops are so heavily outnumbered that only the utmost daring offers any prospect of success.

It must be remembered that all these cases only hold good on condition that the enemy army is in full view, and not covered by a vanguard.

Most night operations, incidentally, are planned to end at daybreak, darkness being used to cover only the approach and the first assault. This enables the attacker to make better use of the confusion into which he throws the enemy. Engagements that do not start till dawn, and in which darkness is used merely for the approach, are not true night operations at all.

BOOK FIVE

Military Forces

CHAPTER ONE

General Survey

Military forces will be examined from the following points of view:

1. Their numerical strength and organization
2. Their state when not in action
3. Their maintenance
4. Their general relationship to country and terrain.

This book will deal not with combat itself, but with those aspects of the armed forces that must be regarded as *conditions necessary to military action.* They are more or less closely related to fighting and interact with it, so they will be frequently mentioned in our discussion of the uses of combat. But first each must be examined as a separate entity with its own characteristics.

CHAPTER TWO

The Army, the Theater of Operations, the Campaign

The very nature of the question makes it impossible to give an accurate definition of these different factors of space, mass, and time; but so as not to be misunderstood, we shall try to clarify the common usage of these terms, which in most cases we like to follow.

1. Theater of Operations

By "theater of operations" we mean, strictly speaking, a sector of the total war area which has protected boundaries and so a certain degree of independence. This protection may consist in fortifications or great natural barriers, or even in a substantial distance between it and the rest of the war area. A sector of this kind is not just a part of the whole, but a subordinate entity in itself—depending on the extent to which changes occurring elsewhere in the war area affect it not directly but only indirectly. A definitive criterion might be found by imagining an advance in one theater simultaneous with a withdrawal in the other, or a defensive action in one simultaneous with an offensive in the other. We cannot always be so precise; we merely wish to indicate the essential point here.

2. The Army

It is easy to define an army by using the concept of "the theater of operations"—that is, all the forces located in a given theater. Yet this obviously does not cover all the common uses of the term. Blücher and Wellington each commanded a separate army in 1815, even though they were in the same theater of operations; so supreme command is another criterion in defining an army. Nonetheless, the two are closely related: where matters are properly arranged, there will be only one supreme commander in a single theater. And a general in control of his own theater of operations will never lack a suitable degree of independence.

In establishing the meaning of the term the army's actual strength matters less than one might at first suppose. Where several armies operate in a *single theater* under a combined command, the term does not derive from their numbers but from their previous history. (In 1813, for example, there were the Silesian Army, the Army of the North, and so forth.) Large numbers of men that are destined to remain in a given theater of operations will certainly be formed into different corps, but never into separate armies. That, at any rate, is not the term that would be used, and usage seems to stick closely to practice. On the other hand it would be sheer pedantry to claim the term "army" for every band of partisans that operates on its own in a remote part of the country. Still, we must admit that no one thinks it odd to talk of the "army" of the Vendée during the French Revolutionary Wars, though it was frequently little more than a band of partisans. The terms "army" and "theater of operations," then, normally go hand in hand, each confirming the other.

3. The Campaign

It is true that the term "campaign" is often used to denote all military events occurring in the course of a calendar year in all theaters of operations, but normally and more accurately it denotes the events occurring in a *single* theater of war. The notion of a single year is harder to dispose of, for wars are no longer broken into annual campaigns by long fixed periods in winter quarters. Events in a given theater of operations tend to group themselves into sections of a certain magnitude; when, for instance a catastrophe of more or less major proportions ceases to produce direct results, and fresh developments start taking shape. These natural divisions must be borne in mind if a year or campaign is to be given its full quota of events. No one will think of the campaign of 1812 as having ended at the Memel just because this is where the armies happened to be on 1 January 1813, and count the subsequent French retreat across the Elbe as part of the next year's campaign: it was clearly part of the whole retreat from Moscow.

The fact that these concepts cannot be more accurately defined should not be considered a disadvantage. Unlike scientific or philosophical definitions, they are not basic to any rules. They are merely meant to serve as an approach to greater clarity and precision of language.

CHAPTER THREE

Relative Strength

In Chapter Eight of Book Three we pointed to the great importance of superior numbers in an engagement and, concomitantly, of superior numbers in general from the point of view of strategy. In its turn that implies the importance of relative strength, on which we must now add a few detailed observations.

An impartial student of modern war must admit that superior numbers are becoming more decisive with each passing day. The principle of bringing the maximum possible strength to the decisive engagement must therefore rank rather higher than it did in the past.

The courage and morale of an army have always increased its physical strength, and always will. But there are periods in history when great psychological advantage was gained by superior organization and equipment; others where the same result was achieved by superior mobility. Sometimes it was a matter of novel tactics; at other times the art of war revolved around efforts to exploit terrain skillfully on large and comprehensive lines. On occasion generals have managed to gain great advantages over one another by such means. But efforts of this type have declined, making way for simpler and more natural procedures. If we take an unbiased look at the experiences of the recent wars, we must admit that those means have almost disappeared, both from the campaign as a whole and the decisive engagements, and particularly from the major battle—as already explained in Chapter Two of the previous book.

Today armies are so much alike in weapons, training, and equipment that there is little difference in such matters between the best and the worst of them. Education may still make a considerable difference between technical corps, but what it usually comes down to is that one side invents improvements and first puts them to use, and the other side promptly copies them. Even the senior generals—divisional and corps commanders—have, as far as their efficacy is concerned, pretty much the same views and methods. The only remaining factor that can produce marked superiority, aside from familiarity with war, consists of the talents of the commander-in-chief, which hardly bear a constant relationship to the cultural standards of the people and the army, and are, indeed, completely left to chance. The decisive importance of relative strength increases the closer we approach a state of balance in all the above factors.

The character of modern battle derives from this state of balance. The battle of Borodino, objectively studied, pitted the French army, the finest in the world, against the Russian army, which, in much of its organization and individual training, was probably the least advanced. In the whole of the battle, there was not a trace of superior skill or intelligence: it was simply a test of strength, and in this the two armies were almost equal. What ensued in the end was merely a slight tipping of the balance in favor of the side that was led with greater vigor and was more familiar with war. We choose Borodino as an illustration since it is a rare example of almost equal numbers being involved.

We do not contend that all battles are like that, but it is typical of most of them.

In a battle consisting of a slow and methodical trial of strength, greater numbers are bound to make a favorable outcome more certain. In fact in modern war one will search in vain for a battle in which the winning side triumphed over an army twice its size. In earlier days this happened now and then. With the sole exception of Dresden in 1813, Bonaparte, the greatest general of modern times, always managed to assemble a numerically superior, or at least not markedly inferior, army for all the major battles in which he was victorious; and where he failed to do so—as at Leipzig, Brienne, Laon, and Belle-Alliance—he lost.

But in strategy, absolute strength is usually a given quantity which a general cannot change. Yet it does not follow that war is impossible for an army whose strength is markedly inferior. War is not always the result of a voluntary policy decision—least of all in instances where there is a great disproportion of forces. So one must admit any kind of relative strength: it would be a peculiar theory of war if it broke off just where the need for it was greatest.

No matter how desirable adequate numbers may be for the purposes of theory, it is not possible to reject even the least adequate as useless. No absolute limits can be set.

The more restricted the strength, the more restricted its goals must be; further, the more restricted the strength, the more limited the duration. These two directions afford escape routes, so to speak, for the weaker side. Any changes in the conduct of the war that are brought about by the degree of strength can be discussed only in their turn; it is enough at this stage to indicate the overall point of view. But for the sake of completeness one more point must be added.

Where the weaker side is forced to fight against odds, its lack of numbers must be made up by the inner tension and vigor that are inspired by danger. Where the opposite occurs, and despair engenders dejection instead of heroism, the art of war has, of course, come to an end.

If an increase in vigor is combined with wise limitation in objectives, the result is that combination of brilliant strokes and cautious restraint that we admire in the campaigns of Frederick the Great.

The less moderation and caution can achieve, the greater must be the dominance of vigor and tension. Where the disparity of strength is so overwhelming that no limitation of one's objectives will provide protection from failure, or where the period of danger threatens to be so extended that not even the greatest economy of strength can lead to success, the tension will, or should, build up to one decisive blow. The hard-pressed army, not expecting help where none can be forthcoming, can only trust to the high morale that despair breeds in all courageous men. At that point the greatest daring, possibly allied with a bold stratagem, will seem to be the greatest wisdom. Where success is out of reach, an honorable defeat will at least grant one the right to rise again in days to come.

CHAPTER FOUR

Relationship between the Branches of the Service

Here we shall only discuss the three main branches: infantry, cavalry, and artillery.

We trust we may be forgiven the following analysis, which really belongs more under the heading of tactics. It is needed here in the interest of clarity.

An engagement consists of two essentially different components: the destructive power of firearms, and hand-to-hand, or individual, combat. The latter in turn can be used for either attack or defense (words employed here in an absolute sense, for we are speaking in the broadest terms). Artillery is effective only through the destructive power of fire; cavalry only by way of individual combat; infantry by both these means.

In hand-to-hand fighting, the essence of defense is to stand fast, as it were, rooted to the ground; whereas movement is the essence of attack. Cavalry is totally incapable of the former, but preeminent in the latter, so it is suited only to attack. Infantry is best at standing fast, but does not lack some capacity to move.

This distribution of elementary military strengths among the three main arms demonstrates the superiority and versatility of infantry in comparison with the other two: it alone combines all three qualities. This also explains how in war a combination of the three arms leads to a more complete use of all of them. It enables the combatant to reinforce at will any one of the functions which, in the infantry, are inseparably united.

In recent wars the major role has undoubtedly been played by the destructive power of firearms: but it is no less clear that the true, the actual core of an engagement lies in the personal combat of man against man. An army composed simply of artillery, therefore, would be absurd in war. An army consisting simply of cavalry is conceivable, but would have little strength in depth. An army consisting simply of infantry is not only conceivable, but would be a great deal stronger. The degree of independence of the three branches, then, is infantry, cavalry, artillery.

But their order of importance is quite different when each is cooperating with the other two. Destruction being a more effective factor than mobility, the complete absence of cavalry would prove to be less debilitating to an army than the complete absence of artillery.

An army consisting only of infantry and artillery would, to be sure, find itself at a disadvantage when faced with one composed of all three arms. But if it were to make up for the missing cavalry by a *proportionately* stronger infantry force, a change in its tactical dispositions would enable it to manage fairly well. Outposts would pose some difficulties: there could be no brisk pursuit of a defeated enemy; and its own retreat would cause greater hardships and exertions. But such difficulties alone would hardly be sufficient to drive it off the field. If, on the other hand, such an army were faced with one composed only of infantry and cavalry it would stand up very well indeed. It is, in turn, almost inconceivable that the latter type could hold out at all against an army composed of all three arms.

It is understood that these reflections on the importance of each arm of the services are derived from the whole mass of military data, where one instance is analogous to another. It cannot be our intention to apply the facts we have discovered to every single phase of any given engagement. A battalion retreating or doing outpost duty would probably prefer some cavalry to a few guns. A body of cavalry and horse artillery with the task of pursuing a retreating enemy or cutting off his escape will find infantry completely useless, and so forth.

Let us recapitulate the results of these reflections:

1. Infantry is the most independent of the arms.
2. Artillery has no independence.
3. When one or more arms are combined, infantry is the most important of them.
4. Cavalry is the most easily dispensable arm.
5. A combination of all three confers the greatest strength.

Since maximum strength derives from a combination of all three arms, the question naturally arises what the optimum proportions would be. An answer is almost impossible.

If one could compare the cost of raising and maintaining the various arms with the service each performs in time of war, one would end up with a definite figure that would express the optimum equation in abstract terms. But this is hardly more than a guessing game. The first part of the equation alone is hard enough to estimate, except for the purely monetary factor; but the value of human life is another matter—one on which no one would be willing to set a price in cold figures.

There is also the fact that each arm really depends on a different sector of the national economy: infantry on the human population, cavalry on the equine, and artillery on finance. That fact introduces an outside determinant, which we clearly see to be dominant in the general historical phases of different peoples at different times.

But since for other reasons we cannot quite dispense with all standards of comparison, instead of taking the first part of the equation as a whole, we shall simply make use of the only ascertainable factor: the monetary cost. For our purposes it will suffice to state that, according to common experience, a squadron of 150 horses, a battalion of 800 men, and a battery of eight six-pounders cost approximately the same both for equipment and maintenance.

So far as the second part of the equation is concerned, it is even more difficult to work out definite figures. It might conceivably be possible if destructiveness were all that had to be measured; but each branch has its own particular use and thus a different sphere of effective action. But the spheres are by no means fixed; they could be expanded or contracted, and the consequence would merely be to modify the conduct of the war without incurring any special disadvantage.

People often talk of the lessons of experience in this context, in the belief that the history of war provides sufficient grounds for a definite answer. But those are obviously empty phrases, which, since they cannot be traced back to any fundamental and compelling basis, are not worth considering in a critical investigation.

In theory, then, there is an optimum proportion between the arms, which in practice remains the unknown X, a mere figment of the imagination. But it is possible to calculate what would happen if one arm were greatly superior or inferior to the same arm on the other side.

Artillery intensifies firepower; it is the most destructive of the arms. Where it is absent, the total power of the army is significantly weakened. On the other hand, it is the least mobile and so makes an army less flexible. What is more, it must always be covered by infantry, since in itself it is unable to engage in hand-to-hand combat. If there is too much artillery, and the troops detailed to cover it are in consequence not strong enough at every point to beat off the enemy, guns are easily lost. This points up an additional disadvantage: artillery is the only one of the three arms whose main equipment—guns and carriages—can be promptly used by the enemy *against* its original owner.

Cavalry increases the mobility of an army. Where there is not enough of it the rapid course of war is weakened, since everything proceeds more slowly (on foot) and has to be organized more carefully. The rich harvest of victory has to be reaped not with a scythe but with a sickle.

An excess of cavalry should never be considered a direct impediment to an army, an organic disproportion. But it does weaken the army indirectly, because of the problems of maintenance and because we must recognize that at the cost of an extra cavalry force of 10,000 men we could maintain an additional 50,000 foot soldiers.

The peculiarities that arise from the predominance of one particular arm are the more relevant to the art of war in the narrower sense, since it is concerned with the use of available forces. These are usually assigned to the commander in proportion to their availability without his having much say in the matter.

Assuming therefore that the character of war is modified by the predominance of one of the arms, it will be in the following manner.

An excess of guns will impose a more passive and defensive character on operations. Greater reliance will be placed on strong positions, major natural obstacles, and even on positions in mountainous areas. The idea will be to let terrain difficulties take care of the defense and protection of the guns and to let the enemy court his own destruction. The whole war will proceed at the solemn, formal tempo of a minuet.

Shortage of artillery will have the opposite effect. It will bring attack to the fore—the active principle of movement. Marching, exertion and continuous effort will become arms in themselves, and war will be a brisker, rougher and more variegated business. Great events will be broken down into small change.

Where cavalry is plentiful, wide plains will be sought out and *sweeping* movements preferred. With the enemy at a distance, we can enjoy greater peace and comfort, without his being able to do the same. Since we are the masters of space, we can be daring in the use of bold flanking movements and generally more audacious maneuvers. Diversions and invasions, insofar as they constitute valid expedients in war, are easily executed.

A serious lack of cavalry impairs the mobility of an army, but without increasing its destructive powers as an excess of artillery does. The war will then be marked by prudent and methodical proceedings. In such a case, the natural tendencies are to stay close to the enemy so as to be able to keep an eye on him; never to make a sudden, or worse, a hasty movement; always to advance one's forces gradually, keeping them well together; and to favor defensive operations and those in rough country. If an attack is necessary, it should be made on the enemy's vital point by the shortest route.

These are the ways in which preponderance of one arm or another will affect the operational conduct of a war; yet they are seldom so complete or decisive that they play the only, or the principal, part in determining the nature of the whole operation. Whether one selects the instrument of strategic attack or of defense, one theater of operations or another, a major battle or some other method of destruction, will probably depend on other, weightier arguments. Where this is not so, we are afraid that nonessentials have taken the place of essentials. But even in cases where the major issues have already been decided on the basis of other reasons, a certain amount of latitude remains in which preponderance in one branch can exert its influence. It is possible to be prudent and methodical in attack, and bold and enterprising in defense, and so on through every possible phase and nuance of military activity.

Conversely, the nature of a war can greatly affect the proportions of the arms of the service.

First, a peoples war, based on militia and home guard, will naturally involve large numbers of infantry. This means a shortage of equipment rather than of men, and equipment will be limited to the barest necessities. It is therefore quite possible to raise not one, but two or three battalions for every eight-gun battery.

Second, where the opposing sides are unevenly matched and the weaker is unable to resort to arming the people, or, what amounts to almost the same thing, raising a militia, an increase in artillery is certainly the fastest means of bolstering its forces and bringing about some sort of balance. One can thereby save manpower while intensifying the principal element of the forces, that is their destructive power. In any case such operations will probably be limited to a small theater, for which artillery will also be most appropriate. Frederick the Great relied on this means in the latter part of the Seven Years War.

Third, cavalry is suited to movement and major decisions. Therefore, its preponderance is important in operations over great distances, and in cases where one expects to carry out major and decisive blows. Bonaparte will serve as an example.

When we come to analyzing attack and defense, we shall see more clearly that no direct influence is exercised by these two forms of warfare as such. For the moment all we wish to point out is that, as a rule, both attacker and defender will operate in the same terrain, and that, in at least a great number of cases, their final intentions may be similar. The campaign of 1812 is relevant here.

It is a common view that in the Middle Ages the proportion of cavalry to infantry was far higher than now, and has gradually declined ever since. To some degree at least this is a misconception. On the average the proportion of cavalry in absolute numbers was probably not significantly larger; and one can easily confirm this by studying the actual figures for armed forces throughout the Middle Ages. We need only mention the masses of foot soldiers that made up the armies of the Crusaders or followed the German Emperors into Italy. It was the *importance* of cavalry that was much greater. Cavalry was the *more effective* arm, consisting of the elite; this made such a difference that, although cavalry was always smaller by far, it was always considered to be the decisive element; while foot soldiers were in low esteem and hardly ever mentioned. Hence the idea that their numbers were comparatively small. No doubt in some minor local incursions in Germany, France, and Italy, a small force consisting of cavalry alone was more common than it would be today; since it was the principal arm, this is not inconsistent. But such cases are not conclusive when one considers the general picture in which they are greatly outnumbered by cases in which larger armies were involved. The custom of using large masses of relatively inefficient foot soldiers came to an end only when the feudal system of military service was replaced by that of hired mercenaries, and the conduct of war became dependent on money and recruiting—as it did during the Thirty Years War and the wars of Louis XIV. There might have been a general return to cavalry at that time if developments in firearms had not given fresh importance to the infantry. One effect was that infantry remained superior in number to cavalry. Even when infantry was weak, its ratio to cavalry during this period was one to one; when it was strong, the ratio was three to one.

As firearms developed further, cavalry steadily continued to lose importance. This is clear enough; but it must be understood that this development related not only to the weapons as such, and to skill in using them, but to the ability to employ troops thus equipped. At the battle of Mollwitz, the Prussians had reached a level of perfection in the use of firepower that has still not been surpassed. On the other hand, the deployment of infantry in rough country and the use of firearms in skirmishing developed only later, and must be considered a major advance in destructive power.

In our view, then, the relationship of cavalry to infantry has changed little in terms of numbers, but greatly in terms of importance. This may seem self-contradictory, but in fact is not. In the armies of the Middle Ages we find great masses of infantry, which, however, stood in no organic relation to the cavalry; foot soldiers were plentiful merely because cavalry was so expensive that all those who could not be equipped as cavalry automatically became infantry. Infantry was therefore merely making the best of necessity: if the quantity of cavalry had been determined only by its intrinsic value, no amount of it would have been too much. This explains why cavalry, though declining in importance, may still have enough significance to maintain that proportion in the armed forces which it has kept until our own times.

It is in fact remarkable that, at least since the War of the Austrian Succession, the ratio of cavalry to infantry has undergone no change at all and has remained between one-quarter and one-sixth. This seems to indicate that these proportions meet some natural need, thereby revealing a ratio that cannot be directly ascertained. But we doubt if this is really so, and believe that in all important instances other reasons are evident for maintaining such large numbers of cavalry.

Russia and Austria, for example, are inclined in this direction because they still maintain fragments of Tartar institutions in their political structures. Bonaparte could never be strong enough to suit his purpose: once he had exhausted the use of conscription, the only means to strengthen his army that remained open to him was to increase the auxiliary arms, which called more for money than for men. Besides, it is plain that the enormous extent of his military operations would place a greater emphasis than usual on cavalry.

It is well known that Frederick the Great took pains to recruit not a single man more than he reckoned his country could afford; his chief concern was to maintain the strength of his army as far as possible at the expense of other countries. It is easy to see that he had good reasons for this: his limited territory at that time did not even include West Prussia or Westphalia.

Cavalry not only required less manpower; it was also more easily recruited. His method of warfare, too, was based entirely on superior mobility. As a result while his infantry declined in numbers his cavalry kept increasing right up to the end of the Seven Years War. Yet even then it hardly amounted to more than a quarter of the infantry in the field.

Nor are examples lacking, during the same period, of armies taking the field exceptionally short of cavalry and still being able to emerge victorious. The outstanding case in point is the battle of Gross-Görschen. Counting only the divisions that took part in the battle, Bonaparte had 100,000 men—5,000 of them cavalry and 90,000 infantry. The allies had 70,000 men, of whom 25,000 were cavalry and 40,000 infantry. Bonaparte's cavalry was thus 20,000 short, and he had only 50,000 more infantry than his adversaries, when he ought to have had a superiority of 100,000. Since he won the battle in spite of the smaller margin, one may well ask whether he could possibly have lost it if he had had 140,000 infantry against the allies' 40,000.

After the battle, to be sure, the allied superiority in cavalry proved to be most valuable: Bonaparte captured hardly any trophies. Victory alone is not everything—but is it not, after all, what really counts?

These considerations make it hard to believe that the ratio of cavalry to infantry which was established eighty years ago and has persisted ever since is the normal one, arising out of the intrinsic value of both arms. We are more inclined to think that, after various fluctuations, the present tendency will continue, and that the constant number of cavalry will eventually be much lower than it is today.

Since the invention of cannon, and as cannon have been improved and reduced in weight, their number has naturally increased. Even so, since the time of Frederick the Great, the proportionate strength of artillery has remained fairly constant: two or three guns per thousand men—that is at the outset of a campaign. In the course of operations guns are not lost as fast as men, and so their proportion is a good deal higher by the end; possibly arriving at a ratio of three, four or five guns per thousand men. Only experience will determine whether these are the normal proportions, or whether guns can go on being increased without encumbering the whole conduct of war.

Let us now summarize the conclusions to which these arguments have led:

1. Infantry is the main branch of the service; the other two are supplementary.

2. A high degree of skill and vigor in the conduct of war can to some extent make up for a lack of the supplementary branches—assuming great numerical superiority in infantry. The higher the quality of the infantry, the easier this will be.

3. It is harder to do without artillery than without cavalry: artillery is the principal agent of destruction, and its use in action is more closely coordinated with the infantry's.

4. In general, artillery being the strongest agent of destruction and cavalry the weakest, one is always confronted with the question of how much artillery one can have without it being a disadvantage, and with how little cavalry one can manage.

CHAPTER FIVE

The Army's Order of Battle

By order of battle we mean the distribution and composition of arms as individual parts of the whole, as well as the disposition which will serve as a standard form during the whole campaign or the duration of the war.

Thus in a sense the order of battle consists of an arithmetical and a geometrical component: *organization* and *disposition.* The former emanates from the army's normal organization in peacetime; certain parts, such as battalions, squadrons, regiments, and batteries, are treated as units that serve as building-blocks for larger structures, which in turn form the whole, depending on the requirements of the moment. Similarly, the army's disposition starts from the basic tactics in which it has been instructed and trained in time of peace—characteristics not susceptible to basic change once war has broken out. This, together with the conditions requiring the use of the troops in war and on a large scale, sets the standards according to which the army is deployed for battle.

This has always been the practice where large armies took the field, and there have even been times when the order of battle was considered the most important part of the action.

In the seventeenth and eighteenth centuries, when the development of firearms caused a great expansion of the infantry and made it possible to deploy soldiers in long, thin lines, the order of battle was certainly simplified, but its handling also called for greater skill. Furthermore, since the only place for cavalry seemed to be on the wings—out of the range of fire, and with room to deploy—the army, once arranged in order of battle, became a solid, indivisible whole. If such an army was split in the middle, it was like an earth-worm cut in half: both ends were still alive and able to move, but they had lost their natural functions. The fighting forces were thus held in what amounted to a bondage of coherence: a minor feat of dislocation and reorganization became necessary whenever a segment had to be deployed separately. When the army as a whole had to undertake a march, it found itself, so to speak, outside its proper element. When the enemy was close at hand, the marching-order called for the utmost ingenuity in order to keep one line or one wing at the proper distance from the other no matter what obstacles it encountered. It was forever necessary to steal marches on the enemy; and this kind of theft was able to escape punishment only because the enemy lay in the same bondage.

In the second half of the eighteenth century it was discovered that cavalry could protect the wings just as well by being posted behind the army as by forming an extension of the line; and, furthermore, that it could be used for purposes other than merely fighting duels with the enemy's cavalry. This was a great step forward, if only for the reason that the whole of an army's front—the width of its disposition—was now composed of homogeneous units: it could be dissected into any desired number of units which were similar to each other and to the whole. At this point the army ceased to be a monolith and became a many-jointed entity which was pliant and flexible. Units could be easily detached and reattached without disturbing the order of battle. This was the beginning of corps made up of all arms—or rather, this was what made them feasible; the need for them had long been felt.

It is understood that all this originated in battle, which used to be the whole of war and will always be its main element. But order of battle is really more a matter of tactics than strategy, and our only point in tracing its development here is to show how tactics, by reorganizing the whole army into smaller units, has paved the way for strategy.

With the increased size of armies and their deployment over wider areas, and the more their individual parts could integrate effectively, the more the scope for strategy expanded. The order of battle, as we have defined it, was, therefore, bound to interact with strategy; and the interaction is most marked at those points where strategy and tactics meet—in other words, where general deployment of armies passes into actual dispositions for battle.

We now turn to the subjects of *organization, combination,* and *deployment of forces,* as seen from the point of view of strategy.

1. *Organization.* From a strategic point of view one should never ask what the strength of a division or corps ought to be. The proper question is how many divisions or corps an army should have. There is nothing more unwieldy than an army split in three parts, save possibly an army split in two. In the latter case the general in command will be practically paralyzed.

To determine the strength of major or minor corps on tactical or operational grounds leaves incredible scope for guesswork, and the most fantastic arguments have been let loose on it. By contrast, it is clear and accepted that an independent whole such as a corps must have a certain number of parts. That fact permits the use of genuinely strategic reasons for settling the number of units that a major force should contain, and what their strength should be; meanwhile, the strength of smaller units, such as companies, battalions, and so forth, can be based on tactical grounds.

It is hard to conceive of even the smallest independent whole without three distinguishable parts: one to be sent ahead of the main force and another to bring up the rear. Four parts would obviously be even better, since the middle part, being the main body, ought to be stronger than either of the other two. One can proceed to eight, which, in our opinion, is the optimum number for an army, assuming that one part will always be needed for the vanguard, three for the main body—a right wing, center, and left wing—two for the reserve, and one each for dispatching to right and left.

It would be pedantic to insist on these numbers and figures; still, in our view, they reflect the normal, most common strategic pattern of deployment, and consequently a convenient system of articulation.

There is no denying that the supreme command of an army (and the command of any independent force) is markedly simpler if orders only need be given to three or four other men; yet a general has to pay dearly for that convenience in two ways. First, an order progressively loses in speed, vigor, and precision the longer the chain of command it has to travel, which is the case where there are corps commanders between the divisional commanders and the general. Second, a general's personal power and effectiveness diminishes in proportion to the increase in the sphere of action of his closest subordinates. A general can make his authority over 100,000 men felt more strongly if he commands by means of eight divisions than by means of three divisions. There are various reasons for that; the most important being that a subordinate commander thinks he has a kind of proprietary right over every part of his corps, and will almost invariably object to any part being withdrawn for however short a time. Anyone with any experience of war will be able to understand this.

On the other hand, the total number of parts must not become so large that confusion will result. It is hard enough to manage eight subdivisions from one headquarters; ten is probably the limit. In case of a division, however, in which there are far fewer means for transmitting orders into action, four, or at the most five subunits, must be considered the appropriate figure.

If these figures of five and ten do not work out—in other words, if the brigades become too large—then corps headquarters must be inserted. But one must bear in mind that this adds another *power* to the chain of command, while simultaneously reducing all the others.

And anyway, when is a brigade too large? Its usual size is two to five thousand men, and there seem to be two reasons for this upper limit. The first is that a brigade is meant to be a unit that one man can directly command by the power of his own voice. The second is that a large body of infantry should not be left without artillery. A combination of these two factors will automatically produce a special unit.

We do not want to get too deeply involved in these tactical subtleties, nor do we mean to go into the controversial question where, and in what proportions, all three arms should be combined—whether in divisions of 8,000 to 12,000 men, or in corps of 20,000 to 30,000. But surely even the strongest opponents of such combinations will not deny that only a combination of the three arms can *make a unit of the army independent*. Combined arms are therefore desirable, to say the least, for any unit that frequently finds itself operating in isolation.

An army of 200,000 men in ten divisions, with five brigades to a division, will have 4,000 men to a brigade. There is no disproportion in this. Of course the army may be divided into five corps, each with four divisions of four brigades, and the brigades would then be 2,500 strong. But considered purely in the abstract the first arrangement seems preferable. Not only does the second contain an extra level of command, but five subdivisions are too few for an army. They make it unwieldy, and the same is true of four units per corps. Also, 2,500 men would make a weak brigade, and there would be eighty of these as against fifty under the first arrangement, which is therefore simpler. All these advantages are abandoned for the sake of reducing by half the number of generals to whom one has to give orders. In the case of smaller armies, it must be obvious that dividing them into corps would be even less suitable.

This is the abstract view of the matter. A given case may suggest different decisions. It may be entirely possible to command eight or ten divisions concentrated on a plain, but it may well be impossible if they are widely scattered in mountainous terrain. An army cut in two by a broad river needs a commander for each half. In short there are hundreds of cogent local and special conditions to which the abstract rule must yield.

Experience shows nevertheless that abstract reasons are used more frequently and thrust aside less often than one might suppose.

We should like to clarify the scope of these observations by means of a simple outline, which will summarize the salient points side by side.

If the term "subdivisions of the whole" is taken to mean only the *first*, or *immediate*, components, we argue that:

a. The whole will be unwieldy if it has too few subdivisions.

b. If the subdivisions are too large, the commander's personal authority will be diminished.

c. Every additional link in the chain of command reduces the effect of an order in two ways: by the process of being transferred, and by the additional time needed to pass it on.

It follows that the number of subdivisions with equal status should be as large as possible, and the chain of command as short as possible; the only qualification being that command is difficult to exercise over more than eight to ten subdivisions in an army, and over more than four to six of them in smaller units.

2. *Combinations of Arms of the Service.* From the strategic point of view, a combination of forces in the order of battle is important only for those parts of the whole that, under ordinary conditions, might be stationed separately, and forced into a separate engagement. It is in the nature of things that it is the major units, and substantially *only those*, that are deployed separately. The reason, which we shall demonstrate elsewhere, is that detached positions are in most cases based on the idea of, and the need for, an independent whole.

In a strict sense, therefore, strategy would require that only corps, or, in their absence, divisions should be made up of a permanent combination of all arms. In the case of less significant units, temporary combinations made to meet the needs of the moment should suffice.

But it is clear that a substantial corps, say 30,000 to 40,000 men, will seldom be found operating as an undivided whole. Any corps of that strength will need a combination of weapons within its divisions. Anyone who would deny the delays that occur not to speak of the confusion that results, when in order to aid the infantry a cavalry unit has to be sent from somewhere else—possibly a good way off—would betray a total lack of operational experience.

The actual combination of all three arms, to what lengths it should go, how close it should be, what proportions it should take, and how much of each should be held in reserve—all these are purely tactical problems.

3. *Deployment*. The arrangements determining the relative position of each formation in the order of battle are also a completely tactical matter, relating only to the battle itself. There is, of course, a strategic deployment, but that depends almost entirely on the dispositions and requirements of the moment. If a rational basis does exist, it is not included in the meaning of the phrase "order of battle." It will therefore be examined elsewhere under the title "Disposition of the Army."

The battle order of an army is thus its organization and disposition *in a body ready for battle*. Its parts are organized in such a way that they can easily be detached from the body and used to meet the demands of the moment, tactical as well as strategic. When these momentary needs no longer exist, the parts will return to their original positions. Thus the order of battle is the first stage and the main basis of that salutary routine which, like a pendulum, regulates the mechanism of war, and which has already been discussed in Chapter Four of Book Two.

CHAPTER SIX

General Disposition of the Army

In most cases there is a great interval between the initial mobilization of the forces and the moment when all decisions have been made; when strategy has brought the army to the crucial point, and tactics has assigned to each unit its place and role. The same holds true of the time between one decisive catastrophe and another.

In the past these intervals did not really seem to be part of war. Take for example Marshal Luxembourg and his methods of encamping and marching. We single him out because this is what he is famous for; he can thus be considered typical of his era. Besides, we have more information about him, from the *Histoire de la Flandre militaire,*[1] than about any other general of that period.

Camps used to be pitched with their rear close to a river, a swamp or a deep ravine—a practice we should consider madness today. The enemy's position did not generally determine the orientation of the encampment, so that frequently the rear was turned toward the enemy, and the front toward one's own territory. This type of arrangement, which now strikes us as incredible, can be understood only where the principal, indeed practically the only factor in the choice of a campsite is that of convenience. Troops in camp were not considered to be in a state of war; they were behind the scenes, so to speak, and entitled to relax. The only security provided was to make sure that the rear was covered by some obstacle—security only in the sense understood at the time. It did nothing to meet the possibility of having to fight an engagement within the camp. But this risk was minimal; engagements were based on a kind of mutual understanding, much like a duel that has been arranged at a meeting place convenient to both parties. Not every type of terrain was suitable for the battles of those days—partly because of the large numbers of cavalry which, though in the twilight of its splendor, was still considered (especially by the French) as the principal arm; and partly because of the armies' unwieldy order of battle. So rough country was almost as safe for troops as neutral territory. An army could make little use of it, and it was considered preferable to go out to meet an enemy who was advancing for battle. We know of course that the battles of Fleurus, Stenkerken, and Neerwinden were conceived by Luxembourg in a different spirit; but under this great commander that spirit was then only in the process of evolving from the earlier methods, and it was not yet reflected in the system of encampment. Indeed changes in the art of war

[1] I.e., de Beaurain, *Histoire militaire de Flandre* (Paris, 1755). Eds.

always emanate from decisive actions, which, little by little, tend to modify the rest. The state of encampment was considered to have little connection with the actual state of war, as is proved by the expression, *il va à la guerre,* which was used to describe a patrol setting out from camp to observe the enemy.

It was much the same with marches. The artillery went its own way in order to travel on safer and better roads, while the cavalry generally alternated on the flanks in order to give every unit in turn the honor of riding on the army's right.

In our day, and particularly since the Silesian wars, the condition of the troops when not in action is permeated by the conditions of combat, and the mutual relationship of the two is so close that one condition can no longer be considered complete without the other. Time was when the engagement was the sword proper, while the period between engagements was the hilt; the former was the steel blade and the latter the wooden handle glued to it. The whole was made up of unrelated parts. In our day the engagement is the edge of the sword, and time out of action is its reverse edge. The whole is so thoroughly welded together that it is not possible to distinguish where the steel starts and the iron ends.

Nowadays the existence of an army between engagements is governed partly by its peacetime rules and regulations, and partly by the tactical and strategic dispositions of the moment. There are three conditions possible for military forces: in billets, on the march, or in camp. Each can be as much a part of tactics as of strategy. In this context, tactics and strategy often come so close that they seem to, or actuallv do, overlap. Consequently there will be many dispositions that are simultaneously both tactical and strategic.

We propose to discuss the three conditions between engagements in a general way before moving on to their specific possibilities. We must first examine the general disposition of the forces, since this provides the context for billets, camps, or marches.

In discussing the disposition of forces in general—that is, without assigning particular purposes, we have to consider them as a single entity—that is, a whole *meant to fight together.* Any deviation from that primary form must necessarily imply some particular purpose. From this arises the idea of an army, no matter what its size.

Further, if no particular purpose is assigned to an army its sole concern will be its own *self-preservation* and consequently its *security.* An army must be able to exist without any particular difficulty, and to fight as a unit. These are its two requirements. In practice they lead to the following considerations concerning the existence and security of the army:

1. Ease of supply
2. Ease of quartering
3. Security of the rear
4. Open ground to its front

5. The position itself on rough ground
6. Strategic points of support
7. Appropriate subdivision of forces.

Our comments on these considerations are as follows.

The first two render it desirable to look for agricultural areas, large towns, and main roads. Their influence is general rather than particular.

The meaning of "security of the rear" will be explained in Chapter 16, which deals with lines of communication. The first and foremost requirement here is that the position form a right angle with the direction of a nearby main line of retreat.

On the fourth point, it is not really possible for an army to command a whole stretch of territory in the same way as it commands the battlefield when tactically deployed. It must use the vanguard as its strategic eyes, sending out individual detachments, spies, and so forth, whose task of observation will naturally be easier in open than in undulating country. The fifth point is merely the reverse of the fourth.

Two features differentiate strategic points of support from tactical ones. First, the army need not be in direct contact with them; and second, they must be of much greater extent. The reason for this lies in the fact that strategy, by its very nature, moves in dimensions of space and time that are greater than those of tactics. If an army takes up a position a mile away from a coastline or a major river, it is considered to rest on them strategically, for the enemy will have no room to make a strategic turning movement. He will not commit himself to prolonged marches lasting for days and weeks in this space. On the other hand, a lake several miles in circumference hardly poses a strategic obstacle. A few miles to one side or the other makes little difference to strategy. Fortresses are strategic points of support in proportion to their size and effective sphere of operations.

The deployment of the army in separate sections may reflect either special purposes and requirements, or general ones; only the latter will be considered here.

The first general requirement is to push out an advance guard, together with other units required to observe the enemy.

The second is that in very large armies reserves have to be stationed several miles in the rear, leading to deployment in separate sections.

Finally, separate corps usually have to be detached to cover the flanks.

This need for cover does not of course mean that part of the army has to be detached in order to protect the area on its flanks—the so-called weak point—from the enemy. In that case, who would defend the flank of the flank? Popular though it is, this type of notion is utter nonsense. Flanks are not in themselves weak points, for the simple reason that the enemy has flanks as well, and cannot endanger ours without incurring a similar risk to his. The flanks will only become weak points where circumstances no longer balance each other, where the enemy army is superior in strength, and where the enemy's lines of communication are more secure (see Chapter 16, on Lines of Communication). But these are special cases, which do not concern us here; neither does the case in which the duties of a flanking corps include the actual defense of the space to our flank. That does not fall within the category of general requirements.

While the flanks may not be particularly *weak* points, they are particularly *important* ones. Defense against flanking movements is not so simple a matter as defense to our front; the necessary measures become more complicated and demand more time and preparation. This is why one usually has to take particular care to guard one's flanks against unforeseen operations by the enemy. It can be done by placing stronger forces on the wings than are needed simply to observe the enemy. To dislodge these forces, even when they do not offer any serious resistance, the enemy will have to spend time and deploy his forces and intentions in proportion to the size of the force to be dislodged—and our purpose will then have been achieved. Further steps will depend on the actual plans of the moment. Corps posted on the flanks may thus be seen as lateral vanguards, charged with delaying the enemy's penetration of the areas beyond the flanks, and thereby providing time for countermeasures.

If these units are to fall back on the main force when the latter is not withdrawing at the same time, they must obviously not be posted in line with the main body, but somewhat ahead of it. Even where their retreat involves no serious fighting, these units should not fall back simply in line with the main body.

These inherent reasons for adopting a divided disposition, then, give rise to a natural system of four or five separate parts, according to whether or not the reserve is stationed with the main force.

Not only do supply and quartering influence dispositions; they are also a contributory factor to a division of forces. The consideration of these two points connects with what we have already said. An effort must be made to meet the one without sacrificing too much of the other. In most cases, subdivision into five separate corps will in itself eliminate the difficulties of providing shelter and supplies, and will therefore necessitate no major changes.

We must still consider the distances at which these separate formations have to be stationed if they are to serve the aim of mutual support—in other words to fight as one. We should here like to recall what we said in the chapters on duration and decision of the engagement. So much depends on absolute and relative strength, on weapons and terrain, that no iron-clad regulation can be formulated; only a general one, a sort of average measure.

The distance of the advance guard is easiest to determine. Since it has to fall back on the main body of the army, it can be placed at most a good day's march away without incurring the risk of having to fight a separate battle. But it should not be posted any further ahead than necessary for the security of the army, since the further it has to retire, the more casualties it will suffer.

As to the flanking corps, it has already been pointed out that an engagement fought by an ordinary division of 8,000 to 10,000 men usually lasts several hours or even half a day before a decision is reached. Therefore such a division may without hesitation be placed at a distance of a few hours' march—say five to ten miles away. For the same reasons, a corps consisting of three or four divisions may be stationed a day's march away—about fifteen to twenty miles.

The very nature of things, then, determines a general disposition of the main body in four or five parts and at given distances. This creates a certain routine method which automatically determines the subdivision of the army, provided no special objectives intervene.

Even though we have stipulated that each of these separate parts must be able to fight on its own, and that they may be forced to do so, it does not necessarily follow that the *actual point* of a divided disposition is that isolated actions should be fought. The necessity for such a distribution usually only arises from the conditions imposed by the factor of time. When the enemy approaches to seek a decision by means of a general engagement, the strategic phase is over, and everything focuses on the moment of battle. All reason for divided disposition is now at an end. Once the battle is joined, questions of quartering and supply are suspended. Observation of the enemy on the front and flank and reducing his impetus by moderate resistance have served their purpose. Everything becomes part of the greater whole, the main battle. The best criterion of the value of a divided disposition is that the separate distribution is a conditional state and a necessary evil, while fighting with all forces combined is the true purpose.

CHAPTER SEVEN

Advance Guard and Outposts

Advance guards and outposts belong to the category of measures where the threads of tactics and strategy are interwoven. On the one hand, they shape the engagement and ensure that the tactical plan is carried out; on the other, they often lead to separate engagements. Besides, since these units are stationed at some distance from the main body of the army, they can be considered as links in the chain of strategy. Their isolation prompts us to give them a moment's study in order to supplement the preceding chapter.

Any force that is not completely ready for battle needs an advance guard to detect and reconnoiter the enemy's approach before he comes into view. After all, a troop's range of vision does not usually extend much beyond its range of fire. How unfortunate it would be if our eyes could see no further than our arms can reach! It has been said that outposts are the eyes of the army. But the need for them varies. It is affected by strength and extent, time and place, circumstances and the type of war being fought, even by chance. We should, therefore, not be surprised if military history yields no definite or simple rules on the use of outposts and advance guards, but rather a jumble of diverse examples.

Thus we find cases in which an army entrusts its safety to one particular corps of the advance guard or to a long line of isolated outposts. The two may be combined, or neither may exist; one guard may be shared by several advancing columns, or each may have its own. We shall try to set out the matter clearly, and then find out whether it can be reduced in practice to a few principles.

Troops on the move are preceded by a force of varying strength as vanguard, or advance guard, which will become the rear guard should the line of march be reversed. When the troops are in billets or in camp, the advance guard takes the form of an extended line of lightly held pickets—the outposts. Essentially, a larger area can and must be covered when the army is halted than when it is on the move. So the natural result in the former case is a line of posts, and in the latter a concentrated corps.

The individual strength of advance guards and outposts can range from a substantial corps, composed of all three arms, to a regiment of hussars; from a strongly entrenched defensive line held by a combined force to mere outlying pickets and scouts sent out from the camp. Consequently their effectiveness ranges from simple observation to actual resistance. The point of such resistance may not be simply to give the main force the time it needs to prepare for battle, but also to make the enemy disclose his dispositions and intentions prematurely. In that case the value of observation is substantially increased.

Consequently, the strength of advance guards and outposts will vary according to whether the main body needs to gain more or less time, and according to whether its resistance is more or less dependent on the particular disposition the enemy may make.

Frederick the Great, who always had his forces ready for action and could practically lead them into battle by word of command alone, did not require strong outposts. That is why he would camp right under the enemy's eyes, without any elaborate security system. He would rely on a regiment of hussars, a light battalion or some scouts and pickets dispatched from camp. On the march his vanguard consisted of a few thousand horse, usually part of the cavalry on the flanks of the first line, which would rejoin the army when the march was over. He seldom kept a permanent corps for vanguard duty.

Where a small army wants to make use of its concentrated weight and greater elasticity and take advantage of its superior training and more determined leadership, everything must be done as did Frederick, when he was facing Daun, practically *sous la barbe de l'ennemi*. His superiority would have been rendered ineffective by deployment of forces far to the rear, and a cumbersome system of outposts. That errors and excesses could lead to defeat as at Hochkirch proves nothing against this type of procedure; on the contrary, the King's mastery of it is demonstrated by the fact that in all the Silesian wars only one battle of Hochkirch occurred.

Bonaparte, on the other hand, who can hardly be said to have been lacking in a steady army and resolute leadership, almost always used a strong advance guard. Two reasons account for this.

The first was the change that had occurred in tactics. Nowadays an army is no longer led into battle as a compact whole and by word of command alone, in order to settle the score more or less by dint of skill and bravery, like a large-scale duel. Today the peculiarities of the terrain and the general circumstances are taken more into account. The battle is composed of a number of different parts. What used to be a simple decision has become a complex plan, and the word of command has turned into lengthy dispositions, based on time-tables and other data.

The second reason is the increased size of modern armies. Frederick went into action with 30,000 to 40,000 men; Bonaparte with 100,000 or 200,000.

We have chosen these two examples because one can assume that generals of that caliber would not have adopted such systems unless they had good reason. On the whole, the use of advance guards and outposts has developed more broadly in our day. But the Austrian practice shows that Frederick's method was not universal in the Silesian wars. Their outpost system was much stronger and they far more frequently used a corps as advance guard—which their situation and resources amply warranted. The most recent wars also provide numerous variations. Even French Marshals—Macdonald in Silesia, for example, or Oudinot and Ney in Brandenburg—advanced with armies of 60,000 to 70,000 men, but there is no mention of any corps serving as vanguard.

So far, we have discussed advance guards and outposts from the point of view of their strength. But there is another distinction that we must make clear. An army advancing or retreating over a certain front may have a single vanguard and rear guard for all its parallel columns, or each column may have its own. In order to elucidate, let us consider the matter in the following way.

When a corps is designated as a vanguard, its basic task is to cover only the advance of the main force in the center. If the main force uses several adjacent roads which may also be taken by the vanguard, and consequently covered by it, the flanking columns will of course need no special measures of protection.

By contrast, the corps that advance at a greater distance and are operationally independent must provide their own vanguard. The same applies to any corps that form part of the central main force if, because of the lie of the land, they find themselves too far from the center. Therefore, there will be as many vanguards as there are separate bodies advancing along parallel lines. If individually they are much less effective than one general vanguard would have been, they really belong in the general category of tactical dispositions, and the vanguard will vacate its place on the strategic board. If, however, the central main force has a significantly larger corps for a vanguard, this will appear to be the vanguard of the whole army, and will in many ways act as such.

We can find three reasons for assigning a much stronger vanguard to the center than to the flanks.

1. The center usually consists of a larger body of troops.
2. The central point of the tract of country over which an army is spread will always be the most important part; all plans will tend to concentrate on it, with the result that the battlefield is usually closer to it than to the flanks.
3. Although a corps stationed ahead of the center cannot really be an advance guard for the wings as well, it can give them a good deal of cover indirectly. Normally the enemy cannot pass within a certain distance to strike a serious blow at one of the flanks, for he will risk an attack on his own rear or flank. The constraint on the enemy that is caused by a corps pushed ahead from the center may not be sufficient to assure the complete security of the lateral corps, but it does eliminate a number of threats to them.

So in cases where the vanguard covering the center is much stronger than that of the flanks—when, in other words, it consists of a special advance corps—it no longer simply serves the purpose of a vanguard, which is to shield the troops immediately behind it from a surprise attack: as an advance corps, it will have a wider strategic role.

The use of such a corps is based on the following objectives, which therefore also determine its application:

1. In cases where our dispositions require a lot of time, it can put up a stronger degree of resistance, and so impose greater caution on the advancing enemy. In this way it augments the effect of an ordinary vanguard.

2. Where the main body is very large and unwieldy, it can to some extent be kept in the background, while contact with the enemy is maintained by means of a more mobile corps.

3. Though other reasons may necessitate keeping a considerable distance between the main body and the enemy, it is useful to keep a corps close to him for purposes of observation.

The idea that observation could be carried out just as well by a minor reconnaissance post or mere patrols will not hold up when one considers how easily these can be dislodged and how inferior are their means of observation in comparison to those of a large corps.

4. In pursuit of the enemy a corps of advance guard to which the bulk of the cavalry should be attached can move much faster on its own; it can keep going until a later hour of the night, and be ready for action earlier in the morning than could the whole army.

5. Finally where, in retreat, it forms the rear guard, it can utilize major natural obstacles for defense. Here too, the center is exceptionally important. True, at first sight one might suppose that a rear guard of that sort would be in constant danger of being outflanked. But we must remember that even if the enemy is pressing hard on the flanks, he still has to cover the distance to the center if he wants to threaten it seriously; which allows the rear guard covering the center some extra time for resistance and for delay in its own withdrawal. If on the other hand the center falls back sooner than the flanks, things look grave indeed; it will seem that the enemy had broken through the line, and one must avoid giving this impression. The need for unity and cohesion is never greater or more acutely felt by everyone than during a retreat. The flanks are ultimately meant to reconverge with the center; even if problems of supply and terrain impose the necessity of spreading out during a withdrawal, the maneuver usually ends with a concentration of forces in the center. If we add the fact that the enemy normally advances on the center with his main force, and consequently exerts the greatest pressure there, it will be plain that the center's rear guard is of particular importance.

It follows that the use of a separate corps for the advance guard is suitable whenever one of the above situations arises. Most of them will hardly ever occur where the center is no stronger in manpower than the wings. This was the case when Macdonald advanced against Blücher in Silesia in 1813, and the latter moved toward the Elbe. Both had three corps which usually moved side by side in separate columns by different roads; which explains why there is no mention of a vanguard on either side.

Nevertheless partly for this very reason an arrangement in three columns of equal strength is anything but advisable, just as it is very awkward to divide an army into three parts, as we argued in Chapter Five of Book Three.[1]

In the previous chapter we maintained that, in the absence of special orders to the contrary, the most natural way of disposing a force is a concentration in the center, with two separate wings. In that case, the simplest scheme will place the advance guard in front of the center, and so also ahead of wings. But the basic function of the lateral corps in relation to the wings is the same as the advance guard's function in relation to the front: and consequently, the lateral corps may often be in line with the vanguard, if not actually ahead of it, depending on circumstances.

Not much can be said about the absolute strength of the advance guard. These days it has justifiably become general usage to assign its task to one or more of the main subunits of the whole army, which are reinforced by part of the cavalry. It will thus consist of a corps if the army is divided into corps, and of one or more divisions if the army is divided into divisions. It will be obvious that in this respect as well, the advantage lies where the army is subdivided into the larger number of units.

The distance at which the vanguard should be stationed depends entirely on circumstances. There may be cases of its being more than a day's march ahead of the main force; in others, it may be immediately in front of it. In the majority of cases it will be found between five and fifteen miles away. While this proves to be the distance most frequently required by circumstances, it should not become a rule for universal application.

Our discussion has lost sight of the subject of *outposts*, to which we shall now return.

When we stated at the beginning that outposts will be required for stationary troops and vanguards for troops on the move, it was in order to determine the origin of the concepts and to keep them separate for the time being; but strict adherence to terms would clearly result in little more than pedantic distinctions.

When an army on the march rests for the night, its vanguard must of course do the same, and must always set up outposts for its own security as well as for that of the whole army. That does not mean it is no longer an advance guard and has been reduced to nothing more than outposts. The only time when outposts may be considered to be the opposite of the concept of the vanguard is when most of the troops making up the advance guard are split up into separate posts with little or nothing of a concentrated corps remaining—in other words, where the concept of a long line of outposts is dominant over that of a concentrated corps.

The briefer its rest, the less the army needs to be completely covered: a single night is too short, at any rate, for the enemy to find out what is covered and what is not. The longer the rest, the more completely each point of access must be watched and covered. So a substantial halt will tend

[1] Book Five is meant. Eds.

as a rule to spread the vanguard into a line of outposts. Two circumstances chiefly decide whether it will do so completely, or whether the principle of a concentrated corps will be retained. The first is the distance between the hostile armies; the second the nature of the terrain.

If the distance between the opposing armies is short in comparison with the extension of their fronts, there will frequently be no room to position an advance guard corps. Each army must then rely for its security on a series of small posts.

In any case a concentrated corps, which is less able to cover the approaches directly, will need more time and space to become effective. Consequently, wherever the army is considerably spread out, as in billets, it must be a good distance away from the enemy if all approaches are to be properly covered by a concentrated corps. That is why winter quarters, for example, are normally covered by a chain of outposts.

The second factor is the nature of the terrain. For instance where a natural formation of the land affords the opportunity to form a strong line of posts with relatively few troops, such a line will be used.

In winter quarters, the severity of the weather may also be a reason for dividing the advance guard into a line of posts: shelter is more easily found for it that way.

The peak of development in making use of a line of reinforced outposts was reached by the Anglo-Dutch army in the Netherlands during the winter campaign of 1794-1795. There the defensive line consisted of brigades of all arms forming separate outposts, the whole system being supported by a general reserve. Scharnhorst, who was with the army, introduced this practice into the Prussian army on the Passarge in East Prussia in 1807. With this exception, it has rarely been used in modern wars—mainly because they have been too full of movement. But it has even been neglected where it might have come in useful: by Murat, for example, at Tarutino. If his defensive line had been extended, he would not have lost some thirty guns in an attack on his outposts.

There is no denying that this procedure offers great advantages under the right conditions. We shall revert to it at a later occasion.

CHAPTER EIGHT

Operational Use of Advanced Corps

We have just explained how an army's security is affected by the way the approaching enemy reacts to its advance guard and flanking corps. In a possible conflict with the enemy's main force, these corps must be considered very weak. It requires a special discussion to show how they can serve their purpose without suffering heavy casualties in consequence of being so greatly outnumbered.

The task of these troops is to observe the enemy and slow down his advance.

Not even the first of these tasks would be served by a small detachment, partly because it would be more easily driven back than a large one, and partly because its means, its tools of observation, would not be sufficiently powerful.

Moreover, observation is meant to serve an ulterior purpose; namely, to induce the enemy to deploy his full forces and so reveal not just his strength but his intentions.

To do no more than that, the mere presence of such a corps would be enough: it would need only to await the enemy's offensive deployment and then fall back.

But it is also meant to delay the enemy's advance, and that involves real resistance.

How is it possible to think of waiting till the last moment and of putting up resistance without the corps running the continual risk of heavy casualties? Chiefly because the enemy will approach with his own advance guard, rather than with the crushing and overwhelming power of his whole army. It is conceivable that from the outset the enemy's advance guard is superior to ours—as it is presumably intended to be—and his main force nearer to it than we are to our own; and, since his main force is already on the move, it will soon be in a position to give full support to the attack of the advance guard. Nevertheless, the opening phase, during which our advance guard has to deal with that of the enemy—its equal, so to speak—will gain some time and permit observation of the enemy's approach for a while without jeopardizing its own retreat.

Even some resistance put up by such a corps in a suitable position does not entail all the disadvantages that one might at other times expect from its numerical inferiority. The chief danger arising from resistance to a superior enemy is the possibility of being outflanked and placed at a definite disadvantage by an enveloping attack. But this risk is usually diminished by the fact that the advancing enemy can never be quite sure how near the support of the main army may lie, and that he therefore has to consider the possibility of getting his columns caught between two fires. An advancing army will, as a result, keep the heads of its columns more or less abreast; only after thoroughly reconnoitering his opponent's position will he start cautiously and watchfully to turn one flank or the other. This phase of wary probing, then, enables our advanced corps to withdraw before any serious threat to it develops.

For how long such a corps may resist a frontal attack and the start of a turning movement will depend mainly on the nature of the terrain and the proximity of support. If resistance is stretched over a longer period than normal—whether as a result of poor judgment or of self-sacrifice in order to buy time for the army—heavy casualties are bound to follow.

Only in the rarest of cases, when a major natural feature can be utilized, will the actual defensive fighting be of any consequence; by itself, the minor engagement that such a corps is able to fight can rarely gain enough time. Rather this gain will result from three circumstances inherent in the situation:

1. The enemy's more cautious and therefore slower advance
2. The duration of actual resistance
3. The withdrawal itself.

The withdrawal must be made as slowly as safety will permit. Any good natural position that is available should be used. It will compel the enemy to work out fresh attacks and turning-movements, and so gain more time. Even a real engagement may prove acceptable in a new position.

It will be obvious that the delaying action is closely linked to the withdrawal. The frequency of engagements will have to make up for the shortness of their duration.

This is the way an advance corps can resist. Its effectiveness depends primarily on its own numerical strength and the terrain; also on the distances it has to cover, and the support and protection it receives.

Even on equal terms, a small body of men cannot hold out so long as a large corps: the more numerous the forces involved, the longer it takes them to complete any action of whatever type. In a mountainous area, marching itself becomes slower, individual positions may be held longer and at a smaller risk, and opportunities to assume such positions present themselves at every turn.

The farther forward an advanced corps is posted, the farther it has to retreat, and therefore the greater the absolute gain in time resulting from its resistance. On the other hand, its isolation limits its capacity to resist and its prospects of support. In that case its withdrawal will take relatively less time than if it had been closer to the main force.

The support and protection that such a corps receives is bound to affect the duration of its resistance. Caution and watchfulness take a toll that must be subtracted from its effectiveness.

The time gained by the resistance of an advanced corps may vary considerably. If the enemy does not appear until after midday this is so much more time gained, for armies seldom advance by night. Thus, in 1815, the first Prussian Corps, under General Ziethen, numbering about 30,000 men, faced Bonaparte with 120,000 men; yet, on the short stretch from Charleroi to Ligny—a bare ten miles—it was able to gain more than twenty-four hours for the concentration of the main Prussian army. General Ziethen was first attacked about nine on the morning of 15 June, and the battle of Ligny did not begin until about two in the afternoon of the sixteenth. General Ziethen admittedly did suffer heavy casualties—5,000 or 6,000 men killed, wounded and taken prisoner.

In terms of practical experience, the following may serve as a guidepost for calculations of this kind.

Take a division of 10,000 or 12,000 men, augmented by cavalry, which has been sent a day's march of fifteen to twenty miles ahead. In ordinary, not particularly difficult country, it will be able to delay the enemy for about half as long again (including the time of his own retreat) as would otherwise be required to cover the distance. But if the unit has been advanced for only five miles, the enemy may well be delayed for twice or three times as long as he would otherwise have spent on the march.

So a distance of twenty miles, which would ordinarily amount to a march of about ten hours, can be expected to take the enemy fifteen hours from the time he appears in strength before one's advanced division until he is in position to attack our main force. But if the advance guard is posted only five miles ahead, more than three or four hours, or, indeed, twice that time, can elapse before the enemy can attack the army. For the time it takes the enemy to deploy against the advance guard is the same in either case; while the period of resistance put up by the advance guard in its original position will be even longer than if it had been posted farther forward.

The upshot is that in the first case the enemy will be hard put to dislodge an advance guard and attack the main force, all on the same day, as experience tends to bear out. Even in the second case the enemy will have to dispose of the advance guard before noon in order to have enough time for the main battle.

Since in the first case night comes to our aid, it becomes clear that much time can be gained by posting the advance guard farther ahead.

The point of posting troops on an army's flank has already been explained, and their operations will usually depend on the particular circumstances. It is simplest to consider them as an advance guard stationed on the flank; and since they are posted slightly ahead, they will fall back toward the army in a diagonal direction.

Since such corps are not directly in front of the main force and are therefore harder to support than an ordinary advance guard, they would be in greater danger were it not that as a rule the impact of the enemy diminishes somewhat toward the extreme ends of the line. And even in the worst of situations, these corps would have room to maneuver without endangering the main force so directly as would the flight of the advance guard.

The reception of a retreating advance corps is best and most frequently done by a considerable cavalry force. This is a good reason for stationing one's reserve cavalry between the main force and the advance corps wherever the distance between them calls for it.

So our final conclusion is that an advanced corps derives its operational value more from its presence than from its efforts; from the engagements it might offer rather than from those it actually fights. It is never intended to stop the enemy's movements, but rather, like the weight of a pendulum, to moderate and regulate them so as to make them calculable.

CHAPTER NINE

Camps

We shall examine the three conditions of the army when not in action only from a strategic point of view—insofar as they concern place, time, and the strength of the forces. The transition to battle conditions, and everything pertaining to the actual course of the action, is a matter of tactics.

The disposition in camp—by which we mean all temporary quarters such as tents, huts, or bivouacs in the open—is, from the strategic viewpoint, completely identical with any engagement contingent on it. That may not always be so from the tactical viewpoint, for there are a number of reasons for choosing somewhat different locations for camping and for the intended battlefield. Since everything necessary has already been said about the disposition of the army, or rather the places to be allocated to its various parts, it remains only to consider the subject of camps from the point of view of the historian.

From the time when armies had again reached a considerable size, wars become longer and their separate actions more integrated up till the French Revolution, armies always camped in tents. That was their normal state. When spring arrived, they left their quarters and did not return before winter. Winter quarters should in a sense be considered as a state of non-war, in which the forces were in a neutral condition and the whole mechanism was for the time being arrested. Other types of quarters such as recuperation camps (which preceded winter quarters), or short-term, cramped billets, were transitional and must be considered as unusual.

This is not the place to inquire how such a regular, voluntary neutralization of power could have been, and in some cases still is, considered to be in keeping with the nature and aims of war. We shall address that question later on. Enough to say it was so.

Since the wars of the French Revolution armies have given up tents because of the mass of baggage they involve. It is now thought more advantageous for an army of 100,000 men to have another 5,000 cavalry or several hundred extra guns instead of 6,000 tent horses. Besides, a large baggage train is a great encumbrance and of little use in extensive and rapid operations.

But this change has resulted in two drawbacks: the increase, both in wear and tear on the troops, and in the devastation of the countryside.

The protection afforded by a roof of cheap canvas may not be much, but over a period of time it is a relief that will be missed by the troops when it is not there. For a single day the difference is slight; a tent gives little shelter against wind and cold, and is far from rainproof. But a slight difference becomes a major one when the situation recurs two or three hundred times a year. Increased losses due to sickness will naturally result.

There is no need to explain the way in which the absence of tents contributes to increased devastation of the countryside.

One would, therefore, think that because of these two drawbacks, doing without tents would have in its own way weakened the energy of warfare by forcing troops to stay in billets longer and more frequently, or, for lack of equipment, to forego the occupation of a number of positions that could have been occupied if tents had been available.

This would indeed have been true were it not for the fact that during that very period warfare underwent an enormous transformation that completely obliterated such minor and less significant effects.

The elemental fire of war is now so fierce and war is waged with such enormous energy that even these regular periods of rest have disappeared, and all forces press unremittingly toward the great decision. This fact will be treated more thoroughly in Book Nine.[1] Under those conditions there can be no question of a diminution in the use of the troops caused by the absence of tents. Troops now camp in shacks or in the open air, heedless of weather, season, or terrain, according to the general plan and purpose of the campaign.

Whether war will retain this energy forever and under all conditions is a question we shall deal with later. Wherever it does not, lack of tents will certainly have some bearing on its conduct, but we would doubt whether such a reaction would ever be strong enough to bring back the custom of camping in tents. The bounds of military operations have been extended so far that a return to the old narrow limitations can only occur briefly, sporadically, and under special conditions. The true nature of war will break through again and again with overwhelming force, and must, therefore, be the basis of any permanent military arrangements.

[1] There is no Book Nine. Eds.

CHAPTER TEN

Marches

Marches are a mere transition from one position to another. Two primary conditions are involved.

The first is the welfare of the men, so that strength that could have been usefully employed should not be wasted. The second is the organization of the march to ensure the punctual arrival of the troops. If 100,000 men were to march in a single column on one road without breaks in time, the tail of the column could never reach its destination on the same day as the head. Progress would either be extremely slow, or the column would, like a falling jet of water, break up into drops. Such a dispersion, together with the extra effort that the length of the column would impose on those at the tail, would soon result in general disorder.

If we can leave this extreme, the smaller the number of troops in a single column, the easier and more orderly the march will be. It follows that there is a need for *subdivision,* which is not the same as the subdivision stemming from divided tactical dispositions. Separation into marching columns will generally be based on the disposition of the force, but not necessarily in each specific case. If one wants to concentrate large numbers in one single area, they have to split up on the march. But even where a divided disposition brings about a divided march, the conditions of either the disposition or the march may predominate. For instance if no battle is expected and the troops are stationed merely with a view to rest, the predominant conditions will be those of the march, and these consist chiefly in the choice of good, well-surfaced roads. With this in mind, roads will in the former case be chosen with an eye to camps and billets, and in the latter, camps and billets will be chosen from the point of view of the roads. When a battle is expected, and the important thing is to get a number of troops to a particular point, even the worst by-road may be used without a moment's hesitation. On the other hand, if an army is still, so to speak, in transit to the theater of operations, one will choose the nearest main roads for the columns, and look for the best quarters and camps that happen to be available in the area.

Wherever there is a possibility of an engagement—in other words, within the whole theater of war—it is a general principle of modern war to organize the columns in such a way as to have their troops ready for independent action, no matter what the type of march. This condition will be met by combining all three arms, by an organic subdivision of the whole, and by a suitable allocation of command. Marches, then, are the principal basis for the modern order of battle, as well as being its principal beneficiaries.

CHAPTER TEN

About the middle of the eighteenth century, particularly in Frederick II's theater of operations, movement itself began to rank as an autonomous principle of fighting. Victories were being won by means of unexpected movements, but the lack of an organic order of battle necessitated the most complicated and laborious arrangements for the army on the march. In order to carry out a movement close to the enemy one had to be constantly ready for battle; but one could not be ready unless the army was concentrated, for only the army formed a self-sufficient whole. When the reserve had to keep in line with the main force, keeping its proper distance never more than a mile from the first, it had to go over hill and dale, with sweat and strain, involving a considerable knowledge of local conditions; for where will you find two paved roads running parallel only a mile apart? The same conditions obtained for the cavalry on the wings when one marched on the enemy in column. The artillery presented a further problem, for it needed a road to itself, with infantry protection; so artillery made infantry lines, which ought to be continuous but were long and irregular enough already, still longer. None of the proper distances could ever have been kept. One only need consult the march tables in Tempelhoff's *History of the Seven Years War* to gain an insight into these conditions and the constraints they imposed on warfare.

Modern war has given armies an organic system of divisions, in which major units form secondary wholes that are as effective in action as the whole army, except that they cannot operate for so long. Nowadays, even if the army is to fight as a whole, columns need no longer be kept together so as to be able to join up before the action begins. They can do so while the engagement is in progress.

The smaller a unit, the easier it is to move, the less it requires the kind of subdivision that results not from a divided order of battle but from the unwieldiness of its own mass. A small body of troops can march along a single road, and if it does have to advance along several lines sufficient adjacent paths can easily be found. The larger the body of troops, the greater will be the need to divide it, the greater the number of columns and the call for well-surfaced roads or even actual highways, and consequently the greater the distances between columns. Arithmetically speaking, the dangers of subdivision are in inverse ratio to its necessity The smaller the units, the more likely they are to have to go to one another's help. The larger they are, the longer they can look out for themselves. One should bear in mind what has been said on the subject in the previous book: in settled areas, fairly well surfaced parallel roads will be found at only a few miles' distance from the main road. It therefore becomes evident that, in planning a march, there will be no major problems that could render *speed and punctual arrival* incompatible with a *proper concentration of strength.* In mountainous areas, parallel roads are rare, and it is very difficult to make connections between them; on the other hand, a single column's power of resistance is much greater.

The matter will be clearer if considered briefly in concrete form.

Experience shows that a division of 8,000 men with its artillery and some other transport normally takes an hour to pass a given point. If two divisions move along a single road, the second will arrive about an hour after the first. As was pointed out in the sixth chapter of Book Four, a division of that size can usually hold its own for several hours, even against superior odds; so the second division ought to arrive in time even in the worst possible situation—that in which the first had been forced to go into action immediately on its arrival. Moreover, in the settled areas of Central Europe one can usually find usable *side roads* within an hour's distance of the main road. They may be used for the march without one's having to march cross-country, as so often happened in the Seven Years War.

Experience also teaches that the head of a column of four divisions with a cavalry reserve can generally do fifteen miles in eight hours, even on inferior roads. At the rate of an hour per division, and the same for the cavalry and artillery reserves, the march will take thirteen hours to complete. This is not too long a time, and yet enough for a total of 40,000 men to march along one road. With a force of this size, one can, in addition, seek out and use the side roads, and so easily shorten the march. If even more troops had to use the road, one could assume that they would not all have to arrive on the same day. Armies of that size do not nowadays give battle at the moment they meet, but usually wait till the next day.

We did not cite these cases in the belief that they cover all possible examples; we did so simply in order to make things clearer and to show, by examining actual experience, that in modern war it is no longer very difficult to organize a march. Prompt and rapid marches no longer require the special skill and the wealth of local knowledge which Frederick the Great, for instance, had to apply during the Seven Years War. Rather, with the present organic division of the army, marches almost organize themselves; at least they do not call for complicated planning. In contrast to the days when battles were conducted by word of command alone while marches called for formal plans, the battle is now what calls for planning, while marching hardly calls for more than the word of command.

Marches are either at right angles to the front or parallel to it. The latter are also known as "flank marches" and alter the geometrical pattern of the army's parts. Those that had been disposed in line will then be marching one behind the other, and vice versa. While the direction of the march can be at any angle inside of ninety degrees, the marching order must be of one definite kind or the other.

Such a geometric alteration can be perfectly accomplished only in the realm of tactics, and then only by using a so-called deployment in file, which is impossible where great numbers are involved. In the realm of strategy it is still less feasible. In the old order of battle, the parts that changed their geometrical relationship were only the center and the wings; today they are units of the first magnitude—corps, divisions, or brigades—depending on the distribution of the whole. Here too, the effects of the modern order of battle will make themselves felt; since it is no longer essential to assemble the whole army before the start of the action, it has become necessary to take greater care that the troops that are assembled will constitute a whole. If two divisions were to be disposed one in reserve behind the other, and were then to advance toward the enemy along two separate roads, we would not dream of sending part of each division by a different road. Each would naturally be given a road to itself: they would be told to keep abreast, and if fighting should occur it would be up to each divisional commander to provide his own reserve. Unity of command is much more important than the original geometrical relationship. If both divisions reach their destinations uneventfully, they can resume their original disposition. If two divisions stationed side by side are to march *parallel* to each other on two roads, it is even less likely to occur to anyone to send the second line, or reserve, of each division along the rearmost road; one road would be given to each of the divisions, and one of them would act as the reserve for the period of the march. If an army of four divisions were to advance against the enemy, with three divisions in the front and the fourth in reserve, it would be natural to assign one road to each of the three divisions and to make the reserve follow the middle one. But if one cannot find three roads close enough to be convenient, one need not hesitate to advance along two; no serious disadvantages are to be expected.

The same applies in the opposite case of a flank march.

A further point concerns columns that march off from the right or left flank. In the case of flank marches this works out automatically. One would certainly not march off from the right flank in order to move leftward. In a march to the front or rear, the marching order should be chosen with an eye to the direction of the road relative to the future line of deployment. This can indeed often be done in tactics, for the scale here is smaller and the geometrical relationships are easier to keep in view. From the strategic point of view it is completely impossible, notwithstanding certain analogies with tactics that have occasionally been suggested: these are sheer pedantry. Formerly, it is true, the whole order of march was purely an affair of tactics, since even on the march the army remained an undivided whole with a view to fighting only as a unit. For instance Schwerin, when he marched off from the area of Brandeis on the fifth of May, had no idea whether his battlefield would be to his right or to his left; hence he had to make his famous countermarch.

If in the old order of battle an army advanced against the enemy in four columns, the outer pair of columns were formed by the cavalry wings of the first and second lines, while the infantry wings of both lines formed the two center columns. All these columns could march off from the right or from the left, or the right wing from the right and the left wing from the left, or the right from the left and the left from the right. The name of the movement in the last of these cases would have been "double column from the center." Even though all these movements should have stood in direct relation to the subsequent deployment, basically they did nothing of the kind. When Frederick the Great went into battle at Leuthen, his army had marched off in four columns from the right wing. This made the transition to marching in line (which has been greatly admired by historians) very easy, since it happened to be the Austrian left wing that Frederick wanted to attack. Had he wanted to turn the right wing, he would have had to make a countermarch as he had done at Prague.

While such maneuvers were unsuited to their purpose even then, they would be downright frivolous today. The relation of the future battlefield to the road along which the army moves is as little known now as it was then, and the short time lost by marching off in the wrong order has become incomparably less important. Here too the new order of battle proves beneficial: it is quite immaterial which division is the first to arrive or which brigade is the first to come under fire.

Under these circumstances it has ceased to make a difference whether you march off to the left or to the right, except that it will balance out the troops' fatigue if it is done alternately. This is the only reason, though a most important one, for retaining the dual method of marching off, even where major bodies of troops are involved.

Under such conditions, marching off from the center automatically disappears as a distinct maneuver, and will only happen by accident. A march from the center in which the two middle columns move as one is strategic nonsense anyway, for it assumes the availability of twin highways.

The order of march is a tactical rather than a strategic matter, for it involves dividing the whole into parts that are to reform as a whole when the march is over. In modern war it is no longer thought necessary to keep the parts close to one another: during a march they are allowed to spread out further and to look out for themselves. This can easily result in engagements in which each of the parts fights alone, and which must therefore be considered complete in themselves. This is why we have found it necessary to go into such detail.

Since incidentally (as was pointed out in Chapter Two[1] of this book) a disposition into three parts in line will prove to be the most natural wherever there are no overriding special purposes, three columns will also tend to be most natural for the order of march.

There is one thing we wish to add: the concept of a column is not based merely on the line of march of one body of troops. In strategy the term also applies to bodies of troops marching along the same road on different days. The fact is that troops are divided into columns mainly in order to shorten and facilitate the march, for a small force marches faster and more easily than a large one. This purpose is of course also served if the troops move on different days rather than along different roads.

[1] Chapter Five is meant. Eds.

CHAPTER ELEVEN

Marches—Continued

Experience is the best guideline for the length of a march and the time it will require.

Modern armies have long been accustomed to consider a fifteen-mile march as a day's work. In extensive operations it must be reduced to an average of ten miles in order to allow for the requisite days of rest on which necessary repairs and maintenance can be carried out.

A division of 8,000 men takes eight to ten hours for such a march in level country and on ordinary roads. In mountainous country, it will take ten to twelve. If a column consists of a number of divisions, a few hours longer will be required, even discounting the delayed starting time of the later divisions.

It is clear that the day is pretty well filled by such a march, and that one cannot compare the strain on a soldier loaded with his pack for ten or twelve hours with an ordinary fifteen-mile walk which would not take an individual more than five hours on a decent road.

Forced marches, if undertaken one at a time, may cover twenty-five miles, or thirty at the most; if they continue, only twenty.

A march of twenty-five miles will call for a rest stop of several hours, and a division of 8,000 men will not manage it in less than sixteen hours even on good roads. If the distance to be covered is thirty miles and several divisions are involved, one has to allow a minimum of twenty hours.

What concerns us here is a march by several complete divisions from one campsite to another, since this is the most common type that occurs in a theater of operations. Where several divisions are to form a single column, the first should be assembled and marched off in advance, and will consequently reach camp that much earlier. But this difference in time can never be so great as the time a division takes to pass a given point—the time required for what the French so aptly describe as its *découlement* (run-off). Thus the soldier is spared but little exertion, and each of the marches will take longer because of the larger number of troops involved. Moving a division by similar methods of assembling and marching off its brigades one at a time is only rarely practicable; that is why the division has been treated as a unit.

On long route marches, with troops transferring from one billet to another, marching in small detachments and without points of assembly, they may indeed cover longer distances. In fact they will be longer because of the detours needed to reach their billets.

The maximum amount of time is taken up by marches on which troops have to be reassembled daily by divisions or even by corps, and must then still go to their billets. This type of march is advisable only with a relatively small body of troops and in areas rich in resources; in that case easier means of obtaining provisions and shelter will compensate sufficiently for a longer period of exertion. There can be no doubt that the Prussian army, on its retreat in 1806, was mistaken in the practice of putting up the troops in billets every night in order to feed them. Supplies could just as well have been procured in bivouacs, and the army need not have made immense exertions to cover some 250 miles in no less than fourteen days.

All such standards of time and distance will undergo so many changes whenever one encounters poor roads or mountainous country that it will be difficult to estimate accurately the time a certain march should take—let alone to set up any general rule. The best a theorist can do is to point out the pitfalls that beset the problem. In order to avoid them the most meticulous calculations, as well as a large margin for unforeseen delays, are necessary. Weather conditions and the state of the troops must also be taken into account.

Once tents had gone out of use and troops began to be supplied by requisitioning food on the spot, an army's baggage shrank considerably. One would expect the most important result to be an acceleration of mobility and, as a consequence, an increase in the range of a day's march. But this will only occur under certain circumstances.

The change did little to accelerate marches in the theater of operations. The reason is the well-known fact that whenever in earlier times the situation called for an exceptional amount of marching, the baggage had always been left behind or sent ahead, and, in general, separated from the troops for as long as movements of this kind were still in progress. Baggage, in point of fact, rarely had any influence on movements; once it had ceased to be a positive encumbrance, no more notice was taken of it—regardless of how much damage it might suffer. So the Seven Years War produced marches that have still not been surpassed: Lacy's, for instance, in 1760, in support of the Russian diversion toward Berlin. He covered the 220 miles from Schweidnitz through Lusatia to Berlin in ten days—a rate of 22 miles a day, which would be astounding even nowadays for a corps of 15,000 men.

The very change in the method of supplying the troops, on the other hand, has tended to retard a modern army's movements. Troops that have to do part of their foraging for themselves, as they often must, spend more time on it than they would need if they only had to get their rations from the bread-cart. Besides, on marches of substantial length one cannot allow large numbers of troops to encamp all in one spot; divisions have to be dispersed to make their feeding easier. Finally, it usually happens that some part of the army, particularly the cavalry, has to be put up in billets. Taken altogether, all this causes considerable delay. Hence the fact that Bonaparte, when pursuing the Prussians and trying to cut off their retreat in 1806, and Blücher, intending to do the same to the French in 1813, both required ten days to cover only 150 miles or so. That was a rate which Frederick the Great achieved, baggage and all, when marching from Saxony to Silesia and back.

On the other hand, both large and small units of troops have increased considerably in mobility and flexibility because the amount of baggage has decreased. For one thing, while cavalry and artillery remain at the same level the number of horses is reduced, thereby reducing the need for forage. For another there is less constraint in occupying positions, since one no longer has to worry all the time about the safety of an endless train of baggage in the rear.

After lifting the siege at Olmütz in 1758, Frederick the Great moved with 4,000 wagons, which were covered by half of his army split into battalions and even companies. A march like that would be impossible today, even when facing the most timid of adversaries.

On long route marches—for example, from the Tagus to the Niemen—the benefit is, of course, more perceptible; while a normal day's march is about the same because of the number of wagons still required, in cases of urgency it may be increased without the same degree of sacrifice.

In general, the reduction of baggage will result in a saving of effort rather than an acceleration of movement.

CHAPTER TWELVE

Marches—Concluded

At this point we must examine the damaging effects of marches on the fighting forces. These are so great that they must rank as a distinct active factor, comparable to the engagement.

A single moderate march will not blunt the instrument; but a series of moderate marches will begin to tell, while a series of strenuous marches will naturally do much greater harm.

In the zone of operations the lack of provisions and shelter, badly rutted roads and the need to be constantly prepared for battle, are the causes of the disproportionate exertions which take their toll of man and beast, wagons and clothing.

It is commonly said that a long period of rest is not good for the physical health of an army, and that there is more sickness at such times than during periods of moderate activity. It may well be that sickness does occur when soldiers are crowded together in cramped quarters, but it can occur just as easily in billets along the march. The cause of such sickness should never be attributed to a lack of fresh air and movement, since these may so easily be provided by exercises.

Consider the difference to a man's unstable and upset organism between falling ill indoors and falling ill on the open road, mired in mud and rain and loaded down by his pack. Even if he is taken ill in camp, he can soon be sent to the nearest village where medical help of some sort will be found; but if stricken on a march he lies by the road for hours on end without any relief whatsoever, and then must drag himself along as a straggler for miles. How many minor ailments this will aggravate; how many serious ones will end in death! Consider also the dust and burning heat of summer, when even a moderate march may cause heat exhaustion. Tortured by parching thirst, the soldier will rush to any cold spring, only to catch some disease and his death.

None of this is meant to say that there should be any less activity in warfare. Tools are there to be used, and use will naturally wear them out. Our only aim is clarity and order; we are opposed to bombastic theories that hold that the most overwhelming surprise, the fastest movement or the most restless activity cost nothing; that they are rich mines which lie unused because of the generals' indolence. The final product may indeed be compared to that of gold and silver mines: one looks only at the end result and forgets to ask about the cost of the labor that went into it.

Lengthy marches outside the theater of war are normally made under easier conditions, and daily casualties are fewer. On the other hand, the slightest sickness usually keeps a man away from his unit for a long time: it is difficult for the convalescent to catch up with the advancing army.

In the case of cavalry there is a steady increase in lame and sore-backed horses; vehicles tend to break down, and confusion results. A march of 500 miles or more will always cause an army to arrive at its destination in a highly weakened condition, especially where horses and wagons are concerned.

If a march of that sort must be made within the theater of war under the enemy's eyes, the disadvantages add up. Where substantial numbers are involved and the general conditions are adverse, losses can mount to unbelievable proportions.

Let us give a few examples to illustrate our point.

When Bonaparte crossed the Niemen on 24 June 1812, his enormous center, which he subsequently led to Moscow, numbered 301,000 men. At Smolensk on 15 August he detached 13,500 men, so 287,500 men should have been left. The actual strength of his army, however, was only 182,000 men—which means that 105,500 had been lost.[1] Bearing in mind that only two engagements worth the name had so far taken place—one between Davout and Bagration and the other between Murat and Tolstoy-Ostermann—the French battle casualties may have been 10,000 men at most. The losses due to sickness and stragglers for this period of 52 days and an advance of about 350 miles would thus number 95,000 men, or about one-third of the whole army.

Three weeks later, at the battle of Borodino, losses (including those in action) had reached 144,000 men, and at Moscow, a week later, they came to 198,000 men. Overall, in the first of the above periods, the army's daily rate of loss was 1 in 150 of the original total strength; in the second period, 1 in 120; and in the third, 1 in 19.

Bonaparte's advance was indeed unrelenting, from the crossing of the Niemen up to Moscow; but one must bear in mind that it took 82 days to cover only about 600 miles, and that the army twice stopped altogether—once for some 14 days at Vilna and the other time for some 11 days at Vitebsk—which must have given many stragglers time to catch up. This fourteen weeks' advance was not made at the worst time of year nor on the worst of roads: it was made in summer and the roads were mostly sandy. The impeding factors were the enormous masses of troops moving along a single road, the shortage of supplies, and an enemy who, though in retreat, was not by any means in flight.

We shall not even mention the French retreat—or, more accurately, the army's *advance* from Moscow to the Niemen—but perhaps we should observe that the pursuing Russian army left the Kaluga area with 120,000 men and arrived in Vilna with 30,000. How few were lost in actual fighting during that time is common knowledge.

Let us take one more example, this one from Blücher's campaign of 1813 in Saxony and Silesia, which was notable not for any length of march, but for a series of movements to and fro. York's corps opened the campaign on

[1] All of these figures are taken from Chambray. Cl.

16 August with about 40,000 men and by 19 October numbered a mere 12,000. The main engagements that it fought—at Goldberg, Löwenberg, on the Katzbach, at Wartenburg and the battle of Möckern (Leipzig) cost, according to the best authorities, 12,000 men. Consequently, in eight weeks its losses from other causes came to 16,000 men—two-fifths, that is, of its initial strength.

Great wear and tear on one's own forces, therefore, must be expected if one intends to wage a mobile war. All other plans must be adjusted to that fact; and above all, replacements must be provided for.

CHAPTER THIRTEEN

Billets

In modern war, billets have once more become indispensable: neither tents nor an ample military train enable an army to do without them. Camps in huts or in the open (so-called bivouacs), no matter how elaborately arranged, cannot be accepted as the regular way of sheltering troops: sooner or later, depending on the vagaries of the climate, sickness will gain the upper hand and prematurely exhaust the strength of the forces. The Russian campaign of 1812 is one of the few in which, during the whole of its six months' course despite the rigors of the climate, the troops were almost never in billets. But then look at the consequences of this effort, which one would be tempted to call an extravaganza if the political concept of the whole undertaking did not deserve that term even more!

Two factors will prevent an army from taking up billets: the proximity of the enemy and the rapidity of movement. Billets are evacuated as the decision draws near, and not reoccupied until after it has been reached.

In the more recent wars—that is, in all the campaigns of the last twenty-five years—the elemental force of war has been unleashed with all its energy. In most cases activity and effort have been pushed to their very limits. But all these were campaigns of short duration. They rarely took more than six months, and frequently less, to reach their goal—that is, the point where the defeated party felt ready to seek an armistice, or even peace, or where the winning side no longer had the momentum required for victory. During such times of major exertion billets could hardly be considered. Even at the time of a victorious pursuit, when there was no longer any question of danger, movements were too rapid for this type of relief.

Where for whatever reason the sequence of events is less impetuous, where there is a balanced suspension of forces, the billeting of troops in solid shelter must be given first consideration. This necessity may affect operations in two ways: one may attempt to gain extra time and security by means of a stronger system of outposts and a larger advance guard placed further forward; and the degree of wealth and cultivation of the area may take precedence over considerations of its tactical advantages and geometric patterns of lines and points. A commercial town with twenty or thirty thousand inhabitants, or a road which runs through an area of substantial villages and prosperous towns, will greatly facilitate the concentration of large forces; and such a concentration in turn confers so much freedom and ease of movement that these advantages will amply compensate for those that a better tactical position might have provided.

We shall comment only briefly on the form to be followed in arranging for billets, since that is chiefly a subject of tactical concern.

In billeting troops one has to determine whether their accommodation is a chief or a secondary consideration. During a campaign the disposition of troops may depend on purely tactical and strategic needs, and their comfort will be best served by putting them in billets near the point of concentration. This is particularly true of cavalry. In that case, billets are a secondary consideration. They are a substitute for camp, and must thus be located within a radius that permits the troops to reach their positions in good time. If, on the other hand, troops are quartered with a view to recuperation, housing becomes the primary consideration that governs everything else, including the actual choice of the point of concentration.

The first problem here is the shape of the billeting area as a whole. The usual shape is an elongated oval, a mere broadening, as it were, of the tactical order of battle. The point of assembly is in front of it, and the headquarters behind. These three determinants, as it happens, constitute an obstacle to, and are practically in conflict with, the safe concentration of the army before the arrival of the enemy.

The more the billets form a square or, better yet, a circle, the quicker the troops can be assembled at a given point, which is the center; the farther back the concentration point is located, the longer the enemy will take to reach it, and the longer the time available. A concentration point to the rear of the billets cannot possibly be threatened. Conversely, the farther forward the headquarters is located, the faster it will receive reports, and the better informed the commander will be. But there are reasons for the first of the arrangements which have to be considered to some extent.

The point of widening the billeting areas is to cover the countryside in order to prevent the enemy from receiving supplies from it. But this reasoning is neither completely sound nor very important. It is sound only in regard to the outermost wings, and does not apply to the gap that develops between two sections of the army whose respective billets are grouped around their concentration point. No enemy unit would venture into such a gap. It is not very important, because there are simpler means of protecting the countryside from enemy requisitions than scattering the army too thinly.

By advancing the concentration point, one intends to cover the billets. The argument goes as follows. In the first place troops who are suddenly called to arms will always leave a wake of stragglers, sick, baggage, supplies, etc., in the billets, which if the concentration point is in the rear the enemy can easily capture. In the second place, one must prevent the enemy from attacking detached regiments and battalions one by one in cases where units of his cavalry have bypassed the advance guard or scattered it altogether. An assembled unit, though it may be weak and may have to succumb in the end, will bring him to a stop, and time will have been gained.

Concerning the location of headquarters, it has always been assumed that the aim should be maximum security.

After weighing all these factors, it is our opinion that the optimum arrangement of the billeting area would be an oblong that comes close to being a square or a circle; the concentration point should be in the center, and wherever troops are at all numerous headquarters should be in the front lines.

So far as the protection of the wings is concerned, the points we made when discussing the general disposition of the army remain applicable. Thus, any corps that are detached on either side of the main force should have their own concentration point in line with that of the main force, even when a joint battle is intended.

For the rest, if one remembers that usually the concentration point is determined by favorable features of the terrain while cities and villages determine the area of the billets, it will easily be seen how seldom matters of this sort are governed by geometric laws. But the point does deserve attention because, like all general laws, it does to some degree affect the general run of cases.

There are some other factors that might be mentioned in connection with favorable locations for billets. One is the choice of a natural feature behind which billets may be occupied; then a number of small detachments will be able to observe the enemy. Billets may also be located to the rear of fortresses; if the enemy has no means of estimating the strength of the garrisons, he will treat them with much more caution and respect.

The subject of fortified winter quarters will be dealt with in a separate chapter.

Billets occupied by troops on the march differ from those occupied by stationary forces in that the former should not be far from the road, but strung out along it, in order to avoid detours. Provided the spread does not exceed a short day's march, this will be anything but harmful to rapid concentration.

In "the presence of the enemy," to use the technical term—that is, in all cases where there is no significant distance between the two advance guards—the strength and position of the advance guards, along with that of the outposts, will be governed by the size of the billeting area and the time needed for the concentration of the troops. Where, on the other hand, the strength and position of advance guard and outposts are determined by the enemy and the general situation, the size of the billeting area will depend on the amount of time which the resistance of the advance guard will afford.

The type of resistance which can be rendered by an advanced corps has been discussed in Chapter Three[1] of this book. Its duration is reduced by the time it takes to alert and call out the troops; only the time which remains is available for their concentration.

Once again we should like to consolidate our ideas into a proposition that will prove valid under most ordinary circumstances. If the billets are spread out within a radius measured by the distance to the advance guard, with

[1] Chapter Eight is meant. Eds.

the point of assembly lying approximately at the center of the billeting area, the time gained by delaying the enemy's advance is the time available for alerting and calling out the troops. In most cases this will be enough, even if the reports are not transmitted by signal lights or shots etc., but merely by relays of messengers, which is the only reliable method.

So with an advance guard lying 15 miles away, billets could cover an area of about 700 square miles. In a moderately populated region, such an area would contain about 10,000 houses. Discounting the advance guard, an army of 50,000 men would find this very comfortable at about four men per house. Even an army twice that size would not be seriously overcrowded at nine men per house. Where, on the other hand, the advance guard cannot be posted more than five miles out, the area will be limited to 80 square miles. Even though the time gained does not diminish in exact proportion to the distance of the advance guard, and a distance of five miles means a six-hour span, extra precautions must be taken where the enemy is this close. An army of 50,000 men would be able to find adequate billets in an area that size only if it were very thickly populated.

This points up the significance of large or at least fair-sized towns in which 10,000 or 20,000 men can be concentrated.

This proposition might be taken to prove that if one is not too close to the enemy and has a sizable advance guard, troops may be left in billets even when facing a concentrated enemy force, as Frederick the Great did at Breslau early in 1762, and Bonaparte at Vitebsk in 1812. When facing the concentrated enemy army, the proper distance and suitable arrangements will ensure the safety of the troops while assembling; but one must remember that an army engaged in rapid assembly is in no position to do anything else during that time. In other words, it will be unable immediately to exploit any opportunities that arise, and this will deprive it of a large part of its effectiveness. Consequently, an army may be completely quartered in billets under only the three following conditions:

1. If the enemy does the same
2. If the condition of the troops makes it absolutely essential
3. If the army's immediate task is limited to the defense of a strong position, and so nothing matters but the prompt concentration of the troops at that point.

The campaign of 1815 provides a remarkable example of the concentration of an army quartered in billets. General Ziethen, commanding Blücher's advance guard of 30,000 men, lay at Charleroi, a mere ten miles from Sombreffe where the army was to assemble. The farthest billets of the main army, however, were forty miles from Sombreffe, beyond Ciney on one side, and as far as Liège on the other. Yet the troops from the Ciney area arrived at Ligny several hours before the battle there began, and those close to Liège (Bülow's corps) would have as well, but for chance and faulty communications.

There is no doubt that the safety of the Prussian army had been neglected; but in explanation one must say that all the arrangements had been made while the French were also still widely scattered in billets. The mistake lay simply in not changing the arrangements as soon as it was learned that the French were on the move and that Bonaparte himself was with them.

It remains noteworthy that the Prussian army might have been assembled at Sombreffe before the enemy attacked. Blücher, it is true, got news of the enemy's advance and began to concentrate his troops on the night of the fourteenth of June—that is, twelve hours before Ziethen was actually attacked. But by nine the following morning Ziethen was under fire, and it was not until that very hour that Thielmann at Ciney received orders to move toward Namur. He thus had to assemble his troops into divisions, and then march thirty-two miles to Sombreffe, which he did in twenty-four hours. General Bülow could have been there at the same time if he had been properly impressed by the order.

Yet it was not until two in the afternoon of the sixteenth of June that Bonaparte made his attack at Ligny. One of the things that held him back was his apprehension at having Wellington on one side and Blücher on the other. In other words, the inequality between his strength and theirs contributed to his hesitation. It only goes to show how even the most resolute commander can be thwarted by having to grope his way carefully, as is inevitable in all complicated situations.

Some of the considerations we have put forward here are obviously more of a tactical than a strategic nature; but we thought it better to stray into the tactical field than to take the risk of not making ourselves clear.

CHAPTER FOURTEEN

Maintenance and Supply

For two reasons the problem of supply has assumed much greater importance in modern warfare. First, armies in general are now much larger than those of the Middle Ages, or even of the *ancien régime*. Those that did approach or even surpass modern ones in size were rare and short-lived. In more recent wars, since the time of Louis XIV, armies have always been very large. The second reason is still more important, and more characteristic of our own day: a war now tends to be more of one piece, and the fighting forces are in constant readiness for action. Earlier wars consisted largely of single, unconnected operations, separated by intervals in which either the war was literally dormant and the only activity was political, or else the hostile armies kept so far apart that each was able to attend to its own needs without regard to its opponent.

More recent wars, by which we mean those since the Peace of Westphalia, have, through the efforts of governments, assumed a more regular, interconnected character. Operational requirements are dominant and even in the sphere of maintenance and supply call for appropriate arrangements. Admittedly, seventeenth and eighteenth century wars had long quiescent periods when fighting practically stopped—we refer to regular winter quarters—yet these too were subject to their operational purpose. These intervals were caused not by problems of supply but by bad weather. Since they always ended with the approach of summer, continuous military activity was the rule for as long as the season allowed.

Transition from one condition, or method, to another has always been gradual, and this case is no exception. In the wars against Louis XIV, it was still the allies' practice to send their troops to winter quarters in distant provinces where it was easier to supply them. By the time of the Silesian wars this was no longer done.

Regulated and coordinated military action did not really become possible until states replaced feudal levies with mercenary troops. Feudal obligations were transformed into money payments, and liege service either vanished altogether in favor of recruitment, or fell only on the lower classes. The nobility considered the furnishing of recruits as a kind of tribute, a human tax, as they still do in Russia and Hungary. At any rate, as we have stated elsewhere, armies now became instruments of the central government, and their cost was borne mainly by the treasury or public revenue.

The very same circumstances that changed methods of recruitment and regular replacement of the fighting forces were to change methods of maintenance and supply. Once the Estates had been exempted from the former in exchange for monetary tribute, they could not very well be burdened with the latter by stealth. So the government, the treasury, had to carry the burden of subsistence: the army could not be allowed to live off the land while it was stationed on its own territory. Governments, therefore, had to treat the maintenance of the army as their responsibility alone. In this way, maintenance presented increased difficulties, for two reasons: the government had to assume responsibility for it, and the fighting forces were required to remain permanently in the field.

Not only an independent military class but also an independent system for supplying it were thus created, and were developed to the fullest possible extent.

Stocks of provisions had to be amassed, either by purchase or from the state's demesnes possibly some distance away, and stored in depots; they had also to be taken from these depots to the troops by the army's own transport, baked nearby by their own bakeries, and finally picked up again for distribution by the units' own vehicles. We have considered this system, not only because it explains some of the characteristics of the wars in which it was employed but also because it will never fall completely into disuse. Some aspects of it will always recur.

Military institutions thus tended to become more and more independent of the country and the people.

In consequence warfare became more regular, better organized, and more attuned to the purpose of war—that is, to its political objective. Movement, on the other hand, was much more limited, much more constrained, and war was waged with far less vigor. Now one was tied to depots and bound by the effective range of transport, with the inevitable result that rations were cut to the barest minimum. Nourished often by only a meager crust of bread, soldiers tottered about like shadows, and there was not even the prospect of better things to come to comfort them in their privations.

Anyone who tries to maintain that wretched food makes no difference to an army, and cites Frederick the Great's accomplishments with ill-fed soldiers, is not taking a dispassionate view of the subject. Ability to endure privation is one of the soldier's finest qualities: without it an army cannot be filled with genuine military spirit. But privation must be temporary; it must be imposed by circumstances and not by an inefficient system or a niggardly abstract calculation of the smallest ration that will keep a man alive. In the latter case it is bound to sap the physical and moral strength of every man. We cannot use Frederick the Great's military accomplishments as a standard. For one thing his enemies used the same system; and for another, who can tell how much more he might have attempted if he could have fed his troops as well as Bonaparte did whenever circumstances allowed?

This artificial system of obtaining supplies was, however, never extended to the feeding of horses, since forage, because of its bulk, is much more difficult to procure. A horse's ration weighs about ten times as much as a man's, and the number of horses per man is not one to ten, but one to four or one to three even today. In the past it was one to three or one to two, so that the weight of the horses' rations was three, four, or five times that of the men's. Therefore, the simplest possible way was found to meet the army's needs: foraging expeditions. These imposed other major limitations on warfare. First of all it became of prime importance to fight on enemy soil, and second it became impossible to stay in any area for long. But by the time of the Silesian wars foraging expeditions had become far less common; they were found to place a far greater burden of devastation and strain on a region than a system of satisfying the needs of the army by local requisitions and deliveries.

When the French Revolution suddenly brought a national army back on the stage of war, the governments' means were no longer adequate. The whole military system, which had been built on these limited means and in turn found security in them, broke up, and that included the sector that concerns us here, the supply system. The French Revolutionary leaders cared little for depots and even less for devising a complicated mechanism that would keep all sections of the transport system running like clockwork. They sent their soldiers into the field and drove their generals into battle—feeding, reinforcing, and stimulating their armies by having them procure, steal, and loot everything they needed.

The Napoleonic Wars were waged by all belligerents in a manner that fell between these two extremes: they chose the most suitable methods from those available; and this is likely to remain so in the future.

This modern way of provisioning the troops—using everything available in the locality, no matter to whom it belongs—falls into the following four categories: supplies furnished by local households; requisition by the troops themselves; general requisition; and depots. Normally all four are used simultaneously, with one predominating; but it can happen that only a single one is used.

1. *Living off local households or the community, which amounts to the same thing.* Supplies for several days will always be available in any community, even if it consists only of consumers, as in the case of large towns. It will thus be understood that even the most populous town can furnish food and lodgings for a day for about as many soldiers as there are inhabitants without any special preparations, and for longer if their number is substantially smaller. In major towns this works out very well, for it means substantial quantities of troops can all be fed and housed in a single place. It would not be satisfactory in smaller towns, and even less so in villages. A population of 3,000 to 4,000 per 25 square miles is pretty dense, and it could only feed some 3,000 or 4,000 troops. If greater numbers were to be fed, they would have to be dispersed so far afield that other requirements would be hard to meet. On the other hand, in the country, and even in small towns, the sort of supplies that matter most in war are much more plentiful. A farmer's stock of bread is usually enough to feed his family for a week or two. Meat can be come by every day, and there is generally a big enough stock of vegetables to last till the next harvest. As a result, in billets that have

not been previously occupied, one can generally find food for three or four times the number of inhabitants for several days, which again works out extremely well. Accordingly, where there is a population of 2,000 to 3,000 per 25 square miles (no substantial town being occupied) a force of 30,000 men would take up 100 square miles or so—requiring a width of 10 miles. An army 90,000 strong (say 75,000 fighting men) marching in three parallel columns, would thus need a front of only 30 miles, provided three roads were available within that space.

If the area is occupied by several columns in succession, special arrangements on the part of the local authorities will be required, but that should be no problem for the needs of a day or two at a time. Consequently, if the 90,000 men were followed the next day by another 90,000, the second force need suffer no privation; and 150,000 fighting men is a considerable army.

The question of forage for the forces is even easier, since forage need not be either ground or baked. Feed enough until the next harvest is bound to be in store for the local horses, so that even where there is little stall-feeding, no shortage should arise. Forage of course should be requisitioned from the community rather than the individual household. It goes without saying that in organizing a march one will have to keep certain considerations in mind concerning the nature of the country, so as not to billet the cavalry in industrial or other areas where forage will be scarce.

The result of this brief survey is that, in an area of average population density—say 2,000 to 3,000 per 25 square miles—a force of 150,000 combatants can live off the local inhabitants and communities within a very small area for a day or two, which will not preclude its fighting as a unit—in other words, it is possible to provision such a force without depots and other preparations on an uninterrupted march.

French operations in the Revolutionary wars and under Bonaparte were based on this conclusion. The French marched from the Adige to the lower Danube and from the Rhine all the way to the Vistula without substantial means of provisioning other than living off the land, and never suffered want. Since their operations were based on physical and moral superiority and were undeniably successful, certainly never being delayed by indecision or timidity, their victorious course was for the most part moving as an uninterrupted march.

Under less favorable conditions—where the population is smaller, or consists more of tradespeople than of farmers, where the soil is poor, or the area has already been occupied several times—the yield will naturally also be lower. But one must bear in mind that by expanding the front of a column from 10 to 15 miles, one more than doubles the area—from 100 to 200 square miles. This is an extent which will normally still permit the column to fight as a unit. So even where conditions are unfavorable, this kind of subsistence is possible provided the troops remain uninterruptedly on the move.

But as soon as a halt of several days occurs, severe shortages would arise unless arrangements had been made in advance. There are two such arrangements without which a substantial army cannot survive even today. The first is to equip the troops with a wagon train that will carry enough bread or flour—the most essential part of their provisions—to last for three or four days. Add the three or four days' ration that a soldier carries in his pack, and bare subsistence for week will be assured.

The second arrangement is an efficient commissariat that can procure supplies from distant areas whenever there is a halt. In that case it is possible to switch at any time from being locally supplied to any other system.

Local supply offers the immense advantage of being the quickest, and needing no means of transport; but it does presuppose that under ordinary circumstances all troops are quartered in local households.

2. *Supply by requisitions is carried out by the troops themselves.* A single battalion can generally camp close to a few villages, and these can be directed to furnish supplies. This type of subsistence is essentially no different from the first. Usually, however, a far larger force is likely to camp in a single area, and in that case there is no alternative but for this major unit—a brigade or a division—to requisition whatever is needed from the surrounding area, and then distribute it.

One can see at a glance that this is no way of raising enough food for a sizable army. The yield from stocks in the countryside will fall far short of what the troops would get if they were quartered in the villages themselves; for when 30 or 40 men move into a farm, they will clean it out if need be. But an officer who is sent out with a few men with orders to exact provisions has neither the time nor the means to find everything. There may also be a shortage of transport, and so they will get only a fraction of what is available. Moreover, troops in camp are so crowded into one place that the area from which food can be gathered rapidly will not be able to supply enough. What, after all, can be expected where 30,000 men are extorting food within a range of 5 miles, or 15 to 20 square miles? Even that little will often be impossible, since most of the nearby villages will have troops billeted in them who will not give anything up. Finally, this method is the least economical. Some units will get more than they can use, much of it will go to waste, and so forth.

In effect then, provisioning by requisitions will be successful only where troops are not too numerous—say in the case of a division of 8,000 to 10,000 men. Even then it should be used only as a necessary evil.

This method is unavoidable for all units directly facing the enemy, such as advance guards and outposts when their own side is advancing. They will reach a point where no preparations could possibly have been made, and the stocks collected for the rest of the army are usually too far away. The same applies to mobile columns operating on their own, and to all cases where neither time nor means for any other type of provisioning happen to be available.

The better the troops are able to manage by means of regular requisitions, and the more time and circumstances permit a transition to this method of supply, the better the results will be. Usually the problem is lack of time: what the troops can get for themselves, they can get more quickly.

3. *Regular requisitions.* Beyond all question, this is the simplest and most efficient way of feeding troops. It has been the basic method in all recent wars.

The difference between this and the previous method lies mainly in the cooperation of the local authorities. Food is no longer seized by force wherever it is found, but is delivered in an orderly way, with the burden reasonably distributed. Only the local authorities can do this.

Here time is of the essence: the more time, the wider the distribution, the lighter the burden, and the more successful the operation. Supplies may even be bought for cash, which comes close to the system of provisioning that will be discussed next. Feeding troops by regular requisitions is no problem when they are concentrated in their homeland; the same, as a rule, is true of an army moving to the rear. On the other hand, all advances into a region that is not one's own will leave very little time for making such arrangements—seldom more than the single day by which the advance guard is ahead of the main force. The former presents the local authorities with the requisitions, specifying the amounts and rations to be provided, and where to deliver them. Since these can only be gathered from the immediate area—that is, within a radius of a few miles from each delivery point—stocks assembled in this hasty manner would not possibly suffice for a good-size army unless it brought with it enough food for several days. It is, therefore, the job of each commissariat to make do with what it receives, and only issue food to units that have none. But every day will ease the shortages. As the distances over which supplies are hauled daily grow greater, so does the territory drawn upon, in terms of square miles. If only 100 square miles can be drawn upon the first day, there will be 400 the second, and 900 the third; to put it another way, the second day's area will be larger than that of the first by 300 square miles, that of the third larger still by another 500.

This of course is only a rough estimate of the circumstances. There may be many limiting factors, the main one being that the area which the army has just left cannot make so significant a contribution as the rest. It must be remembered, on the other hand, that the radius of food collection can be extended by more than 10 miles every day—possibly by 15 or 20, or in some areas even more.

The delivery of requisitions, or at least of most of them, is assured by the executive power of individual detachments which are placed at the disposal of the officials. Even more effective is the fear of being held responsible, of being punished or maltreated—which, under such circumstances, acts as a collective burden that weighs on the whole population.

Since, for our purposes, the result is all that matters, we do not intend to go into details or examine the whole mechanism of commissariat and supply.

This result, derived from a common-sense view of the general situation and confirmed by the experiences of the wars since 1792, is that even the largest army can safely rely on requisitions, provided it carries a few days' rations with it. The requisitions are delivered upon arrival of the army; at first only from the immediate neighborhood, as time goes on from an expanding radius, the activity being organized and controlled by increasingly senior authorities.

This method knows no limits other than the complete exhaustion, impoverishment and devastation of the country. If an army stays for a considerable length of time, the highest civil authorities will become involved in running the system. They will, of course, do whatever they can to spread the load and reduce the burden by purchases. Even belligerent foreign forces that occupy a country for any length of time will hardly be so harsh and pitiless as to place the whole burden of subsistence on the land. Requisitioning, therefore, gradually tends to become increasingly like the depot system, although that does not mean it ever ceases altogether, or that its influence on military movements undergoes any notable change. It is one thing to supplement local resources by supplies imported from a distance, while the region itself remains the army's chief supplier; it is quite another to proceed as in eighteenth-century wars, when the army carried all provisions with it and the countryside was as a rule left alone.

The main difference lies in two factors—the use of local transport and bakeries. They eliminate the enormous burden of wagon trains, which in almost all cases tended to destroy its own labors.

Even today, of course, an army cannot do without some supply wagons of its own, but the need is infinitely smaller. Such wagons are really used only for making the surplus of one day's food available for the next. Special conditions, such as those of 1812 in Russia, will even now call for a colossal wagon train, as well as the transport of field bakeries; but these are to some extent exceptions. It is rare, after all, for an army of 300,000 men to advance for 650 miles on practically a single road, to do it in countries such as Poland and Russia, and just before the harvest. And even in such cases the army's own resources are considered to be only supplementary, while local requisitions are treated as the basis of supply.

In fact the requisitioning system has been the basic method used by all French armies since the first campaigns of the Revolutionary wars. Their enemies were forced to adopt it as well, and one can hardly expect it ever to be abandoned. No other system is as satisfactory, both in regard to the vigor with which war can be waged, and to the ease and flexibility it confers. Subsistence rarely causes any difficulty for the first three or four weeks, no matter where the army goes, and after that supplementary depots will be available; so one can truly say that war has gained the utmost liberty by these arrangements. While there may be difficulties of one kind or another that may influence planning, we will never be faced with an absolute impossibility, and policy can never be dictated by considerations of supply alone.

The one exception is a retreat through hostile country. Here a number of features coincide that have an adverse effect on subsistence. Movements are continuous, usually without definite halts, and so there is no time for gathering supplies. The very conditions that dictate the retreat will normally be highly adverse, making it imperative to keep the troops continually concentrated, which generally rules out any distribution in billets or any considerable extension of the columns. In hostile territory it is not possible to gather supplies by simply issuing orders; and finally, the situation particularly favors resistance and ill will on the part of the local inhabitants. The effect of all this is that in such cases one is strictly limited to established lines of communication and retreat.

When Bonaparte began his retreat in 1812, problems of supply restricted him to the road by which he had advanced: on any other he would undoubtedly have come to grief even earlier. All the blame that has been heaped upon him on this account, even by French writers, completely misses the point.

4. *Subsistence by means of depots.* If a broad distinction were to be made between this and the method previously discussed, we could only do so by referring to the system that was used from the last thirty years of the seventeenth to the end of the eighteenth century. Will such a system ever be used again?

Admittedly any other kind is hard to imagine with the type of warfare in which large armies stayed in the same place for seven, ten, or twelve years—as was the case in the Netherlands, the Rhineland, Northern Italy, Silesia and Saxony. No country could remain the chief supply-agent for the opposing armies for that length of time without being utterly ruined, and gradually failing in its obligations.

This of course leads to the question whether war governs the supply system or is governed by it. We would answer that at first the supply system will govern war insofar as the other governing factors will permit; but where these start to offer too much resistance, the conduct of war will react on the supply system and so dominate it.

Warfare based on requisition and local sources of supply is so superior to the kind that relies on depots, that the two no longer seem to be the same instrument. No government would dare oppose the first kind of warfare with the second; if any minister of war were hidebound or ignorant enough to misjudge the general needs of the situation and, at the opening of hostilities, to provision his army by the old system, force of circumstances would soon overpower his general. Requisitioning would automatically impose itself. One has to remember that since no state ever has more money than it needs, the high cost of maintaining depots will necessarily cut into expenditure on the armament and the size of the army. It follows that there is practically no chance of such arrangements being used, unless the two belligerents were to arrive at a mutual agreement on the subject through diplomatic channels—a possibility that must be considered pure fantasy.

Thus in future, all wars are more than likely to start out with a system of requisitioning. How much either government will do to supplement it by other arrangements so as to spare its rural population and so forth, may be left open. It will probably not be much; priority is always given to the most urgent needs at such a time, and a special supply organization is no longer considered urgent.

Where, on the other hand, a war is not so decisive in its results, or not so extensive in its movements, as its true nature would imply, requisitioning will start to exhaust the area very seriously, to the point where one will be forced either to make peace or to make arrangements for easing the burden on the country by creating an independent supply system. The latter was done by Bonaparte in Spain, but the former is more usual. In most wars the exhaustion of the belligerents increases to such an extent that, instead of making the war more expensive, they are driven to make peace. Here, too, contemporary practice tends to shorten wars.

Still, one cannot altogether deny that war could be conducted under the old method of supply. It may possibly occur again where it is indicated by conditions on both sides, and where other favorable circumstances are present. But its form can never be considered natural; it is an exception that happens to be feasible, but that can never spring from the true concept of war. Even less can it be considered a perfecting of warfare simply because it is more humane. War itself is anything but humane.

Whichever method of supply is chosen, it will, of course, work better in a rich, densely populated area than in a poor and uninhabited one. Density of population affects the size of stocks in an area in two ways. First, where consumption is high, it calls for large reserves; and second, a larger population as a rule means greater production. An exception must of course be made for areas inhabited chiefly by industrial workers, especially when they happen to be in mountain valleys, surrounded by infertile areas, as is not uncommon. Broadly speaking, however, it is far easier to meet an army's needs in a populous area than in a sparsely populated one. No matter how rich the soil, an area of 10,000 square miles with 400,000 inhabitants cannot support an army of 100,000 men so comfortably as one in which the population numbers 2,000,000. Furthermore, in thickly populated areas communications by land and water are better and more plentiful, means of transport more abundant, and regular commercial links simpler and more dependable. In short, it is far easier to feed an army in Flanders than in Poland.

It follows that war, with its numerous tentacles, prefers to suck nourishment from main roads, populous towns, fertile valleys traversed by broad rivers, and busy coastal areas.

All this will indicate the general influence that questions of supply can exert on the form and direction of operations, as well as the choice of a theater of war and the lines of communication.

How far their influence will extend, and how much weight should in the final analysis be attached to the ease or difficulty of supply—those are questions that will naturally depend on how the war is to be conducted. If war is to be waged in accordance with its essential spirit—with the unbridled violence that lies at its core, the craving and need for battle and decision—then feeding the troops, though important, is a secondary matter. On the other hand, where a state of equilibrium has set in, in which troops move back and forth for years in the same province, subsistence is likely to become the principal concern. In that case, the quarter-master-general becomes the supreme commander, and the conduct of war consists of organizing the wagon trains.

Thus there are countless campaigns in which nothing happened, which missed their goal and squandered their resources to no purpose, and the excuse for it all was difficulties of supply. Bonaparte, on the other hand, used to say, *qu'on ne me parle pas des vivres!*

In the Russian campaign, to be sure, he proved that such neglect can go too far. We are not suggesting that as the only reason why the campaign came to grief—that must be a matter of opinion. But it is undeniable that the lack of care over supplies was responsible for the unprecedented wastage of his army on the advance, and for its wholly calamitous retreat.

While one cannot deny that Bonaparte was a passionate gambler who frequently took reckless risks, one must grant that he and the Revolutionary generals before him disposed of some powerful prejudices about maintaining an army in the field. They demonstrated that maintenance had to be regarded as a *condition* of war, and never as its object.

Moreover, privation in time of war may be compared to danger and physical exertion. There are no definite limits to the demands a general can make of his troops. A strong-willed commander will ask more than one ruled by delicate emotions; and an army's performance will also depend on the degree to which its willpower and endurance have been steeled by familiarity with war, military spirit, trust in and devotion to the general, and enthusiasm for the cause. Yet one can take it as a fundamental rule that hardship and privation, no matter how extreme, must always be treated as a temporary condition, which has to lead to a state of plenty—even at times luxury. What can be more moving than the thought of thousands of soldiers, poorly clad, their shoulders bent under thirty or forty pounds of equipment, plodding along for days on end in every kind of weather and on every kind of road, continuously endangering their health and their lives, without even a crust of bread to nourish them? When one knows how often this happens in war, one must marvel at the fact that heart and strength do not give out more often, and at the way in which the power of an idea can, by its lasting effect, summon up and support incredible exertions in human beings.

So if for the sake of great issues one imposes great privations on the troops, one must bear in mind, whether prompted by sympathy or by prudence, the reward owed them at a later time.

Finally, we have to examine the different ways in which attack and defense will affect the problems of supply.

A defending army can always use supplies that it has been able to stockpile in advance; so the defenders will not be lacking in necessities. This is so especially for troops stationed in their own country, but it holds true even in enemy territory. The attacker, on the other hand, leaves his sources of supply behind, and so long as he is advancing and even for some weeks after he has come to a halt, he has to shift for himself from one day to the next. Under these conditions shortages and difficulties will be the rule.

There are two periods when this problem is generally at its worst. The first occurs on the advance, before a decision has been reached. The defender is in full possession of his stocks, whereas the attacker has had to leave his own behind. He must keep his forces concentrated and therefore cannot cover a wide area of country. Even his transport can no longer follow him once movement into battle has begun. Unless thorough preparations have been made by then, the troops may well begin to suffer from shortages and actual hunger several days before they fight the decisive battle. That is hardly a healthy way to lead them into action.

The second crisis most commonly occurs at the end of a victorious campaign when the lines of communication have begun to be overstretched. This is especially true when the war is conducted in an impoverished, thinly populated and possibly hostile country. How vast a difference there is between a supply line stretching from Vilna to Moscow, where every wagon has to be procured by force, and a line from Cologne to Paris, via Liège, Louvain, Brussels, Mons, Valenciennes and Cambrai, where a commercial transaction, a bill of exchange, is enough to produce millions of rations!

Often the finest victory has been robbed of its glory as a consequence of this problem. Strength ebbs away, retreat becames unavoidable, and gradually the signs of genuine defeat appear.

While fodder for the horses is the least scarce of commodities at the outset, as we have pointed out, it is the first to run short when the countryside begins to be exhausted. Its bulk is so great that it is hardest to procure from a distance, and a horse will perish from want much sooner than a man. That is one of the reasons why too much cavalry and artillery can be a real burden, and an actual source of weakness to an army.

CHAPTER FIFTEEN

Base of Operations

When an army begins an operation, whether it is to attack the enemy and invade his theater of war or to take up positions along its own borders, it necessarily remains dependent on its sources of supply and replenishment and must maintain communications with them. They constitute the basis of its existence and survival. As the army grows in size, its dependence on its base increases in intensity and scale. But it is not always either possible or necessary for an army to maintain direct communications with the whole of its own country. What is essential is that part which lies immediately to the army's rear, and is therefore protected by its position. This is where the necessary supply depots will be set up, and arrangements made for supplies and reinforcements to be regularly forwarded. That area, then, is the base for the army and for all its operations; army and base must be conceived as a single whole. If in the interests of security stocks are kept in fortified places, the concept of a base becomes strengthened; but frequently this is not done, and fortifications are not essential to a base.

A stretch of enemy territory may also constitute an army's base, or at any rate part of it, for the area held by an invading force will meet a good many of its needs; provided that the army truly controls the area and can be sure that its orders will be obeyed. The range of that certainty seldom extends beyond the rather limited methods of overawing the inhabitants by means of small garrisons and mobile detachments. In enemy territory, therefore, the area from which supplies can be obtained is small in relation to an army's needs, and seldom adequate. A great deal therefore has to be supplied from home. Once more this means the area closest to the rear of the army, which must be considered as an essential component of its base.

An army's needs fall into two categories: those that any agricultural area can provide, and those that can only be obtained from sources located to the rear. The former are mostly provisions; the latter consist of replacements. The former may to some degree be secured from occupied territory, but the latter—men and weapons, for example, and usually munitions—can, for the most part, only come from home. Exceptions may occur, but they will be rare and insignificant; the distinction remains of great importance and proves again that communications with the homeland are essential.

Food stores are normally kept in undefended towns, both in occupied territory and at home, for there are never enough fortresses to hold these very bulky stocks, which are rapidly consumed and needed in different places at different times. Besides, food is easier to replace than stocks of the army's other needs such as weapons, ammunition, and equipment, which are generally not kept in undefended places near the scene of war, but preferably brought up from a distance. In enemy territory they have to be kept in fortresses. This is further proof that the importance of the base rests more on the need for replacements of weapons and equipment than on food supplies.

The larger the scale on which supplies of both kinds are collected into depots, and the more the sources combine into great reservoirs, the more these can be considered as substitutes for the country as a whole, and the more the idea of the base will primarily refer to the places where these great stores are kept. These locations themselves, on the other hand, must never be mistaken for the base.

Where these sources of replacements and supplies are ample—in other words, where there is a wide expanse of fertile land, where supplies have been stored for greater efficiency in large, secure depots near the army, linked to it by good roads, when depots are spaced over an extensive area to the rear or even to some degree on the flanks of the army—then a more vigorous existence will be generated for the army, as well as a great deal more freedom for its movements. Attempts have been made to unite the advantages of such a situation into a single concept: the dimensions of the operational base. The attempt has been made to express the sum total of advantages and disadvantages arising from the character of an army's sources of supply and replacements, by the relationship between the base and the operational objective and the angle which the outer limits of the base form with this objective (which was assumed to be a point).[1] But it is evident that this elegant piece of geometry is nothing but a toy: it rests on a series of substitutions at the expense of truth. As we have shown, an army's base of operations is composed of the threefold gradations of its situation: the local resources, depots at various points, and the *area* from which supplies are drawn. These three factors are spatially distinct: they cannot be reduced to one. Least of all can they be represented by a line which is supposed to show the width of the base but which, as a rule, is arbitrarily drawn between two fortresses or two provincial capitals or along a country's frontiers. Nor is it possible to define a fixed relationship between the three tiers of the infrastructure, for in actual fact their natures to some extent overlap. In one case, equipment that would otherwise have to be brought from a great distance, can be procured locally; in another, even foodstuffs have to be shipped from a distance. Sometimes the nearest fortresses are major ports, arsenals, or commercial centers that contain the war-potential of an entire State; at other times they consist merely of primitive earthworks, barely adequate for their own defense.

In consequence all that has been deduced from the dimensions of the operational base and the operational angle, and the whole theory of war that was founded on them, insofar as it was geometrical, has been totally ignored

[1] This refers to the work of Heinrich von Bülow. See above, Peter Paret, "The Genesis of *On War*", p. 10. Eds.

in actual warfare, and has given rise to misleading efforts in the realm of theory. Since the basic facts were correct, and only the reasoning from them was spurious, the same notions are occasionally likely to recur.

One cannot then do more, in our view, than recognize the general influence of a base on military operations—the *fact that it may be weak or strong*, and the *factors that make it so*. But one must also admit that there is no way of reducing it to one or two simple propositions that will amount to a useful rule; rather, each separate case requires that all the factors we have mentioned must be *simultaneously* borne in mind.

Once a supply and maintenance depot for an army has been set up in a given area, and for a definite operation, then even in one's own country that area alone can be considered as the army's base. Since changes always call for expenditure of time and effort, an army cannot shift its base overnight, even in its own country. The directions of its operations are, therefore, to some extent restricted. During operations in enemy territory one could treat the common frontier in its entirety as the army's base. This assumption would be valid in a general sense since special arrangements could be made everywhere along the frontier, but not for every particular instance because such special arrangements are not in fact made everywhere. When the Russian army retreated before the French at the start of the campaign of 1812, it could certainly regard the whole of Russia as its base, the more so as the vast extent of Russia gave the army ample room in all directions. This was in no way illusory; the concept came to life later, when other Russian armies converged on the French from several sides. At any given moment of the campaign, however, the base of the Russian army was not so extensive as that. Actually, it was present mainly in the roads on which a tremendous mass of army transport moved in both directions. This type of limitation, for instance, prevented the Russian army, after three days' fighting at Smolensk, from making their further retreat in a different direction rather than toward Moscow. A move toward Kaluga, which had been proposed in order to draw the French away from the capital, would have constituted a change of plan that was impossible to execute without having been prepared in advance.

We have pointed out that an army's dependence on its base increases in intensity and scale with any increase in the army's size; as is only natural. An army is like a tree that draws its sustenance from the ground in which it grows. A mere sapling is easy to transplant, but the taller it grows, the harder this will become. A small detachment also has its channels of sustenance, but unlike a large army it takes root easily wherever it may be. So when one talks about the influence that the base exerts on operations, the size of the army must be the standard by which each factor is measured.

There is a further point inherent in the nature of the subject. Food is important for immediate needs; but, for an army's general existence over a period of time, the flow of men and equipment is more important. The latter can only come from certain sources, while the former may be obtained in a number of ways—a fact that further explains the influence that a base exerts on operations.

No matter how great this influence may be, one must never forget that it is among those that take time to produce a decisive effect. There is always the question, what is to be done in the meantime. The value of an operational base, therefore, will rarely predetermine the choice of an operation: it will do so only where the impossible is demanded. The ordinary difficulties that may arise in this respect will have to be considered in conjunction and comparison with the other means available. Obstacles of this nature tend to vanish in the face of decisive victories.

CHAPTER SIXTEEN

Lines of Communication

Roads that lead from an army's position back to the main sources of food and replacements, and that are apt to be the ones the army chooses in the event of a retreat, have two purposes. In the first instance they are *lines of communication* serving to maintain the army, and in the second they are *lines of retreat.*

In the previous chapter we pointed out that an army, despite the present system of procuring supplies chiefly from the area in which it is stationed, must still be regarded as forming a unity with its operational base. The lines of communication are part of that unity. They link the army to its base, and must be considered as its arteries. The roads are in constant use for all sorts of deliveries, for ammunition convoys, detachments moving back and forth, mail carriers and couriers, hospitals and depots, reserve munitions, and administrative personnel. All this together is vital to the army.

These arteries, then, must not be permanently cut, nor must they be too long or difficult to use. A long road always means a certain waste of strength, which tends to cripple the condition of the army.

In their other role, as lines of retreat, they are in effect the strategic rear of the army.

For both purposes, the value of the roads depends on their *length,* their *number,* and their *orientation* (their general direction and the direction in which they run where they are close to the army); their *state;* the *difficulties of the terrain;* the *condition and temper of the local inhabitants;* and finally the amount of *cover* given them by fortresses or natural barriers.

Not all the roads or tracks that lead from the army to the sources of its life and strength can be counted as lines of communication. No doubt they can all be used as such and ranked as supplementary to the system; but the system itself comprises only roads on which military services have been established. The only true lines of communication are those on which depots and hospitals, relay points and postal services, as well as commandants, field police and garrisons have been installed. At this point the very important and frequently ignored difference must be noted that exists between an army at home and one in enemy territory. Lines of communication will, of course, have been set up at home, but the army is not necessarily tied to them; if need be, it can leave them and use any road available. After all, it is at home everywhere; it can everywhere rely on its own officials, and is everywhere received with goodwill. While other roads may not be so good, and may not suit the army's needs so well, they can at least be used; if the army's position is turned, or if it has to change its front, these roads will not be considered *impossible* to use. In enemy territory, on the other hand, the only lines of communication on which an army can usually rely are the roads on which it advanced in the first place, and small and almost insignificant causes may here make a considerable difference.

As it advances into enemy territory, an army establishes and protects its essential lines of communication. Its presence may inspire fear and terror, but its measures may assume the character of inexorable necessity in the eyes of the inhabitants, who may even be persuaded to consider them as an amelioration of the general evil of war. Small garrisons set up here and there will maintain and reinforce this general system. If, on the other hand, commissaries, commandants, police, security posts and administrative personnel were to be dispatched along some distant road not used by the army, the local inhabitants would regard this as an unnecessary burden. Unless some shattering defeat or disaster had reduced the country to a state of panic, these officials would be treated with hostility, beaten up, and thrown out. In order to secure the new road, one would above all need garrisons, and these would have to be larger than normal; and the risk would still remain that the local population would try to resist. In brief, an army advancing into enemy territory is not equipped to command obedience. It has first to install its own administration and must do so by the authority of its arms. This is not possible everywhere and at once; it demands sacrifices and poses difficulties. It follows that in enemy territory an army is even less able to switch from one base to another by changing its communications system than it is at home, where this is at least possible. The effect, in general, will be a greater limitation on mobility and a greater vulnerability to envelopment.

Even the original selection and organization of lines of communication is limited by a number of conditions. Not only must they as a rule follow major roads, but in general they will be better the wider the roads, the larger and wealthier the towns they link, and the more numerous the fortresses that cover them. The choice is also much affected by rivers as a means of transport, and bridges as points of passage. Thus, the position of lines of communication and, hence, the route an invading army can use are matters of free choice only up to a certain point; their exact location is determined by the facts of geography.

It is the sum of these factors that renders the army's communications with its base strong or weak. The result compared with the enemy's situation determines which of the two sides is better placed to cut the other's lines of communication or even of retreat, or, to use the technical phrase, to *envelop* him. Quite apart from psychological or physical superiority, it can, in effect, only be done by the side that has the superior lines of communication. Otherwise the enemy will soon retaliate.

Just as the roads are made to serve two purposes, the enveloping or turning movement may have two objectives. It may aim at disrupting, or cutting, communications, causing the army to wither and die, and thus be forced to retreat; or it may aim at cutting off the retreat itself.

As to the first objective, one should bear in mind that, with armies supplied as they are today, a brief disruption is rarely serious; indeed, a certain amount of time will elapse before minor individual losses will amount to anything significant. In the past, the artificial system of supply meant thousands of wagons moving to and fro. A single flanking operation could shatter it. Today an operation of that sort, however well it might succeed, would not have any noticeable effect. At best it might capture a single convoy and cause some local inconvenience; but it would not necessitate a retreat.

Flanking operations, which have always been more popular in books than in the field, have thus become still less practicable. They can be considered dangerous only to very long and vulnerable lines of communication, whose chief weakness, however, lies in their being always and everywhere exposed to attacks by an insurgent population.

As for an army's retreat being cut off, the threat of narrowed or endangered lines of retreat should likewise not be overrated. Recent experience has made it plain that where the troops are good and their commanders bold they are more likely to break through than be trapped.

There are only very limited means of shortening and protecting extended lines of communication. One can mitigate the situation somewhat by taking some fortresses near the army's position and on the roads that lead back from it, or, where there are no fortresses, by fortifying appropriate points; by treating the population well, keeping strict discipline on military roads, policing the area thoroughly, and constantly keeping the roads in repair. But the risks can never be entirely eliminated.

Incidentally, what we said, in connection with supply, about the roads on which armies should travel if at all possible, applies particularly to the lines of communication. The widest roads running through the wealthiest cities and the richest agricultural areas are the best lines of communication. They are preferable even where they involve considerable detours, and in most cases it is they that determine the specific dispositions of the army's deployment.

CHAPTER SEVENTEEN

Terrain

Quite apart from their influence on sources of supply, which constitutes a separate aspect of the matter, geography and the character of the ground bear a close and ever-present relation to warfare. They have a decisive influence on the engagement, both as to its course and to its planning and exploitation. We shall now examine these factors in the fullest sense of the meaning of the French term *terrain*.

Their principal effect lies in the realm of tactics, but the outcome is a matter of strategy. An engagement in the mountains is in itself and in its consequences quite different from one on the plains.

So long as we have not defined the difference between attack and defense and examined both closely, we cannot properly consider the salient features of terrain as to their effect; for the present we shall have to confine ourselves to their general characteristics. Geography and ground can affect military operations in three ways: as an obstacle to the approach, as an impediment to visibility, and as cover from fire. All other properties can be traced back to these three.

Undoubtedly this threefold effect of terrain tends to make military activity more varied, complex and skillful, because it introduces three additional elements into the combination.

The idea of an absolute and completely level plain—in other words completely passive ground—actually exists only for very small detachments, and even then only for the duration of a particular episode. But in the case of larger units and of longer periods of time, the physical characteristics of the ground influence the action. When it comes to full-scale armies one can hardly think of a single phase—a battle, for example—where the influence of geography would not be felt.

This influence is thus always active; its degree varies according to the nature of the country.

We shall find, as we examine the data as a whole, that there are three distinct ways in which an area may differ from the concept of a flat and open plain: first in the contours of the countryside, such as its hills and valleys; second in such natural phenomena as forests, swamps and lakes; and third in the factors produced by agriculture. Each of these ways contributes to the influence that geography exerts on military operations. If we further analyze these three types, we shall be able to define them as mountainous country, sparsely cultivated forests and marshlands, and agricultural areas. All three will tend to make warfare more complicated and ingenious.

Not all types of cultivation, naturally, have the same effect. It is strongest in Flanders, Holstein, and other areas, where the land is cut up by numerous ditches, fences, hedges, and walls, and dotted with great numbers of single houses and clumps of trees.

It is therefore easiest to wage war in flat and only moderately cultivated areas. But this holds true only in general and altogether disregards the value of natural obstacles to the defense.

Each of the three kinds of terrain, then, may have a threefold effect: as an obstacle to access and to visibility, and as a means of cover.

In wooded country the dominant impediment is to visibility; in mountainous country it is to access; in intensively cultivated areas, both may exist to some degree.

Military movements are practically impossible in heavily wooded country, since the difficulty of access as well as the complete lack of visibility will not permit all modes of traversing it. Thus in one way it simplifies the very action that it complicates in another. For while it is more difficult to concentrate one's forces for an engagement in such terrain, it will not be necessary to split up into so many parts as it would be in mountains or in heavily intersected areas—in other words, in wooded areas a division of forces will be harder to avoid, but it will not have to be so extensive.

In mountainous areas the problem of access dominates, and it is effective in two ways: one can only penetrate at certain points and not at others; and where it is possible at all, movement is slower and more laborious. Thus, the impetus of all movements will diminish in mountains, and any maneuver will take more time. Moreover, mountainous terrain has the particular characteristic that one point dominates another. The command of high ground in general is a subject we shall deal with in the following chapter. At this stage, we only wish to point out that it is this peculiarity of mountains that causes troops to be split up to so great an extent. High points are important not only for their own sake, but for the effect they have on one another.

As each type of terrain approaches its extreme it will, as we have observed elsewhere, tend to reduce a general's influence on events to the same degree to which it tends to emphasize the personal resources of the ranks, down to the private soldier. The more the forces are divided, the less they can be controlled, and obviously the more each man is on his own. It is of course true that as operations become more and more fragmented, more diversified and specialized, the role of intelligence will in general have to increase, and the supreme commander himself will have more chance of showing superior ability. But at this time we must repeat a point we have made before: in war, the sum total of individual successes is more decisive than the pattern that connects them. Thus if we pursue the present argument to its extreme and imagine an army deployed in a single firing line so that every soldier fights as it were a private battle, more will clearly depend on the sum of individual victories than on their pattern. After all, the effectiveness of favorable combinations can only be the result of success, and never of failure; so in this case the courage, skill and spirit of the individual will be the decisive factor. Only in cases in which the two armies are of equal quality, or where their special virtues cancel each other out, can the talent and insight of the supreme commander again become paramount. It follows that truly national wars with a population in arms, and similar circumstances that generally raise the military spirit of the individual (if not necessarily his courage and skill) will tend to succeed where their forces have to be scattered—in other words, where they are favored by heavily uneven and obstructed terrain. But they can exist only in this type of country; their nature usually denies them the qualities and virtues that are vital to concerted action by even moderately substantial forces.

Between the extremes there are many gradations in the nature of the fighting forces. Even a standing professional army will develop something of a national army's qualities when defending its native soil, which will permit greater independence of action.

The more these characteristics and circumstances are missing in one army, and the more pronounced they are on the other side, the more this army will fear fragmentation and the more it will tend to avoid rough country. But this is seldom a matter of choice: one cannot choose a theater of operations by trying it out as if it were merchandise. Hence troops that are by nature at an advantage when fighting as a concentrated mass, will exert their efforts to the utmost in order to use this system as far as it is at all possible, *in spite of the nature of the terrain.* They will thereby be exposed to other disadvantages, such as difficulties of supply and poor billets; and in action they will be exposed to frequent flank attacks. But they would pay an even greater penalty if they were to give up their own special advantages.

These two opposite tendencies—toward concentration and toward dispersal—normally follow the natural bias of the troops themselves. Still, even in the most unequivocal of cases, it will be as impossible for one side always to remain concentrated, as it will be for the other to count on success simply by always operating in open order. In Spain even the French were forced to disperse their strength, and although the Spaniards fought as insurgents in defense of their native soil they were still forced to hazard some of their strength in major battles.

The influence that the terrain brings to bear upon the general, and particularly the political composition of the fighting forces, is closely followed in importance by its influence on the ratio between the arms of the services.

In all areas that are inaccessible—whether because of mountains, forests, or the type of cultivation—a sizeable cavalry is plainly useless. The same applies to artillery in wooded areas, which seldom offer enough space for the effective use of guns or roads to move them on or forage for the horses. Closely cultivated areas are less of a handicap for artillery, and mountains least of all. Both, of course, provide cover from its fire, and are therefore not favorable to an arm whose chief effect is fire; what is more, heavy guns are frequently endangered by enemy infantry, for infantry can penetrate anywhere. On the other hand, neither type of terrain will be exactly lacking in the space required by a large amount of artillery, and in the mountains artillery also has the enormous advantage of the slowness of the enemy's movements, which adds to its effectiveness.

But it cannot be doubted that in difficult terrain of any kind infantry is the clearly superior arm. In such areas therefore its numbers should be permitted considerably to exceed the normal ratio.

CHAPTER EIGHTEEN

The Command of Heights

In the art of war, the word "dominate" has a charm all its own. In fact this element accounts for a substantial part, and possibly for most of the influence that terrain exerts upon the use of forces. This is at the root of many a sacred convention of military erudition, such as "commanding positions," "key positions," "strategic maneuvers," and the like. We propose to show the facts as clearly as we can without prolixity, and shall examine truth and falsehood, reality and fantasy, one by one.

Physical force is always harder to exert in an upward than in a downward direction, and this must also hold true of an engagement. We can cite three obvious reasons. First of all, high ground always inhibits the approach; second, though it does not add perceptibly to range, shooting downward, considering all the geometrical relations involved, is perceptibly *more accurate* than shooting upward; and third, heights command a wider view. How all these factors combine during the engagement will not concern us here. We shall simply take the sum of the tactical advantages conferred by height and call it the first strategic benefit.

The first and third of these advantages will also arise in a strategic context. Marching and reconnaissance are as much a part of strategy as of tactics. Hence, if a position on higher ground prevents the approach of a force located lower down, it constitutes the second advantage that accrues to strategy. The wider view that high ground affords is the third.

In these elements lies the strength of positions that dominate, survey, command. They are the source of superiority and security for the side that looks down from a mountain ridge at the enemy below, and of inferiority and apprehension for the latter. Possibly this overall impression is greater than it ought to be: the advantages of higher ground impress the mind more acutely than the circumstances that modify them. The impression may, therefore, exceed the facts, in which case this trick of the imagination must be counted as an additional element that reinforces the effect of holding high ground.

The advantage of increased mobility is of course not absolute, nor does it always lie with the side that holds the higher ground. It only does so if the other side intends to attack. Height is no asset if a wide valley lies between the two, and indeed it can favor the side on lower ground if both sides intend to fight an action in the plains—as they did at the battle of Hohenfriedberg. The advantage of a wider view also has definite limitations; it will do no good where the lower ground is wooded, or where part of the view is blocked by the very mountain range on which one is located. There are innumerable cases where one searches the area in vain for the advantageous high points one had located on the map; instead, one often seems to get involved in their very disadvantages. Still, these limitations and conditions will not cancel out the advantages that derive from higher ground, for both defense and attack. We shall briefly show how this applies in either case.

High ground offers three strategic assets: *greater tactical strength, protection from access, and a wider view*. The first two, by their nature, accrue only to the defense; they can be used only by the side that stands still—the side that moves cannot take advantage of them. The third advantage may be used by the attack as well as by the defense.

This suggests the important part that high ground plays in defensive situations. Since high ground is attainable only by taking up positions in mountainous areas, one might conclude that these would constitute a great advantage for the defense. It may turn out to be different in reality, as the chapter on defense in mountainous areas will show.

We must in any case make a general distinction. When it is simply a matter of high ground at a single point—at one position, for example—the strategic benefit tends to merge with the simple tactical benefit of a good position for a battle. If, on the other hand, one imagines a substantial tract of land—a province, say—as a sloping plain, a general watershed, one expands the strategic benefit by being able to move in several directions while still retaining one's elevation above the surrounding country. In that case, one will profit by the advantage of higher ground, not only through a combination of forces in the single engagement but also through a combination of a series of engagements which form a whole. So much for defense.

As for the attack, to some extent it enjoys the same advantages that height confers on defense, for a strategic attack, unlike a tactical one, is not a single action. The attacker's progress is not continuous, like that of a machine, but consists of a number of separate marches, with intervals of varying length between them. At every rest stop the attacker is as much on the defensive as his opponent.

The advantage of a better view means that higher ground confers a certain offensive power on both attack and defense, which calls for some attention: it is the ease with which one can operate in isolated detachments. The advantages that an elevated location give to the whole will also accrue to each of its parts. An isolated group, whether large or small, can wield more power than it would otherwise possess, and is less exposed to risk in taking up a position than it would be without the benefit of height. How such isolated detachments can best be used will be discussed elsewhere.

If a location on high ground is combined with other geographical advantages in relation to one's opponent, if the latter is limited in his movements by other factors, such as the proximity of a major river, for instance, the drawbacks of his situation will appear so decisive as to make him want to withdraw as quickly as possible. No army is capable of maintaining a position in the valley of a major river if it does not command the surrounding heights.

The occupation of high ground can thus mean genuine domination. Its reality is undeniable. But when all is said and done, such expressions as "a dominating area," "a covering position," and "key to the country" are, insofar as they refer to the nature of higher or lower ground, for the most part hollow shells lacking any sound core. These elegant elements of theory have been used above all as seasoning for the apparently overly plain military fare. They are the favorite topics of academic soldiers and the magic wands of armchair strategists. Neither the emptiness of such fantasies nor the contradictions of experience have been able to convince these authors and their readers that they were, in effect, pouring water into the leaky vessel of the Danaïdes. Conditions have been mistaken for the thing itself, the tool for the hand that wields it. Mere occupation of such an area and position is taken for a show of strength, for a thrust or a blow, and the area and the position for an active element. In reality, the occupation is nothing but a raised arm, and the position itself only a lifeless tool, a mere potentiality that needs an object for its realization, a simple plus or minus sign without any value attached. The real thrust and blow, the object, the value is *victory in battle.* It is the only thing that really counts and can be counted on, and one must always bear it in mind, whether it be in passing judgment in books or in taking action in the field.

If the decision depends only on the number and scale of victories, it becomes obvious that the first consideration is the relative quality of the two armies and their commanders. Terrain can only play a minor role.

BOOK SIX

Defense

CHAPTER ONE

Attack and Defense

1. The Concept of Defense

What is the concept of defense? The parrying of a blow. What is its characteristic feature? Awaiting the blow. It is this feature that turns any action into a defensive one; it is the only test by which defense can be distinguished from attack in war. Pure defense, however, would be completely contrary to the idea of war, since it would mean that only one side was waging it. Therefore, defense in war can only be relative, and the characteristic feature of waiting should be applied only to the basic concept, not to all of its components. A partial engagement is defensive if we await the advance, the charge of the enemy. A battle is defensive if we await the attack—await, that is, the appearance of the enemy in front of our lines and within range. A campaign is defensive if we wait for our theater of operations to be invaded. In each of these cases the characteristic of waiting and parrying is germane to the general idea without being in conflict with the concept of war; for we may find it advantageous to await the charge against our bayonets and the attack on our position and theater of operations. But if we are really waging war, we must return the enemy's blows; and these offensive acts in a defensive war come under the heading of "defense"—in other words, our offensive takes place within our own positions or theater of operations. Thus, a defensive campaign can be fought with offensive battles, and in a defensive battle, we can employ our divisions offensively. Even in a defensive position awaiting the enemy assault, our bullets take the offensive. So the defensive form of war is not a simple shield, but a shield made up of well-directed blows.

2. Advantages of Defense

What is the object of defense? Preservation. It is easier to hold ground than take it. It follows that defense is easier than attack, assuming both sides have equal means. Just what is it that makes preservation and protection so much easier? It is the fact that time which is allowed to pass unused accumulates to the credit of the defender. He reaps where he did not sow. Any omission of attack—whether from bad judgment, fear, or indolence—accrues to the defenders' benefit. This saved Prussia from disaster more than once during the Seven Years War. It is a benefit rooted in the concept and object of defense: it is in the nature of all defensive action. In daily life, and especially in litigation (which so closely resembles war) it is summed up by the Latin proverb *beati sunt possidentes*. Another benefit, one that arises solely from the nature of war, derives from the advantage of position, which tends to favor the defense.

Having outlined these general concepts, we now turn to the substance.

Tactically, every engagement, large or small, is defensive if we leave the initiative to our opponent and await his appearance before our lines. From that moment on we can employ all offensive means without losing the advantages of the defensive—that is to say the advantages of waiting and the advantages of position. At the strategic level the campaign replaces the engagement and the theater of operations takes the place of the position. At the next stage, the war as a whole replaces the campaign, and the whole country the theater of operations. In both cases, defense remains the same as at the tactical level.

We have already indicated in general terms that defense is easier than attack. But defense has a passive purpose: *preservation*; and attack a positive one: *conquest*. The latter increases one's own capacity to wage war; the former does not. So in order to state the relationship precisely, we must say that *the defensive form of warfare is intrinsically stronger than the offensive*. This is the point we have been trying to make, for although it is implicit in the nature of the matter and experience has confirmed it again and again, it is at odds with prevalent opinion, which proves how ideas can be confused by superficial writers.

If defense is the stronger form of war, yet has a negative object, it follows that it should be used only so long as weakness compels, and be abandoned as soon as we are strong enough to pursue a positive object. When one has used defensive measures successfully, a more favorable balance of strength is usually created; thus, the natural course in war is to begin defensively and end by attacking. It would therefore contradict the very idea of war to regard defense as its final purpose, just as it would to regard the passive nature of defense not only as inherent in the whole but also in all its parts. In other words, a war in which victories were used only defensively without the intention of counterattacking would be as absurd as a battle in which the principle of absolute defense—passivity, that is—were to dictate every action.

The soundness of this general idea could be challenged by citing many examples of wars in which the ultimate purpose of defense was purely defensive, without any thought being given to a counteroffensive. This line of argument would be possible if one forgot that a general concept is under discussion. The examples that could be cited to prove the opposite must all be classed as cases in which the possibility of a counteroffensive had not yet arisen.

In the Seven Years War, for instance, Frederick the Great had no thought of taking the offensive, at least not in its final three years. Indeed, we believe that in this war he always regarded offensives solely as a better means of defense. This attitude was dictated by the general situation; and it is natural for a commander to concentrate only on his immediate needs. Nevertheless one cannot look at this example of defense on a grand scale without speculating that the idea of a possible counteroffensive against Austria may have been at the root of it, and conclude that the time for such a move had not yet come. The peace that was concluded proves that this was not an empty assumption: What else could have induced the Austrians to make peace but the thought that their forces could not on their own outweigh the genius of the King; that in any case they would have to increase their efforts; and that any relaxation was almost bound to cost them further territory? And, indeed, is there any doubt that Frederick would have tried to crush the Austrians in Bohemia and Moravia again if Russia, Sweden, and the Army of the Empire had not diverted his energies?

Now that we have defined the concept of defense and have indicated its limits, we return once more to our claim that defense is *the stronger form of waging war.*

Close analysis and comparison of attack and defense will prove the point beyond all doubt. For the present, we shall merely indicate the inconsistencies the opposite view involves when tested by experience. If attack were the stronger form, there would be no case for using the defensive, since its purpose is only passive. No one would want to do anything but attack: defense would be pointless. Conversely, it is natural that the greater object is bought by greater sacrifice. Anyone who believes himself strong enough to employ the weaker form, attack, can have the higher aim in mind; the lower aim can only be chosen by those who need to take advantage of the stronger form, defense. Experience shows that, given two theaters of operations, it is practically unknown for the weaker army to attack and the stronger stay on the defensive. The opposite has always happened everywhere, and amply proves that commanders accept defense as the stronger form, even when they personally would rather attack.

Some related points remain to be discussed in the following chapters.

CHAPTER TWO

The Relationship between Attack and Defense in Tactics

First let us examine the factors that lead to victory in an engagement.

At this stage we are not concerned with numerical superiority, courage, training, or other qualities of an army. All of these as a rule depend on matters beyond that part of the art of war we are concerned with here; in any case their bearing would be the same on attack and defense. Even *general superiority of numbers* is not relevant, since numbers, too, are usually a given quantity in which a commander has no say. Moreover, these matters have no special bearing on attack and defense. Only three things seem to us to produce decisive advantages: *surprise, the benefit of terrain, and concentric attack.*

Surprise becomes effective when we suddenly face the enemy at one point with far more troops than he expected. This type of numerical superiority is quite distinct from numerical superiority in general: it is the most powerful medium in the art of war. The ways in which the advantage of terrain contributes toward victory are fairly obvious. But it should be noted that this is more than a matter of obstacles to an attack—steep slopes, high mountains, marshy streams, hedges, and the like. Terrain may be just as useful by enabling us to hold a concealed position; even a featureless landscape can provide some advantage to those familiar with it. Concentric attack comprises all tactical envelopment, great or small; its effectiveness is produced partly by the double effectiveness of cross fire, and partly by the fear of being cut off.

What is the relationship of attack and defense to these matters?

Bearing in mind the three elements of victory already described, the answer must be this: the attacker is favored by only a small part of the first and third factors while their larger part, and the second factor exclusively, are available to the defender.

The one advantage the attacker possesses is that he is free to strike at any point along the whole line of defense, and in full force: the defender, on the other hand, is able to surprise his opponent constantly throughout the engagement by the strength and direction of his counterattacks.

For the attacker it is easier to surround the whole opposing force and cut it off than it is for the defender: the latter is tied to his position and has thereby presented the attacker with an objective. But the attacker's envelopment and its advantage is applicable only to the whole position, for in the course of the engagement, it is easier for the defender to attack segments of the opposite force concentrically; for *as we have already said, the defender is better placed to spring surprises by the strength and direction of his own attacks.*

It is self-evident that it is the defender who primarily benefits from the terrain. His superior ability to produce surprise by virtue of the strength and direction of his own attack stems from the fact that the attack has to approach on roads and paths on which it can easily be observed; the defender's position, on the other hand, is concealed and virtually invisible to his opponent until the decisive moment arrives. Ever since the right method of defense was adopted, reconnaissance has gone out of fashion—or, rather, it has become impossible. Some reconnaissance is still carried out now and again, but as a rule nothing much comes of it. And yet, no matter how great the advantage of being free to choose the ground for a position and become familiar with it before the action, and no matter how plain it is that the defender in the concealed position he has selected is bound to cause far more surprise than the attacker, the older view persists: a battle accepted is regarded as already half lost. This stems from the type of defense that was practiced twenty years ago and to some extent in the Seven Years War as well. In those days the only kind of advantage looked for in terrain was that the front had to be difficult to approach (because of steep slopes, etc.). The shallowness of the positions and the difficulties of maneuvering one's flanks made armies so weak that they had to dodge each other from hill to hill and so made things even worse. Once some kind of support had been secured, the army was stretched as tight as on an embroidery frame: all depended on its not being pierced. The defended terrain was considered valuable for its own sake, and so had to be defended at all points. Thus movement or surprise in battle were out of the question. This was the complete opposite of what a good defense can be, and in fact has been in the recent past.

Defense appears to fall into disrepute whenever a particular style of it has become obsolescent; that is what happened in the case described above. In its day this method of defense really had been superior to the attack.

If we survey the development of modern war, we find that at the beginning—in the Thirty Years War and the War of the Spanish Succession—an army's deployment and disposition was one of the main elements in a battle. It was the most important part of the plan of action. This normally worked to the advantage of the defender because his forces were deployed and in position from the start. With the troops' increased ability to maneuver, this advantage was lost, and for a time the attack gained the upper hand. The defender now sought protection behind rivers or deep valleys, or on mountains. He thus recovered a distinct advantage, which lasted until the attacker became so mobile and so skilled that he could venture even into rough country and attack in separate columns; which enabled him to *turn* the enemy. This led to greater and greater extension of the line of battle until it naturally occurred to the attacker to concentrate on a limited number of points and pierce the enemy's shallow position. Thus the offensive gained the upper hand for the third time, and once again the defensive had to change its methods. That is what happened in the recent wars. Forces were kept concentrated in large masses, most of them not deployed and, wherever possible, in concealed positions. The object was simply to be ready to deal with the attack as soon as its intentions became clear.

This does not entirely preclude defending one's ground in a partly passive manner, for to do so offers such decisive advantages that it is frequently done in the course of a campaign. But usually the passive defense of terrain is no longer dominant—which is all we are concerned with here.

If the offensive were to invent some major new expedient—which is unlikely in view of the simplicity and inherent necessity that marks everything today—the defensive will also have to change its methods. But it will always be certain of having the benefit of terrain, and this will generally ensure its natural superiority; for today the peculiarities of the topography and the ground have a greater effect on military action than ever.

CHAPTER THREE

The Relationship between Attack and Defense in Strategy

Let us again begin by examining the factors that assure strategic success.

As we have said before, in strategy there is no such thing as victory. Part of strategic success lies in timely preparation for a tactical victory; the greater the strategic success, the greater the likelihood of a victorious engagement. The rest of strategic success lies in the exploitation of a victory won. The more strategy has been able, through its ingenuity, to exploit a victorious battle; the more that it can wrest out of the collapsing edifice whose foundations have been shattered by the action; the more completely the fruits of the hard-won victory can be harvested; then the greater the success. The main factors responsible for bringing about or facilitating such a success—thus the main factors in strategic effectiveness—are the following:

1. The advantage of terrain
2. Surprise—either by actual assault or by deploying unexpected strength at certain points
3. Concentric attack (all three as in tactics)
4. Strengthening the theater of operations, by fortresses, with all they involve
5. Popular support
6. The exploitation of moral factors.[1]

What is the relationship of attack and defense with regard to these factors?

In strategy as well as in tactics, the defense enjoys the advantage of terrain, while the attack has the advantage of initiative.[2] As regards surprise and initiative, however, it must be noted that they are infinitely more important and effective in strategy than in tactics. Tactical initiative can rarely be expanded into a major victory, but a strategic one has often brought the whole war to an end at a stroke. On the other hand, the use of this device

[1] Anyone who has learned his strategy from Herr von Bülow will not understand how it is that we have simply left out the whole of Bülow's teaching. But it is not our fault if Bülow deals with minor matters only. An office boy would be just as puzzled if he searched the index of an arithmetic book and found no entry for such practical rules as the rules of three or five. But Herr von Bülow's opinions can hardly be counted as practical rules. We have made the comparison for other reasons. Cl.

[2] The German term is *Überfall* (surprise attack), which Clausewitz here employs in a general sense. Eds.

assumes *major, decisive, and exceptional* mistakes on the enemy's part. Consequently it will not do much to tip the scales in favor of attack.

Surprising the enemy by concentrating superior strength at certain points is again comparable to the analogous case in tactics. If the defender were compelled to spread his forces over several points of access, the attacker would obviously reap the advantage of being able to throw his full strength against any one of them.

Here too the new system of defense has, by its new approach, imperceptibly introduced new principles. Where the defender has no reason to fear that his opponent will be able by advancing along an undefended road to seize an important depot or munitions dump, or take a fortress unawares, or even the capital unawares; where, therefore, he is not forced to attack the enemy on the road chosen by the latter in order to avoid having his retreat cut off; then there is no reason for him to split his forces. If the attacker chooses a road on which he does not expect to meet the defender, the latter can still seek him out there with his entire strength a few days later. Indeed he can be sure that in most cases the attacker himself will oblige him by seeking him out. But if for some reason the attacker has to advance with divided forces—and problems of supply often leave him little choice—the defender obviously reaps the benefit of being able to attack a part of his opponent with his own full strength.

In strategy, the nature of flank and rear attacks on a theater of operations changes to a significant degree.

1. The effect of cross fire is eliminated, since one cannot fire from one end of a theater of operations to the other.
2. There is less fear of being cut off, since whole areas cannot be sealed off in strategy as they can in tactics.
3. Because of the greater areas involved in strategy, the effectiveness of interior and therefore shorter lines is accentuated and forms an important counterbalance against concentric attacks.
4. A new factor emerges in the vulnerability of lines of communication, that is, in the consequences of their being cut.

Because of the greater areas involved in strategy, envelopment or concentric attack will of course only be possible for the side which takes the initiative—in other words, the attacker. The defender cannot, as he can in tactics, surround the surrounder in turn, for he cannot deploy his troops in the relative depth required, nor keep them sufficiently concealed. But what use to the attack is ease of envelopment if its rewards do not materialize? In strategy, therefore, there would be no justification at all in putting forward the enveloping attack as a means of victory, were it not for its effect on lines of communication. Yet this is seldom an important factor at the earliest stage when attack is first confronted by defense, and the two sides face each other in their opening positions. It only begins to tell in the course of a campaign, when the attacker, in enemy territory, gradually becomes the defender. At that point the new defender finds his lines of communication weakening, and the original defender can exploit that weakness once he has taken the offensive. But it must be obvious that as a rule the defender deserves no credit for this advantage, since it really derives from the principles inherent in the defense itself.

The fourth element, *the advantages of the theater of operations*, naturally benefit the defender. By initiating the campaign, the attacking army cuts itself off from its own theater of operations, and suffers by having to leave its fortresses and depots behind. The larger the area of operations that it must traverse, the more it is weakened—by the effect of marches and by the detachment of garrisons. The defending army, on the other hand, remains intact. It benefits from its fortresses, nothing depletes its strength, and it is closer to its sources of supply.

The support of the population, the fifth principle, will not necessarily apply to every defense; a defensive campaign may be fought in enemy territory. Still, this element derives from the concept of defense alone, and it is applicable in the vast majority of cases. What is meant is primarily (but not exclusively) the effectiveness of militia, and arming the population. Furthermore, every kind of friction is reduced, and every source of supply is nearer and more abundant.

The campaign of 1812 will here serve as a magnifying glass, for it clearly reveals how the third and fourth of these factors can operate. Half a million men crossed the Niemen; only 120,000 fought at Borodino, and still fewer reached Moscow.

One may say indeed that the outcome of this enormous effort was so great that even if the Russians had not followed it up with their own counteroffensive, they would have been secure from any fresh invasion for a long time to come. Of course no European country, except for Sweden, is in a similar position to Russia's; but the principle is universal and differs only in degree.

As to the fourth and fifth factors, one might add that these assets pertain to the basic case of defense in one's own country. If defense is moved to enemy soil and gets involved in offensive operations, it will be transformed into a further liability of the offensive, in much the same way as with the third element mentioned above. The offensive is not composed of active elements alone, any more than the defensive is made up solely of passive elements. Indeed, any attack that does not immediately lead to peace must end on the defensive.

Thus, if all elements of defense that occur during an offensive are weakened by the very fact that they are part of the offensive, then we must regard this as another general liability pertaining to it.

This is not simply hairsplitting. Far from it: this is the greatest disadvantage of all offensive action. Hence when a strategic attack is being planned one should from the start give very close attention to this point—namely, the defensive that will follow. The matter will be discussed in greater detail in the book on strategic planning.

The important moral forces that sometimes permeate war like a leaven may occasionally be used by a commander to invigorate his troops. These forces may be found on the side of defense as well as that of attack; at least one can say that the ones which especially favor attack, such as panic and confusion in the enemy's ranks, do not normally emerge until after the decisive blow has been struck, and so seldom have much bearing on its course.

All this should suffice to justify our proposition that *defense is a stronger form of war than attack*. But we still have to mention a minor factor that so far has been left out of account. It is courage: the army's sense of superiority that springs from the awareness that one is taking the initiative. This affinity is a real one, but it is soon overlaid by the stronger and more general spirit that an army derives from its victories or defeats, and by the talent or incompetence of its commander.

CHAPTER FOUR

Convergence of Attack and Divergence of Defense

These two concepts, these two ways of using one's forces in attack and defense, crop up so frequently in theory and in practice that one unconsciously comes to regard them as inherent forms, almost as indispensable elements, of attack and defense. The slightest reflection will show that this is not actually the case. That is why we should like to examine them as early as possible and see them for what they really are. Once this is done we can leave them out of account in our further analysis of the relationship between attack and defense without being constantly distracted by any apparent advantage or disadvantage that seems to attach to them. We will consider them here as pure abstractions and distill their essence; reserving to a later stage any comment on their role.

It is assumed in both tactics and strategy that defense is a state of expectation—that is to say, immobile—whereas the attacker is in motion—relative, that is, to the defender's immobility. It follows automatically that turning and enveloping are methods available to the attacker only so long as his movement and the immobility of the defense persist. This option of moving convergently or not—depending on whether or not it is in the attacker's interest—would have to be counted as one of his advantages. But the freedom of choice is available only in tactics, not always in strategy. In the former the points on which the flanks rest hardly ever provide the kind of absolute security that they frequently do in strategy, where the line of defense may run from sea to sea or from one neutral country to another. In this case, there can be no convergent attack; freedom of choice is limited. It is even more awkwardly limited where the attack *has* to be convergent. Russia and France cannot attack Germany in any other way than by convergent movements; they can never attack with united forces. So if the convergent use of forces is assumed to be the less effective in the majority of cases, the benefit the attacker usually gets from greater freedom of choice is probably nullified by the fact that in other cases he is forced to use the less effective form.

Let us examine the effect of these forms in tactics and strategy more closely.

When forces operate convergently, from the circumference toward the center, the fact that they converge as they advance has been considered a major advantage. The fact is true enough, but the presumed advantage is not; because convergence takes place on both sides, and they therefore cancel each other out. Where movement is from the center outward, the same is true of divergence.

The real advantage is a different one, in that forces operating on convergent lines direct their effectiveness toward a *common destination,* while divergent lines do not. What, then, are these effects? Here we must distinguish between tactics and strategy.

We do not want to push our analysis too far, and shall therefore cite only the following points as the advantages of these effects:

1. The doubled, or at least intensified, effect of cross fire, once a given degree of convergence has been reached
2. Convergent attack on a single force
3. Cutting off the retreat.

Cutting off a retreat is possible in strategy as well, but it is obviously much harder because the greater areas involved are not very easily sealed off. Generally speaking a concentric attack on part of a force will gain in weight and decisiveness the smaller the force attacked and the closer it comes to the absolute minimum—viz., a single soldier. An army can fight quite well on several fronts; a division less so; and a battalion only in close order. A single man can do nothing at all. Strategy is concerned with major bodies of troops, wide areas and substantial lengths of time; tactics with the opposite. It follows that a convergent attack cannot have the same effect in strategy as in tactics.

The effect of cross fire is not relevant to strategy at all; its place is taken by another factor—the alarm that every army experiences to some degree when the enemy is victorious in its rear, however far or near he may be.

The convergent use of forces has, then, this undoubted merit: its effect on A works also upon B without detracting from the effect on A; while its effect on B works likewise upon A. The sum is therefore not A + B, but something more; and this is an advantage that occurs both in tactics and in strategy, though rather differently in each.

What is it in the divergent operations of the defense that offsets these advantages? Plainly, it must be the fact that the troops are closer together and operating on interior lines. There is no need to demonstrate how this can multiply strength to the point where the attacker dare not expose himself to it unless he is greatly superior.

Once the defense has embraced the principle of movement (admittedly starting later than the attacker, but still in time to burst the numbing bonds of inactivity) the benefit of greater concentration and interior lines becomes a decisive one which is more likely as a rule to lead to victory than a convergent pattern of attack. But victory is the prerequisite for success: one must vanquish the enemy before one can think of cutting off his retreat. In short it becomes evident that the relationship is similar to that of attack and defense in general. The convergent form pays dazzling dividends, but the yield of the divergent form is more dependable. The former is the weaker form with the active purpose; the latter the stronger form with the passive purpose. That being so, it seems to me that they almost balance one another. One might add that defense, not being absolute in every case, will not always find it impossible to use its strength convergently. There is then no longer a valid basis for the belief that the convergent form alone provides attack with a universal advantage over defense. One's judgment is thus freed from the influence which that fallacy is apt to exert on it every time the point comes up.

The foregoing remarks apply to tactics and to strategy alike. A particularly important point, concerning strategy only, remains to be made. The advantage of interior lines increases with the distances to which they relate. Where distances measure a few thousand paces or a few miles, less time will obviously be saved than over a march of several days or a distance of a couple of hundred miles. The former, shorter, distances are a matter of tactics; the longer ones, of strategy. While it is true that in strategy one needs more time to reach an objective than in tactics, and it takes longer to defeat an army than a battalion, even in strategy this time will increase only up to a certain point—that is, until a battle has been fought, and possibly also for the few days in which a battle can be avoided without serious consequences. What is more an even greater difference lies in the absolute lead gained in each case. The distances in tactics or in a battle are so slight that an army makes its moves almost in full view of the enemy: therefore anyone stationed on the outermost lines will soon be aware of the enemy's movements. With the greater distances involved in strategy, the movements of the one will nearly always remain concealed from the other for at least a day; and there are many instances of moves involving only part of an army, or the despatch of a considerable force, that were kept from the enemy for weeks. The considerable advantage of concealment to the side that, because of its natural position, can best profit by it, must be obvious.

This concludes our discussion of the convergent and divergent effect of forces and of their relation to attack and defense; we shall return to both aspects at a later stage.

CHAPTER FIVE

The Character of Strategic Defense

We have already stated what defense is—simply the more effective form of war: a means to win a victory that enables one to take the offensive after superiority has been gained; that is, to proceed to the active object of the war.

Even when the only point of the war is to maintain the *status quo*, the fact remains that merely parrying a blow goes against the essential nature of war, which certainly does not consist merely in enduring. Once the defender has gained an important advantage, defense as such has done its work. While he is enjoying this advantage, he must strike back, or he will court destruction. Prudence bids him strike while the iron is hot and use the advantage to prevent a second onslaught. How, when, and where that reaction is to begin depends, of course, on many other conditions which we shall detail subsequently. For the moment we shall simply say that this transition to the counterattack must be accepted as a tendency inherent in defense—indeed, as one of its essential features. Wherever a victory achieved by the defensive form is not turned to military account, where, so to speak, it is allowed to wither away unused, a serious mistake is being made.

A sudden powerful transition to the offensive—the flashing sword of vengeance—is the greatest moment for the defense. If it is not in the commander's mind from the start, or rather if it is not an integral part of his idea of defense, he will never be persuaded of the superiority of the defensive form; all he will see is how much of the enemy's resources he can destroy or capture. But these things do not depend on the way in which the knot is tied, but on the way in which it is untied. Moreover, it is a crude error to equate attack with the idea of assault alone, and therefore, to conceive of defense as merely misery and confusion.

Admittedly, an aggressor often decides on war before the innocent defender does, and if he contrives to keep his preparations sufficiently secret, he may well take his victim unawares. Yet such surprise has nothing to do with war itself, and should not be possible. War serves the purpose of the defense more than that of the aggressor. It is only aggression that calls forth defense, and war along with it. The aggressor is always peace-loving (as Bonaparte always claimed to be); he would prefer to take over our country unopposed. To prevent his doing so one must be willing to make war and be prepared for it. In other words it is the weak, those likely to need defense, who should always be armed in order not to be overwhelmed. Thus decrees the art of war.

When one side takes the field before the other, it is usually for reasons that have nothing to do with the intention of attack or defense. They are not the motives, but frequently the result of an early appearance. The side that is ready first and sees a significant advantage in a surprise attack, will for *that* reason take the offensive. The side that is slower to prepare can to some degree make up for the consequent disadvantage by exploiting the advantages of defense.

Generally speaking, however, the ability to profit from being the first to be ready must be considered an advantage to the attacker, as we have acknowledged in Book Three. Still, this general advantage is not essential in every specific case.

Consequently, if we are to conceive of defense as it should be, it is this. All means are prepared to the utmost; the army is fit for war and familiar with it; the general will let the enemy come on, not from confused indecision and fear, but by his own choice, coolly and deliberately; fortresses are undaunted by the prospect of a siege; and finally a stout-hearted populace is no more afraid of the enemy than he of it. Thus constituted, defense will no longer cut so sorry a figure when compared to attack, and the latter will no longer look so easy and infallible as it does in the gloomy imagination of those who see courage, determination, and movement in attack alone, and in defense only impotence and paralysis.

CHAPTER SIX

Scope of the Means of Defense

In Chapters Two and Three of this book we have shown that defense has a natural superiority in the use of the means—other than the absolute strength and quality of the forces—that determine tactical and strategic success. Among them are advantage of terrain, surprise, concentric attack, advantages of the theater of operations, support of the populace, and the harnessing of moral forces. It may be useful to cast another glance at the range of resources that are preeminently at the disposal of the defender. They may be compared to the various types of pillar on which his edifice rests.

1. *Militia.* In recent times the militia has been employed not only at home but also to invade enemy territory, and there is no denying that its organization in some countries—Prussia, for instance—is such that it can almost be considered part of the regular army. In such cases it is not purely an instrument of defense. Still, one should not forget that its vigorous use in the years 1813, 1814, and 1815 originated in defensive warfare; that in only the minority of countries is it organized as in Prussia; and that wherever its organization is imperfect it will be better suited to defense than to attack. Quite apart from this, the concept of a militia embodies the idea of an extraordinary and largely voluntary participation in the war by the whole population, with its physical strength, its wealth, and its loyalty. The less the institution resembles this model, the more a militia will become a regular army under another name. It will then have the advantages of a regular army, but it will also be lacking in the advantages of a genuine militia: a reservoir of strength that is much more extensive, much more flexible, and whose spirit and loyalty are much easier to arouse. These factors are the essentials of a militia. Its organization must leave scope for the participation of the populace. If it does not, any great hopes one may have from it are mere delusions.

The close relationship between the popular nature of a militia and the concept of defense is unmistakable, and thus also the fact that such a militia is likelier to be a part of defense than of attack. The qualities in which it is superior to the aggressor will mainly show in the course of defensive action.

2. *Fortresses.* The part played by the attacker's fortresses is limited to those that are closest to the border and is not very important. The influence of the defender's fortresses extends more deeply into his country; therefore more of them are involved, and the contribution made by each is incomparably greater. A fortress that attracts, and holds out against, a full-scale siege will naturally weigh much more in the scales of war than one that is so strongly fortified that it is clearly impregnable, and therefore neither really engages nor destroys the enemy's forces.

3. *The People.* Although one single inhabitant of a theater of operations has as a rule no more noticeable influence on the war than a drop of water on a river, the *collective influence* of the country's inhabitants is far from negligible, even when we are not dealing with popular insurrection. At home, everything works more smoothly—assuming the public is not wholly disaffected. Nothing, major or minor, is done for the enemy save under *force majeure,* which the troops must apply at the expense of their own strength and exertions. The defender can get all he wants. It may not be freely given, as the fruit of enthusiastic loyalty; usually it is due to a long tradition of civil obedience which is the citizen's second nature, and also to orders from the government, and other constraints not originating with the military. But voluntary collaboration born of genuine attachment is also always of great value; particularly it will never be wanting when no actual sacrifices are involved. Let us mention just one example, which is of great importance for the conduct of operation: information. We refer not so much to the single outstandingly significant report, but to the countless minor contacts brought about by the daily activities of our army. Here the defender's close relations with the population give him a general superiority. The smallest patrol, every picket, every sentry, every officer on a mission, all have to turn to the local inhabitants for news of friend or foe.

If we proceed from these general conditions which always apply, to the special cases in which the population begins to participate in the fighting itself, until we reach the highest level at which, as in Spain, the war is primarily waged by the people, it will be understood that we are dealing not simply with an intensification of popular support but with a genuine new source of power; which entitles one to say that:

4. A *people in arms,* or a home guard, may be listed as a specific means of defense.

5. Finally, a defender's *allies* can be cited as his ultimate source of support. By this we do not mean the ordinary type of ally such as the aggressor also possesses, but the kind who have a *substantial interest* in maintaining the integrity of their ally's country. If we consider the community of states in Europe today, we do not find a systematically regulated balance of power and of spheres of influence, which does not exist and whose existence has often been justifiably denied; but we certainly do find major and minor interests of states and peoples interwoven in the most varied and changeable manner. Each point of intersection binds and serves to balance one set of interests against the other. The broad effect of all these fixed points is obviously to give a certain amount of cohesion to the whole. Any change will necessarily weaken this cohesion to some degree. The sum total of relations between states thus serves to maintain the stability of the whole rather than to promote change; at least, that *tendency* will generally be present.

This, we suggest, is how the idea of the balance of power should be interpreted; and this kind of balance is bound to emerge spontaneously whenever a number of civilized countries are in multilateral relations.

To what extent this tendency of the common interest helps maintain existing conditions is another question. One can certainly imagine changes in relations between individual states that would strengthen this effect, and others that could weaken it. The first kind are attempts to perfect the political balance, and as their aim reflects that of the common interest, the majority of the parties would be in favor. The other kind, however, are deviations, hyperactivity of individual states, actual cases of disease; one should not be surprised that diseases occur in a loosely constituted polity such as a multitude of states of various sizes: after all, they also occur in the marvelously structured organic whole of all living nature.

It may be objected, of course, that history offers examples of single states effecting radical changes that benefit themselves alone, without the slightest effort by the rest to hinder them. There have even been cases in which a single state has managed to become so powerful that it could virtually dictate to the rest. We would reply that this does not disprove the tendency on the part of common interests to support the existing order; all it shows is that at the moment the tendency was not sufficiently effective. Aspiration toward a goal is not the same as motion, but that is not to say that it is a nullity—witness the dynamics of the heavens.

We therefore argue that a state of balance tends to keep the existing order intact—always assuming that the original condition was one of calm, of equilibrium. Once there has been a disturbance and tension has developed it is certainly possible that the tendency toward equilibrium will shift direction and try to bring about a particular change. It lies in the nature of things, however, that such a change can affect only a few states, never the majority. Most states will certainly assume that the collective interest will always represent and assure their stability. It is thus also certain that in defending itself every individual state whose relations with the rest are not already strained will find that it has more friends than enemies.

One may laugh at these reflections and consider them utopian dreams, but one would do so at the expense of philosophic truth. Philosophy teaches us to recognize the relations that essential elements bear to one another, and it would indeed be rash from this to deduce universal laws governing every single case, regardless of all haphazard influences. Those people, however, who "*never rise above anecdote*" as a great writer said, and who would construct all history of individual cases—starting always with the most striking feature, the high point of the event, and digging only as deep as suits them, never get down to the general factors that govern the matter. Consequently their findings will never be valid for more than a single case; indeed they will consider a philosophy that encompasses the general run of cases as a mere dream.

If it were not for that common effort toward maintenance of the *status quo*, it would never have been possible for a number of civilized states to coexist peacefully over a period of time; they would have been bound to merge into a single state. The fact that Europe, as we know it, has existed for over a thousand years can only be explained by the operation of these general interests; and if collective security has not always sufficed to maintain the integrity of each individual state, the fact should be ascribed to irregularities in the life of the system as a whole which instead of destroying were absorbed into it.

There is no need to review the countless instances in which changes that might have upset the balance too severely were prevented or reversed by the more or less overt reaction of the other states. The briefest glance at history will reveal them. One case, however, calls for mention—one which is always trotted out by those who ridicule the very idea of political balance—because it seems to be an extremely relevant example of how a harmless, unaggressive country perished without any other coming to its assistance. We refer to Poland. The fact that a state of eight million inhabitants could vanish, partitioned by three others, with none of the remaining states resorting to arms, will seem at first glance to be a case that either proves the political balance to be generally ineffective or at least shows how ineffective it can be in given circumstances. The fact that so large a state could vanish and fall prey to others among which were already numbered some of the most powerful (Russia and Austria) seemed an extreme case. If such an event was unable to arouse the common interest of the community of nations, one could argue that the effectiveness of the common interest in assuring the survival of single states is an illusion. We insist, however, that a single case, however striking, cannot vitiate a general principle; and further contend that the demise of Poland is not as strange as it appears. Could Poland really be considered a European state, an equal among equals in the European community of nations? She could not: she was a Tartar state. But instead of lying on the Black Sea, like the Tartars of the Crimea, on the fringe of the European community, she was located in the midst of it on the Vistula. In saying this we do not wish to slight the Poles or justify the partition of their country. Our only concern is to face the facts. Poland had not really played a political part for a century or so; she had merely been a cause of dissension among other states. Given her condition and the kind of constitution she had, she could not possibly maintain her independence. A radical change from these Tartar-like conditions could have been accomplished in the space of fifty or a hundred years, provided her leaders had been willing. They, however, were too much Tartars themselves to desire such a change. Their chaotic public life and their boundless irresponsibility went together, and thus they were swallowed up by the abyss. Long before the country was partitioned, the Russians were doing what they liked there. The idea of Poland as an independent state with meaningful frontiers no longer corresponded to the facts and nothing was surer than that Poland would have become a Russian province if she had not been partitioned. Had things been otherwise, had Poland been a country able to defend itself, the three powers would not have so lightly undertaken to partition it, and the powers most interested in maintaining its independence (France, Sweden and Turkey) would have been able to collaborate in its survival. But it is asking too much when a state's integrity must be maintained entirely by others.

Partitioning Poland had been under discussion for more than a century. Since then, the country had lost the character of a private home and had become more like a public highway on which foreign armies could disport themselves whenever and however they pleased. Were the other states supposed to put a stop to that? Were they supposed to be continually in arms in order to guard the political sanctity of the Polish frontier? That would have been to ask for the morally impossible. Poland in those days was, politically speaking, little better than an uninhabited steppe. This open prairie land, located in the midst of other states, could neither be shielded from their encroachments, nor could its political integrity be guaranteed by others. For all these reasons one should find the silent demise of Poland no more strange than that of the Crimean Tartar state. The Turks were certainly more concerned about the Crimea than any of the European states were in saving Poland; but they realized that it would simply be a waste of effort to try and protect an unresisting steppe.

To return to our subject: we believe we have shown that as a rule the defender can count on outside assistance more than can the attacker; and the more his survival matters to the rest—that is, the sounder and more vigorous his political and military condition—the more certain he can be of their help.

Naturally, the factors listed here as being the real means of defense will not all be available in every case. Some may be missing in one case, some in another; but they all come under the general heading of defense.

CHAPTER SEVEN

Interaction between Attack and Defense

The time has come to consider defense and attack separately, insofar as they can be separated. We shall start with defense for the following reasons. While it is quite natural and even indispensable to base the principles of defense on those that govern attack and vice versa, there must be a third aspect to one of them that serves as a point of departure for the whole chain of ideas and makes it tangible. Our first question, therefore, concerns this point.

Consider in the abstract how war originates. Essentially, the concept of war does not originate with the attack, because the ultimate object of attack is not fighting: rather, it is possession. The idea of war originates with the defense, which does have fighting as its immediate object, since fighting and parrying obviously amount to the same thing. Repulse is directed only toward an attack, which is therefore a prerequisite to it; the attack, however, is not directed toward defense but toward a different goal—possession, which is not necessarily a prerequisite for war. It is thus in the nature of the case that the side that first introduces the element of war, whose point of view brings two parties into existence, is also the side that establishes the initial laws of war. That side is the *defense*. What is under discussion here is not a specific instance but a general, abstract case, which must be postulated to advance theory.

We now know where to find the fixed point that is located outside the interaction of attack and defense: it lies with the defense.

If this argument is correct, the defender must establish ground rules for his conduct even if he has no idea what the attacker means to do, and these ground rules must certainly include the disposition of his forces. The attacker, on the other hand, so long as he knows nothing about his adversary, will have no guidelines on which to base the use of his forces. All he can do is to take his forces with him—in other words, take possession by means of his army. Indeed, that is what actually happens: for it is one thing to assemble an army and another to use it. An aggressor may take his army with him on the chance that he may have to use it, and though he may take possession of a country by means of his army instead of officials, functionaries, and proclamations, he has not yet, strictly speaking, committed a positive act of war. It is the defender, who not only concentrates his forces but disposes them in readiness for action, who first commits an act that really fits the concept of war.

We now come to the second question: what in theory is the nature of the underlying causes that initially motivate the defense, before it has even considered the possibility of being attacked? Obviously, it is an enemy's advance with a view to taking possession, which we have treated as extraneous to war but which forms the basis for the initial steps of military activity. This advance is meant to deter defense, and it must, therefore, be thought of in relation to the country; and this is what produces the initial general dispositions of the defense. Once these have been established, the attack will be directed toward them, and new ground rules of defense will be based on an examination of the means used by the attack. At this point the interaction has become evident, and theorists may continue to study it as long as new results appear and make the study seem worthwhile.

This brief analysis was necessary to provide somewhat greater clarity and substance to our subsequent discussion; it is not intended for the battlefield, nor for any future general, but for the legions of theorists who, up to now, have treated such questions far too lightly.

CHAPTER EIGHT

Types of Resistance

The essence of defense lies in parrying the attack. This in turn implies waiting, which for us is the main feature of defense and also its chief advantage.

Since defense in war cannot simply consist of passive endurance, waiting will not be absolute either, but only relative. In terms of space, it relates to the country, the theater of operations, or the position; in terms of time, to the war, the campaign, or the battle. True these are not unalterable units, but the central points of certain areas that overlap and merge with one another. In practice, however, one must often be satisfied with merely arranging things into categories rather than strictly separating them; and those terms, in general usage, have become clearly enough defined to serve as nuclei around which other ideas may conveniently be gathered.

The defender of a country, therefore, merely awaits the attack on his country, the defender of a theater of war awaits the attack on that theater, and the defender of a position awaits the attack on that position. Once the enemy has attacked, any active and therefore more or less offensive move made by the defender does not invalidate the concept of defense, for its salient feature and chief advantage, *waiting*, has been established.

The concepts characteristic of time–war, campaign and battle—are parallel to those of space–country, theater of operations and position—and so bear the same relation to our subject.

Defense is thus composed of two distinct parts, waiting and acting. By linking the former to a definite object that precedes action, we have been able to merge the two into one whole. But a defensive action—especially a large-scale one such as a campaign or a war—will not, in terms of time, consist of two great phases, the first of which is pure waiting and the second pure action; it will alternate between these two conditions, so that waiting may run like a continuous thread through the whole period of defense.

The nature of the matter demands that so much importance should be attached to waiting. To be sure, earlier theorists never gave it the status of an independent concept, but in practice it has continuously served as a guideline, though for the most part men were not consciously aware of it. Waiting is such a fundamental feature of all warfare that war is hardly conceivable without it, and hence we shall often have occasion to revert to it by pointing out its effect in the dynamic play of forces.

We should now like to elucidate how the principle of waiting runs through the entire period of defense, and how the successive stages of defense originate in it.

In order to establish our ideas by means of a simpler example, we shall defer (till we reach the book on war plans) the defense of a country, a more diversified subject, and one that is more strongly influenced by political circumstances. On the other hand, defense in a position or in a battle is a tactical matter; only when it is *completed* can it serve as the starting point of strategic activity. Therefore, we shall take the defense of a *theater of operations* as the subject that will best illustrate the conditions of defense.

We have pointed out that waiting and acting—the latter always being a riposte and therefore a reaction—are both essential parts of defense. Without the former, it would not be defense, without the latter, it would not be war. This conception has already led us to argue that *defense is simply the stronger form of war, the one that makes the enemy's defeat more certain.* We must insist on this interpretation—partly because any other will eventually lead to absurdity, partly because the more vivid and total this impression, the more it will strengthen the total act of defense.

It would be contrary to this interpretation to discuss reaction, the second necessary component of defense, by making a distinction between its parts, and considering that phase which, strictly speaking, consists in warding off the enemy—from the country, the theater of operations, the position—as the only *necessary* part, which would be limited to what is needed to achieve those purposes. The other phase, the possibility of a reaction that *expands into the realm of actual strategic offense,* would then have to be considered as being foreign to, and unconnected with, defense. Such a distinction is basically unacceptable: we must insist that the idea of *retaliation* is fundamental to all defense. Otherwise, no matter how much damage the first phase of reaction, if successful, may have done to the enemy, the proper balance would still be wanting in the restoration of the dynamic relationship between attack and defense.

We repeat then that defense is the stronger form of war, the one that makes the enemy's defeat more certain. It may be left to circumstances whether or not a victory so gained exceeds the original purpose of the defense.

Since defense is tied to the idea of waiting, the aim of defeating the enemy will be valid only on the condition that there is an attack. If no attack is forthcoming, it is understood that the defense will be content to hold its own; so this is its aim, or rather its primary aim, during the period of waiting. The defense will be able to reap the benefits of the stronger form of war only if it is willing to be satisfied with this more modest goal.

Let us postulate an army that has been ordered to defend its theater of operations. It may do this in the following ways:

1. It can attack the enemy the moment he invades its theater of operations (Mollwitz, Hohenfriedberg).

2. It can take up position near the frontier, wait until the enemy appears and is about to attack, and then attack him first (Czaslau, Soor, Rossbach). Such an attitude is obviously more passive; it calls for a longer

period of waiting; and though little or no *time* may be gained by the second plan as compared to the first if the enemy really does attack, still, the battle that was certain in the first case will be less certain in the second, and it may turn out that the enemy's determination will not extend as far as an attack. The advantage of waiting, therefore, has become greater.

3. It can wait, not merely for the enemy's decision to attack—that is, his appearance in full view of the position—but also for the actual attack (as at Bunzelwitz, to take another example from the campaigns of the commander we have been referring to).[1] In that event the army will fight a true defensive battle, but one, as we have said before, that may include offensive moves by some part of the army. Here too, as in the previous case, the gain in time is immaterial, but the enemy's determination will be tested once again. Many an army has advanced to the attack but refrained at the last moment, or desisted after the first attempt on finding the enemy's position too strong.

4. It can withdraw to the interior of the country and resist there. The purpose of this withdrawal is to weaken the attacker to such an extent that one can wait for him to break off his advance of his own accord, or, at least, be too weak to overcome the resistance with which he will eventually be confronted.

The simplest and most outstanding example would be the case in which the defender is able to leave one or more fortresses behind, which the attacker must invest or besiege. It is obvious that this will weaken his forces and provide an opportunity for an attack by the defender at a point where he has the upper hand.

Even where there are no fortresses, such a retreat to the interior can gradually restore to the defender the balance or superiority that he did not have on the frontier. In a strategic attack, every advance reduces the attacker's strength, partly as an absolute loss and partly because of the division of forces which becomes necessary. We shall discuss this in greater detail in connection with the attack. For the present, we shall simply assume this statement to be correct, since it has been sufficiently demonstrated in past wars.

The main advantage of this fourth case lies in the time that is gained. If the enemy lays siege to our fortresses, we have gained time until their surrender (which is probable, but which may take several weeks, and in some cases months). If, on the other hand, his loss of strength, the exhaustion of the momentum of his attack, is caused simply by his advance and by having to leave garrisons at vital points, and thus only by the distance he has covered, the amount of time gained will usually be even greater and we are not so strongly compelled to act at any given moment.

Not only will the relative strength of defender and attacker have changed when this action has run its course, the former will also have to his credit

[1] Frederick the Great. Eds.

the increased benefit of waiting. Even if the attacker has not been weakened enough by his advance to prevent him from attacking our main force where it has come to rest, he may lack the determination to do so. This determination must be stronger here than it would have had to be at the frontier: the reason is partly that his forces are reduced and no longer fresh while his danger has increased, and partly that irresolute commanders will completely forget about the necessity of a battle once possession of the area has been achieved; either because they really think it is no longer necessary, or because they are glad of the pretext. Their failure to attack is not, of course, the adequate negative success for the defender that it would have been on the frontier, but the time gained is substantial nonetheless.

In all four cases cited, it goes without saying that the defender has the benefit of terrain, and that the support of his fortresses and the populace are favorable to his action. With each successive stage of defense these elements become more significant, and in the fourth stage they are particularly effective in weakening the enemy. Since the advantages of waiting also increase with each phase, it follows that each successive stage of defense is more effective than the last, and that this form of warfare gains in effectiveness the further it is removed from attack. We are not afraid of being accused on this account of believing that the most passive kind of defense is the strongest. Each successive stage, far from being intended to weaken the act of resistance, is meant merely to *prolong and postpone* it. Surely there is no contradiction in saying that one is able to resist more effectively in a strong and suitably entrenched position, and that, after the enemy has wasted half his strength on it, a counterattack will be that much more effective. Daun could hardly have won at Kolin without his strong position. If his pursuit of the mere 18,000 men whom Frederick was able to lead from the field had been more energetic, this victory could have been one of the most brilliant in the annals of war.

What we do maintain is that with each successive stage of defense the defender's predominance or, more accurately, his counterweight will increase, and so in consequence will the strength of his reaction.

Can we say that the advantages that derive from an intensified defense are to be had without cost? Not at all: the sacrifices with which they must be purchased will increase equally.

Whenever we wait for the enemy inside our own theater of operations, no matter how close to the frontier the decisive action may be fought, the enemy's forces will enter our theater of operations, which will entail sacrifices in this area. If we had attacked him first, the damage would have been incurred by him. The sacrifices tend to increase whenever we fail to advance toward the enemy in order to attack him; the area he occupies and the time he takes to advance to our position will continue to increase them. If we intend to give defensive battle and thus leave the initiative and the timing up to the enemy, the possibility exists that he may well remain for a considerable time in the area he holds. So the time we gain by his postponement of the decision has to be paid for in this manner. The sacrifices become even more noticeable in the case of a retreat into the interior of the country.

However, the reduction of the defender's strength that is caused by all of these sacrifices will usually affect his fighting forces only later, not immediately: it is frequently so indirect as to be barely noticeable. Thus the defender tries to increase his immediate strength by paying for it later—in other words, he borrows like anyone else who needs more than he has.

In order to assess the results of these various forms of resistance, we have to examine *the purpose of the enemy's attack*. It is to gain possession of our theater of operations, or at least a substantial part of it; for the concept of the whole implies at least the greater part, and a strip a few miles wide is seldom of independent strategic importance. Therefore, so long as the attacker is not in possession, so long, in other words, as fear of our strength has prevented him from entering our theater of operations or seeking out our position or has caused him to avoid the battle we are prepared to give, the objects of the defense have been accomplished. Our defensive dispositions have proved successful. Admittedly, this is only a negative success which will not directly produce enough strength for a real counterattack. But it may do so in an indirect way, gradually: the time that passes is *lost to the aggressor*. Time lost is always a disadvantage that is bound in some way to weaken him who loses it.

Thus, in the first three stages of defense (in other words, those taking place at the border) *the very lack of a decision constitutes a success for the defense*.

That, however, is not the case in the fourth stage.

If the enemy lays siege to our fortresses we must relieve them in good time—in other words, it is up to us to bring about a decision by positive action.

This is also the case where the enemy has pursued us into the interior without besieging any of our fortresses. While we may have more time and can wait until the enemy is at his weakest, the assumption will remain that we shall have to take the initiative in the end. Indeed, the enemy may by then have taken all of the area that was the object of his attack, but he holds it as a loan. The tension continues to exist, and the decision is still to come. So long as the defender's strength increases every day while the attacker's diminishes, the absence of a decision is in the former's best interest; but if only because the effects of the general losses to which the defender has continually exposed himself are finally catching up with him, the point of culmination will necessarily be reached when the defender must make up his mind and act, when the advantages of waiting have been completely exhausted.

There is of course no infallible means of telling when that point has come; a great many conditions and circumstances may determine it. We should note, however, that the approach of winter is usually the most natural turning point. If we cannot prevent the enemy from wintering in the area he has occupied, we might as well give it up for lost. Even so there is the example of Torres Vedras to remind us that this is not a universal rule.

What is it that, broadly speaking, constitutes a decision?

In our discussion, we have always assumed decision to occur in the form of battle, but that is not necessarily so. We can think of any number of engagements by smaller forces that may lead to a change in fortune, either because they really end in bloodshed, or because the probability of their consequences necessitate the enemy's retreat.

No other kind of decision is possible in the theater of operations itself: that necessarily follows from the concept of war we have proposed. Even where the hostile army is forced to retreat because of lack of food, this factor does after all arise out of the limitations that our forces impose. If our army were not present, the enemy would surely find ways of helping himself.

So even when his offensive has run its course, when the enemy has become the victim of the difficult conditions of the advance and has been weakened and reduced by hunger, by sickness and the need to detach troops, it is really the fear of our fighting forces alone that makes him turn about and abandon all he has gained. Nevertheless there is a vast difference between such a decision and one that has been reached at the border.

At the border, his arms are faced by our arms—they alone hold him in check or do him damage. But when his offensive has run its course, he has been worn out largely by his own efforts. This will impart a completely different value to our arms, which are no longer the only factor in the decision though they may be the ultimate one. The ground has been prepared for it by the breakdown of the enemy's forces during their advance to the extent where a retreat, and a complete reversal of the situation, can be caused by the mere possibility of our reaction. In such a case one is bound to be realistic, to give credit for the decision to the difficulties of the offensive. Admittedly, one will not be able to find an example in which the defending forces were not also a factor, but for the sake of practical considerations, it is important to distinguish which of these two factors was dominant.

In the light of these ideas we think it is fair to say that two decisions, and therefore two kinds of reaction, are possible on the defending side, depending on whether the attacker is to *perish by the sword* or *by his own exertions*.

It is obvious that the first type of decision will predominate in the first three stages of defense, and the second type in the fourth. Indeed, the latter can essentially only take place where the retreat penetrates deeply into the interior of the country. It is in fact the only reason that can justify such a retreat and the great sacrifices it entails.

Two basically different types of resistance have now been identified. There are cases in military history where they stand out as clearly and distinctly as any abstract concept ever can in practice. In 1745 Frederick the Great attacked the Austrians at Hohenfriedberg as they descended from the Silesian mountains—at a time when their strength could not have been sapped either by exertions or by the detachment of troops. Wellington, on the other hand, stayed in the fortified lines of Torres Vedras until cold and hunger had left Masséna's army so depleted that it withdrew of its own accord. In that case the defender's forces took no part in the actual process of wearing down the enemy. At other times where both types are closely linked, one will still be distinctly dominant. Take, for instance, 1812. In that famous campaign so many savage engagements were fought that, under different circumstances, an absolute decision might well have been reached by the sword alone; yet there is probably no other case in which the evidence is so clear that the invader was destroyed by his own exertions. Only about 90,000 of the 300,000 men who made up the French center reached Moscow. Only 13,000 men had been detached. Casualties, therefore, numbered 197,000, of which fighting certainly cannot account for more than a third.

All campaigns that are known for their so-called temporizing, like those of the famous Fabius Cunctator, were calculated primarily to destroy the enemy by making him exhaust himself.

In general there have been many campaigns that were won on that principle without anyone explicitly saying so. One will arrive at the true cause of many decisions only by ignoring the far-fetched explanations of historians, and instead closely examining the events themselves.

We believe that we have thus adequately described the considerations that underlie defense and its various phases. By pointing out these two chief means of resistance, we hope we have explained clearly how the principle of waiting runs through the whole system and combines with the principle of positive action in such a way that the latter may appear early in one case and late in another. After which, the advantages of waiting will be seen to be exhausted.

We believe that we have now surveyed as well as delimited the whole field of defense. True, there still are aspects that are important enough to deserve a chapter on their own, points on which a series of reflections could be based and which ought not to be overlooked: for instance, the nature and influence of fortresses and entrenched camps, the defense of mountains and rivers, flanking operations, and so forth. We shall deal with these in the chapters that follow. Still, none of these subjects seems to fall outside the scope of the ideas explained above, but merely constitute their application to specific places and circumstances. The foregoing sequence of ideas has developed out of the concept of defense and its relation to attack. We have linked those simple ideas with reality, and so demonstrated how to move from reality to these simple ideas and achieve a solid analytic base. In the course of debate we will therefore not need to resort to arguments that themselves are ephemeral.

But armed resistance, by its diversity of possible combinations, can so change the appearance and vary the character of armed defense, especially in cases where there is no actual fighting but the outcome is affected by the fact that there could be, that one is almost tempted to think some new effective principle here awaits discovery. The vast difference between savage repulse in a straightforward battle, and the effect of a strategic web that prevents things from getting that far, will lead one to assume that a different force must be at work—a conjecture somewhat like that of the astronomers' who deduced from the enormous void between Mars and Jupiter that other planets must exist.

If an attacker finds the enemy in a strong position that he thinks he cannot take, or on the far side of a river that he believes to be impassable, or even if he fears he will jeopardize his food supply by advancing any further, it is still only the force of the defender's arms which produces these results. What actually halts the aggressor's action is the fear of defeat by the defender's forces, either in major engagements or at particularly important points; but he is not likely to concede this, at least not openly.

One may admit that even where the decision has been bloodless, it was determined in the last analysis by engagements that did not take place but *had merely been offered*. In that case, it will be argued, *the strategic planning* of these engagements, rather than the tactical decision, should be considered the operative principle. Moreover such strategic planning would be dominant only in cases where defense is conducted by some means other than force of arms. We admit this; but it brings us to the very point we wanted to make. What we say in fact is this: where the tactical results of the engagement are assumed to be the *basis* of all strategic plans, it is always possible, and a serious risk, that the attacker will proceed on that basis. He will endeavor above all to be tactically superior, in order to upset the enemy's strategic planning. The latter, therefore, can never be considered *as something independent*: it can only become valid when one has reason to be confident of tactical success. To illustrate briefly what we mean, let us recall that a general such as Bonaparte could ruthlessly cut through all his enemies' strategic plans in search of battle, because he seldom doubted the battle's outcome. So whenever the strategists did not endeavor with all their might to crush him in battle with superior force, whenever they engaged in subtler (and weaker) machinations, their schemes were swept away like cobwebs. Schemes of that sort would have been enough to check a general like Daun; but it would have been folly to oppose Bonaparte and his army in the way the Prussians handled Daun and the Austrians in the Seven Years War. Why? Because Bonaparte was well aware that everything turned on tactical results, and because he could rely on them, while Daun's situation was different in both respects. *That is why* we think it is useful to emphasize that all strategic planning rests on tactical success alone, and that—whether the solution is arrived at in battle or not—this is in all cases the actual fundamental basis for the decision. Only when one has no need to fear the outcome—because of the enemy's character or situation or because the two armies are evenly matched physically and psychologically or indeed because one's own side is the stronger—only then can one expect results from strategic combinations *alone*.

When we look at the history of war and find a large number of campaigns in which the attacker broke off his offensive without having fought a decisive battle, consequently where strategic combinations appear effective, we might believe that such combinations have at least great inherent power, and that they would normally decide the outcome on their own whenever one did not need to assume a decisive superiority of the offensive in tactical situations. Our answer here must be that this assumption, too, is erroneous in situations that arise in the theater of operations and are therefore part of war itself. The reason for the ineffectiveness of most attacks lies in the general, the political conditions of war.

The general conditions from which a war arises, and that form its natural basis, will also determine its character. This will later be discussed in greater detail, under the heading of war plans. But these general conditions have transformed most wars into mongrel affairs, in which the original hostilities have to twist and turn among conflicting interests to such a degree that they emerge very much attenuated. This is bound to affect the offensive, *the side of positive action*, with particular strength. It is not surprising, therefore, that one can stop such a breathless, hectic attack by the mere flick of a finger. Where resolution is so faint and paralyzed by a multitude of considerations that it has almost ceased to exist, a mere show of resistance will often suffice.

We can see that in many cases the reason for the defender's being successful without having to fight does not lie in the fact that he occupies many impregnable positions, nor in the size of the mountain ranges that lie across the theater of operations, nor in the broad stream that traverses it, nor in the ease with which the threatened blow can be paralyzed by a well-planned series of engagements. The real reason is the faintness of the attacker's determination, which makes him hesitate and fear to move.

Countervailing forces of this kind can and must be reckoned with; but they must be recognized for what they are, rather than having their effects attributed to other causes—those that alone concern us here. We must state emphatically that, in this respect, military history can well become a chronic lie and deception if critics fail to apply the required correctives.

At this point let us examine, in their most common form, the vast number of offensive campaigns that failed without a decisive battle being fought.

The aggressor marches into hostile territory; he drives the enemy back a little, but then begins to have doubts about risking a decisive battle. He halts and faces his opponent, acting as if he had made a conquest and was interested only in protecting it—in short, he behaves as if it were the enemy's affair to seek a battle, as if he himself were ready to fight at any time, and so forth. All of these are mere *pretexts*, which a general uses to delude his army, his government, the world at large, and even himself. The truth of the matter is that the enemy's position has been found too strong. Here we are not talking of a case in which the aggressor fails to attack because a victory would be of no use to him, because his advance having run its course he does not have enough resiliency to start a new one. This would assume that a successful attack had already taken place and resulted in a genuine conquest; rather, we have in mind a case in which the aggressor gets bogged down in the middle of an intended conquest.

At that point the attacker will wait for a favorable turn of events to exploit. There is as a rule no reason to expect such a favorable turn: the very fact that an attack had been intended implies that the immediate future promises no more than the present. It is therefore a fresh delusion. If, as is usual, the operation is a joint one timed to coincide with others, the other armies will then be blamed for his failures. By way of excusing his inaction he will plead inadequate support and cooperation. He will talk of insuperable obstacles, and look for motives in the most intricately complicated circumstances. So he will fritter his strength away in doing nothing, or rather in doing too little to bring about anything but failure. Meanwhile the defender is gaining time—which is what he needs most. The season is getting late, and the whole offensive ends with the return of the invader to his winter quarters in his own theater of operations.

This tissue of falsehoods ends by passing into history in place of the obvious and simple truth: that failure was due to *fear of the enemy's forces.* When the critics begin to study a campaign of this sort they tend to get lost in argument and counterargument. No convincing answer will be found, because everything is guesswork and the critics never dig deep enough to find the truth.

That sort of fraudulence is not merely a matter of bad habit; its roots lie in the nature of the case. The counterweights that weaken the elemental force of war, and particularly the attack, are primarily located in the political relations and intentions of the government, which are concealed from the rest of the world, the people at home, the army, and in some cases even from the commander. For instance no one can and will admit that his decision to stop or to give up was motivated by the fear that his strength would run out, or that he might make new enemies or that his own allies might become too strong. That sort of thing is long kept confidential, possibly forever. Meanwhile, a plausible account must be circulated. The general is, therefore, urged, either for his own sake or the sake of his government, to spread a web of lies. This constantly recurring shadow-boxing in the dialectics of war has, as theory, hardened into systems, which are, of course, equally misleading. Only a theory that will follow the simple thread of internal cohesion as we have tried to make ours do, can get back to the essence of things.

If military history is read with this kind of skepticism, a vast amount of verbiage concerning attack and defense will collapse, and the simple conceptualization we have offered will automatically emerge. We believe that it is valid for the whole field of defense, and that only if we cling to it firmly can the welter of events be clearly understood and mastered.

Let us now examine the employment of these various methods of defense.

They are all intensifications of the same thing, each one exacting increased sacrifices on the part of the defender. A general's choice, all other things being equal, would largely be determined by this fact. He would choose the method he considered adequate to give his forces the necessary degree of resistance; but to avoid unnecessary losses he would not retreat any further.

It must be admitted, however, that the choice of different methods is already severely limited by other major factors that play a part in defense and are bound to urge him to use one method or another. A withdrawal to the interior calls for ample space, or else it requires conditions like those that obtained in Portugal in 1810: one ally (England) provided solid support to the rear, while another (Spain), with its extensive territory, reduced a great deal of the enemy's impact. The location of fortresses—close to the border or farther inland—may also decide for or against a certain method. Even more decisive are the nature of the terrain and the country, and the character, customs, and temper of its people. The choice between an offensive and defensive battle may be determined by the enemy's plans or by the characteristics of both armies and their generals. Finally, the possession or lack of an outstanding position or defensive line may lead to one method or the other. In short, a mere listing of these factors is enough to indicate that in defense they are more influential on the choice than is relative strength alone. Since we shall become more familiar with the most important factors that have here only been touched upon, we will later be able to demonstrate in greater detail what influence they exert on the choice. Finally, the implications will be treated in a comprehensive analysis in the book on war plans and campaign plans.

That influence, however, will normally become decisive only if the relative strengths are not too disproportionate. Where they are (and therefore in the majority of cases), relative strength will prevail. The history of war is full of proof that this has actually occurred—quite apart from the chain of reasoning developed here—*through the hidden processes of intuitive judgment*, like almost everything that happens in war. It was the same general, with the same army, who, in the same theater of operations, fought the battle of Hohenfriedberg and also moved into camp at Bunzelwitz. Thus, even Frederick the Great, who when it came to a battle was the most offensive-minded of generals, was finally compelled to resort to a strict defensive when the disproportion of strength became too great. Indeed did not Bonaparte, who used to rush at his enemies like a wild boar, twist and turn like a caged animal when the ratio of forces was no longer in his favor in August and September 1813, without attempting a reckless attack on any one of his enemies? And do we not find him at Leipzig, in October of the same year, when the disparity of forces had reached its peak, taking refuge in the angle made by the Parthe, Elster, and Pleisse rivers, as if he were cornered in a room with his back to the wall, waiting for his enemies?

We should like to add that this chapter, more than any other of our work, shows that our aim is not to provide new principles and methods of conducting war; rather, we are concerned with examining the essential content of what has long existed, and to trace it back to its basic elements.

CHAPTER NINE

The Defensive Battle

In the last chapter we stated that in the course of his defense the defender can fight a tactically offensive battle by seeking out and attacking the enemy as soon as he invades his theater of operations. Alternatively, he may await the enemy's appearance and then attack him in which case the battle is still offensive in a tactical sense, though somewhat modified in form. Finally he may actually wait for the enemy to attack his position and then strike back; not only using part of his force to hold the enemy locally, but also attacking him with the rest. Naturally various degrees and stages are possible, running gradually from positive counterattack to local defense. We cannot here enter into a discussion as to how far this should go, and what would be the most favorable ratio between the two elements for the purpose of winning a decisive victory. Where this is the object, we do insist that the offensive element must never be completely absent. And we are convinced that all the consequences of a decisive victory can and do result from this offensive phase, just as they do in a purely offensive battle.

As the battlefield in strategy is simply a point in space, so the duration of a battle is strategically only a moment in time, and a battle's strategic significance lies not in its course but in its outcome and its consequence.

If it were true that total victory could be linked to the offensive elements that are present in every defensive battle, no basic strategic difference between an offensive and a defensive battle would exist. That is in fact our own belief, although appearances seem to contradict it. To examine the matter more closely, to clarify our views and so eliminate that contradiction, let us briefly outline our idea of a defensive battle.

The defender waits for the attack in position, having chosen a suitable area and prepared it; which means he has carefully reconnoitered it, erected solid defenses at some of the most important points, established and opened communications, sited his batteries, fortified some villages, selected covered assembly areas, and so forth. The strength of his front, access to which is barred by one or more parallel trenches or other obstacles or by dominant strong points, makes it possible for him, while the forces at the points of actual contact are destroying each other, *to inflict heavy losses on the enemy at low cost to himself* as the attack passes through the successive stages of resistance until it reaches the heart of the position. The points of support on which his flanks rest secure him against sudden attacks from several directions. The covered ground on which the defender has taken up his position will make the attacker wary, even timid. It will enable the defender to slow down the general retrograde movement by means of small successful counterattacks as the area of action steadily narrows. In this way the defender can confidently survey the battle as it smolders before his eyes. But he knows that he cannot hold his front forever, that his flanks are not invulnerable, and that he cannot change the whole course of the battle by the successful counterattack of a few battalions or squadrons. He holds his position *in depth*, for at every level, from division to battalion, his order of battle has reserves for unforeseen events and to renew the action. A substantial reserve, however—perhaps one quarter or one third of his whole force—is kept far to the rear, far enough to avoid any casualties from enemy fire and, if possible, far enough to remain outside any possibility of envelopment. This reserve is meant to cover his flanks against any wider and larger turning movement, and to protect him against the unexpected. In the final third of the battle, when the enemy has revealed his whole plan and spent the major part of his forces, the defender intends to fling this body against a part of the enemy forces, thus opening a minor offensive battle of his own, using every element of attack—assault, surprise, and flanking movements. All these pressures will be brought to bear on the battle's center of gravity while the outcome still hangs in the balance, in order to produce a total reversal.

This is the normal course of a defensive battle as we conceive of it, based on contemporary tactics. In such a battle the attacker's turning movement, intended to give his attack a better chance and his victory a greater scale, is countered by a subordinate turning movement, which is aimed at that part of the enemy's force that has executed the original envelopment. This subordinate turning movement may suffice to cancel out the effect of the enemy's, but it cannot be expanded into a similar general envelopment of the enemy's forces. So, the difference between the forms of victory will always lie in the fact that in an offensive battle the attacker does the turning and then converges on the center, while in a defensive battle the movement is more likely to fan out from the center toward the periphery.

On the battlefield, and in the first stage of pursuit, it must be acknowledged that a turning movement is the more effective form. This is not due to the form of envelopment as such; rather it holds true only where the envelopment can be pushed to an extreme, when it can severely restrict the enemy's chances of retreat while the battle is still in progress. This is the very situation that the defender's positive counterattack is designed to prevent. In many cases where a counterattack is not enough to win a victory, it may still suffice to provide protection in that extremity. At all events, we must admit that, in a defensive battle, the danger of having one's retreat severely restricted is preeminently present; where it cannot be averted, the impact of defeat and of the early stage of pursuit is aggravated.

Normally, however, this holds true only in the first stage of pursuit—that is, until nightfall. By the following day the envelopment will have reached its limit, and in this respect both sides will again be in balance.

The defender's principal line of retreat may of course be lost. If so, he will thereafter be at a strategic disadvantage. But encirclement itself will, with few exceptions, be at an end, since it was intended only for the battlefield, and cannot go much beyond it.

What happens, on the other hand, if *the defender* is victorious? The defeated army will split up. At first, this will make the retreat easier, but by the *following* day the *reunion of all the parts* will become its primary objective. Where the victory was decisive and the defender pursues it energetically, concentration is frequently impossible. The fragmentation of the beaten army can thus lead to the gravest consequences, which may by stages end in complete disintegration. If Bonaparte had won the battle of Leipzig, the allied armies would have been cut off from one another and the effect on their strategic cohesion would have been serious. At Dresden, where Bonaparte admittedly did not fight a true defensive battle, the attack still maintained the geometric form under discussion—from the center toward the periphery. The embarrassment suffered by the allies as a consequence of operating in separate columns is well known. Their embarrassment was ended only by the victory on the Katzbach, the news of which caused Bonaparte to return to Dresden with the Guards.

The battle on the Katzbach is itself a case in point. The defender took the offensive at the last possible moment, and his attack therefore had a diverging effect. The French corps were forced apart. Several days after the battle, Puthod's division fell into allied hands as a fruit of victory.

We therefore conclude that if the attacker can use the convergent form, which is natural for him, to enhance his victory, the defender has in the divergent form, which is natural for him, a means of making his victory more effective than it would be in the simple case of parallel positions and vertical operations. We believe that one form is worth at least as much as the other.

In the history of war major victories are less often the consequence of defensive battles than of offensive ones, but that does not prove that defensive battles are inherently less likely to be victorious: rather, the defender simply finds himself in markedly different circumstances. In most cases, he is the weaker belligerent—not only in numbers but also in terms of his entire situation. Usually he is not, or does not believe himself to be, able to follow up a victory, and is, therefore, satisfied with having repulsed the danger and saved the honor of his arms. There is no question that the defender may be handicapped by his numerical weakness and his circumstances; but frequently what should be seen as the result of necessity has been interpreted as the result of defense as such. In this absurd manner it has become a basic assumption that defensive battles are meant merely to repulse the enemy, and not to destroy him. We consider this as a most damaging error, in fact a confusion between form and substance. We maintain unequivocally that the form of warfare that we call defense not only offers greater probability of victory than attack, but that its victories can attain the same proportions and results. Moreover, this applies not only to the *aggregate success* of all the engagements that make up a campaign, but to *each individual* battle, provided there is no lack of strength and determination.

CHAPTER TEN

Fortresses

In former times, before the days of great standing armies, fortresses—castles and walled towns—existed simply for the protection of their inhabitants. A noble who was hard pressed on all sides fled to his castle in order to gain time and wait for a better turn of events. By their fortifications towns sought to ward off the storm clouds of war.

These simple and basic purposes of fortresses did not remain the only ones. The relation in which fortresses stood to the country as a whole and to the troops who fought across the land soon extended their importance. Their significance was felt beyond their walls; it contributed to the conquest or retention of the country, the successful or unsuccessful outcome of the whole struggle, and thus tended to give the war itself greater coherence. In this way, fortresses attained a strategic significance that for a time was considered so important that they formed the basis of strategic plans, which were more concerned with capturing a few fortresses than with destroying the enemy's armies. People examined the root cause of this significance, the connection between a fortified point and its surroundings and the army, and believed that no amount of care, ingenuity, and abstract thought could be too great when it came to choosing the points to be fortified. These abstractions tended to obscure the original purposes altogether, and eventually someone conceived of the idea of fortresses without towns and inhabitants.

But the time is past when mere enclosure and fortification, without any other military preparations, protect a locality against a tide of war that inundates the whole country. This used to be possible, partly because of the petty states into which nations were split, partly because of the periodic nature that then characterized invasions. Their duration used to be limited like the seasons of the year, either because the feudal levies would hurry home or because the cash with which condottieri were paid would regularly run out. Since great standing armies mechanically flattened the resistance of individual strong points with their powerful artillery no city or minor township felt like gambling with its strength, only to be taken a few weeks or months later, and then to receive harsher treatment. It is still less in the interest of the army to be split up into a large number of small fortified garrisons. While this may delay the enemy's progress, it will undoubtedly end with the fortresses being taken. Enough strength must always be left to meet the enemy in battle on equal terms, unless one can count on the arrival of an ally who will relieve the fortresses and set the army free. As a result, fortresses have had to be heavily reduced in number. This in turn has made us discard the notion of using fortresses for the immediate protection of the population and property of the towns, and led us to the idea of treating fortresses as an *indirect protection* of the country, by means of their strategic value as knots that hold the web of strategy together.

Such has been the development of ideas not only in books but also in practice—though the books have admittedly spun them out, as they always do.

While this was the right course to take, the ideas led too far: artifice and fancy crowded out the healthy core of natural and major necessities. We shall consider only these requirements when listing the purposes and conditions of fortresses, proceeding from the simple to the complex. The following chapter will indicate the deductions concerning location and number of fortresses that can be drawn from their purposes.

The effectiveness of a fortress is obviously composed of two distinct elements, one active and the other passive. The first appears in the protection that a fortress gives to the area and everything in it; by the second, it exercises a certain influence on the countryside beyond the range of its artillery.

The active element lies in the garrison's ability to attack any enemy who approaches. The larger the garrison, the larger the units that can be sent out for this purpose; and as a rule, the larger they are, the farther they can be sent. It follows that the active sphere of influence of a large fortress, as compared to a small one, is not only stronger but also more extensive. In turn the active element itself consists, so to speak, of two additional elements: operations carried out by the garrison itself, and operations carried out by independent forces, large or small, which are keeping in touch with it. In fact, forces that would be too weak to face the enemy alone can keep the field and control the area to some extent if they are sure they can take refuge inside its walls in the last resort.

The operations that the garrison of a fortress can risk are always fairly limited. Even in the case of large fortresses and numerically strong garrisons, the units that can make a sortie are not significant compared to the forces in the field, and their average range of operations is rarely more than that of a few marches. Minor fortresses can only send out very small detachments, and those can affect only the nearest villages. Independent corps, however—troops that are not part of the garrison and therefore do not necessarily have to fall back upon the fortress—are much less constrained. In the right conditions, they can greatly extend a fortress' active sphere of influence. Therefore, when we talk about the active effectiveness of fortresses in general, this is the part to bear particularly in mind.

Yet, slight as the active influence of the smallest garrison may be, it is still essential to every function that a fortress must fulfill. Strictly speaking, even the most passive function of a fortress, defense against assault, cannot, after all, be imagined without this active element. It is quite obvious, however, that among the various ways in which a fortress may be significant—whether it be intrinsic or dependent on the circumstances of the moment—

some will tend to involve the more passive effect, and others the more active. Sometimes the significance of a fortress may be said to be simple, in which case its effectiveness is direct; sometimes it is complex, and its effectiveness will be more or less indirect. We shall proceed from the former to the latter. Let us state at the outset, however, that any fortress may of course be of significance in several or even all of the ways listed simultaneously, or at any rate at different times.

We suggest that fortresses constitute the first and foremost support of defense, in the following ways:

1. *As secure depots.* During an offensive, the attacker lives from hand to mouth; the defender, as a rule, has to make his preparations well ahead, and therefore he cannot live only off the country he holds—which in any case he would rather spare. Consequently, he is in great need of depots. As the attacker advances, he leaves his supplies behind and so protects them; the supplies of the defense lie in the middle of the operational zone. Unless these supplies are stored *in fortified places,* operations in the field will suffer badly. The most extensive and inconvenient positions often have to be set up solely for the purpose of protecting them.

Without fortresses, an army on the defensive is vulnerable everywhere. It is a body without armor.

2. *As protection for large and prosperous towns.* This function is closely related to the first, because large and prosperous towns, especially commercial ones, are an army's natural sources of supply, which is, therefore, immediately affected by their possession or loss. Besides, it is always worth an effort to retain this component of the national assets, partly because of all that can be drawn from it indirectly, and partly because an important town weighs heavily in the scales when it comes to negotiating peace.

In recent times, this function of fortresses has been underrated; yet it is among the most obvious, highly effective, and least susceptible to misuse. Imagine a country where not only the large and prosperous town, but every sizable one is fortified and defended by its citizens and the farmers of the surrounding areas. The speed of military operations would be so reduced, and so much weight thrown into the scale by the defending inhabitants, that the skill and determination of the enemy commander would dwindle almost to insignificance. We only mention this ideal scale of fortification in order to do justice to this particular role of a stronghold, and to make sure that the value of the *direct* protection it provides will not for a moment be ignored. This mental image, by the way, will not affect our inquiry, for among so many towns, some will always be more strongly fortified than others, and those are the ones that must be considered the real support of the armed forces.

These first two functions almost exclusively involve the passive purposes of fortresses.

3. *As real barriers.* These fortresses block the roads, and also, in most cases, the rivers on which they are located.

It is not so easy as one might think to find a serviceable detour to bypass the fortress. It must not only be out of artillery range but also more or less beyond the reach of possible sorties.

If the terrain is at all difficult, the slightest deviation from the road can often cause delays worth a whole day's march. This can be of great importance if the road is used repeatedly.

The extent to which a blockage of river traffic will affect operations is clear enough.

4. *As tactical points of support.* The range of fire of a not entirely insignificant fortress will normally have a diameter of several miles, and in any case its offensive effectiveness will extend still farther. Therefore fortresses should be considered the most favorable points of support for the flanks of a position. A lake that is several miles long may be of value as a point of support, but one can accomplish more with a moderate-sized fortress. The flank need never be close by: the attacker cannot enter the area between the two, for he would have no line of retreat.

5. *As a staging post.* Where fortresses are located along the defender's lines of communication, as they are in most cases, they are convenient stopping places for the traffic moving up and down the line. The danger to which the lines of communication are exposed stems mainly from periodic raids by partisans. If an important convoy is threatened by this type of action and can reach a fortress by increasing its speed or doubling back, it is safe and can wait until the road is clear. A fortress also gives all troops in transit a chance to rest for a day or two, and so to complete their march in better time. Since rest days are the ones that pose the greatest danger to the troops, a fortress that is situated at the midpoint of a line of communications 150 miles long will, in a sense, cut it by half.

6. *As a refuge for weak or defeated units.* Under the protection of the guns of a fortress of reasonable size, troops are safe from enemy action, even without an especially entrenched camp. Granted, a unit that wants to stay there has to give up the idea of further retreat; but there are times when this is no great loss—when a continued withdrawal would only end in complete destruction.

In many cases, however, a fortress can offer a few days' rest without the retreat having to be abandoned. It serves as a refuge, particularly for the lightly wounded, the dispersed, and so forth, who have hurried ahead of the defeated army: a place for them to wait until they can rejoin their units.

If Magdeburg had lain in the direct line of the Prussian retreat of 1806, and if that line had not already been lost at Auerstädt, the army could easily have halted there for three or four days, using the time to rally and reform. Even under the circumstances that prevailed, Magdeburg served as a rallying point for what was left of Hohenlohe's corps; it was only there that this corps took visible shape again.

Only actual experience in war will enable one to conceive of the stabilizing influence that a nearby fortress exerts under adverse conditions. There one can find stocks of arms and powder, forage and bread, shelter for the sick, safety for the able-bodied, and comfort for the panic-stricken. A fortress is an oasis in the desert.

It is obvious that the last four functions begin to involve the active influence of fortresses to a somewhat greater degree.

7. *As an actual shield against enemy attack*. Fortresses that cover a defensive line may be likened to blocks of ice in the course of a river's flow. The enemy must invest them, and, if the garrisons do their best, this will require twice their number. On top of that, however, half a garrison can and should consist of troops which, but for the fortress, could never have gone on active service: half-trained militia, convalescents, armed civilians, home guard, etc. In such a case, the drain on the enemy may be four times as much as the drain on our side.

This disproportionate debilitation of the enemy's strength is the first and greatest merit of a fortress that holds out against a siege, but there are others. Once an attacker has pierced the line of the defender's fortresses, he has far less freedom of movement: his lines of possible retreat are limited, and he must constantly attend to the need for direct coverage of any siege he undertakes.

With this fortresses begin to play a great and decisive part in the conduct of defense. It should be considered the foremost role a fortress can fill.

Nonetheless, this use of fortresses, far from being employed with regularity, is comparatively rare. The reason lies in the nature of most wars. For many, it is in a sense too drastic, too decisive a method to employ, as we shall explain in detail later.

This role of a fortress involves mainly its offensive power; at least, that is what produces its effectiveness. If a fortress were merely a point that the attacker was unable to occupy, it would admittedly be an obstacle to him, but never enough to make him feel that he has to lay siege to it. However, he cannot leave six, eight or ten thousand men to do as they like behind his back; he must therefore invest the place in adequate strength. In order not to be forced to do this repeatedly, he must assault the fortress and take it. Once the siege begins, it is mainly the passive role that comes into play.

Each of the foregoing roles of fortresses are quite direct in action and simple in form. The two purposes that follow, on the other hand, are more complex in their effect.

8. *As protection for extensive camps*. For a fortress of moderate size to cover a billeting area 15 to 20 miles wide behind it is a simple factor of its existence; but if this kind of place—as military history frequently tells us—has been privileged to cover a line of billets extending over 75 to 100 miles, an explanation is required, as well as a special comment on the cases where it was merely an illusion.

The following points should be considered:

1. The fortress itself must block one of the main roads and must effectively cover an expanse of 15 to 20 miles.

2. It must either be considered as an exceptionally strong advanced

post or it must offer a full view of the countryside, heightened by secret intelligence obtained through the ordinary intercourse between an important town and its surrounding area. It is to be expected that in a place of six, eight or ten thousand inhabitants, more news will be available than in a mere village where an ordinary outpost might be based.

3. It is a source of support for minor units, which can find safety and cover there. From time to time they can make a sortie against the enemy, collect intelligence, or attack his rear if he happens to pass by. In brief, then, while a fortress is not itself mobile, it can to some extent have the effect of an advanced corps (Book Five, Chapter Eight).

4. After having assembled his troops, the defender must be able to take up his position immediately to the rear of the fortress. The attacker will then not be able to reach him without the fortress becoming a danger to his own rear.

Of course, any attack on a line of billets should in itself be considered as a surprise; or rather, that is the only aspect that concerns us here. A surprise attack will obviously take effect more quickly than a major offensive against a theater of operations. In the latter case, of course, if one has to bypass a fortress, one has no choice but to invest and hold it in check; but in a surprise attack on a line of billets this is not so necessary, and a fortress will therefore not hamper it to the same degree. This is certainly true; besides, a fortress cannot directly cover the flanks of a line of billets thirty to forty miles away. On the other hand, a surprise attack of that kind is not just aimed at a few billets. Later, in the book on attack, we will be able to explain more fully what such an attack really aims at and what can be the expected result. This much, however, can be said at once: its chief results will not derive from the actual attack on individual billets, but from the engagements it forces upon isolated, unprepared units that are less concerned with fighting than with getting to their destinations in a hurry. This type of attack and pursuit, however, must always be more or less directed against the center of the enemy's quarters; if an important fortress were located in front of the center, it would pose a serious obstacle to the attacker.

We maintain that the combined effect of these four points constitutes evidence that a major fortress can at any rate provide a direct and indirect degree of cover for a far greater billeting-area than one might suppose at first glance. We say "a degree of cover," for all these indirect effects together do not make the enemy's advance impossible; they only make it more difficult and more hazardous—in other words, less likely, and less dangerous to the defender. Indeed, that is all one asks of it, and all that the term implies in this particular connection. Real, direct protection must come from outposts and the way the billets are arranged.

There is some sense, then, in regarding a great fortress as a source of cover for a widespread billeting-area in its rear; but there is no denying, either, that one can find empty phrases and sophistical opinions on the matter in actual war plans, and even more frequently in military histories. After all, if that amount of cover is the result only of a combination of a number of circumstances, and then does no more than reduce the danger, one can see that in individual cases the cover may turn out to be quite unreal—because of unusual circumstances, or, more likely, because of the boldness of the attacker. In wartime then, one should never take it for granted that a given fortress will produce a given effect; the matter calls for careful study in each case.

9. *As cover for an unoccupied province*. If a province is left unoccupied, or occupied only by a token force, and at the same time is more or less exposed to enemy raids, a reasonably important fortress located in the area can be considered as its cover or even as its security. It may certainly be considered as protection, because the enemy cannot become master of the province until he has taken the fortress, which gives the defender time to come to its aid. The actual cover, though, can only be described as indirect, or *symbolic*, since only the active effectiveness of the fortress can keep enemy raiders within bounds. If this effectiveness is limited to its garrison, the result will hardly be noticeable: the garrisons of these fortresses are usually weak—infantry alone and not of the best at that. The idea of cover will have more substance if the fortress is in touch with small units that use it as a base and for support.

10. *As the focal point of a general insurrection*. In a guerrilla war, to be sure, provisions, arms, and munitions are not issued on a regular basis. Indeed, it is in the nature of that type of war that one must get along as best one can, and thus discover countless minor sources of resistance, which would otherwise remain untapped. Still, an important fortress stocked with substantial amounts of such supplies can obviously intensify the whole resistance movement, give it greater solidity, and lead to better cohesion and results.

Besides, the fortress serves as refuge for the wounded, seat of the civil authorities, and treasury, as well as a point of assembly for larger operations, and so forth. Above all, it will be the focal point for resistance; during a siege the enemy's forces are placed in a situation that facilitates and favors attacks by local partisans.

11. *As a defense of rivers and mountainous areas*. Nowhere can a fortress serve so many purposes or play so many parts as when it is located on a great river. Here it can assure a safe crossing at any time, prevent the enemy from crossing within a radius of several miles, command river traffic, shelter ships, close roads and bridges, and make it possible to defend the river indirectly—that is, by holding a position on the enemy's bank. It is clear that this versatile influence greatly facilitates the defense of a river and must rank as one of its essential elements.

In mountainous areas, fortresses are of similar importance. Here they can open or close whole networks of roads that converge upon them, and thereby dominate the whole area to which the roads give access. Thus, they serve as veritable buttresses for their whole defensive system.

CHAPTER ELEVEN

Fortresses—Continued

Having examined the roles of fortresses, let us now consider their location. At first glance, the matter looks extremely complicated, given the wide variety of determinants, every one of which is modified by its locality. But this concern is unfounded if we can keep to the essentials of the matter and guard against unnecessary refinements.

It will be clear that all requirements can be satisfied simultaneously if all the largest and wealthiest towns in the area of the operational theater are fortified—those that lie on the main highways linking the two countries, and especially those located at ports and gulfs, on major rivers and in mountains. Major towns and major roads always go together, and both have a natural affinity for great rivers and the seacoast. These four requirements will, therefore, coexist easily and cause no conflicts. Mountains are another matter; there one will rarely find important cities. Therefore, if the location and direction of a chain of mountains make it a suitable line of defense, it will be necessary to block its roads and passes with minor forts especially built for that purpose, and at the lowest possible cost. Large, elaborate fortifications should be reserved for large towns in the plains.

So far we have paid no attention to frontiers, and have said nothing about the geometric pattern of the line of fortification as a whole; nor have we mentioned any other geographic aspects of their location. In our opinion, the requirements already named are the essential ones, and in many instances, particularly those of minor states, they will suffice. However there are cases where other determinants are permissible and even necessary. This is true of countries that extend over a larger area and contain a number of important towns and roads, or, on the other hand, are almost completely devoid of them; rich countries that desire to add to the many fortresses they already possess, or, conversely, that are very poor and are forced to make the best of very few; in short, cases where the fortresses are disproportionate in number to the roads and cities, being either substantially more or substantially fewer. Let us briefly consider these additional factors.

The principal questions that remain are the following:

1. The choice of a main road, where there are more roads linking the two countries than one would wish to fortify
2. Whether fortresses should be located only on the frontier or distributed throughout the country
3. Whether to distribute them evenly or in groups

4. The geographic characteristics of the area that have to be taken into account.

Several other points may arise from the geometric pattern of a line of fortresses. Should they be placed in one or in several rows? In other words, will they be more effective behind each other, or side by side? Should they be arranged as on a checkerboard, or would it be better to place them in a straight line with salients and reentrants as in individual fortresses? All these questions, in our opinion, are sheer sophistry, considerations so trivial that they are not worth mentioning when more important ones are discussed. The only reason we touch on them at all is that they are not only cited in so many books, but are given a great deal more weight than such trivial matters deserve.

To clarify the first question, let us call to mind the relationship of southern Germany to France—in other words, to the upper Rhine. If we think of this area as a whole, to be fortified on strategic lines without considering the individual states of which it is made up, we shall encounter great problems because so many well-made roads run from the Rhine into Franconia, Bavaria, and Austria. True, there is no lack of major towns, such as Nuremberg, Würzburg, Ulm, Augsburg, and Munich; but if you cannot fortify them all, you will have to make a choice. Further, while, in our opinion, it is most important to fortify the largest and wealthiest towns, one cannot deny that Nuremberg and Munich, being so far apart, will have significantly different strategic values. One might, therefore, entertain the question whether, instead of Nuremberg, one should not fortify a second place, even if it is of lesser importance, in the vicinity of Munich.

The decision in such a case—in other words, the answer to the first question—must relate to what we have discussed in the chapters on defensive plans in general and on the choice of objectives for an attack. The points in immediate danger of attack are the ones that have to be fortified.

Thus, if there are a number of main roads that run from enemy territory to our own, we shall preferably fortify the road that leads straight to the heart of our country, or else the one that offers the greatest advantages to the enemy in the way of fertile areas, a navigable river, and so forth. One may then be certain that the enemy will either run up against one's fortifications, or, if he elects to circumvent them, we shall have the means to execute a natural and favorable flanking operation.

Vienna is the heart of southern Germany, and Munich or Augsburg would clearly be better places for the principal fortress than Nuremberg or Würzburg, even against France alone—assuming Switzerland and Italy to be neutral. This will become even more evident when one studies the roads that come from Italy and lead from Switzerland and through the Tyrol; for these, Munich and Augsburg retain some value, while Würzburg and Nuremberg might as well be nonexistent in that respect.

Now let us turn to the second question, whether fortresses should be located only on the frontier, or distributed throughout the country. To start with, it must be noted that this is an academic question for small countries, in which the frontier, from the strategic point of view, extends pretty much over the whole area. The larger the state involved, the clearer the necessity of posing this question.

The obvious answer is that the place for fortresses is on the frontier. They exist to protect the country, and the country is protected if its frontiers are. This answer may be valid in general, but the following observations will show how subject it is to limitations.

Any defense that relies chiefly on outside help will set great store by gaining time. It is not a vigorous counterattack, but rather a drawn-out process in which the advantage lies more in gaining time than in reducing the enemy's strength. Other things being equal, it lies in the nature of the case that if fortresses are spread throughout the country and cover a wide area between them, they will take longer to capture than if they are densely packed along the frontiers. Further, wherever the aim is to defeat the enemy by the length of his line of communications and the difficulties of maintenance—in countries, naturally, where these things can be relied on—it would be absurd to build fortifications only on the frontiers. Finally, one must remember that the fortification of the capital must receive priority wherever circumstances permit; that, in accordance with our principles, provincial capitals and commercial towns should also be fortified; that rivers which traverse the country, mountain ranges, and other natural obstacles will serve as additional defensive lines; that many towns with a strong natural position are worth fortifying; and lastly, that certain military installations such as munitions factories are situated better in the interior than on frontiers and surely deserve to be protected by a fortress. Thus we find that in some cases there may be more and in some cases less reason for building fortifications in the interior of the country. Our feeling is, therefore, that while states that have many fortresses will be correct in locating most of them on the frontiers, it would still be a great mistake to leave the interior completely unfortified. In France, for example, this mistake is in our opinion astonishingly common. Some confusion may understandably arise in this respect if, in the border areas, there are no major towns at all, or where these are located further inland. This is the case in southern Germany: there is hardly a major town in Swabia, whereas Bavaria is full of them. We do not think it is necessary to solve this problem once and for all, but believe that such cases must be determined by allowing for the particular circumstances of each case. All the same, we would draw attention to the closing remarks of this chapter.

The third question—whether fortresses should be grouped or spread out evenly—is one that seldom arises, all things considered. Still, we would not, for that reason, treat it as idle speculation: a group of two, three, or four fortresses only a few days' march away from a common center will give great strength to that center and to the army based on it. Therefore, if other conditions permit at all, one would be very tempted to create such a strategic bastion.

The final point concerns the natural setting of the locations to be chosen. A fortress will be twice as useful if it lies on the coast, on a stream or great river, or in the mountains. As we have already pointed out, one of the main requirements is then fulfilled. Still, there are several other aspects to be considered.

If a fortress cannot be located directly on a river, it is better not to place it in the immediate vicinity, but some fifty to sixty miles away; otherwise the river will cut through and interfere with its sphere of influence with respect to all the points mentioned above.[1]

This does not hold true in mountains: they do not limit the movement of armed forces, large or small, to given points, as rivers do. But a fortress on the enemy's side of a mountain range is poorly placed because it is difficult to relieve. Conversely, if it is located on the near side, the enemy will be hard put to besiege it, for the mountains will obstruct his lines of communication. Olmütz in 1758 is an example.

It will be self-evident that large impenetrable forests and marshes have the same effect as rivers.

The question frequently arises whether towns in inaccessible locations will make better or poorer fortresses. Since it costs little to fortify and defend them, while an equal expenditure of strength will make them much stronger and often impregnable, and since the services of a fortress are always more passive than active, the objection that they are easy to blockade need not in our opinion be taken very seriously.

In conclusion, let us review our straightforward system of fortifying a country. We feel justified in claiming to have based it on important and permanent considerations directly related to the vital affairs and interests of the state. It is, in consequence, immune from transient military fashions, flights of ingenious strategy, or the special needs of a given case—any one of which could have unhappy consequences for a fortress built to last five hundred or even a thousand years. Silberberg, built by Frederick the Great on a ridge of the Sudeten mountains, has, under radically changed conditions, lost practically all its purpose and importance; whereas Breslau, if it had been and had remained a proper fortress, would have kept its value under all conditions—against the French no less than against the Russians, Poles, or Austrians.

The reader should not forget that our arguments are not intended to be valid for a state that wants to build an entirely new set of fortifications. Since this seldom, indeed practically never, happens, our arguments would have been a useless exercise; but they might apply when any single fortress is being planned.

[1] Philippsburg was a perfect example of how not to site a fortress. Its location was that of an idiot standing with his nose against the wall. Cl.

CHAPTER TWELVE

Defensive Positions

Any position in which one accepts battle and makes use of terrain to protect oneself is a defensive position, and it makes no difference whether one's general attitude is mainly passive or mainly active. This follows from our general view of the defensive.

One could go a step further and call all positions defensive in which an army advancing toward the enemy would be ready to accept battle if the enemy were to seek it there. This, in the last analysis, is how most battles come about, and throughout the Middle Ages no other way was known. However, that is not our present subject. Most positions are of this type, and the concept of a *position*, as opposed to a *temporary camp*, would suffice for present purposes. A position specifically designated as a *defensive position* must, therefore, have additional characteristics.

It is obvious that decisions arrived at in ordinary positions are dominated by the element of time. The armies march toward one another: where they meet is a minor matter; all that counts is that the place be fairly suitable. In true defensive positions, however, the element of *place* is dominant: the decision must be reached at that place, or rather mainly because of that place. This is the kind of position we are discussing here.

The element of place, then, has a dual aspect: first of all, a force in a given position has a certain influence upon the whole; and second, the position itself serves to protect the force and adds to its potentialities. In brief, it will have strategic as well as tactical pertinence.

To be quite accurate, only the tactical aspect justifies the term "defensive position." The strategic aspect—the fact that a force positioned at that point protects the country by its mere presence—could just as well be applied to an attacking force.

The first of these two aspects—the strategic influence of a position—can only be fully demonstrated when we come to discuss the defense of a theater of operations; here we shall consider it only so far as can now be done. For this purpose we must more closely examine two somewhat similar concepts that are often confused: turning a position and by-passing it.

A position is turned in relation to its front, and this is done either to attack it from the flank, or even from the rear, or to cut its lines of retreat and communication.

The first of these—an attack on flank or rear—is tactical in nature. Nowadays, when troops are extremely mobile and all operational plans are aimed to some extent at turning or enveloping the enemy, any position must be prepared for this eventuality; and a position that deserves to be called strong must not only be strong in front but must also allow adequate scope for fighting on the flanks and in the rear, if either should be threatened. Thus, if the position is *turned* with the intention of attacking its flank or rear, it will not be rendered useless. The battle taking place here was determined by the position, and the defender should be able to benefit from all the advantages for which he chose the position in the first place.

If the attacker turns the position so as to threaten its line of retreat and communication, he brings the strategic aspect into play, and the question then is how long the position can be held and whether the enemy can be beaten at his own game. Both depend on the actual location of the place—chiefly that is, on the relation of the lines of communication of each side to one another. Any good position should be sure of giving the defending force superiority in that respect. At any rate, the position will not be rendered useless by being turned; indeed, an enemy who attempts this type of undertaking will at least be neutralized by it.

If the attacker, on the other hand, ignores the force that lies in wait for him in a defensive position, and pursues his aim by leading his main force forward by another route, he *by-passes* the position: and if he can do this without loss, and really carries it off, he forces the defender to abandon the position instantly—in other words, he makes it useless.

There is hardly a position in the world that is impossible to by-pass in the literal sense of the term. Instances such as the Isthmus of Perekop are too rare to deserve detailed consideration. What is meant by impossibility, therefore, refers to the disadvantages that accrue to the attacker from by-passing the position. Chapter Twenty-seven will offer a more suitable opportunity for discussing these disadvantages; whether they be great or small, they are the equivalent of the unexploited tactical effectiveness of the position, and together with it they fulfill its function.

Two strategic features of a defensive position have thus emerged:

1. That it may not be by-passed
2. That it gives the defender an advantage in the struggle for lines of communication.

Two other strategic features will now be added:

3. That the relation of the lines of communications also have a favorable effect on the pattern of the engagement
4. That the general influence of the terrain be favorable.

The relationship of the lines of communication affects not only the possibility of by-passing a position and possibly of cutting off its supplies but also the whole course of the battle. An oblique line of retreat facilitates a tactical turning movement by the attacker, and inhibits tactical movements by the defender while the battle is in progress. An oblique disposition relative to the line of communication is not, however, necessarily due to faulty tactics, but may well be the result of a strategic error in the choice of the position. For example, it cannot be avoided if the road changes direction near the position (as at Borodino, 1812). The attacker is then headed in the right direction for a turning movement *without having to diverge from his perpendicular disposition.*

Moreover, if the attacker is able to retreat along a number of roads, while the defender only has a single one, the former enjoys far greater tactical freedom. In all such cases, the defender will struggle in vain to overcome the disadvantages of the strategic error. He will not succeed.

Finally, with regard to the fourth point, other aspects of the terrain can be so disadvantageous that the most careful choice and utmost ingenuity of tactics will be of no avail. In that case, the following will be the main considerations:

1. Above all, the defender must seek to keep the enemy under observation, and be able to fling himself upon him within his own immediate area. The defender will reap the benefit of the terrain only where the natural obstacles of the environment combine with these two conditions.

There is a disadvantage then, to all places that are dominated by generally higher ground; all or most positions in mountainous areas (which will be more specifically considered in the chapters on mountain warfare); all positions with one flank protected by a mountain (while the attacker will find these hard to *by-pass,* he will find them easy enough to *turn*); any position with a mountain immediately in front of it; and, in general, all cases that derive from the above circumstances and their relation to the general conditions of the terrain. Among the opposites of these disadvantageous conditions, we shall only mention the case of a position backed by mountains. Here the advantages are so many that it may be ranked among the best conceivable places for the purpose.

2. The terrain may, to a greater or lesser degree, be suited to the army's nature and composition. Superiority in cavalry should cause us, with good reason, to look for open ground. The use of rough and very difficult terrain is indicated where there is a shortage of cavalry and perhaps also of artillery, but where the infantry consists of brave men who have experience in war and know the surrounding countryside well.

There is no need to discuss in detail the tactical effect that the location of a defensive position may have on the fighting forces. We shall simply consider the result as a whole, for this alone has strategic significance.

Any position in which an army means to await the enemy's attack should obviously be one that offers solid advantages of terrain, which will in turn serve to multiply the army's strength. Where Nature helps a good deal, but not as much as one could wish, entrenchment must come to one's aid. By this one can frequently make individual sectors—indeed, at times the whole position—*impregnable.* In the latter case, the purpose of the measures that will be taken has obviously changed completely. We are no longer seeking a battle under favorable conditions—a battle intended to make a success of the whole campaign. Our aim now is to succeed without any battle at all.

By keeping our fighting forces in an impregnable position we actually refuse battle, and force the enemy to seek other ways of reaching a decision.

These two cases, therefore, must be clearly distinguished from one another. The second will be discussed in the following chapter under the heading of "strong positions."

The defensive position with which we are concerned here is meant to be simply an exceptionally advantageous battlefield. But if it is intended to be a battlefield, the defensive advantages must not be *too great*. How strong, then, should such a position be? Obviously, the more aggressive our opponent, the stronger the position needs to be, and therefore each case must be judged on its own merits. Against a man like Bonaparte, one can and should retreat behind stronger ramparts than against a Daun or a Schwarzenberg.

If sections of a position are impregnable, such as, for instance, its front, this should be considered as a single factor of its total strength, for it means that forces not needed here may be used elsewhere. However, one should not fail to recognize the fact that the enemy, in avoiding the unconquerable parts, will alter the whole pattern of his attack. It remains to be seen whether the new pattern suits our purposes.

For instance, consider a position taken up directly behind a major river, so that the river serves to strengthen the front, as must have happened on occasion. In effect the river has become a point of support for either the right or the left flank, for the enemy naturally must make his crossing further to the left or right and change his front in order to attack. The principal question is, therefore, what advantages or disadvantages this has for the defender.

In our opinion, a defensive position approaches the ideal the more its strength is masked, and the more it lends itself to taking the enemy by surprise in the course of the action. One always attempts to deceive the enemy as to the true numerical strength of one's fighting forces and their true direction. By the same token, then, one should not let him see how one intends to take advantage of the terrain. This is only possible up to a certain point, of course, and perhaps calls for a special technique that has not yet been attempted.

Proximity to a substantial fortress, no matter in which direction it lies, lends superiority to any position by increasing the mobility and usefulness of its forces. Appropriate use of individual fieldworks can make up for lack of natural strength at some point, permitting one at will to determine the broad outlines of the engagement in advance. These are the reinforcements that art can provide. Combining them with the correct choice of natural obstacles (impeding the enemy forces without rendering them completely ineffective) and the advantages that derive from knowing the battlefield while the enemy does not, with our ability to conceal our arrangements better than he can, and, in general, with our superiority in means of surprise in the course of the action, can make the influence of the terrain itself overpowering and decisive, so that the enemy will succumb without ever knowing the real cause of his defeat. This is how we conceive of a defensive position; it is in our opinion one of the greatest advantages of defensive war.

Without making allowances for any special circumstances, one may assume that undulating country, which is neither too closely nor too sparsely cultivated, will afford the greatest number of positions of this type.

CHAPTER THIRTEEN

Fortified Positions and Entrenched Camps

In the last chapter we stated that a position which nature and skill have made so strong that it must be considered unassailable is distinct from the category of advantageous battlefields. It is in a class by itself. In this chapter, we shall discuss its peculiarities; and, because its characteristics are much like those of a fort, we shall call it a *fortified position.*

Fortified positions are not easily created by simple entrenchment, except for entrenched camps close to fortresses. Still less are they created by exploiting natural obstacles. Nature and skill usually work hand in hand in their creation; therefore they are often referred to by the term entrenched camps or positions. That name, however, can apply just as well to any position improved by a few fieldworks—which is not the same as the subject being discussed here.

The function of a strong position, then, is to make the forces holding it practically unassailable. By doing this, it will either serve as a direct protection of the area *itself,* or merely of the *fighting forces* stationed in the area. They, in turn, will cover the country indirectly. The former was the function of the lines in earlier wars, especially on the frontiers of France, while the latter is the purpose of entrenched camps facing in all directions, and of those in the vicinity of fortresses.

If, for instance, the front of a position has been made so strong by fieldworks and obstacles to the approaches that no attack is possible, the enemy will be forced to turn it and attack in flank or rear. In order to make this more difficult, however, one would look for points of support for these lines that could protect their flanks, such as the Rhine and the Vosges Mountains for the lines in Alsace. The longer the front of such a line, the easier it was to prevent its being turned, for a turning movement always implies a certain risk to the force executing it, and the risk increases in proportion to the amount of deviation from the original direction of the forces. An extended length of front that could be made impregnable, therefore, together with good points of support, would usually provide the opportunity of directly protecting a considerable area from invasion by the enemy. That, at least, was the original idea, and that was the significance of the lines in Alsace, with their right flank on the Rhine and their left flank on the Vosges, and of the lines in Flanders extending seventy-five miles, their right flank resting on the fortress of Tournai and the Scheldt, and their left flank on the sea.

When the means are not available for so long and strong a front and well-supported flanks, if the area is to be defended at all by a well entrenched force, the latter will have to use all-round defense to protect itself from being turned. The concept of an effectively covered area is lost: such a position, strategically speaking, is only a point. The fighting forces are all that is covered, and it is therefore up to them to hold the country, or rather to *maintain themselves in the country*. It will be impossible to *turn* such a camp, because it has no *weak* parts in flanks or rear that can be attacked: it is all front, and equally strong everywhere. But a camp of that sort can be *by-passed*—and much more easily than an entrenched line at that, for it has next to *no* extension.

Basically, entrenched camps in the vicinity of fortresses belong to the latter category: they are meant to protect their forces. Their broader strategic significance, namely the employment of these protected forces, differs somewhat from that of other entrenched camps.

Having traced their origins and development, we now propose to assess the value of each of these three methods of defense. To distinguish them we shall call them "fortified lines," "fortified positions," and "entrenched camps near fortresses."

1. *Lines*. Lines constitute the most ruinous form of cordon-warfare. The obstacle they offer the attacker is worthless without powerful fire to support it. Otherwise it is good for nothing. Moreover, the distance an army can extend and still use its firepower effectively is minimal in relation to the country; the lines will, therefore, have to be very short and thus cover very little of the country, or the army will not be able to defend all points effectively. The suggestion has, of course, been made that not all points along the line need necessarily be occupied: they could merely be kept under observation and defended by means of available reserves, much like the defense of a moderate-sized river. But this device is inappropriate to the means employed. Where the natural obstacles are so great that it could be used, entrenchments would not be needed and would, in fact, be dangerous. Lines are not appropriate to local defense, but entrenchments are. Where the entrenchments themselves are supposed to be the main obstacle to access, it will be plain that they are of little use if they are *not defended*. How much use is a twelve- or fifteen-foot ditch, or a rampart ten or twelve feet high, against a combined assault by several thousand men undisturbed by enemy fire? The conclusion is that lines of that sort can be *turned* if they are short and fairly well defended; if they are long and insufficiently defended, they can be taken from the front without trouble.

Since lines of this type tie down the forces to local defense and rob them of all mobility, they are a poorly designed device against an enterprising enemy. If they nevertheless persisted in recent wars, the reason lies in the nature of the wars, when apparent difficulties were often treated as real. In most campaigns these lines were in any case only used as a supplementary defense against raiders. As such they may have had some value; but one must remember that the troops needed for their defense could often have accomplished more valuable things at other points. They were out of the question in the most recent wars, nor is there any sign that they were used. One may doubt that they will ever reappear.

2. *Positions.* As we shall show in detail in Chapter Twenty-seven, the defense of an area exists so long as the force entrusted with it can maintain itself there. It does not end until that force withdraws from the area and abandons it.

Consequently, if a force is to hold its ground in a country that is attacked by a much more powerful enemy, one means of protecting this force against overwhelming odds is to place it in an impregnable position.

As has already been pointed out, such a position must have all-round defense. The *normal* width of a tactical disposition, where the force is not *very large* (which would be contrary to the whole nature of the case) will take up *very little space*—a space so small that in the course of the engagement it would suffer from innumerable disadvantages; no matter how much it was strengthened by ramparts of all kinds, there would be little hope of making a successful stand. In a camp that is to present a front on every side, all sides must be extensive as well as virtually impregnable. The art of entrenchment alone will not be enough to provide for such strength where extension is so great. It is therefore a fundamental condition that such a camp should look for strength in natural obstacles that will make some parts of it unapproachable and others very hard to reach. If this method of defense is to be used, therefore, the right position must be found; one cannot do it merely by entrenchments. These remarks apply to tactical results and are meant to establish the existence of this strategic means. By way of illustration, we would mention Pirna, Bunzelwitz, Kolberg, Torres Vedras, and Drissa. Now let us consider its strategic qualities and effects.

The first condition, of course, is that the troops detailed to hold this camp should have their food supplies assured for some time—that is, for as long as one relies on the effectiveness of the camp. This can only be the case where the rear of the position rests upon a port, as at Kolberg and Torres Vedras; or, if like Bunzelwitz and Pirna, it is in close communication with a fortress; or else if stocks have been built up in the camp or close at hand, as was the case at Drissa.

This condition will have been adequately met only in the first case, and only partially in the second and third. It will always be a source of risk. At the same time it is clear that problems of supply rule out a good many points that would otherwise have served for an entrenched position. Suitable ones are consequently *rare.*

In order to gauge the effectiveness of such a position, to balance its assets and liabilities, we must ask ourselves how an attacker is likely to react to it.

a. The attacker can by-pass the strong position, pursue his own designs, and place it under observation by a suitable number of troops.

Here it makes a difference whether the entrenched position is held by the main force or only by a subsidiary one.

If it is held by the main force, the attacker will only benefit from by-passing it if there is an additional accessible and *decisive object for him to attack*, such as a fortress, the capital, or the like. Even assuming that this exists, he can pursue it only if the strength of his base and the location of his lines of communication are such that his strategic flanks are in no danger.

The conclusion to be drawn concerning the propriety and effectiveness of a strong position for the main force of the defender, is that it will be found only where it is such a threat to the attacker's flank that one can be sure to pin him down where he can do no harm; or else if there is no objective in his reach which might be of concern to the defense. Where there is such an objective, and where, at the same time, the strategic flank of the enemy cannot be seriously threatened, the position should either not be taken up at all, or, if it is, only by a token force, simply to see if the enemy will respect it. In that case the danger still exists, however, that if he does not do so, one will no longer be able to relieve the threatened position.

If the fortified position is held only by a subsidiary force, the attacker will never lack for another objective for his attack, which may be the main army itself. The value of the position, in that event, is confined to the threat it represents to the enemy's strategic flank, and will be confined to this condition.

b. If the attacker dare not by-pass the position, he can lay siege to it and starve it into surrender. However, there are two prerequisites for this. The first is that the position must not be open to the rear; and the second is that the attacker must be strong enough for such an investment. If both conditions are met, the position will, of course, neutralize the attacker for some time; but the defender will have to pay for this advantage with the loss of the defending force.

It follows, then, that the occupation of such a strong position by the *main defending force* is a measure to be taken only under certain circumstances: (aa) if its rear is absolutely safe (as at Torres Vedras); (bb) if one can predict that the enemy's superiority will not suffice for a siege. If he does attempt it without sufficient superiority, the defender will be able to make successful sorties, and defeat the enemy in stages; (cc) if one can count on relief, as the Saxons did at Pirna in 1756. Basically this was also the case after the battle of Prague in 1757: Prague may be considered as an entrenched camp in which Prince Charles would never have let himself be bottled up had he not known that the army from Moravia could release him.

One of these three conditions must absolutely be met if the choice of a fortified position for the main defending force is to be justified. It must be admitted, however, that the second and third of these conditions come close to forcing the defender to accept a serious risk.

Where a subordinate corps is involved, however, one that can, if necessary, be sacrificed for the benefit of the main force, these conditions disappear. The only question then is whether the sacrifice will in fact avert a greater evil. No doubt this will only rarely be the case, but it is not impossible. It was the fortified camp at Pirna that prevented Frederick the Great from invading Bohemia as early as 1756. Austrian unpreparedness was so great at that time that Bohemia would surely have been lost, and the resulting casualties might well have gone beyond the 17,000 men who surrendered when Pirna fell.

c. The situation may not afford the attacker either of the possibilities outlined at a. and b. above. In that event, the conditions we prescribed for the defender are fulfilled, and the attacker frankly has no choice but to come to a halt facing the position, like a setter pointing at a covey. The most he can do is to send out detachments and thereby spread his forces over the area, contenting himself with that advantage, insignificant and indecisive though it be, and wait for the future to decide the possession of the area. In that case, the position has discharged its function to the full.

3. *Entrenched Camps near Fortresses.* These camps fall, as we have said, within the general category of fortified positions to the extent that their purpose is to protect not an area but a force from enemy attack. The only difference from other positions lies in the fact that they form an inseparable whole with the fortress—a fact that, of course, adds immensely to their strength.

They give rise to the following features:

a. They may serve the special purpose of making the siege of the fortress either very difficult or impossible. This may be worth substantial casualties if the fortress is a port that cannot be blockaded; in any other case, however, the risk exists that the place may be starved into surrender too soon to justify the sacrifice of a considerable number of troops.

b. Near a fortress, it is possible to establish an entrenched camp for a smaller body of troops than would be feasible in the field. Within the walls of a fortress, four or five thousand men may be invincible, whereas no camp in the world would be strong enough to save them in the open field.

c. They may be used for the assembly and training of troops whose morale is not strong enough for them to be exposed to the enemy save under the protection of a fortress: recruits, militia, home guard, and the like.

Camps of this sort might thus be recommended for a number of useful purposes, if they did not have the exceptional disadvantage of putting the fortress in jeopardy when they cannot be occupied. On the other hand, it would be much too heavy a commitment for a fortress to maintain a garrison large enough to suffice also for the occupation of the camp.

We are therefore inclined to recommend their use for coastal fortresses only, but to consider them as more of a liability than an asset in all other instances.

In conclusion, we summarize our views on fortified and entrenched positions, as follows:

1. The smaller the country, and the less room for evasive movements, the harder it is to do without such positions.

2. The greater the certainty with which help and relief can be relied upon from other forces, the onset of bad weather, a popular insurrection, shortages, or the like, the less are they a source of risk.

3. Their effectiveness will become greater, the weaker the impetus of the enemy's attack.

CHAPTER FOURTEEN

Flank Positions

The only reason we are devoting a separate chapter to flank positions is to provide easy reference to it, in the manner of a dictionary. While it is a prominent subject in the world of orthodox military ideas, it is not in our opinion an independent topic.

A flank position is any position that is meant to be held even though the enemy may pass it by: once he has passed it, the only effect it can have is on his strategic flank. Every *fortified position* is therefore a flank position: since it is impregnable, the enemy has to by-pass it, and thereafter its only value lies in the effect it has on his strategic flank. It does not matter which way the real front of the position faces—whether parallel to the enemy's strategic flank as at Kolberg, or at right angles to it, as at Bunzelwitz and Drissa. A fortified position must face in all directions.

But one may intend to hold a position that is *not* impregnable, even if the enemy has passed it by. This would be the case if it were so well placed with respect to the enemy's lines of communication and retreat that not only would an effective attack on his strategic flank be possible as he advanced; in addition the enemy, being concerned about his own retreat, would not be able to cut ours off altogether. If this were not the case, we would, since the position is not impregnable, run the risk of fighting without any chance of retreat.

In 1806, for example, the position of the Prussian army on the right bank of the Saale could have become in every sense a flank position in relation to Bonaparte's advance through Hof, simply by having the front face the Saale, and then waiting on events.

If there had been less of a disproportion of physical and psychological forces, if the French army had been led by a man like Daun, the Prussian position would have proven brilliantly effective. It was impossible to by-pass: even Bonaparte admitted as much when he decided to attack it. Not even he could cut it off entirely; and if the disproportion of physical and moral forces had been less severe, trying to cut off the Prussians would have been as impracticable as to move past them, because defeat of its left wing constituted less of a danger to the Prussian army than the same thing did to the French. In spite of the imbalance of both physical strength and morale, determined and intelligent leadership would have provided sufficient reason to hope for a victory. There was nothing to prevent the Duke of Brunswick giving orders on the thirteenth for his 80,000 men to confront, on the morning of the fourteenth, the 60,000 with whom Bonaparte crossed the Saale at Jena and Dornburg. While his superiority in numbers, together with the deep valley of the Saale in the rear of the French, would not have sufficed for a decisive victory, one must recognize that it was a very advantageous combination. If it would not have led to success, then there should have been no thought of gaining a decision in this area in the first place. The Prussians should have fallen back still further, thereby adding to their own strength and reducing the enemy's.

To sum up, the Prussian position on the Saale, vulnerable as it was, could be considered as a flank position toward the road coming from Hof; but, because of its vulnerability, it could not become one in an absolute sense until the enemy made it so by not daring to attack it.

It would be still less consistent with a clear conception if we were to apply the term to positions that *cannot* maintain themselves when they are bypassed, and from which the defender must therefore attack the enemy in the flank; they would be called *flank positions* simply because that is where the enemy has been attacked. This type of flank attack has little connection with the position in itself; at least, it does not primarily result from the nature of the position, as would be the case in an action against the enemy's strategic flank.

At any rate, it follows that there is nothing new to be said about the attributes of a flank position. But this may be a suitable place for a few words about the character and implications of this expedient.

We shall not include genuine fortified positions, which have previously been discussed at length.

A flank position that is not impregnable is a very effective instrument, but because of its vulnerability, it is a risky one. If it gives an attacker pause, it produces a great effect with little effort, much like the pressure of a rider's little finger on the curb rein of the bit. If, however, the effect is insufficient and the attacker is not checked, the defender will pretty well have lost his chances of retreat. He must attempt a rapid escape by roundabout routes, that is under most unfavorable conditions, or he will be in danger of fighting without retreat being possible. If the enemy is daring, superior in morale, and looking for a sweeping decision, this is a very hazardous course, and far from suitable—as our example of 1806 has shown. With a cautious opponent, on the other hand, and in a war of observation only, it is one of the best means the defense can use. One will find examples in Duke Ferdinand's defense of the Weser River, taking a position on its left bank, and in the well-known positions at Schmottseifen and Landeshut. In the latter case, however, the disaster that befell Fouqué's corps in 1760 also shows the danger of using it incorrectly.

CHAPTER FIFTEEN

Defensive Mountain Warfare

Mountainous terrain exercises a strong influence on warfare; the subject is therefore significant for the theorist. Since that influence consists in introducing a retarding element into operations, it belongs primarily to the realm of defense. We shall, therefore, discuss the matter here without confining ourselves to the narrower subject of the defense of mountains. In the course of studying this subject, since our analysis has resulted in some respects in rather unorthodox conclusions, we shall have to go into the matter in some detail.

We shall first consider the tactical aspects of the matter, from which we can proceed to its links with strategy.

There is no doubt that the reputation for effectiveness and strength enjoyed by defensive mountain warfare has traditionally derived from two main factors: first, the difficulty of moving long columns over mountain roads; and second, the extraordinary degree of strength attained by a small post whose front is covered by a steep mountainside while its flanks are supported by deep ravines. Only the characteristics of armaments and tactics at certain periods have prevented major forces from making use of this effectiveness and strength.

A column toils at snail's pace up a mountain through narrow gorges; gunners and teamsters yell and swear as they flog their weary beasts along the rocky tracks; each broken-down wagon has to be removed at the cost of indescribable effort while behind it the rest of the column stops, grumbles, and curses. At such a moment, each man secretly thinks that in this situation a few hundred of the enemy would suffice to cause a total rout. Here one can see the origin of the expression used by historians who speak of a gorge so narrow that a handful of men could hold off an army. Still, anyone who has experience in war will know, or should know, that such a march through the mountains has little, if anything to do with an *attack* in them. It would be quite wrong to infer from *this particular* difficulty that an attack would be even more difficult.

A novice will naturally jump to this conclusion, and it is equally natural that at certain periods military practice got entangled in the same error. At the time, the phenomenon was almost as new to the veteran as it was to the novice. Before the Thirty Years War the deep order of battle, the swarms of cavalry, the crudeness of the firearms and various other factors made any exploitation of the major obstacles of the terrain highly unusual. Methodical defense of mountain areas, at least with regular troops, was practically impossible. Not until a more extended order of battle came into use, and infantry, and with it firearms, became the dominating component of an army, did it dawn on anyone that good use could be made of heights and valleys. Even then, it took another hundred years—approximately till the middle of the eighteenth century—for it to be developed to its full potential.

The second aspect—the strong resistance that a small post can put up at an almost inaccessible point—reinforced the belief in the greater effectiveness of defense in mountains. It seemed as if all one had to do was to increase such positions in order to transform a battalion into an army, and a mountain into a mountain range.

Undeniably, in a mountainous area a small post in a favorable position acquires exceptional strength. A unit that on open ground can be dispersed by a couple of cavalry squadrons, and will think itself lucky if it can escape capture or annihilation by rapid retreat, can face an army in the mountains. By a kind of tactical effrontery, one might say, it can exact from a whole army the military tribute of a full-scale attack, an envelopment, and so on. The manner in which it adds to its power of resistance through obstacles, flank-support and new positions taken up as it retreats is a question of tactics; we accept it here as a matter of fact.

It was only natural to assume that a series of strong posts of this sort would result in a strong, almost impenetrable front. One only had to guard against being outflanked by extending the position to right and left until it reached adequate points of support, or until one believed that the extension alone was enough to prevent the position from being turned. In that respect a mountain area is very tempting; it offers such a wealth of defensive positions, each better than the last, that it is hard to know where to stop. One ended up by occupying and defending every point of access within a given area, and believed that if one occupied a space of fifty miles or more with ten or fifteen posts, one would at last be safe from the horror of envelopment. Since these positions seemed firmly linked by inaccessible terrain (columns not being able to march across broken country), one appeared to confront the enemy with a wall of bronze. For extra safety, one retained a couple of infantry battalions, some horse artillery and a dozen cavalry squadrons in reserve, just in case the enemy should by fluke succeed in breaking through.

No one will deny that this is an accurate historical depiction; and it cannot be said that people have completely shed this error.

The course of tactical development since the Middle Ages, as armies increased in size, also contributed to strengthen the apparent defensive significance of mountainous terrain for military action.

The leading feature of defensive mountain warfare is its decisively passive character. It was therefore natural enough to have recourse to it before armies achieved their present state of mobility. The number of troops had steadily increased, and, to utilize their firepower they were increasingly deployed in long thin lines, elaborately linked and very difficult, if not impossible, to maneuver. Deploying an intricate machine of this kind could take half a day; it made up half the battle and included practically everything that can be found in a modern battle plan. Once all this had been accomplished, it was difficult to make changes if different circumstances arose. It followed that an attacker who delayed drawing up his line of battle could then do so on the basis of the defender's position, and the latter was unable to respond. The attack thus acquired an overall superiority, and all the defense could do was to protect itself by means of natural obstacles. Nothing served that general purpose more effectively than mountains. One therefore aimed, so to speak, at a union of the army and suitable terrain; united they made common cause. The battalion defended the mountain, and the mountain defended the battalion. A mountain area thus conferred a high degree of strength on passive defense, and was not in itself a disadvantage, except that it resulted in an additional loss of mobility—which no one had been adept at using anyway.

In a clash of two opposing systems the exposed side, that is the weaker side, always attracts the enemy's attack. If a defender's posts are strong and impenetrable, transfixed, so to speak, and immobile, the attacker will be emboldened to turn them, because he will not have to worry about his own flanks. That indeed is what happened, and soon it was the order of the day. In response, positions became more and more extended, and their front became proportionately weaker. Then the attacker changed his method: he no longer tried to outflank the enemy by outextending him, but massed his strength against a single point and pierced the line. This was roughly the stage reached by defensive mountain warfare at the end of the recent wars.

Thus, attack had again attained complete superiority, due to its constantly increasing mobility. Only mobility could strengthen the defense, but mobility is inhibited by mountainous terrain. As a result defensive mountain warfare has (if we may use such an expression) suffered a defeat; just as armies did so frequently when they attempted this kind of defense during the Revolutionary Wars.

In order not to throw the baby out with the bath-water and be swept along by the stream of platitudes into assertions that can be disproved in practice a thousand times by actual experience, we must distinguish the effects of defense in mountains according to the nature of the individual case.

The central question to be decided here, and the one that sheds the greatest light on the whole subject, is whether resistance in defensive mountain warfare is intended to be relative or absolute. Is it meant to last only a certain time, or to end in definite victory? Mountains are eminently suited to defense of the first type, because of the increased factor of strength. For the second type, on the other hand, they are, except for a few special cases, generally not suited at all.

In mountains, any movement is slower and more difficult; it takes more time, and if it is made within range of the enemy, it also costs more lives. The resistance encountered by the attacker is measured in expenditure of time and lives. Therefore the defender has a clear advantage so long as movement is only up to the attacker; the advantage vanishes as soon as the defender has to move as well. It is basic, and tactically necessary, that limited resistance allows for far more passivity than an aim of outright victory. Passivity, moreover, can continue indefinitely—right to the very end of the engagement. That would be impossible in the case of absolute resistance. The impeding characteristic of mountain country, a sort of viscous element that clogs and stultifies initiative, is therefore the ideal setting for this purpose.

We have already stated that a small post can acquire extraordinary strength in mountainous terrain. While this tactical result needs no further proof, one further explanation is necessary: one has to distinguish whether the size of a unit is small in an absolute or a relative sense. If a force of any given size decides to station a single unit in an isolated spot, that unit may find itself under attack by the whole enemy force—in other words, by superior forces, compared to which it is quite small. In that event, it can, as a rule, merely hope to put up relative, rather than absolute, resistance. The smaller the unit in relation to its own main force and to the enemy's, the more this applies.

But even a post small in an absolute sense, that is a post opposed by an enemy no stronger than itself, that could expect to offer absolute resistance aimed at actual victory, will be infinitely better off in the mountains than would a large army. It will make better use of the terrain, as we shall explain in more detail later.

The conclusion that emerges, then, is that a small post can be very strong in mountain country. It is plain enough that this can be of decisive value whenever *limited* resistance is required; but will it have the same *decisive* value for the absolute resistance of an army? That is the question we must now investigate.

First, we must ask another question: will a front composed of several posts be so strong proportionately as each single one of them, as has hitherto been assumed? Certainly not: that assumption is the result of one of two possible errors.

For one thing, an impassable area is often confused with an inaccessible one. Where one is not able to *march* in a column, or with artillery or cavalry, one can, in most cases, still advance with infantry, or make some use of artillery: the brief exertions involved in movements during battle cannot be measured by the standards of a march. The belief that posts enjoy secure communications with each other therefore rests on a complete illusion, and one that endangers their flanks.

The other error is to think that a line of small posts that are very strong on their front are just as strong on their flanks, because a ravine, a precipice, or the like make good points of support for a small post. But why do they? Not because they make a post impossible to turn, but because they burden a turning movement with a cost in time and effort that has to be measured against the post's significance. The enemy who wants to and has to turn such a post despite the difficulties of terrain because a frontal attack is impossible may easily spend half a day on this maneuver, and may still not be able to accomplish it without casualties. If a post of this type is dependent on help, or if it is intended to hold out for only a limited time, or, finally, if its strength is equal to the enemy's, the points of support of the flanks have done their job, and it would be true to say that the post was strong not only frontally, but on its flanks as well. However, this does not hold in the case of a chain of posts that are part of an extended position in the mountains. Here, none of those three conditions is fulfilled. The enemy attacks a single point in superior strength; support from the rear is negligible, and yet the situation calls for absolute defense. The post's points of support are worthless under such conditions.

This is the weakness against which the attack will direct its blows. An assault by concentrated and therefore vastly superior forces on a single point on the front will meet with fierce resistance *when measured by the strength of that point; but measured by the whole, that resistance is negligible.* Once it has been overcome, the line is pierced and the objective is achieved.

It follows that *limited* resistance is more effective in the mountains than in the plains; that it is most effective, relatively speaking, in the case of small posts; and that it does not grow in proportion to the forces engaged.

We now turn to the true object of every large engagement—*positive* victory. This must be the aim of defensive mountain warfare wherever the whole force or a major part of it is involved. In that event, *defensive mountain warfare automatically turns into a defensive battle in the mountains.* Its form now becomes that of a battle: the whole force is committed to the purpose of annihilating the enemy, and victory is the object of the engagement. The defensive mountain warfare which is here involved is secondary; it is no longer the end, but the means. That being the case, what is the relation of the mountain area to this end?

Characteristically, a defensive battle calls for passive reaction at the front, and increased active reaction in the rear; but mountainous country tends to produce paralysis. Two factors are at work. First of all, there are no roads that allow for rapid marching from the rear to the front. Even a sudden tactical assault is hampered by the uneven nature of the ground. Second, it is impossible to keep the area and the movements of the enemy under observation. Thus the attacker gets the same advantages from the terrain that had accrued to our front, while the better part of the defense is completely paralyzed. Now a third factor comes into play: the risk of being cut off. No matter how much a retreat, in the face of pressure on the front, is favored by mountainous terrain, no matter how much time the enemy loses by attempting a turning movement—these advantages only matter in the case of a *relative* resistance. They have no bearing on a decisive battle in which resistance must be maintained to the last. Granted, it will take a little longer for the enemy's flank columns to occupy the points that threaten or even cut off our retreat; but once he is there, no relief is possible. No offensive from the rear can dislodge him from the points that threaten the retreat; no desperate assault by the entire force can subdue him where he blocks the way. If this seems to be a contradiction, and one believes that the advantages that mountains afford the attacker must also accrue to the force that tries to cut its way out, one should not forget the difference of their circumstances. The corps that attempts to block the way is not meant to render an *absolute* defense; a few hours' resistance may be enough. Its situation therefore is that of a small post. Moreover, its opponent is no longer in control of all his means of operation; he is in disorder, short of ammunition, and so forth. In any case, the prospects for success are slight; it is this danger that causes the defender to fear this situation more than any other. This fear spreads through all phases of the battle and weakens the contestant's every fiber. His flanks become abnormally sensitive; indeed, every handful of soldiers that the attacker deploys within view on a wooded slope in the rear provides new leverage toward his victory.

These disadvantages would vanish to a large extent, while all the advantages would remain, if the defense of the mountains could be conducted by an army concentrated on an extensive mountain plateau. Here it would be possible to envisage a strong front, flanks difficult of access, with, on the other hand, complete freedom of movement within and in rear of the position. Such a position would rank as one of the strongest possible. However, this is little better than an illusion: while most mountain ranges are somewhat more accessible from the rear than up their slopes, most plateaus are either too small for such purposes, or the term plateau does not rightfully apply to them—it may refer to geological rather than topographical considerations.

As has already been pointed out, the drawbacks of a defensive position in the mountains tend to diminish for smaller units. The reason is that smaller forces need less room, fewer lines of retreat, and so on. One mountain does not constitute a range, nor does it possess its disadvantages. The smaller the unit, the more its position will be limited to single mountains and ridges, and the less it will be forced to involve itself in the densely wooded labyrinth of steep and narrow gorges, which is the source of all these troubles.

CHAPTER SIXTEEN

Defensive Mountain Warfare—Continued

We now turn to the strategic uses to be derived from the tactical conclusions developed in the previous chapter.

The following aspects are to be distinguished:

1. The mountain area as a battlefield
2. The effect that its possession has on other areas
3. Its effectiveness as a strategic barrier
4. The problems of supply to which it gives rise.

In the first and most important aspect we must further distinguish between a. a major battle, and b. secondary engagements.

We have pointed out in the previous chapter that in a *decisive battle, mountainous terrain* is of no help to the defender; on the contrary, that it favors the attacker. This is in direct contradiction to the general opinion; but then, general opinion is usually in a state of confusion, and unable to distinguish between diverse aspects of a question. People are so much impressed by the powerful resistance of a minor unit that they assume that all defensive mountain warfare possesses extraordinary strength. They are surprised when the existence of such strength in the core of all resistance, the defensive battle, is denied. On the other hand, they are always ready to blame the incredible mistake of cordon warfare for the loss of any defensive battle in the mountains, completely ignoring the force of circumstances that are inevitably involved. We are not afraid of being in direct conflict with such opinions. On the other hand, we are gratified to have found support in the writings of an author who, for several reasons, commands respect in the matter—Archduke Charles, in his histories of the campaigns of 1796 and 1797. Charles was a sound historian, a shrewd critic, and, what counts even more, a good general.

The defender, though outnumbered, has gathered his forces with deliberation and at great pains, in order to impress the attacker, at the time of the decisive battle, with his patriotism, enthusiasm, and keen intelligence. All eyes are upon him. We therefore cannot help but find it regrettable if he chooses to take up a position in the twilight area of densely wooded mountains, fettered in his movements by the unrelenting terrain, and open to countless forms of attack by his numerically superior opponent. The opportunity to exercise his intelligence is limited to one area only: the full use of natural obstacles. That device, however, brings him dangerously close to cordon warfare, which can be ruinous and should be avoided at any cost. In case of a decisive battle, therefore, we are far from regarding mountainous terrain as a refuge for the defender; on the contrary, we would advise any commander to avoid it if at all possible.

One must admit that it is not always possible. The character of the battle will then be markedly different from what it would have been in the plain. His positions will be far more extended—usually two or three times as long. Resistance will be much more passive, and counterattacks less violent. Such are the unavoidable conditions of mountainous terrain; still, the defensive in such a battle should not fall back on defensive mountain warfare. Rather, its chief characteristic in the mountains should be that of a concentrated disposition of its forces, fighting a *unified* battle under a *single* commander, with enough reserves left over to make the decision count for more than a mere repulse, or shielding action. This is an indispensable condition, but a very difficult one to meet; and it is so easy to slide into defensive mountain warfare that it is not surprising that it should happen so often. But it is so dangerous that the theorist cannot overstate his warnings.

So much for a decisive battle involving the main force.

In the case of engagements of lesser importance and significance, on the other hand, mountains can offer infinite advantages, because absolute resistance is not needed and no decisive results will follow. Let us clarify this by listing the objects of such resistance:

a. Simply to gain time. This is a very common object. It is always present if a defensive position has been formed for gathering intelligence, and also in any case where reinforcements are expected.

b. The repulse of a mere demonstration or minor venture of the enemy. Where a province is protected by a mountain range, no matter how lightly the range is defended, the defense will at any rate suffice to prevent enemy raids and other plundering expeditions. Without a mountain range, such a weak chain would be absurd.

c. For a demonstration of one's own. It will take a long time for the correct view on the subject of mountain country to gain recognition. Until then, one may find opponents who are afraid of it and who are immobilized by it. In cases of this kind, even the main force may be used to defend a mountain range. This is a common situation in a war fought without great energy or movement; but the condition that must remain constant is that one neither has the intention of accepting, nor of being forced into, a major battle in this position.

d. Generally speaking, mountains are well suited for any disposition in which one does not intend to accept a major engagement, because in the mountains each unit is stronger individually; only their aggregate strength will be less. Moreover, it is easier to avoid being taken by surprise there and to be forced into a decisive encounter.

e. Finally, national insurrection thrives in mountains, but always needs support from small regular units. The proximity of the main force, on the other hand, seems to work to its disadvantage. An insurrection therefore seldom justifies leading an army up into the mountains.

So much for mountains from the point of view of battle positions.

2. *The effect that possession of mountains has on other areas.* We have indicated the ease of securing a substantial sector in the mountains by means of minor posts—posts which would be too small to maintain themselves, and so in constant danger if they were more accessible. When the enemy holds the mountains, every advance takes much longer than in the plain, and therefore cannot be expected to keep the same pace. It follows that in mountains possession is a much more important aspect than in any other area of equal size. Open country can change hands from one day to the next. One need only advance some strong detachments in order to make the enemy relinquish the desired area. This is not the case in mountains, where much smaller forces can put up a serious resistance: if one desires to seize a mountainous sector, a special operation is required demanding the expenditure of much time and effort before the area can be taken. Even though a mountain range may not be the arena of the principal action, it should not be treated as being completely dependent on that action—which would be the case with a more accessible region. Its seizure and occupation must not be considered as an automatic consequence of the advance.

Mountain country, then, has far more independence. Its possession is more absolute, and less liable to change. Moreover the outer slopes of a mountain range usually afford a good view of the surrounding countryside, whereas its own interior is shrouded, so to speak, in deepest night. One can understand, then, that for anyone faced with mountains he does not hold, a mountain range represents a perennial source of unfavorable influence and a secret forge of the enemy's strength. That impression is heightened if the enemy not only holds the mountains, but is their rightful owner. This is where the smallest bands of courageous partisans can find refuge from pursuit, only to break through, unharmed, at a different point. The strongest columns can pass through them unobserved; and the attacker's forces must keep at a considerable distance in order to avoid being drawn into their area of dominance, there to be involved in an unequal fight consisting of sudden assaults and blows which they are powerless to return.

This is how any range of mountains exercises a continuous effect, within a certain radius, on the low-lying country around it. It will depend on local conditions whether this effect is direct, for instance on the outcome of a battle (as at Malsch on the Rhine in 1796), or will be felt on the lines of communication only after a certain time. Whether or not it can be overwhelmed and swept away by decisions reached in the valley or on the plain depends on the forces involved.

In 1805 and 1809 Bonaparte marched on Vienna without troubling much about the Tyrol; but Moreau had to abandon Swabia in 1796 mainly because he did not hold the hill country, and had to devote too much of his strength to its observation. In a campaign that sways to and fro, with evenly matched forces, one does not like to be exposed to the constant disadvantage represented by mountains that remain in enemy hands. One will, therefore, try to take and hold the part necessary to the main lines of one's offensive. That is why, in such cases, mountains tend to be the principal scene of minor actions between the two armies. One must be careful, however, not to overrate this by always treating mountains as the key to the whole situation, and their possession as one's chief concern. When victory is at stake, one's chief concern is victory; once it has been won, the rest of the situation can be dealt with as circumstances warrant.

3. *The effectiveness of mountains as a strategic barrier.* Two factors have to be distinguished here. Once again, the first is a decisive battle. One can treat the mountains as a river—an obstacle with certain points of access. It provides an opportunity for a victorious battle by splitting the enemy's advance and confining him to certain roads, thereby enabling us to assault part of his force with all of our own, which has been massed on the other side of the mountains. Even if the attacker disregards all other factors, there is one decisive reason why he cannot march through mountains in a single column: he would be exposed to the fatal risk of having to fight a decisive battle with only a single line of retreat. Certainly, then, this method of defense is based on cogent arguments. However, since the terms "mountains" and "mountain access" are very vague, everything depends on the terrain itself. The method can, therefore, be indicated merely as a possibility, which also involves two disadvantages. The first is that an enemy who has suffered a defeat will easily find refuge in the mountains; and the second, that he will be the one who holds the higher ground. That may not be a decisive factor, but it is a disadvantage for the defender.

We know of no battle that actually took place under such conditions, unless one counts the one Alvinczy fought in 1796. But Bonaparte's crossing of the Alps in 1800 shows that it is possible: Melas could and should have fallen on him in full strength before he had assembled his columns.

The second factor is the influence of a mountain barrier on the enemy's lines of communication where it intersects them. Quite apart from forts that block the passes or the effects of a general insurrection, poor mountain roads in bad weather can be enough to drive an army to despair. More than once they have forced an army to retreat, having first utterly worn it out. If, in addition, there are constant raids by partisans, or even a general insurrection, the enemy will have to send out large expeditions, and, finally, occupy strong points in the mountains. He will then be entangled in the most adverse situation possible in offensive war.

4. *Problems of supply to which mountains give rise.* This is a very simple and straightforward matter. The greatest advantage that can accrue to the defender in this respect arises when the attacker either has to remain in the mountains, or at least has to leave them in his rear.

These reflections on defensive mountain warfare basically apply to mountain warfare in general, insofar as they also illuminate the concept of offensive war. They will surely not be considered incorrect or impractical simply because mountains cannot be turned into plains, or vice versa, or because the choice of a theater of operations is governed by so many other factors that there seems to be little scope for arguments of this sort. That scope, however, is not so limited when it is applied to large-scale operations. When the problem is how to dispose the main force to best advantage—especially at the moment of decisive battle—a few extra marches to the front or rear are enough to bring the army from the mountains to the plains: when the whole force is firmly concentrated there, the neighboring mountains will be neutralized.

Having thrown light on the matter in general, we shall once more bring the picture into sharp focus.

We maintain, and hope that we have proved, that mountains are generally unsuited to defensive warfare, from the point of view of both tactics and strategy. Defense, in this sense, is of the decisive kind that determines the question of possession of the country. Mountains reduce one's control, and impede movement in all directions; they impose passivity, and, by requiring every means of access to be blocked, they almost always lead to some degree of cordon-warfare. Whenever possible one should therefore keep one's main force out of the mountains, leaving them to one side, or taking up a position in front of or behind them.

On the other hand, we think that in the case of minor operations or objectives, mountain country makes for greater strength. Based on what we have previously stated, it will not be inconsistent to call it a true refuge for the weak—for those no longer able to seek an absolute decision. Its very suitability for minor operations is another reason for excluding major forces.

All these considerations, however, will scarcely serve to counterbalance the psychological impact. The imagination, not only of the novice, but also of those who have been trained in the wrong methods, will be overwhelmingly impressed by the difficulties posed by mountainous terrain, which, like a dense, hostile element, impedes all the attacker's movements. Such people will, therefore, be hard put not to consider our opinion as the most fantastic paradox. When one takes a more general view, the history of the eighteenth century, with its peculiar form of warfare, takes the place of such impressions. For instance, some will never be persuaded that it would have been easier for Austria to defend herself on the Rhine than in Italy. The French, on the other hand, having made war for twenty years under vigorous, ruthless leadership, with the successes of this method strongly impressed on their minds, will for a long time to come be foremost in applying sound instinct based on practiced judgment to this as to other situations.

It would seem, then, that a state would find greater protection in open country than in mountains; that Spain would be stronger without the Pyrenees, Lombardy less accessible without the Alps, and a flat country such as northern Germany more difficult to conquer than a mountainous one such as Hungary. Those deductions are false, and that brings us to our concluding observations.

We do not maintain that Spain would be stronger without the Pyrenees, but we do hold that a Spanish army that felt it had the strength to risk a decisive battle would be wiser to make a concentrated stand behind the Ebro than to split up among the fifteen passes of the Pyrenees. This is not going to obviate the effect of the Pyrenees on the war. We believe the same to hold true of an Italian army. Scattered among the Alpine peaks, it would be no match for a resolute opponent, and have no choice of victory or defeat; but on the plains of Turin its claims would be as strong as those of any other army. Still no one will be ready to believe that an attacker likes to march across a mountain massif like the Alps, and to leave it in his rear. Moreover, acceptance of a major battle in the plains does not necessarily imply that there can be no preliminary defensive action by minor units in the mountains. In ranges such as the Alps or the Pyrenees that sort of action would be advisable. Finally, we are far from asserting that a country in the plains would be easier to conquer than a mountainous one unless a single victory sufficed to disarm the enemy completely. After such a victory the conqueror assumes a posture of defense, mountains will then be as awkward for him as they were for the defender—even more so. Should the war go on, should outside help arrive for the defender, should the people be up in arms, each of these reactions will be reinforced by the mountains.

The case may be likened to dioptrics: the image becomes more luminous when it is moved in a certain direction, but only until the focus is reached. Once one has passed that point, the effect is reversed. If defense is weaker in the mountains, it might encourage the attacker to choose a mountain route for his advance. However, this will happen only rarely: problems of supplies and roads, uncertainty as to the enemy's accepting battle in the mountains and deploying his main force there, are enough to offset any possible advantages.

CHAPTER SEVENTEEN

Defensive Mountain Warfare—Concluded

In Chapter Fifteen we discussed the nature of fighting in mountains, and in Chapter Sixteen the strategic uses to which it may be put. In the course of these discussions, the actual concept of defensive mountain warfare kept coming up, but we did not stop to explain what it meant in terms of form and organization. We shall now examine it more closely.

Mountain ranges often stretch across the surface of the earth like bands or belts, forming the divides between entire systems of irrigation. Their basic forms repeat themselves in its lesser parts, with ridges and valleys issuing from the principal range and forming smaller watersheds in turn. Hence it was natural for defensive mountain warfare to be thought of in terms of the principal mountain range, an obstacle of length rather than of breadth, which acted like an extensive barrier. Geologists may not yet be agreed on the origins of mountains or the laws of their development; in any case, the pattern of the watercourses is the most direct and reliable guide to the structure of the system—whether it was itself shaped by their effect, through erosion, or whether the course of the water is the result of the structure. It was natural, therefore, in planning defensive mountain warfare, to be guided by the watercourses. Not only do they provide a natural series of levels that enable one to plot the mountain's general height and profile with accuracy, but the valleys they form will always be the shortest, safest means of access to the heights. In any case, this much is known about erosion: it will always tend to wear uneven slopes into a single, regular curve. The theory of defensive mountain warfare that resulted would treat a mountain range that ran generally parallel to the front as a major obstacle to the approach, a kind of rampart, whose points of entry were formed by the valleys. The actual defense would be conducted on the highest ridge (that is, on the edge of the highest plateau in the range) and its line would cut across the major valleys. If the range were to run more at right angles to the defensive front, the defense would be conducted on one of its major spurs; it would then run parallel to a major valley and up to the principal watershed, which would be considered as its terminal point.

This formalism, which bases defensive mountain warfare on geological structure, has here been outlined because for a time it actually occupied the imagination of theorists; indeed, in the so-called theory of terrain, the laws of erosion had been absorbed into the conduct of war.

All this, however, is so full of erroneous assumptions and of loose analogies that there is not enough left on which to base any serious practical system.

The principal ridges of mountains are in reality much too bare and inaccessible to hold troops in any numbers. The spurs are often just as bad—too short and too irregular in shape. Plateaus do not exist on every mountain ridge, and, where they do, they are generally too narrow and inhospitable. Indeed, on close examination, it is very rare to find a mountain range that rises to a single, unbroken ridge with sides that fall away in anything like regular slopes or a series of terraces. The main ridge twists and turns and forks into mighty spurs, which curve far out into the land and often end in peaks that rise to greater heights than the main ridge itself. Adjacent foothills form great valleys, which will not fit into the system. Add to this that at the points where several mountain ranges meet, or from which they originate, the notion of a narrow band or belt must be discarded altogether and replaced by that of a star-shaped group of watersheds and mountain chains.

It must follow—and anyone who has studied mountains from this point of view will feel it even more strongly—that the idea of a systematic deployment of troops must be rejected as being too unrealistic to serve as a basis for a general plan. But we must note another important point in the area of practical application.

If we look closely at the tactical aspects of mountain warfare, two main problems stand out. They are the defense of steep mountains and the defense of narrow valleys. The latter, which frequently, *indeed usually*, affords greater effectiveness to defense, is not easily combined with positions on the main ridge: often the valley itself must be occupied—usually at the point where it opens out from the mountain massif, rather than higher up where it originates and where its sides are very steep. The defense of valleys, furthermore, provides a method of defending mountain country even when it is out of the question to station troops on the main ridge. The part it plays increases in importance therefore with the height and inaccessibility of the massif.

All these factors make it clear that one must completely give up thinking of a fairly regular defensive line that coincides with a basic geological feature. Mountains should be thought of simply as a flat surface strewn with irregularities and obstacles of every sort, each part of which one seeks to use to the best advantage possible. In short, while knowledge of the geological structure of the terrain is essential to a complete grasp of the shape of the mountain massif, it will rarely be evident in the organization of defense.

No disposition that spanned a whole range of mountains, and in which the defense followed its basic features, will be found in either the War of the Austrian Succession, or the Seven Years War, or the wars of the Revolution. Armies were never found on the principal ridge, but always on the slopes: higher up or lower down; facing this way or that—parallel, at right angles, or oblique; following the watercourse or crossing it; in the higher ranges, such as the Alps, often even continuing along a valley floor; and, greatest anomaly of all, in the minor ones, such as the Sudeten, halfway up the slope opposite the defender, so that the main ridge faced them. This was the position with which Frederick the Great covered the siege of Schweidnitz in 1762, with the heights of the Hohe Eule facing the front of his camp.

The famous positions of Schmottseifen and Landeshut in the Seven Years War were located, for the most part, on the valley floor. The same holds true of the Feldkirch position in the Vorarlberg. In the campaigns of 1799 and 1800, both the French and the Austrians set their main posts up in the valleys proper—not merely spanning them so as to form barriers, but along their entire length, while the ridges were either left unoccupied, or were occupied by only a few isolated posts.

The ridges of the High Alps are, in fact, so inaccessible and inhospitable as to rule out holding them with significant forces. If one insists on stationing troops in mountainous areas in order to control them, one cannot help but station them in the valleys. At first sight, this must seem absurd, for the accepted theory is that valleys are controlled by ridges. But in practice things are not as bad as that. The ridges can only be reached by a few tracks and paths, and usually only on foot. All the roads are in the valleys. Consequently, only at isolated points could the enemy's infantry put in an appearance; but distances in such mountain ranges are too great for small firearms to be effective. Thus, positions in the valleys are less dangerous than they appear. Admittedly, however, defense in valleys is exposed to another serious danger—that of being cut off. The enemy, it is true, can get infantry down to the valleys only at a few points, and even then he can only do so slowly and with great effort. Hence there can be no question of surprise; but if no posts defend the points where these paths open into the valley, the enemy will eventually succeed in descending with superior numbers and fanning out. He will then be able to pierce the fragile line, which has now become quite weak, its only protection being the stony bed of a shallow mountain stream. A retreat that has to take place along the valley floor by bounds until an exit from the mountains has been found, becomes impossible for many parts of the line. That explains why, in Switzerland, the Austrians almost always lost a third or half their men as prisoners.

Now a few more words about the extent to which forces on such a defensive mission are usually divided.

All such dispositions are based on the position taken by the main force along the principal approach, more or less near the center of the general line. To the right and left of that point, other units will be sent to occupy the most important points of access, and the whole will therefore form a position of three, four, five, six or more posts, more or less in a line. How far it is necessary or wise to extend the position will depend on individual requirements. A few marches' length, say thirty to forty miles, is a reasonable extension, and cases are on record where the position ran to 100 or 150 miles.

Between the single posts, which are a few hours' distance from one another, one will easily find other, less important access routes, favorable positions for a few battalions, well suited for liaison between the chief positions. These will accordingly be occupied. One can easily imagine that the forces could be even further subdivided, down to single companies and squadrons—which has in fact often been done. Indeed, there is no general limit to the splintering process. On the other hand, the strength of the individual posts depends on the strength of the whole; and that alone makes it impossible to state the probable or natural degree of strength maintained by the chief positions. For guidance we merely offer a few propositions derived from experience and the nature of the case.

1. The higher and less accessible the mountains, the more the forces may be split: indeed, the more they *must be split,* because the smaller the area that can be secured by combinations based on movement, the more its security must be taken care of by direct coverage. Defense in the Alps requires far greater subdivision, and comes much closer to the cordon method, than defense in the Vosges or in the Riesengebirge.
2. So far, in defensive mountain warfare, forces have been usually divided so as to give a single line of infantry to the principal posts, supported by a few cavalry squadrons. Only the main force, which is stationed in the center, might possibly have a few battalions in the second line of battle.
3. There have only been very few cases of a strategic reserve being kept in the rear to reinforce points under attack: a front so widely stretched was generally thought too weak at all points to start with. Support for posts under attack, therefore, was generally brought up from other posts in the line that had not been attacked themselves.
4. Even where forces had not been greatly subdivided, and individual posts were therefore fairly strong, the main resistance offered by them always took the form of local defense. Once the enemy was in full possession of a post it was considered past recovery by any reinforcements that might reach it.

How much one can expect from defensive mountain warfare; when it should be used; to what lengths one can, and may, go in extending and subdividing the forces—these are matters that the theorist must leave to the discretion of the general. It is enough for a theorist to describe the means and the part they play in military operations.

A general who allows himself to be decisively defeated in an extended mountain position deserves to be court-martialled.

CHAPTER EIGHTEEN

Defense of Rivers and Streams

If we consider the defense of rivers and major streams, they belong, like mountains, in the category of strategic barriers. But they differ from mountains in two ways; one concerns their relative, the other their absolute, defense.

Like mountains, they reinforce a limited defense; but their peculiar characteristic is that they act like a tool made of a hard and brittle substance: they either stand the heaviest blow undented, or their defensive capacity falls to pieces and then ceases completely. If the river is very wide, and all other conditions are favorable, a crossing may be absolutely impossible. But once the defense is breached at any point, the kind of resistance in depth that would occur in mountains does not take place. The matter is settled in this single act, unless the river happens to flow through mountainous terrain.

The other attribute of rivers in their relation to combat is that they generally permit of more favorable, and in some cases excellent tactical possibilities for a decisive battle; usually better ones than do mountains.

What rivers and mountains have in common is that they are dangerous and alluring objects, which have often led to wrong decisions and into dangerous situations. When we come to a more detailed discussion of river defense, we shall call attention to these implications.

Historical examples of the successful defense of rivers are fairly rare, justifying the view that they are not such formidable barriers as people used to think in the days when systems of absolute defense used every means of reinforcement offered by the terrain. Still, a river is undoubtedly an asset to the engagement as well as to the defense of the country in general.

In order to provide some cohesion and perspective, we shall list the various aspects from which the subject will be examined.

First, and generally, the strategic value provided by the defense of rivers must be distinguished from the influence they exert on the defense of the country without themselves being defended.

The significance of the defense proper may be of three different types:

1. Absolute resistance by the main force
2. A mere show of resistance
3. Limited resistance carried out by subordinate elements such as advance posts, covering lines, detached corps, and so forth.

Finally, we have to distinguish three main degrees, or types, that the form of the defense may take:

1. Direct defense intended to prevent a crossing
2. A more indirect form, in which the river and its valley serve only as components for a more favorable tactical development
3. An absolutely direct defense, which consists of holding an unassailable position on the enemy side of the river.

These three degrees will form the framework of our discussion; and when each has been examined in the light of the first and most important consideration, we shall conclude by taking up the other two considerations. First, then, let us look at the direct defense, which attempts to prevent the enemy army from crossing the river.

This can only apply in the case of major rivers—that is, great bodies of water.

The combination of space, time, and strength that must be considered as the basic elements of this theory of defense makes this a fairly complicated matter. Consequently, it is not easy to find a fixed point of departure. Upon careful thought, one will arrive at the following conclusion.

The intervals at which the units defending the river should be stationed are determined by the time required to build a bridge. One must divide the total length of the defensive line by these intervals in order to find the number of units; then divide this number into the total strength available, to find the strength of each individual unit. By comparing that figure with the number of troops with which the enemy can cross the river by using other means while the bridge is being built, one can gauge the chances of a successful defense. Unless the defense is able to attack any enemy units that get across before the bridge is finished, in really superior strength—say two to one—it would be dangerous to assume that the enemy could not force a crossing.

Suppose, for instance, that it will take the enemy twenty-four hours to build his bridge. If he cannot get more than 20,000 men across in that time by using other means, and if the defense can concentrate that number at any point in twelve hours or so, no crossing can be forced; 20,000 men will be there by the time the enemy has ferried across half that number. Allowing for the time that messages will take, one can march twenty miles in twelve hours; 20,000 men would therefore be needed for every forty miles, or 60,000 men for the defense of 120 miles of river front. That would be sufficient for 20,000 men to be sent to any point even if the enemy tried to cross at two points simultaneously, and with twice that number if he did not.

The three governing factors are as follows: (1) the width of the river; (2) the means of crossing it, since both together govern the time it will take to build a bridge and the number of men that can get across while it is being built; (3) the strength of the defending force. The attacker's strength is not relevant at this stage. This theory would lead to the view that there is a point at which a crossing completely ceases to be possible and at which no degree of superior strength can force it.

This is the basic theory of the direct defense of a river—that is to say, a defense intended to prevent the enemy from finishing his bridge and from crossing the river by other means. It does not take into account the effect of any demonstrations that the enemy may employ. We shall now examine the particular circumstances and the measures required by this type of defense.

If, to start with, one disregards all geographical details, it will be enough to state that the units required, according to this view, must be stationed directly on the river bank, each one in concentrated formation. They must be on the river bank because any position further back adds needlessly to the distances that must be traversed. Since the width of the river covers the position against any serious enemy activity, there is no need to keep it at a distance like a reserve force in an ordinary defensive line. Besides, the roads running parallel to a river are generally more passable than those leading down to it. Finally, there is no doubt that this type of position will make possible better observation of the river than would a mere chain of posts, chiefly because all senior officers will be close at hand. Each unit must be kept concentrated, otherwise our calculations would have to be altered. Anyone who knows how long it takes to assemble a unit will recognize that to have the units already concentrated will assure the greatest effectiveness of the defense. At first sight it may be very tempting to set up a line of posts to stop the enemy crossing by boat; but save at the few points, especially suitable for ferrying, such a disposition would be most unwise. Aside from the danger that the enemy can generally reduce such a post by superior fire power from the opposite bank, it is likely to be a total waste of strength: all that is accomplished by such a post is that the enemy will choose a different point for crossing. Unless, therefore, one is strong enough to treat and defend the river like a moat around a fortress—in which case one needs no additional advice—this defense of the river bank itself will necessarily be unproductive.

In addition to these general principles of disposition, we must take into account, first, the individual characteristics of the river; second, the removal of all means of crossing; and third, the effect of fortresses on the river.

If one considers the river as a defensive line, it must have points of support at each end, such as the ocean or neutral territory, or other factors that will prevent the enemy from crossing above or below the defended sector. Such points of support or other conditions will occur only if the line is extremely long, and it becomes evident that the defense of rivers must extend over considerable distances. It is therefore not a practical proposition (and we need not bother with any other kind) to defend a river by massing a large force on a relatively short stretch of it. By *a relatively short stretch of river front,* we mean a distance not much greater than the normal extension of a position where there is no river. We maintain that cases of that sort do not occur; any direct defense of a river must always be extended until it amounts to a kind of cordon system. It is, therefore, ill-advised to counter any enemy envelopment by methods that would be natural in a concentrated deployment. Hence, where an envelopment is possible, the direct defense of a river, however promising under other circumstances, is a very risky affair.

As to the river between these limits, obviously not all points are equally suited to a crossing. We can discuss this further in a general way, but we cannot actually categorize the possibilities, since the slightest local variation often outweighs the most massive arguments in books. Such categorization would in any case be entirely useless; one look at the river, combined with information received from the local inhabitants, will provide guidance, and there is no need to resort to books.

In a general sense, we would say that the features that most favor a crossing are roads running down to the river, tributaries flowing into it, large towns located on its banks, and, above all, its islands. On the other hand, features that tend to be stressed in the literature, such as the greater elevation of one of the river's banks, or a bend in its course at the point of crossing, have seldom proved of great significance. The reason is that the influence exerted by these factors is limited to the narrow concept of an absolute defense of the banks, a matter that rarely, if ever, arises in the case of the largest rivers.

Anything that makes a crossing easier at one point than at another is bound to affect the position, and to modify the general mathematical rule in some respects; but it would not be wise to stray too far from this rule and rely too heavily on the difficulties presented at certain points. The enemy will choose the places least favored by nature if he can be sure that he will be least likely to meet us there.

In any case, one measure that can be recommended is the strongest possible occupation of the river's islands. A serious attack on them is the safest clue to the intended point of crossing.

The units posted on the river bank are expected to move upstream or down as the situation may require. If no existing road runs parallel to the river the improvement of the nearest parallel track, or, alternatively, the construction of short stretches of new roads, can be counted among the most important preparations the defense can make.

The second point under discussion is the removal of the means of crossing. This is not an easy matter on the river proper, and is, at any rate, very time-consuming. On the tributaries, especially those on the enemy's side, it is next to impossible, for they are usually already in the enemy's hands. It is most important, therefore, to seal off the mouth of every tributary with fortifications.

The means of crossing that the enemy brings with him—pontoons, that is—are seldom sufficient for major rivers. Consequently, a great deal will depend on the materials for building boats and rafts that he can find on the river itself and on its tributaries, in the large towns along its banks, and finally in adjacent woods. There have been cases in which all of these circumstances work against him to such an extent as to make a crossing virtually impossible.

Finally, there are the fortresses located on either bank or on the enemy's alone. They not only serve as protection against a crossing in their vicinity, whether up- or downstream, but also as a means of sealing off the tributaries and of storing material that could be used for crossing.

So much for the direct defense of rivers, which presupposes a large body of water. The addition of a deep and narrow gorge, or marshy banks, will, it is true, increase the difficulties of crossing and the effectiveness of the defense; but these can never replace a large body of water, for they do not constitute the major break in terrain which is the *first requirement* for direct defense.

The question arises as to the part played by such direct defense of a river in the strategic plan of a campaign. One must admit that it can never lead to a decisive victory: partly because its intention is not to permit the enemy to cross, but to crush the first substantial force he has landed; partly because the river itself prevents us from exploiting with an energetic counterattack, any advantages gained.

On the other hand, this type of river defense can often gain considerable time—and time, after all, is what the defender is most likely to need. It takes time to assemble the means of crossing. If several attempts at crossing fail, even more time will have been gained. If the enemy changes his direction because of the river, still other benefits will no doubt fall to the defense. Finally, in all cases where the enemy is not determined on an advance, the river will put a halt to his movements and serve as a permanent protective barrier for the country.

Where two substantial forces are involved, the river is broad, and conditions are favorable, the direct defense of a river can be considered an excellent device, and may yield results which, in recent times, have received too little attention due to failures that were caused by insufficient means. The above-mentioned requirements are, after all, easily met by rivers like the Rhine and the Danube. If one can maintain an effective defense against substantially superior forces over 120 miles of river front by means of 60,000 men, one may well consider it a noteworthy achievement.

Let us return once more to the phrase "*substantially superior forces.*" In the theory we have outlined, everything depends upon the means of crossing, and nothing on the force that seeks to cross, provided it is not inferior to the defending force. Strange as this may seem, it is nonetheless true. But one must not forget that most or practically all river defenses have no absolute points of support. They can all be turned; and great superiority in numbers will greatly facilitate the turning operation.

One must also remember that such a direct defense, even if it is overwhelmed by the enemy, cannot be equated with a lost battle. Even less can it lead to complete defeat: only part of our troops will have been involved, and the enemy, delayed by his slow passage across the bridge, cannot immediately follow up his victory. For all these reasons, one should not underrate this method of defense.

What matters in all practical affairs is to find the valid point of view.

Thus, in the defense of a river, it makes a great difference whether we have a correct impression of the whole position: some apparently trivial element may significantly alter the situation. What may have been a sound, effective measure in one case may be a disastrous mistake in another. The difficulty of judging everything correctly and refraining from assuming that one river is like another is perhaps greater in this instance than elsewhere. That is why we must constantly be on guard against the danger of applying the wrong methods or misinterpreting the facts. We must add unequivocally, however, that we consider it beneath our dignity to notice the clamor of those whose vague emotions and still vaguer minds impel them to expect everything from attack and movement, and whose idea of war is summed up by a galloping hussar waving his sword.

Even where they are actually justified, such ideas and feelings are not always enough (we need only cite the once famous "dictator" Wedel at Züllichau in 1759);[1] but what is worse, most of the time they are inapplicable. They leave the commander in the lurch at the very moment when he is beset by a mass of highly complex problems.

In our opinion, then, as long as one aims no higher than a modest negative, the direct defense of a river with a large number of troops and under the right conditions can bring about good results. But this does not apply to minor units. While 60,000 men along a given stretch of river front are able to stop 100,000 from crossing, 10,000 along the same stretch will not be able to stop a corps of 10,000—probably not even half that number provided these are willing to run the risk of placing themselves on the same side of the river with a defender so superior in numbers. The point is clear, for the means of crossing are the same in either case.

So far we have said little on the subject of feints, since they rarely play a role in the direct defense of a river. Part of the reason is that such a method of defense does not require the concentration of an army at one point, but gives each unit its own sector to defend, and partly it is because, under the conditions assumed here, the pretense of a crossing is an extremely difficult affair. Where the means for a crossing are themselves scarce—less than the attacker feels he needs to ensure the success of his operation—he can hardly want, or afford, to earmark a considerable part for a feint. In any case, it would diminish by that much the size of the forces he can get across at the real crossing point. The other side gains thereby in time what it might have lost through uncertainty.

The direct defense of a river is suitable as a rule only for the very largest European rivers, and only on the lower half of their course.

[1] In 1759 Frederick replaced the commander of the Prussian corps operating against the Russians with Karl Heinrich v. Wedel. Since Wedel was junior to other generals in the corps, Frederick wrote a letter that confirmed his supreme authority: "He, Lieutenant-General v. Wedel, represents in the army what a Dictator represented in Roman times." Wedel's position at Züllichau prevented the Russians from crossing the Oder there, but they threatened a crossing further north, and when Wedel, though outnumbered two to one, attacked them at Kay on 23 July, he was badly defeated. Eds.

The second form of defense is suited to minor rivers and deep valleys—frequently even for insignificant ones. It consists in taking up a position farther to the rear. The distance should be such as to make it possible either to catch the enemy army in separated units if it crosses at several points, or, if it crosses at a single point, to catch it close to the stream, where it is confined to a single bridge or road. An army whose rear is up against a river or cramped in a deep valley, which is limited to a single line of retreat, is in a most disadvantageous situation for battle. The defense of all moderate-sized rivers and deep valleys consists in exploiting these circumstances.

The deployment of an army in large units close to a river—which we consider best for direct defense—assumes that the enemy cannot cross by surprise and in great strength; otherwise, the risk of being separated and beaten individually would be too great. Thus, if conditions are not sufficiently favorable to the defense of the river, if the enemy can lay his hands on too many means of crossing, if the river has too many islands or even fords, if it is not wide enough, or if our forces are too weak, this method of defense must not be considered. The troops, in order to stay in close touch with one another, must be withdrawn some distance from the river. What remains to be done is to converge as rapidly as possible on the enemy's crossing point and attack him before he holds enough of the river bank to enable him to cross at several other points. In this case, the river or the valley must be watched and lightly defended by a chain of outposts, while the army, divided into several corps, takes up a position at appropriate points some distance from the river—normally a few hours' march away.

The important feature here is the passage through the narrow river valley. What counts is not only the body of water as such, but the passage as a whole. As a rule, a deep, rocky gorge is of greater significance than a river of considerable width. The difficulties presented by the march of a substantial body of troops through a narrow passage are actually much greater than they appear to be at first glance. The time it takes is considerable, and the risk that the enemy will meanwhile seize the surrounding heights is most disquieting. If the leading units get too far ahead, they will meet the enemy too soon and are in danger of being crushed by a superior force; if they remain near the crossing point, they will be in the worst possible position for fighting. Crossing such a divide with the idea of facing the enemy on the other side is therefore extremely daring or presupposes a great superiority in numbers and self-confidence on the part of the commander.

This sort of defensive line cannot, of course, be extended as far as it would be in the case of the direct defense of a major river: one wants to fight with the total force united, and no matter how difficult the crossing points they cannot be compared to those of a major river. The enemy is therefore in a much better position to turn our line. On the other hand, this will take him away from his real direction (assuming, of course, that it runs approximately at right angles to the divide) and the handicap of a narrowed line of retreat is not overcome all at once, but only by degrees. The defender therefore still retains a few advantages over the attacker even if he does not catch him at the critical stage, but only after his envelopment has given him somewhat greater scope.

When speaking of rivers, we are concerned not only with the body of water, but, almost more to the point, with the deep depressions formed by their valleys. We must, therefore, above all make it clear that we do not mean regular mountain valleys, since in that case everything that has been said about mountain warfare would apply. But there is much open country where even the smallest streams run between high, precipitous banks. Besides, marshy banks and other obstacles to approach belong in this category.

Under such conditions, therefore, the position of a defending army behind a fair-sized river or a deep valley is very advantageous; this type of river defense must be counted among the best strategic devices.

Its weakness, the point on which the defender may easily go wrong, lies in the overextension of his forces. It is only natural, in such a situation, to string out one's forces from one crossing point to the next, and not to know where to stop. But if one cannot fight with the army united, the whole enterprise has failed. A lost engagement, an unavoidable retreat, confusion, and casualties of all kinds may bring the army to the brink of total disaster, even if it does not fight to the last.

It is enough to say that one should not extend one's forces too far, and that, in every case, one must be able to assemble one's troops by the end of the day on which the enemy has crossed. This principle will take the place of all further discussion about time, strength and space, which depend on a variety of local factors.

The battle resulting from such conditions is bound to have one peculiar characteristic: the defender must show the utmost impetuosity. The feints with which the enemy may well have kept him guessing for a time will generally allow him to get to the right place only at the last minute. The special advantages of his situation lie in the difficult position of those enemy troops that are directly opposite him. If additional forces arrive from other crossing points and envelop him, he cannot deal with them in the normal way by sustained counterattacks from the rear. If he did, he would sacrifice the advantages of his position. He must decide the issue before these additional troops begin to press him—in other words, he must attack whatever troops are before him with the utmost speed and vigor, and through their defeat reach a decision for the encounter as a whole.

One must remember that the objective of *this* type of river defense can never be to resist a vastly superior force, as it might perhaps be in the case of the direct defense of a major river. Usually one will have to deal with the largest part of the enemy's force and even if this happens under favorable conditions, it is easy to see that the disparity of strength must be reckoned with.

This holds true in the defense of medium-sized rivers and deep valleys where large forces are involved; forces which seek a decisive victory and for whom the effective resistance that can be sustained on the rim of the valley bear no comparison with the drawbacks of a dispersed position. If all that is needed, however, is the reinforcement of a secondary line of defense, which is meant to resist for a time and depends on the arrival of reinforcements, a direct defense of the ridges and even of the river bank would indeed be in order. While one cannot expect the advantages of a mountain position, resistance here can be kept up longer than it would in ordinary country. The one condition under which it can be really risky, or impossible, is where the river winds in hairpin bends, which is just what rivers in deep valleys are apt to do. (Consider the course of the Moselle in Germany.) In such a case, the units holding the salients formed by the bends would almost certainly be lost in the event of a retreat.

A major river obviously offers the same possibilities of defense as we have attributed to rivers of medium size where the bulk of an army is engaged, and under much more favorable conditions. Such a defense will invariably be employed where the defender aims at total victory. Aspern is a case in point.

A completely different case arises when an army occupies a river, a stream, or a deep valley *immediately* to its front, in order to gain a tactical obstacle to approach, *a tactical strengthening of its front.* A closer study of this belongs to the realm of tactics, but in terms of its effectiveness, we can only call it pure self-delusion. If the divide is great enough, it will make the position's fronts impregnable, but since it is no more difficult to by-pass than any other, the effect is almost as if the defender had evaded the attacker—which was hardly the point of occupying the position in the first place. This type of position, therefore, is of use only where local conditions make the attacker's lines of communication so unfavorable that any departure from the most direct route would involve unacceptable consequences.

In this second form of defense feints constitute a much greater threat. The attacker will find them easier to make, while the defender will still have to concentrate his whole force at the real point of crossing. However, the defender will not be quite so pressed for time, for the advantage will remain with him until the attacking force is fully massed and has taken several crossing points, while enemy feints will never be so effective as they will be with a cordon-defense, where no ground can be yielded at all. When it comes to using the reserve, therefore, the problems are very different. In one case, it is simply a matter of knowing the whereabouts of the main enemy force; in the other, it is the far more difficult problem of guessing which will be the first point to be overrun.

We would add a general comment on the subject of either form of defense of major or minor rivers: if they have been adopted in the hurry and confusion of retreat, without preparation, without taking away the means of crossing, and without familiarity with the terrain, they cannot possibly yield the results described above. Usually nothing of the sort can be expected, and so it will be a grave mistake to spread a force too thin over an extended position.

In any case, since everything is apt to go wrong in war unless it is done with full awareness, firmly and wholeheartedly, the same will hold true of defending a river for fear of meeting the enemy in open battle and in the hope that the width of the river or the depth of the valley will stop him. Such decisions show a lack of confidence in the situation; they often fill the general and the army with dire forebodings, which usually come true only too quickly. After all, a battle in open country is not like a duel that presupposes equal terms: the defender who is unable to find an advantage by exploiting the special nature of defense, or by using rapid marches, or by familiarity with the terrain and freedom of movement, has little to hope for. Least of all can he look to a river or its valley for salvation.

The third form of defense is by means of a strong position that one holds on the enemy's side of the river. Its effectiveness is based on the risk incurred by the enemy that the river traverses his lines of communication, once he had crossed it, and thus would limit him to one or two bridges. Obviously, this will be the case only with major rivers that run broad and deep; it would not apply to a river with a narrow valley, which usually has many crossing points.

The position must be strongly fortified—practically impregnable. Otherwise we would play into the enemy's hands, and our advantage would be lost. If, however, it is strong enough to deter the enemy from attack, the effect may be to tie him down to the bank. If he were to cross, he would expose his lines of communication—though, of course, he would also threaten the defender's. Here, as in all cases where two armies pass each other by, the crucial question is whose lines of communications are the more secure—in number, position, and other respects. In addition, it depends on which side has more to lose, and is therefore more easily outbid by the other; and finally, whose army retains the greater determination on which it can draw as a last resort. The river contributes nothing, except to increase the danger of any such movement for both sides, because both are confined to bridges. Insofar as one can normally assume the defender's crossing points and his various depots to be better fortified than his opponent's, this is a perfectly feasible form of defense which will suffice where other circumstances do not favor a direct defense. Admittedly, it means that the army is not defended by the river, or the river by the army; but the country is defended by the combination of the two, which is what really matters.

We must allow, however, that this form of defense, in which there is no decisive blow, is like the tension set up in the atmosphere between positive and negative electric currents: it will only be able to stop a blow of minor proportions. It might suffice against a cautious, hesitant general who is not compelled to press on even when he has greatly superior strength; it might also do if the armies were already in a state of balance, with neither of them looking for more than minor advantages. But as a means of coping with superior numbers and a dashing general it is a dangerous course, leading close to disaster.

This method of defense carries such an air of boldness and appears so scientific that one might almost call it elegant; but since elegance easily comes close to fatuousness—which is less excusable in war than in society—

few examples of this elegant method exist. It can, however, be developed into a special means of support for the first two methods: by holding a bridge and a bridge-head, one can always threaten a crossing oneself.

Apart from the purpose of absolute defense with the main force, each of these three forms of river defense may have a further one: that of *feigned defense*.

An empty show of resistance can, of course, be used in connection with a number of other measures, and basically with any position that is not simply an overnight camp. But the feigned defense of a great river becomes an effective deception if it involves a number of more or less complex measures. The effect is usually larger in scale, and lasts longer than in other cases. The act of crossing a river in the face of an enemy is always a serious decision for the attacker. He is apt to consider it at length or postpone it until a more favorable time.

A feigned defense requires that the main force should deploy along the river in approximately the same way as it would in the case of a real defense. However, the intention of a mere feint proves that circumstances are not favorable enough for a real defense. It follows that the positions you take up—which are inevitably more or less extended and scattered—may well give rise to serious losses if the units really get involved in resistance, on however limited a scale. That would actually be a half-measure. Therefore, in a feigned defense everything must be calculated in terms of a real concentration of the army at a point considerably further to the rear—frequently as far as several days' march. One can render only as much resistance as is consistent with that plan.

To explain exactly what we mean, and at the same time show the significance that such a show of resistance can have, we recall the final phase of the campaign of 1813. Bonaparte had returned across the Rhine with 40,000 to 50,000 men. With so small a force it would have been impossible to defend the length of this river between Mannheim and Nijmwegen—the stretch where, according to the general direction of its forces, the allied army would be most likely to cross. The only practical thing Bonaparte could do was to plan his first real stand on the French part of the Meuse, where his army could expect reinforcements. If he had withdrawn to that line at once, the allies would have followed hard on his heels; the same would have happened before long if he had sent his troops to rest-camps on his own side of the Rhine. No matter how cautious and faint-hearted the allies might have been, they would have sent swarms of Cossacks and other light troops across, and if these had succeeded other units would have followed. In consequence the French had no choice but to prepare to defend the Rhine in earnest. Since it was to be expected that, as soon as the allies really started to cross, nothing would be accomplished by this defense, the whole maneuver has to be considered as a mere show of resistance in which the French, in fact, were risking nothing, since their point of assembly was located on the upper Moselle. Only Macdonald, stationed at Nijmwegen with 20,000 men, made the mistake of waiting to be driven out. Because of the late arrival of Wintzingerode's corps, this did not happen until the middle of January, and prevented Macdonald from rejoining Bonaparte before the battle of Brienne. The feigned defense of the Rhine, then, sufficed to bring the allies to a halt and make them decide to postpone the crossing until the arrival of reinforcements—a period of six weeks. These six weeks must have been of incalculable value to Bonaparte. Without the show of resistance on the Rhine, the battle of Leipzig would have led the allies straight to Paris; a battle anywhere east of Paris would have been quite beyond the powers of the French at the time.

A demonstration can also be made with the second form of river defense—one that involves a river of medium size. But it will normally be much less effective, because the mere attempts at crossing are easier and therefore the game is given away sooner.

In the third form of river defense, the demonstration would probably be even less effective. It would hardly be of more use than any other temporary position.

Finally, the first two forms of defense are well suited to confer much greater strength and security on a chain of outposts or other defensive line established for some secondary purpose (a cordon), or even on a small observation corps, than these would possess without the river. In all these cases, we are talking only about relative resistance, which will become much more effective wherever a break in the terrain exists. But we must keep in mind not only the fairly long time gained by resistance during the actual engagement but also the many doubts that accompany the planning of the attack, which in ninety-nine cases out of a hundred will cause it to be cancelled unless there are urgent reasons to proceed.

CHAPTER NINETEEN

Defense of Rivers and Streams—Continued

We should like to add a few observations concerning the effect of rivers and streams on the defense of a country, even if they themselves are not defended.

Any important river valley, together with its tributaries, constitutes a substantial natural obstacle, and as such it is generally an asset for the defense; but the salient features of its actual role can be defined in greater detail.

To start with, we must establish whether the river runs parallel, diagonal, or at right angles to the border—that is, to the main strategic front. If it does run parallel, we must distinguish whether it runs behind the defender's army or the attacker's; and in either case, it makes a difference how far away the army is from the river.

A defending army with a major river close (but no less than a normal day's march) behind it, a river on which it has secured a sufficient number of crossing points, is undoubtedly in a much stronger position than it would be without the river. While concern for its crossing points may deprive it of some liberty of movement, it will gain a great deal more through the security of its strategic rear, especially of its lines of communication. It must be understood that we are talking about defense on one's own territory; on hostile territory, even where the enemy army is in one's front, one must expect him in one's rear as well once one has crossed the river. Then the river would be more of a disadvantage than an advantage, because it constricts our communications. The further distant the river is from the army, the smaller its usefulness; and at a certain distance, its value vanishes altogether.

If an advancing army has to leave a river in its rear, the river is bound to impede its movements, because the lines of communication will be limited to a few crossing points. In 1760, when Prince Henry marched against the Russians along the right bank of the Oder near Breslau, the river, which was only a day's march to his rear, clearly gave him a point of support. Later on, by contrast, when the Russians crossed the Oder under Chernichev, they were in a most uncomfortable position simply because of the risk of losing their line of retreat, which depended on a single bridge.

Where a river runs more or less at right angles across a theater of operations, the advantage again lies on the side of the defender. To start with, he will usually have a choice of good positions by using the river as support and the tributary valleys as reinforcements for his front (as the Prussians used the Elbe in the Seven Years War); second, the attacker must either leave one side of the river alone or split his forces. In the latter case, the defender will doubtlessly be favored by virtue of having a larger number of safe crossings than the attacker. A glance at the Seven Years War is enough to demonstrate that the Oder and the Elbe were of great help to Frederick in defending his theater of operations—that is to say, Silesia, Saxony, and the Mark—and, conversely, constituted a decided obstacle to the Austrian and Russian conquest of these provinces. Yet neither river was really defended during the course of the war. Moreover, for the most part, both ran diagonally or at right angles to the enemy's front more frequently than they ran parallel to it.

Generally speaking, a river's role as a means of transportation is its most favorable aspect as far as the attacker is concerned, provided it runs at right angles to the front: his lines of transportation are longer and he therefore has greater problems in moving up his supplies. Water transport will, therefore, come as a relief and be an advantage. It is true that here too the defender has the advantage of being able to close the river to traffic by means of fortresses from the border on; but this does not counteract the benefits that the attacker reaps up to that point. Still, several factors must be called to mind. A river may be wide enough to be of some military significance without necessarily being navigable; it may not be navigable all year round; river traffic upstream is extremely slow and often difficult; frequent bends may more than double the distance to be traveled; highways nowadays serve as the main arteries between two countries; and finally, the bulk of an army's needs is at present raised through local requisitioning rather than through commercial procurement from distant places. These considerations make it clear that water transport plays a much smaller part in the supply of armies than textbooks would have us believe. Its effect on the course of events is therefore quite remote and hard to measure.

CHAPTER TWENTY

A. Defense of Swamps

Great expanses of marsh like the Bourtang Moor in northern Germany are so rare that we need not spend much time on them; but one should not forget that certain types of low-lying country, and minor rivers with marshy banks, occur with greater frequency. These can amount to substantial segments of the terrain suitable for defense and indeed are all too often used for this purpose.

The principles of using them defensively are of course much the same as for rivers; still, they have some special features that must be noted. The first and foremost is that unless there are dikes, a swamp is impassable for infantry. It is far more difficult to cross than any river. For one thing, a dike is not so quickly built as a bridge; for another, there are no temporary means by which troops can get to the far side to cover its construction. One would never start to build a bridge without using boats to ferry over an advance guard, but with a swamp this is impossible. The simplest way of getting infantry across is by means of planks; but that is a tedious business, and if the marsh is of some width, it takes infinitely longer than getting the first boats across a river. If in the middle of the swamp there is also a river that cannot be crossed without a bridge, the task of getting the first troops across becomes even more difficult: planks may be enough for men to cross one by one, but not for transporting heavy bridging material. Under some circumstances this difficulty may prove insurmountable.

Another characteristic of marshes is that one is never able to demolish a means of crossing entirely, as one can in the case of a river. One can dismantle a bridge, or destroy it enough to make it unusable, but the most one can do to a dike is to breach it—which is not doing much. If a stream flows through the swamp, one can indeed demolish the bridge crossing it, but this will not impede the crossing as a whole to the same extent as would the destruction of a bridge over a major river. The natural consequence is that existing dams must always be fairly strongly held and seriously defended if the marsh is to be of any military advantage.

On the one hand, one is thus restricted to purely local defense; this is facilitated, on the other hand, by the difficulties of crossing elsewhere. These two factors combine to make the defense of swamps more local and more passive than that of rivers.

Consequently, one will have to muster greater strength, relatively speaking, than for the direct defense of a river. In other words, one cannot rely on so long a line of defense, especially not in closely settled parts of Europe, where even under the best of circumstances the crossing points tend to be quite numerous.

In that respect, then, marshes are not so useful as great rivers, and this is an important difference; since there is always something very insidious and dangerous about localized defense. Yet one must bear in mind that most marshes and bogs are a great deal wider than the widest rivers of Europe; that therefore there is never any danger that a post set up to defend a crossing point will be neutralized by fire from the opposite bank; that on a long narrow dike the effect of one's own fire is considerably raised; and that, in general, incomparably greater delays will be encountered in filing along a narrow track a mile or two long than in crossing a bridge. It must be admitted then—provided there are not too many crossing points—that such swamps and marshes are among the strongest lines of defense possible.

An indirect defense of the type discussed in connection with rivers and streams—using a natural obstacle for the favorable initiation of a major battle—may just as easily be applied in the case of swamps.

However, the third form of river defense—by means of holding a position on the enemy side—would be too dangerous because it takes so long to cross the swamp.

It would be most hazardous to get involved in the defense of any swamps, water meadows, bogs, or marshes that might possibly be crossed by passages other than dikes. The enemy's discovery of a single such crossing point may be enough to breach the whole line of defense; which in the case of serious resistance will lead to heavy losses.

B. Inundations

We still have to consider inundations, which closely resemble swamp conditions, both as a means of defense and as a natural phenomenon.

Admittedly, inundations are rare. The Netherlands are quite possibly the only country in Europe where they constitute an element that deserves our attention. Indeed, it is this very country which, because of the remarkable campaigns of 1670 and 1787, and because of its geographical relation to Germany and France, obliges us to devote some consideration to the matter.

The characteristics of flooding in the Netherlands differ from ordinary swampy and impassable bogs in the following ways:

1. The country itself is dry, and consists either of dry meadows or of fields under cultivation.

2. The land is intersected by a number of irrigation channels and drainage ditches of varying width and depth, running parallel to one another in certain areas.

3. Major canals for the purposes of irrigation, drainage and navigation traverse the land in all directions. They run between embankments and can only be crossed by bridges.

4. The elevation of the ground in the whole area subject to flooding is well below sea level, and consequently also below the level of the canals.

5. It follows that the whole area can be flooded by breaching the dams and opening or closing the sluice gates. Only the roads that run along the higher dams will remain dry. The rest are either under water or become so waterlogged as to be unfit for use. Even where the flood level is no higher than three or four feet, possibly enabling one to wade short distances, this is prevented by the smaller ditches mentioned at point 2 above, which can no longer be seen. Only where these ditches run in the right direction so that one can wade between them without having to cross any of them, can flooding be considered anything but a total barrier to access. Naturally, one will be able to wade only over very short stretches, and will thus be restricted to specific tactical purposes.

It follows from all this that:

1. The attacker is confined to a relatively small number of access routes, which run along rather narrow dikes; since, in addition, these are usually flanked by ditches, they amount to endlessly long and dangerous defiles.

2. Every defensive measure taken on dikes of that sort can very easily be reinforced to the point where they become impregnable.

3. Because of his limitations, the defender—even at each individual point—must restrict himself to mere passive resistance: it is his only chance.

4. This is not a matter of one long line of defense, enclosing the country like a single barrier. All flanks are equally covered by the difficulty of access, and new posts can always be set up so that any breach in the original line can be sealed off. One might almost say, that, as on a chessboard, the range of combinations is inexhaustible.

5. This entire condition is possible only in a densely cultivated and populated country. It follows automatically that the number of crossings and, consequently, of posts to close them, must be very large in comparison with other strategic dispositions; and from this it follows further that that type of defensive line should not be a long one.

The principal defensive line in Holland runs from Naarden on the Zuider Zee (mostly behind the Vecht) to Gorkum on the Waal—on the Biesbosch, to be accurate—for a length of about forty miles. In 1672 and 1787, a force of 25,000 to 30,000 men was used to defend it. If really insurmountable resistance could be expected, its value would certainly be very great—at least to the province of Holland in its rear. The line held fast in 1672 against a far greater force under eminent generals—first Condé, and then Luxembourg. They could well have attacked with 40,000 to 50,000 men; yet they preferred not to use force but to wait for the winter—which, as it happened, was not severe enough. In 1787, on the other hand, resistance in the first line was nonexistent; and although on a much shorter line, between the Zuider Zee and the Haarlemer Meer, it was somewhat more serious, the Duke of Brunswick broke it in a single day by means of carefully planned tactical dispositions that exactly suited the local conditions; this in spite of the fact that the Prussian force that actually advanced against the line was little stronger, if at all, than the defense.

The difference in the outcome of these two occasions may be ascribed to the difference in the supreme command. In 1672, when Louis XIV surprised the Dutch, they were on a peacetime footing. In the army, as is well known, military spirit was not high. The majority of fortresses were short of ammunition and equipment; they were held by weak garrisons of mercenary troops, and commanded either by foreigners devoid of loyalty, or nationals without ability. Therefore, the fortresses on the Rhine, belonging to Brandenburg but occupied by the Dutch, as well as their own eastern defensive line, except for Groningen, quickly fell to the French and, for the most part, without real resistance. The capture of these numerous fortresses constituted the chief activity of the French army of 150,000 men.

But when in August 1672 the brothers De Witt were assassinated, the Prince of Orange came to power and brought the national defense under unified command. He had just time to seal the defensive line described above, and arrangements were so well coordinated from then on that neither Condé nor Luxembourg (who commanded the forces remaining in Holland after Louis XIV and Turenne had left) dared attack a single post.

In 1787, conditions were quite different. It was not the Republic of the Seven United Provinces that faced the French: the Province of Holland alone opposed the aggressor, and was to put up the chief resistance. It was not a matter of capturing fortresses, which had made up the main activity in 1672; the defense fell back at once on the line described above. For their part, the invaders were only 25,000 strong instead of 150,000; they were not led by the mighty sovereign of a great neighboring power but only by a general, subordinate to a distant prince who himself was bound by various restraints. The people, it is true, were split into two parties everywhere, even in Holland; but the Dutch republicans were in the majority, and in a state of genuine enthusiasm. Under these conditions, resistance should have achieved at least as much in 1787 as in 1672. But there was a fatal difference: there was no unified command. In 1672 William of Orange had been entrusted with the command, and executed it with competence, intelligence and energy. In 1787 reliance was placed on a so-called Defense Commission, which, though consisting of four energetic men, was quite unable to bring about unified direction or to inspire confidence. Hence, the whole apparatus proved deficient in use and unreliable in action.

We have dwelled on this example to clarify this method of defense; at the same time, we wanted to demonstrate how much difference unity and consistency in the leadership can make.

Although the organization and operation of such defensive lines is a question of tactics, the defensive line itself is rather more closely related to strategy, and we should like to make one observation about it, arising from the 1787 campaign. We believe that, passive as the defense of individual posts must inevitably be, a counterattack from some part of the line would be not impossible, and hold out good prospect of success if as in 1787 the enemy is not numerically stronger. Such an attack could only be made along the dikes and could hardly have much freedom of movement or impetus; still, the invader cannot occupy all the dikes and tracks on which he is not advancing. Thus the defender, who occupies the fortresses and knows the country, should be able to launch serious flank attacks against the enemy or cut off his supplies. If we consider the constricted circumstances in which the advance operates, and particularly its unusual dependence on lines of communication, we can understand that any counterattack with even a remote prospect of success must be highly effective even as a demonstration. We greatly doubt whether so cautious a man as the Duke of Brunswick would have dared to advance on Amsterdam if the Dutch had made a single demonstration of that sort, for example from Utrecht.

CHAPTER TWENTY-ONE

Defense of Forests

One must above all distinguish between dense, impenetrable, overgrown forests, and extensive, cultivated woods that may have numerous clearings and be traversed by a large number of paths.

Whenever a defensive line is planned, woods of the latter kind should be kept in rear or strenuously avoided. It is in the interest of the defender, even more than of the attacker, to command an unimpeded view, partly because he is normally the weaker of the two, and partly because the natural advantages of his position lead him to develop his plans later than the attacker. If he were to fight with a forest in front of him, he would become like a blind man fighting a man who can see. If he took up position in the middle of the forest, both of course would be equally blind; but this equality would be detrimental to his interests.

A forest of that sort, therefore, cannot bear any useful relationship to the defense unless it is kept in the defender's rear. Then it can be used as a screen for whatever movements are in train, and to cover and facilitate his eventual retreat.

These remarks apply, of course, only to forests in the plains. Wherever the terrain is mountainous it will predominate in tactical and strategic arrangements. This has already been discussed elsewhere.

Impenetrable forests, on the other hand—or rather, forests where one must keep to the roads traversing them—do present opportunities for indirect defense similar to those offered by mountains: one can initiate a battle when conditions are favorable. The army, in a more or less concentrated position, can wait behind the forest for the enemy to appear and attack him as he emerges from the narrow road. In its effects, such a forest is more like mountains than like a river: it is very slow and difficult to traverse, but, as far as a retreat is concerned, it is rather an asset than a danger.

No matter how impenetrable a forest, however, its direct defense is still a risky matter, even for the lightest chain of outposts. Abattis are psychological obstacles only, and no forest is so impassable that small units cannot infiltrate it in hundreds of places. In a defensive chain, these are like the first few drops of water leaking through a dam: a general breakthrough is sure to follow.

The influence of great forests of all kinds becomes infinitely more important in the case of a national insurrection, for which they are unquestionably the right environment. If a strategic plan of defense can be contrived that makes the enemy's lines of communication run through deep forests, a powerful lever has been added to the machinery of defense.

CHAPTER TWENTY-TWO

The Cordon

By a cordon we mean any system of defense in which a series of interconnected posts is intended to give direct protection to an area. We stress the word direct, since several corps of a large army stationed in line with one another could protect a considerable area from enemy incursions without constituting a cordon. In that case, protection would not be direct but would result from concerted movements and maneuvers.

A defensive line, long enough to cover an extensive area directly, will obviously be able to render only minimal resistance. This would be true even if large numbers were deployed, provided they were opposed by equally numerous forces. The intent of a cordon, then, is to withstand a slight attack—slight either because the attacker is easily discouraged or because the attacking force is small.

This was the function of the Great Wall of China: a protection against Tartar raids. It is also the significance of all the lines and frontier defenses of the European States bordering on Asia and Turkey. In this sense, cordons are neither absurd nor ill-adapted to their purpose. Admittedly, they will not be able to prevent every raid; but all incursions are made more difficult, and they therefore occur less frequently—which is an important consideration in the relations prevailing with Asiatic peoples, where a state of war is virtually permanent.

The lines that were erected in the recent wars between European states—for instance the French lines on the Rhine and in the Netherlands—come close to being cordons in that sense. Basically, they are meant to protect the country from inroads intended merely to levy contributions or live off the enemy. Since they are meant to deal only with minor incursions, they will only have minor forces at their command. Of course, wherever the enemy's main force is directed against these lines, they will have to be held by the defender's main force as well—and that is far from being the best way of organizing a defense. Because of this drawback, and because the protection against raids is of extremely limited importance in a short war in which the existence of such lines may entail an overexpenditure of forces, lines of that sort are nowadays regarded as a disadvantageous arrangement. The greater the fury with which the war is waged, the more dangerous and futile they become.

Finally, an extended line of outposts designed to cover an army's billeting area and to offer a certain degree of resistance, should also be considered as a true cordon.

Resistance of that type is mainly meant to cope with raids and other minor operations aimed at making billets unsafe, and if the terrain is right, it can serve the purpose well enough. Against the main force of the enemy, on the other hand, resistance can be only relative—concerned, in other words, with gaining time: but any time that is gained will usually not be significant. It will, therefore, seldom be considered the purpose of a cordon. The enemy force can never assemble and advance so secretly that the defender's first news of it would come from his outposts. If that were to happen, one could only feel very sorry for him.

So even in this case the cordon is only meant to meet a light attack; as in the other two instances there is nothing contradictory about it.

But for the main force, intended to defend the country, to be strung out in a long series of defensive posts against the enemy's main army—in fact, in a cordon—would be so absurd that one would have to investigate the immediate circumstances accompanying and explaining such an occurrence.

In the mountains, any position, even one occupied with the intention of fighting a battle with the fullest possible strength, is bound to be more extended than it would be in the plains. This is possible because the terrain enormously enhances the defensive potential; it is required because a wider base is needed for retreat, as we have seen in the chapter on defensive mountain warfare. However if no battle is imminent, the enemy is likely to be facing us for a considerable time without making any move unless a favorable opportunity arises—which has been the normal condition in most wars. In such a situation, one will naturally not want to be confined to the occupation of a minimum area: one will want to be in control of as much of the country on all sides as the safety of one's army permits. This gives rise to a number of benefits to be discussed in detail later on. In open, unobstructed country, it is made easier than in the mountains by the element of *mobility*; consequently, there will be less need for the extension and splintering of forces to accomplish this purpose. It would also be much more dangerous because each individual unit would have a reduced capacity to resist.

In the mountains, however, possession of ground turns more on local defense; it is harder to reach a threatened point; and if the enemy gets there first, he is not easily dislodged even by a slightly superior force. Such conditions will tend to lead to arrangements which, while not exactly a cordon, do consist of a series of defensive posts very much like one. From such a disposition, consisting of several detached posts, to a cordon is still a long step; but it is one that is frequently taken by a general without his realizing it—by being lured from one stage to the next. To begin with, the object of such dispersion is the security and possession of the country; then it is the safety of the troops themselves. Each commander will weigh the advantages of occupying various points of access to either side of his post, and so the whole force gradually slides imperceptibly from one degree of dispersal to the next.

It follows that a cordon-war involving the main force should not be seen as a matter of deliberate choice, designed to stop all enemy attacks, but a condition into which one is drawn in the pursuit of a completely different objective—the preservation and security of the country against an enemy who is not intent on major action. That is always wrong, and the arguments by which a general lets himself be coaxed into forming one small post after another are always trivial compared with the importance of keeping his main force intact. Yet the existence of these arguments shows at least the possibility of such confusion. The fact that it really is a mistake—a misjudgment of the enemy and of one's own position—is apt to go *unnoticed*; what is blamed is the faulty *method*. But the method is tacitly approved wherever it has been used to advantage, or at least without doing any harm. Everyone praises the *faultless* campaigns fought by Prince Henry in the Seven Years War, because that is what the King called them; actually, these campaigns include the most extreme and incomprehensible examples of extended chains of posts, which deserve the name of cordon as much as any. One can fully justify these positions by saying that the Prince knew his opponents and was sure that no decisive action was to be expected; moreover, the object of his disposition was always to control the largest possible area, and he therefore went as far as circumstances would possibly allow. Still, if the Prince had once been caught in such a cobweb and had sustained a heavy loss, one would have had to conclude, not that he had followed a faulty system of warfare, but rather that he had used the wrong measure, applying it to a situation to which it was not suited.

So much for our attempt to indicate how a so-called cordon-system can originate with the main force of a theater of operations and be sensible and useful enough not to appear absurd. We must add that there seem to have been occasions when commanders and their general staffs have overlooked the real significance of a cordon-system, treating its relative value as absolute, and believing it capable of holding off all enemy attacks whatever. These were not misapplications of a method, but a total inability to grasp its nature. We admit that this absurdity, among others, seems to have played a part in the defense of the Vosges by the Prussian and Austrian armies in 1793 and 1794.

CHAPTER TWENTY-THREE

The Key to the Country

There is no theoretical concept in the art of war dearer to the hearts of critics than the one under discussion here. It has been the prize exhibit of innumerable accounts of battles and campaigns, the favorite theme of all arguments—one of those pseudoscientific terms with which critics hope to show their erudition. Yet the underlying concept has neither been established, nor even clearly defined.

We shall attempt a clear exposition of the concept and consider what practical value it retains.

We deal with it at this point, since it is closely bound up with defense of mountains and rivers, as well as the concept of fortified and entrenched positions, which we have just discussed.

Behind the hoary martial metaphor of "key to the country" there lurks a vague, confused idea, sometimes denoting the most widely open region of a country, and sometimes the region most strongly defended.

If there is an area *without the possession of which one cannot risk an advance into enemy territory*, it may correctly be designated as the key to the country. But this simple and frankly not very valuable concept has not been enough for the theoreticians; they have raised it to a higher power and used it to denote points that will put you in possession *of the whole country*.

When the Russians set about invading the Crimean peninsula, they had to start by capturing the Isthmus of Perekop and its defensive lines—not because there was no other way of entering (Lacy, in fact, had turned the lines in 1737 and again in 1738) but in order to be tolerably safe once they had established themselves in the Crimea. That is plain enough, though the concept of a key-point hardly adds to our understanding. However, if one were able to say that whoever occupies the area around Langres owns or is in command of all of France as far as Paris—in other words, once Langres is occupied it depends only on him whether or not to take possession of the country—it would obviously be a very different and much more important claim. According to the first view, possession of a country is considered impossible without the point designated as the key. That stands to reason. According to the second view, however, possession of the country is the inevitable consequence of possession of the point designated as the key. That is obviously mysterious, beyond the bounds of normal understanding, requiring the magic power of the occult sciences. This incantation actually began to appear in print about fifty years ago and reached its zenith at the end of the eighteenth century. Despite the overwhelming force, assurance and logic with which Bonaparte's leadership swept away previous military conceptions, that magic formula contrived to keep a tenuous hold on life and continued to spin its fragile thread in the literature.

Let us ignore *our* definition of a key point: it is obvious that in each country there are some points of *exceptional* importance, where a number of roads converge, where it is easy to stockpile supplies, whence one can conveniently move in several directions; in short, whose possession satisfies a number of needs and offers a number of advantages. If generals wanting to describe the importance of such a point in a single word wish to call it a key to the country, it would be pedantic to object; on the contrary, the expression is apt and agreeable. But if this little flower of speech is blown up to form the core of a whole system, branching out in various directions like a tree, sheer common sense should warn you to restrain yourself and keep to the proper value of the term.

The memoirs of wars and campaigns written by generals use the concept of a key to the country in a practical but, on the other hand, very imprecise sense. In order to develop it into a system, one had to transform it into a more specific and consequently more limited term. From all available aspects, that of high ground was chosen.

When a road traverses a mountain ridge, the traveler breathes a sigh of relief once he has reached the summit and the descent begins. This holds true of an individual, and even more so of an army. All difficulties seem to be, and usually have been, overcome. We feel that the descent will be easy, and we shall be able to conquer any obstacle in our path. The country is spread out before us and appears to be at our feet, metaphorically as well as physically. Thus, the highest point of a road across the mountains has usually been considered the decisive one. In the majority of cases, this is what it actually is—though by no means in all. These are the points that are often designated as key points in generals' memoirs—usually, to be sure, in a slightly different sense, and mostly with limited application. This is the concept that has frequently been used as a starting point for the erroneous theory (of which Lloyd may have been the founder); that is why the elevated points from which several roads descend into the country to be conquered have been considered as the keys to that country—in fact as points that *dominate* the country. It followed as a matter of course that this concept merged with a closely related one, that of the *systematic defense of mountain country*; the effect was to drive the matter still further into the realm of fantasy. A mass of tactical elements relevant to mountain warfare became involved, and as a result the concept of the highest *point on a road* being the key to the country was replaced by the highest point in the mountain range—in other words, the *watershed*.

At that very time, however, at the end of the eighteenth century, new theories began to be disseminated concerning the formation of the earth's surface by the process of erosion. Natural science, in the form of this geological system, became the ally of military history. This broke the dam of practical common sense; sensible discussion was swept away in a flood of illusions based on geological analogy. Hence in the late eighteenth century one heard—or rather, read—about nothing but the sources of the Danube and the Rhine. We admit that this nonsense abounded primarily in the literature, and that only a fraction of book-learning will seep into practical life anyhow; and the more foolish the theory, the less of it. Yet this particular theory had some practical effect, at Germany's expense. In order to prove that we are not tilting at windmills, let us cite two actual examples: first, the important but highly doctrinaire campaigns of the Prussian army in the Vosges in 1793 and 1794 (to which the works of Massenbach and Grawert provide the theoretical key); and second, the campaign of 1814 when, in obedience to the theory, an army of 200,000 men was forced to make a senseless march through Switzerland to get to Langres.

The high point of an area, the watershed, is usually nothing more than just that. All that was written at the turn of the century about its influence on military affairs was an exaggeration and a faulty application of basically sound ideas, and was completely unrealistic. If the Rhine, the Danube, and all six rivers of Germany agreed to honor a *single* mountain with a common point of origin, its military value would still not be increased thereby, except possibly as a place for a trigonometrical marker. It would have little value for a signal tower, still less for an observation post, and none at all for a whole army.

To look for a key position in a so-called *key area*—in other words, in the nodal point of several mountain ranges where the highest river sources are located—is nothing but an impractical application of the textbook. It is repudiated by Nature herself: she never makes ridges and valleys as accessible from above as is postulated by what has hitherto been known as the theory of terrain, but scatters peaks and gorges at random, and often puts the lowest lakes among the highest massifs. If we consult military history, we will find that the salient geological features of an area hardly affect its military use with any regularity; what little effect they have is outweighed by other local needs and factors. Military lines will often pass very close by a feature of that sort without its attracting them specifically.

We shall now leave this mistaken idea, with which we have dealt at such length only because a whole elegant system has been developed out of it. Let us return to our own point of view.

To repeat, we maintain that if the term "key position" is to rank as an autonomous strategic concept, all it can mean is an area which one must hold before one can risk an advance into enemy territory. If the idea is to be stretched, on the other hand, to cover any point of easy access to a country, or any convenient central point therein, the term will start to lose its proper meaning, and its value will decline accordingly. It would merely denote something that could be found more or less anywhere, and would become nothing but a convenient figure of speech.

The positions we have in mind are admittedly hard to find. The real key to the enemy's country is usually his army, and if terrain is to have precedence over military force, it must promise some exceptionally advantageous conditions. If these are present, they can usually be recognized by two outstanding characteristics: first, that the powers of resistance of the force deployed in that particular place be notably improved through the support of the terrain; and second, that the position effectively threatens the enemy's lines of communication before one's own are threatened by him.

CHAPTER TWENTY-FOUR

Operations on a Flank

We hardly need stress that the subject of this discussion is the strategic flank—in other words, the side of a theater of operations. It has no connection with a flank attack in battle, which is a tactical matter; and even if a strategic flank attack were in its last phase to coincide with a tactical flank attack, the two can easily be kept distinct, because the one will never automatically develop out of the other.

These flanking operations and the flank positions that go with them are also among those prize exhibits of the theorists that are seldom found in actual war. The reason is not that they are ineffective or illusory as devices, but that both sides normally take precautions against them; the cases in which such precautions cannot be taken are rare. In these rare cases, of course, the measure has often been most effective; for this reason, and because of the constant *vigilance* it causes in wartime, the theorist must define the matter with the utmost clarity. While a strategic flanking operation is possible in attack as well as in defense, it has far greater affinity with the latter, and therefore must take its place among the methods of defense.

One simple principle must be laid down before we go into the matter—one that should never be overlooked in the subsequent discussion: forces sent to operate against the enemy's rear and flank are not available for use against his front. It would therefore be quite mistaken, both in tactics and strategy, to think that *falling on the enemy's rear* is an accomplishment in itself. It has no value in isolation, but will become effective only in conjunction with other factors. Moreover, depending on these other factors, its value may be positive or negative. These now require examination.

First of all, we must distinguish between two aspects of the effect of a strategic flanking operation: the effect on the enemy's *line of communication* only, and the effect on his *line of retreat*—which may in turn have repercussions on communications.

When Daun sent raiding parties in 1758 to capture supply convoys destined for the siege of Olmütz, he was obviously not interested in blocking the King's retreat into Silesia: on the contrary, he hoped to bring it about, and was ready to facilitate it.

In the campaign of 1812 the only purpose of the raiding parties sent out by the main Russian army in September and October was to interrupt communications—not to block the French retreat. But the latter was obviously the intention of the Army of the Moldau, which advanced toward the Beresina under Chichagov; it was also the purpose of the attack that General Wittgenstein was ordered to make on the French corps at the Dwina.

These examples are given simply to clarify the point.

Pressure on lines of communication is aimed at enemy convoys, small detachments in the army's wake, couriers, travelers, minor enemy depots, and so forth—at anything, in fact, that the enemy needs to keep his army in a healthy, vigorous condition. In this manner, pressure is meant to weaken the condition of the enemy forces, and thereby bring about their retreat.

Pressure on lines of retreat is meant to cut off the enemy's withdrawal; but it can only do so if a retreat has really been his intention. However, the mere threat may bring about the withdrawal, and thus, by acting as a demonstration, it may have the same effect as pressure on lines of communication. Still, as we have said, none of these effects can be expected to result from the mere act of turning the enemy's position or from the mere geometrical form of disposition; they can only obtain under the appropriate conditions.

In order to make these conditions easier to understand, we shall keep the two kinds of flanking operations separate, and start by examining those directed at lines of communication.

To begin with, we shall have to postulate two main conditions, one or the other of which must be present.

The first is that effective action against the enemy's flank can be taken by small detachments of troops that are so insignificant in number that they will hardly be missed at the front.

The second is that the enemy offensive will have run its course, and that he is therefore unable to exploit a new victory over our forces, or to pursue us if we were to withdraw.

The latter case is not so rare as might appear; but for the moment we shall disregard it and deal with the additional conditions of the former.

The more important of these preconditions is that the enemy's lines of communication be fairly long—too long to be covered by a small number of good posts; and the other, that their location would expose them to action on our part.

Two factors in turn lend themselves to such exposure: their direction, where it is not perpendicular to the enemy's front; and the fact that the lines of communication may run through enemy territory. If both factors are at work, the lines will be all the more exposed. Each factor calls for closer study.

When there are two or three hundred miles of communications to be covered, one would hardly think it mattered much whether the army at the far end of the line was deployed at right angles to it or not: the total extent of its position is a mere dot compared to the length of the line. But in fact it matters a good deal. If one army is drawn up at right angles to its line of communication, even a numerically superior enemy will find it hard to disrupt the line by means of raiding parties. If one only considers the problem of giving absolute protection to a certain area, this is hard to believe; on the contrary, it would seem that an army would be hard-driven to protect its rear—that is the area behind it—from all the raids a stronger enemy could mount. True enough, if only war were as predictable in practice as it is on paper! In that event, the side which had to cover its lines of communication would be in a constant state of uncertainty as to where to expect the raiding parties—it would be practically blind in comparison with the raiders. But consider the unreliable and fragmentary nature of all intelligence in war. Remember that both sides fumble in the dark at all times. One will quickly realize that a party sent past the enemy's wing to raid his rear is like a man in a dark room with a gang of enemies. They will get him in the end. A similar fate awaits the raiders. Once they have turned the enemy's perpendicular position, he is close upon them, but their friends are far away. Not only is there a risk of considerable losses, but the machinery itself is in danger of breaking down: as soon as one of the parties experiences any trouble, the rest will all lose heart. Instead of a bold attack and a daring provocation, one will witness only constant attempts to escape.

This is how the perpendicular disposition of an army manages to cover the nearest points of its lines of communication for a distance of two to three marches, depending on the size of the army. These nearest points are, however, the ones in most immediate danger, because they are also closest to the enemy.

On the other hand, in case of a decidedly oblique position, no part of the lines of communication can be covered in this way. The slightest presssure, the most faint-hearted attempt on the part of the enemy, constitutes a direct threat to some vulnerable point.

What is it that determines the front of a position if it is not the right angle it makes with its line of communication? It is the enemy's front, of course—although this in turn could be considered as being conditioned by our own. Here an interaction is at work for which we must find the causes.

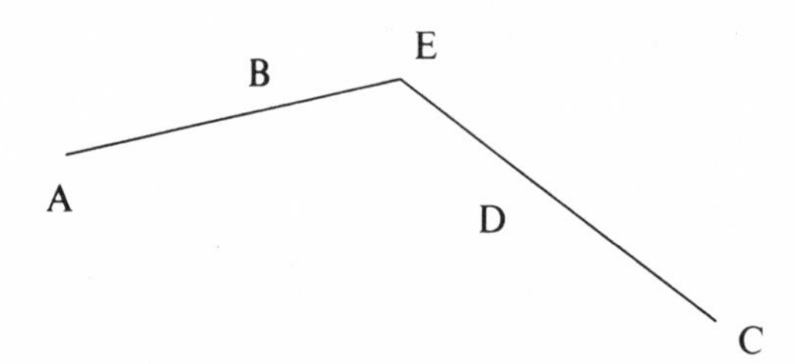

Suppose the attacker's line of communication (A-B) is located so as to make a wide angle with the defender's (C-D). If the defender wanted to take up a position at E, where the two lines meet, it is evident that the attacker could, from position B, compel the defender, by means of the geometrical relation alone, to face him with his front, and consequently expose his line of communication.

The opposite would be the case if the defender took up his position somewhere short of the junction point, say at D. In that event, the attacker's front would have to face him—provided the attacker's line of operations, which is mainly dependent on topography, could not be altered at will and made to run, for example, from A to D. From this one might deduce that the play of interaction benefits the defender, for he need only take up position somewhere short of the junction point. We do not attach much importance to this geometrical argument, and have only brought it in to make the matter absolutely clear. Actually, we are convinced that the position taken by the defender will be strongly influenced by local and other specific considerations. Consequently, one cannot lay down a general rule to show which of the two lines of communication is more likely to be exposed.

If both lines of communication run in exactly the same direction, the side that places its position at an angle to that line will, of course, force the enemy to do the same. But in that case there is no geometric gain. Advantages and disadvantages will be equal for both sides.

We shall therefore limit our further analysis to situations where only one side's line of communication is exposed.

In the second type of vulnerability of a line of communication—when it runs through hostile territory—the extent of its vulnerability where the population is in arms is obvious: one must treat the situation as if enemy forces were stationed all along the line. They may be small in numbers, lacking in depth and in capacity to follow through; but think what constant interference with the line at so many points along its length can mean! There is no need to elaborate. But even if the enemy population is *not* in arms, and there is no militia or any other form of military support in the area—even, indeed, if the population has no stomach at all for war—its mere allegiance to the enemy remains a palpable disadvantage to the other side's lines of communication. The assistance available to a raiding party that speaks the local tongue, knows the country and the people, is able to receive messages, and enjoys the support of the local authorities, can make all the difference to a small detachment, and it is easily obtainable. Furthermore, there are bound to be fortresses, rivers, mountains, and other points of refuge within reasonable range which are still in enemy hands, unless one has taken formal possession and stationed garrisons there.

In such a case, and especially if other circumstances are in our favor, action against the enemy's lines of communication is possible even if they join his front at a right angle; our raiders do not constantly have to fall back on their own forces, but can find plenty of shelter by simply vanishing into the countryside.

This gives us the following:

1. Considerable length
2. An oblique direction and
3. Hostile territory

as the main conditions that expose an army's lines of communication to being severed by relatively small forces. To make this disruption effective, a fourth condition is required—a certain length of time. On that subject, we would ask the reader to refer to our remarks in the fifteenth chapter of Book Five.

But these four conditions are only the outstanding factors that affect the matter: linked to them are numerous local and individual circumstances, often more significant and of wider implication than our four conditions. To mention only the most important, let us list the state of the roads, the nature of the country which they traverse, the cover afforded by rivers, mountains and morasses, the season and the weather, the importance of particular convoys (such as siege-trains), the number of light troops, and so forth.

All these factors, then, will determine the success with which a general can operate against his opponent's lines of communication. Taking the sum of all these parts on both sides and weighing each against the other, one can strike a balance between the two systems of communication. This balance will determine which general can outdo the other in that respect.

While our exposition of the matter may seem rather lengthy, in practice a decision can often be made at first glance. But this calls for seasoned judgment; reflection on all the cases reviewed here will provide the answer to the usual absurd arguments by critics who think they can settle the matter without closer examination merely by mentioning "turning the flank" and "flanking operations."

We have now to deal with the *second main condition* for strategic flanking operations.

If the enemy is blocked from making further progress by something other than our own defense—no matter what it may be—we need no longer be afraid of weakening our forces by sending out strong detachments. Even if the enemy hopes to make us pay by launching an attack, we can simply yield some ground and decline the battle. This is what the main Russian army did before Moscow. But the enormous dimensions and conditions obtaining in that campaign are not essential. During the First Silesian Wars, Frederick the Great found himself constantly in this situation on the frontiers of Bohemia and Moravia. One can postulate a wide variety of reasons, particularly political ones, in the complicated relations of generals and their armies, that would make all further progress impossible.

In such a case, larger forces can be used for flanking operations, and therefore other conditions need not be so favorable. Even the relation between the enemy's system of communication and our own need not be in our favor: the enemy cannot make much use of our withdrawal, nor can he easily retaliate. He will be more concerned with directly covering his own retreat.

This type of situation is, therefore, well suited to produce an effect without a battle, which might be considered too risky: it makes use of a means that is less brilliant and far-reaching, but at the same time less risky than an outright victory.

In such a case, it is not as risky to take up a position on one of the flanks, even if it exposes one's own line of communication. It will always cause the enemy to form front obliquely to his own communications—whereby *this particular* condition from among those listed above will be met almost every time. The likelihood of success will increase in proportion to the influence exerted by the other factors, together with other favorable circumstances; to the degree that these are not available, everything will depend on superior skill in planning and on rapid, accurate execution.

This is the proper setting for strategic maneuvers of the kind so often seen in Silesia and Saxony in the Seven Years War, in 1760 and 1762. The fact that these maneuvers were so common in many wars fought with only moderate intensity should not always be taken as evidence that a general has reached the end of his tether. It may just as well mean that lack of resolution, courage and enterprise, and unwillingness to take responsibility have taken the place of real impediments. We need only cite the case of Field-Marshal Daun.

To sum up the results of this discussion, we would say that flanking operations are most useful:

1. In defense
2. Toward the end of a campaign
3. Especially during a retreat into the interior of the country and
4. In conjunction with armed insurrection.

Concerning the execution of these operations against lines of communication, a few words will suffice.

Their conduct must be in the hands of skillful raiders who must move daringly in small detachments and attack boldly, assaulting the enemy's weaker garrisons, convoys and minor units on the march. They must encourage the local home guards and occasionally join them in operations. The numbers of these units matters more than their individual strength, and they should be so organized that several can link up for a major operation without being too much hampered by the vanity and caprice of individual leaders.

Finally, we must consider the effect on the enemy's line of retreat.

Here one should particularly bear in mind the principle stated at the start, namely, that troops used in the enemy's rear cannot be used against his front; that is to say, that the effect of an action on the rear or flanks will not in itself multiply our forces. Rather it will raise their potential to a higher power—higher as to possible success, but also higher as to possible danger.

All armed resistance, other than the direct and simple kind, tends to raise the stake and raise the risk as well. That rule applies to flanking operations, whether carried out by the whole united force, or with divided forces attacking and enveloping on several sides.

For cutting off the enemy's retreat—if the move is meant seriously and not just as a demonstration—the best solution is a decisive battle, or at least its pre-conditions. In this solution the two elements of higher stakes and higher risks will be re-united. Only favorable conditions, therefore, can justify a commander adopting this method.

In this method one must distinguish between the two forms we have mentioned. The first arises when a general decides to use his full force to attack the enemy in the rear, either from a flank position taken with that end in mind, or by means of a full-scale turning movement. The second arises if he divides his force, assumes an enveloping position, and threatens the enemy rear with one part and his front with the other.

The effect is equally intensified either way: the retreat may be cut absolutely, in which case a large part of the enemy force will be captured or scattered; or the enemy, in order to avoid that danger, will beat a hasty and lengthy retreat.

The risk, however, is different in each case.

If one uses one's whole force to turn the enemy, one risks leaving one's rear exposed. Once again, all depends on the relationship between the two lines of retreat, just as, in the parallel case of the effect on the lines of communication, the deciding factor was their relation to each other.

A defender fighting in his homeland is of course much less constricted than the attacker in terms of lines of communication and retreat, and to that extent he is better placed for a strategic turning movement. But this general relationship is not cogent enough to base an effective system on it. Again individual cases can only be decided in the light of all the circumstances.

We would simply add the following. Favorable conditions are naturally easier to find in large areas than in restricted ones; they also arise more often in truly independent states than in weaker ones that depend on foreign help and whose armies, therefore, must above all secure the point of juncture with their allies. Finally, conditions favor the defender most toward the end of a campaign when the attack has spent its force—again in much the same way as in the case of lines of communication.

A flanking position such as the one the Russians took up so profitably on the Moscow-Kaluga road in 1812, when Bonaparte's assault was slackening, would have been a serious error in the Drissa camp at the start of the campaign, had not the Russians had the wit to change their plan in time.

The other form of envelopment and cutting the line of retreat entails a division of forces. The danger here lies in that division itself, for the enemy has the benefit of the concentration of his internal lines and can thus bring superior numbers against any individual part of his opponent's force. This hazard cannot be eliminated, and only three main causes justify one's exposure to it:

1. A previous division of forces, which makes this type of operation obligatory if one wishes to avoid a major loss of time

2. Great physical and moral superiority, which will justify taking drastic measures

3. A loss of impetus on the enemy's part once the assault has run its course.

Frederick the Great did not intend to combine his frontal attack with another on the enemy's strategic rear when he invaded Bohemia by convergent routes in 1757; at least, that was not his main concern (as we shall show elsewhere in greater detail). It is clear, at all events, that there could be no question of concentrating his forces in Silesia or Saxony before beginning the invasion, since that would have cost him all the benefits of surprise.

When the allies were planning the fall campaign of 1813, their great material superiority entitled them to plan to use the bulk of their strength in an attack on Bonaparte's right flank, which rested on the River Elbe, in order to shift the theater of operations from the Oder to the Elbe. Their reverse at Dresden was not caused by this general plan, but by specific strategic and tactical measures that were taken. After all, the strength with which they confronted Bonaparte at Dresden was 220,000 to 130,000—a ratio that presumably left little to be desired. Anyway, the ratio was not much more favorable at Leipzig: 285,000 to 157,000. It is true that Bonaparte's army was too evenly deployed to operate effectively along a single defensive line, for he had 70,000 men against 90,000 in Silesia, and 70,000 in the Mark of Brandenburg against 110,000; but in any case it would have been difficult for him—short of abandoning Silesia completely—to collect an army on the Elbe that could have dealt the main force of the allies a decisive blow. Likewise, the allies could easily have sent Wrede's army to the River Main with orders to attempt to cut off Bonaparte's road to Mainz.

In 1812 the Russians were able to direct the Army of the Moldau toward Volhynia and Lithuania so as to use it later in the rear of the main French army; since it was absolutely safe to assume that Moscow was to be the culminating point of the French offensive. In that campaign there was no danger to Russia beyond Moscow, and the Russians therefore did not have to worry about their main force being too weak for this operation.

The original plan of defense, worked out by General Phull, had been based on the same disposition of forces: Barclay's army was to occupy the Drissa camp while Bagration's advanced against the rear of the enemy's main force. But what a difference there was between these two instances! In the first, the French were three times as strong as the Russians; in the second, the Russians were distinctly stronger than the French. In the first, Bonaparte's army had an impetus that took it all the way to Moscow, 400 miles beyond Drissa; in the second, it was unable to make another day's march from Moscow. In the first, the line of retreat to the Niemen would not have been more than 150 miles; it was 500 in the second. The same operations, then, that eventually were so effective against the French retreat would have been the wildest folly in the beginning.

Action against an enemy's line of retreat (if it is to amount to more than a demonstration) constitutes an actual attack upon his rear. A good deal more could be said on this subject, but the book on the attack is a more suitable place for that. We shall, therefore, leave it for now, and content ourselves with having shown the conditions under which a counteraction of this sort can be carried out.

When this subject is discussed it is generally more in terms of the demonstration than of the actual attack—the intention being to cause the enemy to retreat. If to be effective each demonstration had to signify that an actual attack was completely feasible—which, at first glance, seems obvious—the two would converge in every detail. But that is not the case: we refer the reader to the chapter on demonstrations, which will show that somewhat different conditions are involved.[1]

[1] No such chapter appears in the book. Eds.

CHAPTER TWENTY-FIVE

Retreat to the Interior of the Country

We regard a voluntary withdrawal to the interior of the country as a special form of indirect resistance—a form that destroys the enemy not so much by the sword as by his own exertions. Either no major battle is planned, or else it will be assumed to take place so late that the enemy's strength has already been sapped considerably.

All attackers find that their strength diminishes as they advance; this point will be discussed in detail in Book Seven. Here we shall anticipate that conclusion; which we can afford to do since military history has shown it to be true in every campaign in which great distances were covered.

Debilitation in the course of an advance is increased if the defender is undefeated and retreats voluntarily with his fighting forces intact and alert, while by means of a steady, calculated resistance he makes the attacker pay in blood for every foot of progress. The advance becomes a steady push forward rather than a mere pursuit.

Conversely, the losses sustained by a retreating defender will be much more serious after a defeat than in a voluntary withdrawal. Even assuming that he can resist the invader day after day, as he could if he withdrew voluntarily, his losses will be *at least as heavy,* and the casualties in battle would have to be added. But such an assumption flies in the face of probability. The best army in the world that is forced to retire to the interior of the country after a defeat in battle will sustain *disproportionate* losses; and if the enemy is considerably stronger—as one can assume in the cases under study—if he presses on with vigor, as has normally been done in recent wars, the outcome will most likely be an actual rout, and will ruin the defending army.

A *calculated* day-by-day resistance means one in which resistance is maintained only so long as the issue can be kept in doubt. It calls for avoiding defeat by yielding the contested ground in time. That sort of fighting will cost the attacker at least as many lives as the defender: the latter's unavoidable losses in men captured will be equaled by the attacker's losses in combat, since he has always to struggle against the defender's advantages of terrain. True, the retreating army loses its severely wounded altogether, but the attacker also loses his, since they usually have to remain in hospital for several months.

One can assume that this continuous attrition will affect the two sides more or less equally.

The situation is completely different when a defeated army is being pursued. Resistance becomes difficult, indeed sometimes impossible, as a consequence of battle casualties, loss of order and of courage, and anxiety about the retreat. The pursuer who, in the former case, had to move with circumspection, almost groping like a blind man, can now advance with the assurance of the victor, the arrogance of the fortunate, and the confidence of the demigod. The faster his pace, the greater the speed with which events will run along their predetermined course: this is the primary area where psychological forces will increase and multiply without being rigidly bound to the weights and measures of the material world.

This should make it clear how the relationship of the two armies will differ according to the manner in which they have reached the point that may be regarded as the terminus of the offensive.

We have considered the result of mutual destruction; but one must also add the debilitation suffered by the attacker in other ways, which we have already said will be treated in Book Seven. The defender, on the other hand, will almost always gather additional strength from reinforcements, whether contributed by outside forces or by his own persistent efforts.

Finally, there is considerable disproportion in supplies. Frequently the retreating army will have more than enough, while the attacker is in dire need.

The retreating army has the means of collecting supplies at prearranged points; the pursuing army is dependent on having its supplies forwarded—a difficult task while it is on the move, no matter how short its lines of communication. It is bound to have shortages from the start.

The retreating army has first call on, and usually exhausts, the local resources. All that remains are devastated towns and villages, harvested and trampled fields, empty wells, and muddied streams.

It is not uncommon, then, for an invader to be faced with serious shortages from the outset. He can never count on finding enemy supplies: if indeed he were to capture some from time to time it would be a matter of pure luck, or of gross neglect on the defender's part.

Thus, there can be no doubt that when the distances involved are long, and the strength of the belligerents not too unequal, a relative state of forces will result that offers the defender far greater prospects of success than a decisive battle on the frontier. But it is not the mere prospect of a victory that is increased through the change in relative strengths; the altered situation will increase the impact of that victory as well. There is an enormous difference between losing a battle on one's own frontier and losing it in the very heart of enemy territory! Indeed, the condition of the invader at the end of his course is often such that even a *victory* can force him to retreat; he may not have enough reserves to follow up his success and make the most of it, nor can he hope to replace his losses.

It makes a vast difference, therefore, whether the decision comes at the start or the end of the offensive.

The great advantages of this method of defense are counterbalanced by two drawbacks. The first consists of the losses the country suffers as a consequence of the enemy invasion; the second is the moral impact.

It cannot be the object of defense to protect the country from losses; the object must be a favorable peace. This is the aim for which one strives, and for which no temporary sacrifice should be considered too severe. Still, while the wastage of the land may not be decisive, it must figure on the balance sheet, because it always affects our interests.

A retreat will directly strengthen the fighting forces, the losses it also involves will not damage the army directly, but rather in a roundabout way. It is difficult to balance the advantages against the disadvantages; they are different in kind, and their effects have no common meeting point. All that can be said is that the loss is increased if a fertile, densely populated area with great commercial cities has to be given up, but that it is greatest if it includes the loss of war material, whether in finished form or in process of production.

The second drawback is psychological. There are times when a general must rise above it, calmly hold to his plans and face the short-term objections that are advanced by fainter hearts. Still, this impression is no mere specter that can easily be dismissed; it is not like a force that affects only a single point, but rather one that spreads instantaneously through every sinew, and paralyzes all military and civil activity. There may be times when the army and the nation fully understand the reasons for withdrawing to the interior, when confidence and hope may even be fortified as a result; but they are very rare. As a rule, the people and the army cannot even tell the difference between a planned retreat and a backward stumble; still less can they be certain if a plan is a wise one, based on anticipation of positive advantages, or whether it has simply been dictated by fear of the enemy. There will be public concern and resentment at the fate of the abandoned areas; the army will possibly lose confidence not only in its leaders but in itself, and never-ending rear guard actions will only tend to confirm its fears. *These consequences* of retreat should not be underrated. Moreover, in the abstract it is, of course, more natural, simpler, nobler, and more in keeping with a nation's moral character to face the challenge squarely, and ensure that an enemy who violates a frontier will be made to pay a penalty in blood.

Such are the pros and cons of this method of defense. Now some words about the conditions and circumstances that will favor it.

The main, the basic need is for ample space, or at least a long line of retreat; just a few days' advance will obviously not weaken the enemy perceptibly. In 1812 Bonaparte's center numbered 250,000 men at Vitebsk; it was down to 182,000 at Smolensk, and not until Borodino had it been reduced to 120,000 men—that is, become equal to the Russian center. Borodino is 450 miles from the frontier, but not until Moscow did the Russians possess a decisive superiority. Once it was established, however, the reversal was so inevitable that even the French success at Maloyaroslavetz had no substantial effect.

No other state in Europe is the size of Russia, and a line of retreat 500 miles long is conceivable in very few. On the other hand, the conditions that produced an army like that of the French in 1812 are unlikely to recur—to say nothing of the disproportion that obtained between the two antagonists at the outset, when the French had more than twice the amount of men, with immense prestige into the bargain. Therefore, what was accomplished here in the space of 500 miles, may in other cases be accomplished in 250 or even 150.

Among the favorable circumstances are:

1. A sparsely cultivated area
2. A loyal and warlike people
3. Severe weather conditions.

All these make it hard to keep an army in the field. They involve massive convoys, numerous detachments, and heavy fatigue. They lead to sickness, and make flanking operations easier for the defender.

The final factor affecting this method of defense is the actual numbers of the fighting forces involved.

It is of course in the nature of things that, apart from the relative strength of the two armies, a smaller force will be exhausted sooner than a larger one; it cannot run so long a course, and therefore the radius of its theater of operations is bound to be restricted. In fact, a fairly constant ratio exists between the size of a force and the area it can occupy. This ratio cannot be expressed in numbers; besides, it is subject to change by other circumstances. Here it is enough to say that the relationship between the two is permanent and fundamental. One may be able to march on Moscow with 500,000 men, but never with 50,000—even if the invader's strength relative to the defender's were greater in the second case than in the first.

Let us assume, therefore, that the absolute size of the force in relation to the area is constant in two different cases: there is no doubt that the effect of our retreat will increase the debilitation of the enemy in proportion to the numbers involved.

1. The invader will find it more difficult to supply and billet his troops. Even if the areas covered expand in proportion to the armies, they can never be the sole source of supplies, and everything sent forward from the base will suffer a higher rate of wastage. Moreover only a small part of the area, and never the whole, will be used for billeting; and the area used will not increase in proportion to the numbers of the troops.

2. Progress will be slower as troops increase in number. Therefore it will take more time for the advance to run its course, and the aggregate of daily losses will be that much greater.

Three thousand men on the heels of two thousand will generally not allow them to retreat by easy marches of five, ten, or even fifteen miles at a time, or come to a halt for a few days every now and then. Only a few hours are needed to catch up, attack, and disperse the enemy. But if we multiply these figures by a hundred, we have another problem altogether. It is no longer a matter of a few hours, but of a day, or even two. Neither side can remain assembled in one place; all plans and movements become intricate, and consequently everything requires more time. The attacker is faced with the additional handicap that problems of supply make him spread out much more than the defender, and in consequence he always runs the risk of being overwhelmed by a superior force at some point—which is what the Russians attempted at Vitebsk.

3. The larger the forces engaged, the greater the effort required by the day-to-day needs of strategy and tactics. Imagine 100,000 men marching to the assembly point and back, day after day, stopping and starting, now taking up arms, now cooking meals or distributing rations, all being kept in the field until the necessary reports have been received. For all these minor efforts, which are only incidental to the march itself, those 100,000 men will as a rule need twice the time that 50,000 would—but the day only has twenty-four hours for each. As for the march itself, we have pointed out in Chapter Ten of the preceding book how much the time and energy consumed will vary with the number of troops involved.

All these problems also of course affect the retreating army, but they bear more heavily on the invader because:

1. He has more men (we assumed the attacker's superiority from the beginning);

2. By yielding ground the defender acquires the right to guide the operation and compel the other side to adjust to his actions. He is able to plan ahead and can usually hold to his intentions. The invader, however, can make his dispositions only after his opponent has taken up positions, which have always first to be reconnoitered.

We must recall that we are discussing the pursuit of an enemy who has not suffered a defeat, who has not even lost a battle; we are not contradicting what was said in Chapter Twelve of Book Four.

The advantage of compelling the attacker to adjust his movements to ours makes a difference in terms of saving time and effort, and in many other minor ways. In the long run this can be of great importance;

3. The retiring army takes great pains to make its retreat easier: roads and bridges are repaired, the most convenient sites are chosen for camps, and so forth. On the other hand, it is trying just as hard to make the pursuit more difficult for the enemy by destroying bridges, making bad roads worse merely by using them, denying the enemy the best camping places and watering points by occupying them itself, and so forth.

Finally, we must mention that a popular insurrection is an exceptionally favorable factor. Since this will be taken up in a separate chapter, we need not go into details here.

Up to now, we have discussed the advantages of this type of retreat, the sacrifices it entails, and the conditions it requires. We shall now look at the way it is carried out.

The first question to be examined is that of the direction taken by the retreat.

It should lead to the *interior* of the country—if possible, in fact, toward a point where the enemy will be surrounded by our territory on all sides, and is completely exposed to its effects. In such a case the defender runs no risk of *being diverted from the heart of his territory,* as he would if he chose a line of retreat too close to the frontier. That could easily have happened to the Russians in 1812 if they had decided to retreat in a southerly rather than an easterly direction.

This condition is implied by the very aim of such a retreat. Which point in the country is the right one, how far it is consonant with the intention of directly covering the capital or some other vital point, or of drawing the enemy away from it—these matters depend on circumstances.

If in 1812 the Russians had planned their withdrawal in advance and carried it out accordingly, they could easily have gone from Smolensk toward Kaluga instead of taking that direction only after leaving Moscow. It is quite possible that under those circumstances, Moscow might have been saved altogether.

At Borodino the French were 130,000 strong; there is no reason to suppose that they would have been any stronger if the Russians had accepted battle half-way to Kaluga. But in that case how many men could the French have spared to march on Moscow? Obviously very few. But Moscow is 250 miles from Smolensk, and one cannot send a small force over that sort of a distance to take a place like Moscow.

The battle at Smolensk left Bonaparte with about 160,000 men. Suppose he had felt he could risk detaching a force to capture Moscow before fighting another major battle, and had sent off 40,000 men. He would then have had 120,000 left to face the main Russian army. But only 90,000 of them could have taken part—40,000 less than fought at Borodino. That would have left the Russians stronger by 30,000. If the course of the battle of Borodino is any guide, that is a margin with which the Russians might have been victorious. At any rate, according to this calculation, their chances would have been better than at Borodino. But the Russian retreat was not the result of advance planning. They retired as far as they did because whenever they had the chance to accept battle, they did not feel strong enough to do so. All their supplies and reinforcements were directed toward the road from Moscow to Smolensk, and nobody at Smolensk would have dreamed of leaving it. Besides a victory somewhere between Smolensk and Kaluga would not, in Russian eyes, have justified the disgrace of leaving Moscow unprotected and exposed to possible capture.

In 1814, Bonaparte could almost certainly have saved Paris from being attacked if he had taken up position obliquely—behind the Canal de Bourgogne, for example—leaving only a few thousand men and the large National Guard to hold the capital. The allies would never have dared to send 50,000 to 60,000 men toward Paris, knowing that Bonaparte was at Auxerre with 100,000 men. Conversely, had the allies been in Napoleon's position, it is certain that no one would have dared to suggest that they leave the road to their capital uncovered while Bonaparte was on the offensive.

With that much superiority he would not have hesitated a moment before marching on the capital. It simply shows the difference moral and psychological factors can make, even where all other circumstances are the same.

We would just add that in such a maneuver the capital, or whatever point one is trying to save, should have sufficient powers of resistance to protect it against capture or plunder by any raiding party that happens to come along. We shall leave this subject, which will be taken up again when we consider the war plans.

In connection with such a line of retreat, one other point deserves attention—that of a sudden *change of direction*. The Russians kept to the same direction as far as Moscow, beyond which it would have led them to Vladimir. Leaving it, they headed first for Riazan, then switched toward Kaluga. If they had been forced to continue their retreat, they could easily have done so in this new direction, which would have led them to Kiev—thus returning much closer to the enemy's frontier. Even if the French had still been superior at that stage, they could obviously not have held the enormous bend in their line of communication, which ran through Moscow. They would have had to abandon not only Moscow but most likely also Smolensk—in other words, they would have had to give up the territory they had occupied at such expense, and content themselves with the theater of operations west of the Beresina.

Admittedly, the Russian army would have been faced with that very risk if it had headed for Kiev in the first place—the danger of being cut off from the main part of the country. But this disadvantage had practically disappeared: the French would have arrived at Kiev in a very different condition if they had not taken the road by way of Moscow.

It is clear that decided advantages can be gained by such a change in the direction of the line of retreat—which is quite feasible if the distances involved are large enough:

1. The turn makes it impossible for the enemy to maintain his existing lines of communication. Setting up new ones is always difficult; what is more, the direction changes only gradually, and therefore new lines of communication will have to be established several times.
2. The result is to bring both armies closer to the frontier again. The invader is no longer able to use his position to cover the territory he has conquered, and he will most likely be forced to abandon it. In a country as immense as Russia, two armies could easily play a regular game of tag with one another.

Under favorable conditions a turn of this sort is possible even where smaller distances are involved. But every case must be decided on its own merits in the light of all the circumstances.

Once the direction in which the enemy is to be led into the country has been determined, the main defending force must move that way. Otherwise, the enemy will not follow it, and even if he did, the defender would not be able to impose all the restrictions which we have assumed above. The only question is whether to take that route with an undivided force, or move sideways with substantial portions of it, thereby retreating along divergent lines.

The answer is that the second course should be avoided for the following reasons:

1. It is a further dispersal of forces, whereas their concentration in one area would present a major obstacle to the invader.

2. The enemy thereby enjoys the advantage of interior lines. He will then be better concentrated than the defender and have greater superiority at any given point. Admittedly, this superiority is to be feared less if from the beginning our plan aims at avoiding battle. But this depends on the enemy continuing to respect our strength and not feeling free to push us around, which could conceivably happen. This type of retreat further implies that one main force will eventually gain the upper hand, so that it can strike a decisive blow. But one cannot count on this once the force has been split up.

3. Concentric operations are simply unsuited to the weaker side.

4. A divergent retreat cancels out part of the enemy's weakness.

The main weak points of an attack that penetrates deeply are, of course, the long lines of communication and the exposed strategic flanks. A retreat that takes a divergent form obliges the attacker to show a front to the flank with one part of his force; and although that part is really only meant to contain the forces facing it, it may incidentally serve an additional purpose—that of covering part of the lines of communication.

In short, so far as the purely strategic value of the retreat is concerned, the divergent form is not advantageous. If it is to pave the way for future action against the enemy's line of retreat, we must refer to the content of the previous chapter.

The *only* purpose that justifies a retreat along divergent lines is one in which it serves to protect provinces that would otherwise be occupied by the enemy.

The areas that the invader will occupy on both sides of his line of march can usually be predicted fairly accurately from the concentration and direction of his forces, and the location of his territory, fortresses and so forth, in relation to our own. It would be a dangerous dissipation of strength if we were to station our fighting forces in areas that will probably not be touched. *Whether one would be able to prevent an occupation, by stationing troops* in areas likely to be occupied, is a great deal harder to predict. It is largely a matter of judgment.

When the Russians retreated in 1812, they left 30,000 men under Tormasov to defend Volhynia from invasion by the Austrians. Considering the size of that province, the number of natural obstacles it presents, and the unimpressive force that was supposed to invade it, the Russians had reasonable hope that they could keep the upper hand on that stretch of their frontier, or, at any rate, maintain themselves nearby. If they were able to hold their position, solid advantages could be expected to accrue in the future, which need not be discussed here. In any case, it would have been almost impossible for these forces to rejoin the main army in time, even if that had been intended. All these factors amply justified the Russians' leaving this force to fight an independent war in Volhynia. On the other hand, according to the plan submitted by General Phull, only Barclay's force of 80,000 men was to retire into Drissa. Bagration with 40,000 men was to operate on the right flank, with the intention of falling on the French rear later. One will realize at a glance that this force had not the slightest hope of holding out in southern Lithuania—holding *additional* territory, that is, and *closer* to the rear of the French. It would have been demolished by their overwhelming numbers.

The defender is interested, of course, in yielding as little ground as possible to the invader, but this always remains a secondary aim. It is also obvious that an attack becomes more difficult the smaller, or rather, the more limited the theater of operations to which the invader can be confined. Yet all these considerations assume the probability of success and that they will not weaken the main force too much—because it is with this force that one must primarily seek the final result. Difficulties encountered by the enemy's main force are most likely to cause him to retreat, and will increase to the greatest degree the loss of physical and moral forces incurred on the retreat.

As a rule, therefore, a retreat to the interior of the country should be undertaken by an unbeaten and undivided force. It should retire immediately ahead of the enemy's main force, as slowly as possible. By maintaining continuous resistance, it should keep the enemy in a constant state of alarm, forcing him into the fatal extravagance, so to speak, of tactical and strategic precautions.

Once both parties have, in this manner, reached the end of the aggressor's offensive, the defender should, if at all possible, take up a position at an angle to the direction of that line and exert pressure on the enemy's rear by all means at his disposal.

All these features and their effects are displayed to a high degree by the Russian campaign of 1812, as through a magnifying glass. Even though it was not a voluntary retreat, it can be regarded as one, since there can be no doubt that if the Russians, knowing what they know now, had to repeat it under the same conditions, they would do systematically what they did, mostly unintentionally, in 1812. One would be quite wrong, however, in assuming that the same effect has never been produced, and never can be, where the Russian dimensions of space are lacking.

The principal effects and conditions of this type of resistance have been present—no matter what modifying circumstances may have attended them in addition—wherever a strategic attack has foundered, not as a consequence of a decisive battle, but because of the sheer problems of existence; which forced the invader to undertake a retreat that, to smaller or greater extent, proved disastrous. Frederick the Great's campaigns of 1742 in Moravia and 1744 in Bohemia, the French campaign of 1743 in Austria and Bohemia, the Duke of Brunswick's of 1792 in France, and Masséna's winter campaign of 1810–1811 in Portugal, are examples that illustrate cases that are similar though much more limited in scale and circumstances. Moreover, there are countless other situations where the principle established here was partly, if not completely responsible for the result. We do not cite them because it would mean going into unnecessary detail.

In Russia, and in all the other cases mentioned here, the tide turned without a victorious battle to provide the decision at the point of culmination. Even where such an effect cannot be expected, it remains sufficiently important to bring about, by means of this type of resistance, a change in the relative strength of both sides, that will make victory a possibility. Once that has been won, one must ensure that it touches off a series of calamities, which, in accordance with the law of falling bodies, will keep gathering momentum.

CHAPTER TWENTY-SIX

The People in Arms

In the civilized parts of Europe, war by means of popular uprisings is a phenomenon of the nineteenth century. It has its advocates and its opponents. The latter object to it either on political grounds, considering it as a means of revolution, a state of legalized anarchy that is as much of a threat to the social order at home as it is to the enemy; or else on military grounds, because they feel that the results are not commensurate with the energies that have been expended.

The first objection does not concern us at all: here we consider a general insurrection as simply another means of war—in its relation, therefore, to the enemy. The second objection, on the other hand, leads us to remark that a popular uprising should, in general, be considered as an outgrowth of the way in which the conventional barriers have been swept away in our lifetime by the elemental violence of war. It is, in fact, a broadening and intensification of the fermentation process known as war. The system of requisitioning, and the enormous growth of armies resulting from it and from universal conscription, the employment of militia—all of these run in the same direction when viewed from the standpoint of the older, narrower military system, and that also leads to the calling out of the home guard and arming the people.

The innovations first mentioned were the natural, inevitable consequences of the breaking down of barriers. They added so immensely to the strength of the side that first employed them that the opponent was carried along and had to follow suit. That will also hold true of the people's war. Any nation that uses it intelligently will, as a rule, gain some superiority over those who disdain its use. If this is so, the question only remains whether mankind at large will gain by this further expansion of the element of war; a question to which the answer should be the same as to the question of war itself. We shall leave both to the philosophers. But it can be argued that the resources expended in an insurrection might be put to better use in other kinds of warfare. No lengthy investigation is needed, however, to uncover the fact that these resources are, for the most part, not otherwise available and cannot be disposed of at will. Indeed, a significant part of them, the psychological element, is called into being only by this type of usage.

When a whole nation renders armed resistance, the question then is no longer, "Of what value is this to the people," but "what is its potential value, what are the conditions that it requires, and how is it to be utilized."

By its very nature, such scattered resistance will not lend itself to major actions, closely compressed in time and space. Its effect is like that of the process of evaporation: it depends on how much surface is exposed. The greater the surface and the area of contact between it and the enemy forces, the thinner the latter have to be spread, the greater the effect of a general uprising. Like smoldering embers, it consumes the basic foundations of the enemy forces. Since it needs time to be effective, a state of tension will develop while the two elements interact. This tension will either gradually relax, if the insurgency is suppressed in some places and slowly burns itself out in others, or else it will build up to a crisis: a general conflagration closes in on the enemy, driving him out of the country before he is faced with total destruction. For an uprising by itself to produce such a crisis presupposes an occupied area of a size that, in Europe, does not exist outside Russia, or a disproportion between the invading army and the size of the country that would never occur in practice. To be realistic, one must therefore think of a general insurrection within the framework of a war conducted by the regular army, and coordinated in one all-encompassing plan.

The following are the only conditions under which a general uprising can be effective:

1. The war must be fought in the interior of the country.
2. It must not be decided by a single stroke.
3. The theater of operations must be fairly large.
4. The national character must be suited to that type of war.
5. The country must be rough and inaccessible, because of mountains, or forests, marshes, or the local methods of cultivation.

The relative density of the population does not play a decisive part; rarely are there not enough people for the purpose. Nor does it make much difference whether the population is rich or poor—at least it should not be a major consideration, although one must remember that poor men, used to hard, strenuous work and privation, are generally more vigorous and more warlike.

One peculiarity of the countryside that greatly enhances the effectiveness of an insurrection is the scattered distribution of houses and farms, which, for instance, can be found in many parts of Germany. Under such conditions the country will be more cut up and thickly wooded, the roads poorer if more numerous; the billeting of troops will prove infinitely more difficult, and, above all, the most characteristic feature of insurgency in general will be constantly repeated in miniature: the element of resistance will exist everywhere and nowhere. Where the population is concentrated in villages, the most restless communities can be garrisoned, or even looted and burned down as punishment; but that could scarcely be done in, say, a Westphalian farming area.

Militia and bands of armed civilians cannot and should not be employed against the main enemy force—or indeed against any sizable enemy force. They are not supposed to pulverize the core but to nibble at the shell and around the edges. They are meant to operate in areas just outside the theater of war—where the invader will not appear in strength—in order to deny him these areas altogether. Thunder clouds of this type should build up all around the invader the farther he advances. The people who have not yet been conquered by the enemy will be the most eager to arm against him; they will set an example that will gradually be followed by their neighbors. The flames will spread like a brush fire, until they reach the area on which the enemy is based, threatening his lines of communication and his very existence. One need not hold an exaggerated faith in the power of a general uprising, nor consider it as an inexhaustible, unconquerable force, which an army cannot hope to stop any more than man can command the wind or the rain—in short, one need not base one's judgment on patriotic broadsides in order to admit that peasants in arms will not let themselves be swept along like a platoon of soldiers. The latter will cling together like a herd of cattle and generally follow their noses; peasants, on the other hand, will scatter and vanish in all directions, without requiring a special plan. This explains the highly dangerous character that a march through mountains, forests, or other types of difficult country can assume for a small detachment: at any moment the march may turn into a fight. An area may have long since been cleared of enemy troops, but a band of peasants that was long since driven off by the head of a column may at any moment reappear at its tail. When it comes to making roads unusable and blocking narrow passes, the means available to outposts or military raiding parties and those of an insurgent peasantry have about as much in common as the movements of an automaton have with those of a man. The enemy's only answer to militia actions is the sending out of frequent escorts as protection for his convoys, and as guards on all his stopping places, bridges, defiles, and the rest. The early efforts of the militia may be fairly weak, and so will these first detachments, because of the dangers of dispersal. But the flames of insurrection will be fanned by these small detachments, which will on occasion be overpowered by sheer numbers; courage and the appetite for fighting will rise, and so will the tension, until it reaches the climax that decides the outcome.

A general uprising, as we see it, should be nebulous and elusive; its resistance should never materialize as a concrete body, otherwise the enemy can direct sufficient force at its core, crush it, and take many prisoners. When that happens, the people will lose heart and, believing that the issue has been decided and further efforts would be useless, drop their weapons. On the other hand, there must be some concentration at certain points: the fog must thicken and form a dark and menacing cloud out of which a bolt of lightning may strike at any time. These points for concentration will, as we have said, lie mainly on the flanks of the enemy's theater of operations. That is where insurgents should build up larger units, better organized, with parties of regulars that will make them look like a proper army and enable them to tackle larger operations. From these areas the strength of the insurgency must increase as it nears the enemy's rear, where he is vulnerable to its strongest blows. The larger groups are intended to harass the more considerable units that the enemy sends back; they will also arouse uneasiness and fear, and deepen the psychological effect of the insurrection as a whole. Without them the impression would not be sufficiently great, nor would the general situation give the enemy enough cause for alarm.

A commander can more easily shape and direct the popular insurrection by supporting the insurgents with small units of the regular army. Without these regular troops to provide encouragement, the local inhabitants will usually lack the confidence and initiative to take to arms. The stronger the units detailed for the task, the greater their power of attraction and the bigger the ultimate avalanche. But there are limiting factors. For one thing, it could be fatal to the army to be frittered away on secondary objectives of that kind—to be dissolved, so to speak, in the insurgency—merely to form a long and tenuous defensive line, which is a sure way of destroying army and insurgents alike. For another, experience tends to show that too many regulars in an area are liable to decimate the vigor and effectiveness of a popular uprising by attracting too many enemy troops; also, the inhabitants will place too much reliance upon the regulars; and finally, the presence of considerable numbers of troops taxes the local resources in other ways, such as billets, transportation, requisitions, and so forth.

Another means of avoiding an effective enemy reaction to a popular uprising is, at the same time, one of the basic principles of insurrection: it is the principle of seldom, or never, allowing this important strategic means of defense to turn into tactical defense. *Insurgent actions* are similar in character to all others fought by second-rate troops: they start out full of vigor and enthusiasm, but there is little level-headedness and tenacity in the long run. Moreover, not much is lost if a body of insurgents is defeated and dispersed—that is what it is for. But it should not be allowed to go to pieces through too many men being killed, wounded or taken prisoner: such defeats would soon dampen its ardor. Both these characteristics are entirely alien to the nature of a tactical defense. A defensive action ought to be a slow, persistent, calculated business, entailing a definite risk; mere attempts that can be broken off at will can never lead to a successful defense. So if the defense of a sector is entrusted to the home guard, one must avoid getting involved in a major defensive battle, or else they will perish no matter how favorable the circumstances. They may and should defend the points of access to a mountain area or the dikes across a marsh or points at which a river can be crossed for as long as possible; but once these are breached, they had better scatter and continue their resistance by means of surprise attacks, rather than huddle together in a narrow redoubt, locked into a regular defensive position from which there is no escape. No matter how brave a people is, how warlike its traditions, how great its hatred for the enemy, how favorable the ground on which it fights: the fact remains that a national uprising cannot maintain itself where the atmosphere is too full of danger. Therefore, if its fuel is to be fanned into a major conflagration, it must be at some distance, where there is enough air, and the uprising cannot be smothered by a single stroke.

This discussion has been less an objective analysis than a groping for the truth. The reason is that this sort of warfare is not as yet very common; those who have been able to observe it for any length of time have not reported enough about it. We merely wish to add that strategic plans for defense can provide for a general insurrection in one of two ways: either as a last resort after a defeat or as a natural auxiliary before a decisive battle. The latter use presupposes a withdrawal to the interior and the form of indirect defense described in Chapters Eight and Twenty-Four of this book. Therefore, we shall add only a few words concerning the calling out of the home guard after a battle has been lost.

A government must never assume that its country's fate, its whole existence, hangs on the outcome of a single battle, no matter how decisive. Even after a defeat, there is always the possibility that a turn of fortune can be brought about by developing new sources of internal strength or through the natural decimation all offensives suffer in the long run or by means of help from abroad. There will always be time enough to die; like a drowning man who will clutch instinctively at a straw, it is the natural law of the moral world that a nation that finds itself on the brink of an abyss will try to save itself by any means.

No matter how small and weak a state may be in comparison with its enemy, it must not forego these last efforts, or one would conclude that its soul is dead. The possibility of avoiding total ruin by paying a high price for peace should not be ruled out, but even this intention will not, in turn, eliminate the usefulness of new measures of defense. They will not make the peace more difficult and onerous, but easier and better. They are even more desirable where help can be expected from other states that have an interest in our survival. A government that after having lost a major battle, is only interested in letting its people go back to sleep in peace as soon as possible, and, overwhelmed by feelings of failure and disappointment, lacks the courage and desire to put forth a final effort, is, because of its weakness, involved in a major inconsistency in any case. It shows that it did not deserve to win, and, possibly for that very reason was unable to.

With the retreat of the army into the interior—no matter how complete the defeat of a state—the potential of fortresses and general insurrections must be evoked. In this respect, it will be advantageous if the flanks of the main theater of operations are bordered by mountains or other difficult terrain, which will then emerge as bastions, raking the invader with their strategic enfilade.

Once the victor is engaged in sieges, once he has left strong garrisons all along the way to form his line of communication, or has even sent out detachments to secure his freedom of movement and keep adjoining provinces from giving him trouble; once he has been weakened by a variety of losses in men and matériel, the time has come for the defending army to take the field again. Then a well-placed blow on the attacker in his difficult situation will be enough to shake him.

CHAPTER TWENTY-SEVEN

Defense of a Theater of Operations

Having discussed *the most important methods of defense,* we could perhaps delay any discussion of the way they fit into an overall plan of defense until the final book, on war plans. A war plan is, after all, the source of all the lesser plans of attack and defense, and determines their main lines; indeed frequently a war plan is nothing more than a plan for attacking or defending the main theater of operations. But at no time have we yet been able to begin our discussion with war as a whole, despite the fact that in war, more than anywhere else, it is the whole that governs all the parts, stamps them with its character, and alters them radically. On the contrary, it seemed necessary to start by thoroughly examining the various parts as separate components. If we had not advanced from the simple to the complex, we should have been swamped by a multitude of vague concepts; more particularly, the variety of interactions that occur in war would have constantly confused our ideas. One more stage remains before we reach the whole: to examine the defense of a theater of war as a subject in itself, and to look for the thread that ties together all the subjects discussed.

Defense, as we see it, is nothing but the stronger form of combat. The preservation of one's fighting forces and the destruction of the enemy's—in a word, victory—is the substance of this struggle; but it can never be its ultimate object.

The ultimate object is the preservation of one's own state and the defeat of the enemy's; again in brief, the intended peace treaty, which will resolve the conflict and result in a common settlement.

But what, in the context of war, is meant by the enemy state? First of all his fighting forces; then his territory. Of course, it means a great many other things as well, which, depending on circumstances, can attain genuine significance. Chief among these are the foreign and domestic political conditions, which are sometimes more decisive than anything else. But although the enemy's fighting forces and his territory may not be the state itself, nor represent all his means of making war, they will always be the *dominant* factors, and usually they *far exceed* all others in importance. The fighting forces are meant to protect their own territory and to seize the enemy's: it is the territory, on the other hand, that sustains them and keeps restoring their strength. Each, then, depends on the other. They give mutual support and are of equal value to each other. But while they interact, they do so with a difference. If the forces are destroyed—in other words, overcome and incapable of further resistance—the country is automatically lost. On the other hand, loss of the country does not automatically entail destruction of the forces; they can evacuate the country of their own accord, in order to reconquer it the more easily later on. Not only the complete annihilation, but any considerable weakening of the fighting forces, will generally lead to a loss of territory. Conversely, however, not every considerable loss of territory automatically leads to a weakening of the forces. It will happen in the long run, of course, but not always within the decisive phase of the war.

It follows that it is always more important to preserve, or, as the case may be, destroy armed forces than to hold on to territory—in other words, the former must be a general's prime concern. The possession of territory will become an end in itself only *where those means are not enough in themselves*.

If all the enemy's forces were united in a *single* army, and if the war consisted of a *single* battle, possession of the country would turn on that battle's outcome. The destruction of the enemy's forces, the occupation of his territory, and the safety of one's own would follow automatically and, in a sense, be identical with it. The question now arises, what will cause the defender to abandon this, the simplest of all forms of war, in the first place, and disperse his forces in space? The answer lies in the inadequacy of the victory that he can achieve with his combined forces. Each victory has its own sphere of influence. If that sphere includes the whole of the enemy state—fighting forces, territory, and all—in other words, if all the components of his strength are carried away in the very torrent that has hit its core, that victory is all that is needed. There will be no need for a division of forces. If, on the other hand, parts of the enemy's forces and of both countries are beyond the scope of our victory, these parts will require special attention. Since one cannot concentrate land as one can an army, it will be necessary to divide the army in order to defend the land.

Only in the case of small and compact states is such a concentration of force possible and probable that its defeat will decide everything. If the area involved is very large and the frontier long, or if one is surrounded on all sides by a powerful alliance of enemies, such a concentration is a practical impossibility. A division of forces then becomes inevitable, and with it, several theaters of operation.

The scale of a victory's sphere of influence depends, of course, on the *scale* of the victory, and that in turn depends on *the size of the defeated force*. For this reason, the *blow* from which the broadest and most favorable repercussions can be expected will be aimed against *that area* where the greatest concentration of enemy troops can be found; the larger the force with which the blow is struck, the surer its effect will be. This rather obvious sequence leads us to an analogy that will illustrate it more clearly—that is, the nature and effect of a center of gravity.

A center of gravity is always found where the mass is concentrated most densely. It presents the most effective target for a blow; furthermore, the heaviest blow is that struck by the center of gravity. The same holds true in war. The fighting forces of each belligerent—whether a single state or an alliance of states—have a certain unity and therefore some cohesion. Where there is cohesion, the analogy of the center of gravity can be applied. Thus, these forces will possess certain centers of gravity, which, by their movement and direction, govern the rest; and those centers of gravity will be found wherever the forces are most concentrated. But in war as in the world of inanimate matter the effect produced on a center of gravity is determined and limited by the cohesion of the parts. In either case, a blow may well be stronger than the resistance requires, and in that case it may strike nothing but air, and so be a waste of energy.

There is a decided difference between the cohesion of a *single* army, led into battle under the personal command of a *single* general, and that of an *allied force* extending over 250 or 500 miles, or even operating against different fronts. In the one, cohesion is at its strongest and unity at its closest. In the other, unity is remote, frequently found only in mutual political interests, and even then rather precarious and imperfect; cohesion between the parts will usually be very loose, and often completely fictitious.

On the one hand then, the force at which our blow is to be aimed requires that our strength be concentrated to the utmost; on the other, any excess is to be regarded as a decided disadvantage, since it involves a waste of energy, which in turn means a *lack of strength* elsewhere.

It is therefore a major act of strategic judgment to distinguish these centers of gravity in the enemy's forces and to identify their spheres of effectiveness. One will constantly be called upon to estimate the effect that an advance or a retreat by part of the forces on either side will have upon the rest.

Far from believing we have discovered a new technique, we are merely providing a rationale for the actions of every general in history, which serves to explain their connection with the nature of the problem.

The last book will describe how this idea of a center of gravity in the enemy's force operates throughout the plan of war. In fact, that is where the matter properly belongs; we have merely drawn on it here in order not to leave a gap in the present argument. Our reflections are intended to demonstrate the general reasons for dividing one's forces. Basically, there are two conflicting interests: one, *possession of the country*, tends to disperse the fighting forces; the other, *a stroke at the center of gravity of the enemy's forces*, tends, in some degree, to keep them concentrated.

This is how operational theaters, or individual armies' zones of operations, are created. A country and the forces stationed there are divided in such a way that any decision obtained by the main force in a particular theater directly affects the whole and carries everything along with it. We say *directly*, since any decision reached in one particular operational theater is also bound to have a more or less remote effect on adjoining areas.

We want to reiterate emphatically that here, as elsewhere, our definitions are aimed only at the centers of certain concepts; we neither wish nor can give them sharp outlines. The nature of the matter should make this obvious enough.

Our position, then, is that a theater of war, be it large or small, and the forces stationed there, no matter what their size, represent the sort of unity in which a *single* center of gravity can be identified. That is the place where the decision should be reached; a victory at that point is in its fullest sense identical with the defense of the theater of operations.

CHAPTER TWENTY-EIGHT

Defense of a Theater of Operations—Continued

Defense, however, consists of two different elements—the *decision* and the *period of waiting*. This chapter deals with the connection between the two.

We must begin by pointing out that the state of waiting is not the sum total of the term "defense." It is, however, the phase by which a defense approaches its goal. So long as a fighting unit has not abandoned the area assigned to it, the tension that an attack creates on both sides will continue. Only a decision can put an end to it; and that decision, whatever it may be, can only be considered a fact after either the attacker or the defender has abandoned the theater of war.

So long as a force maintains itself in its area, its defense of that area continues, and in that sense one can say that defense *of* an operational theater is the same thing as defense *within* it. It is immaterial how much or how little of the area is temporarily occupied by the enemy; it is merely lent to him.

This conceptualization, which is meant to clarify the true relationship between the state of waiting and the whole, is valid only if a decision is really intended, and is regarded as inevitable by both sides. It is the decision that changes the centers of gravity on each side, and the operational theaters they create, into *active agents*. If one drops the idea of a decision, the centers of gravity are neutralized, and so, indeed, in a certain sense, are all the forces. At this point, possession of the country, the next most important component of the theater of war, will become a direct objective—in other words, the importance of possessing the country increases, the less a decision is actively sought by the belligerents, and the more the war becomes a matter of mutual observation. The defender will be more intent on the immediate coverage of all he has, while the attacker increasingly attempts to spread his forces out in his advance.

There is no denying that a great majority of wars and campaigns are more a state of observation than a struggle of life and death—a struggle, that is, in which at least one of the parties is determined to gain a decision. A theory based on this idea could be applied only to the wars of the nineteenth century, which alone have displayed that characteristic to such a high degree. Not every future war, however, is likely to be of this type; on the contrary, one may predict that most wars will tend to revert to wars of observation. A theory, to be of any practical use, must allow for that likelihood. We shall, therefore, start by considering the kind of war that is completely governed and saturated by the urge for a decision—of true war, or absolute war, if we may call it that. In a later chapter we shall consider the modifications that arise from a greater or lesser approximation to a war of observation.

In the first instance, either the attacker is expected to bring about a decision or it is sought by the defender; for present purposes it does not matter which. Here the defense of a theater of war will consist of maintaining the position in such a way as to be able to bring about an advantageous decision at any moment. Such a decision may be made up of a single battle or a series of major engagements; it may also consist, however, of the mere effect of the relationships that arise from the disposition of both forces—in other words, from *possible engagements*.

Even if a battle were not the primary, the most common, the most effective means of reaching a decision (as we think we have already shown more than once) the mere fact that it is one of the means of obtaining a decision should be enough to call for the *utmost possible concentration of strength* permissible under the circumstances. A major battle in a theater of operations is a collision between two centers of gravity; the more forces we can concentrate in our center of gravity, the more certain and massive the effect will be. Consequently, any partial use of force not directed toward an objective that either cannot be attained by the victory itself or that does not bring about the victory should be *condemned*.

The basic condition, however, does not consist merely in the greatest possible concentration of forces; they must also be deployed in a way that enables them to fight under sufficiently favorable circumstances.

The various gradations of defense, which were dealt with in the chapter on types of resistance, are completely consonant with these basic conditions; there can, therefore, be no difficulty in establishing a connection between them according to the needs of the individual case. There is only one point that, at first sight, seems self-contradictory, and that, because it is one of the most important points in defense, is all the more in need of further development: it is how to hit the enemy's exact center of gravity.

If the defender finds out early enough by what roads the enemy will advance, and on which of them the core of his force is to be found, he will be able to confront him there. That is what normally happens; for while defense may anticipate attack by general precautions—such as building fortresses, stocking major arms depots, and peacetime disposition of forces, thus establishing the lines that the attack must follow when actual hostilities begin—the defender also possesses the inherent advantage over the offensive of being able to make the riposte.

An advance into enemy territory by a sizable force calls for considerable preparations such as building up food stocks and supplies or equipment and so forth. This will take long enough to give the defender time to make his own preparations. And one must not forget that the defender normally needs less time than the attacker, since all states are usually better prepared for defense than for attack.

But while this may be perfectly true in the majority of cases, the possibility still remains that in a particular instance the defense may not be sure of the main route of the enemy's advance. This is more probable when the defense relies on measures that themselves take time, such as the preparation of a strong position. Furthermore, even where the defender is blocking the line of advance—provided he does not himself assume the offensive by giving battle—the attacker can avoid the defender's position by a slight change in his original line of march. In the settled parts of Europe one will never be at a loss for roads on which to by-pass a position on one side or another. In that case the defender obviously will not await his opponent in position, not at least with the intention of giving battle there.

Before we discuss the means that remain available to the defender in this situation, we must examine it more closely and consider the probability of its arising.

In every state and therefore in every theater of war (which is our chief concern here) there will, of course, be certain objectives and points that offer the most effective target for an attack. The detailed discussion of this topic belongs in the book on the attack. At this stage we merely want to stipulate the following: if the most advantageous objective and target of attack determines the direction of the offensive, the same reasoning will affect the defender as well, and must guide his dispositions whenever he does not know the intentions of his opponent. If the attacker were to fail to take the most favorable direction, he would forfeit some of his natural advantages. If the defender does lie across that route, the remedy of avoiding and by-passing him is clearly no bargain; it has to be paid for. It follows, therefore, that neither the defender's risk of *misjudging the direction of the attack* nor *the enemy's ability to by-pass him* is so great as might at first appear. In fact, a definite and usually overriding reason for taking one route or the other is already given. Thus, the defender, though his dispositions may be tied to a certain place, is generally in no danger of missing the enemy's main force. In other words, *if the defender has taken up the right position, he can be fairly sure that the attacker will seek him out there.*

Still, there is no denying the possibility that the defender's dispositions may at times fail to connect with the attack. The question therefore arises what to do then, and how much of the original advantage of his situation will still remain.

The choices open to a defender who is being by-passed are the following:

1. He can split his forces in two from the very beginning in order to be sure to catch the enemy with one part, while the other hurries to its assistance.

2. He can concentrate his forces in one position, and if the enemy passes him by, move quickly to the flank. In most cases, such an advance can no longer be made precisely to the flank: the new position will have to be taken somewhat further to the rear.

3. He can throw his full force at the enemy's flank.
4. He can operate against the enemy's lines of communication.
5. He can mount a counterattack on the enemy's theater of operations, thereby producing the same effect on the enemy the latter meant to produce in by-passing him.

The last choice is mentioned here because a case can be conceived in which it would be effective. But basically it conflicts with the intent of the defense, or rather with the reasons for which it was chosen. It must, therefore, be considered as an abnormal situation that can be caused only by major errors on the part of the enemy or by other peculiarities of the individual case.

Operating against the enemy's line of communication presupposes the superiority of our own line of communication, which, indeed, is one of the basic elements of a good defensive position. But while such an action promises the defender some benefits, in defending a theater of operations it seldom leads to the *decision* that we have assumed to be the object of the campaign.

A single theater of operations is rarely large enough to make the attacker's lines of communication critically vulnerable. Even if they were, the effects of an action against them are too gradual to delay seriously the execution of the enemy's plans, which usually do not require much time.

In most cases, therefore, an action against lines of communication will be totally ineffective if the enemy is bent on gaining a decision—nor will it help bring about a decision for the defender.

The three other means left to the defender are more suited to the objective, because they aim at an immediate decision, a confrontation of the two centers of gravity. We shall say at once, however, that we decidedly prefer the third to the other two. While we would not reject these completely, we believe the third to be the appropriate means of resistance in most cases.

By splitting one's forces, one is in danger of getting involved in a war of outposts: against a determined enemy, this will at best result in a *substantial limited resistance*, but never in the decision that was intended. Even if that pitfall can be avoided, the assault will be noticeably weakened because of the temporary split in the defense. Nor can one ever be sure that the advance units will not suffer disproportionate losses. Furthermore, the resistance rendered by these units generally ends in their falling back on the main force, which is hurrying to their support. To the troops this will usually look like a defeat or failure, and will in this way significantly lower their morale.

The second way—intercepting the opponent with combined forces on the very route by which he means to evade us—involves the risk of arriving too late, and thereby falling between two stools. What is more, a defensive battle calls for calm, reflection, knowledge of the area, indeed real familiarity with it; none of these can be expected if we are engaged in rapid movements. Finally, positions that make good defensive battlefields are scarce; one cannot assume that one will find them on every road and at every turn.

Great advantages, however, attach to the third course—that of falling on the attacker from the flank and thereby forcing him to fight a battle with a change of front.

First of all, as we have already seen, this is bound to make him expose his lines of communication—in this case lines of retreat. The advantage that accrues to the defender derives from his condition in general, but more especially from the strategic properties that we have claimed for his position.

Furthermore—and this is the main point—an attacker trying to by-pass his enemy is involved in doing two incompatible things simultaneously. His first concern is to advance and reach his objective; but as he may be attacked from the flank at any moment, he feels he must also be ready to strike back instantly, and with full force. These two aims are mutually exclusive: they create so much confusion and make it so difficult to cover all eventualities that one would be hard put to conceive of a worse strategic situation. If the attacker knew exactly when and where he would be assaulted, he could prepare himself with skill and resource; but in his uncertainty and the necessity of keeping up his advance, a sudden battle can scarcely fail to find him badly concentrated, and thus certainly not in an advantageous position.

If there is ever a suitable occasion for the defender to fight an offensive battle, one must surely expect it under such circumstances. If we further bear in mind that the defender has the advantage in knowledge and choice of terrain, that he is able to prepare his movements and initiate them, there can be no doubt that under these circumstances he will retain definite strategic superiority over his enemy.

We therefore feel that a defender who is located with his full force in a well-placed position can safely wait to be by-passed. If his opponent does not seek him out, and if circumstances prevent the situation from having an effect on his opponent's lines of communication, he still has an excellent means of producing a decision by falling on the enemy's flank.

Historically this seldom occurs. The reason is partly that defenders rarely dare to hold out in such a position; they would rather split their forces or hasten to cut the attacker off by means of oblique or lateral marches. Besides, an attacker will not dare to by-pass a defender under such circumstances, and this usually brings him to a halt.

In such a case, then, the defender is forced to fight an offensive battle and to forego the further benefits of *waiting, of a strong position, and of good entrenchments,* etc. As a rule, the situation in which he finds the advancing enemy will not completely make up for the lack of these advantages; after all, it was to circumvent them that the attacker exposed himself to these conditions. It does afford a *certain amount of compensation,* however. Thus, this is not an instance where the theorist finds that a quantity has suddenly vanished and *pro* and *contra* cancel each other out, as so often happens when military critics introduce a fragment of theory into their work.

We do not mean to imply that this is a matter of logical subtleties; on the contrary, the more one looks at the practical side of the matter, the more one sees that the idea applies to the whole field of defense, ruling and permeating its every aspect.

Only if the defender is determined to attack the enemy in full strength as soon as he has been by-passed can he avoid the two pitfalls that beset his path so closely—a divided position and a hasty advance. Either way he is governed by the conditions of the offensive; either way he must make do with makeshift expedients and dangerous haste. Consequently, wherever a determined adversary, intent on pursuing victory and reaching a decision, has encountered this type of defensive system, he has demolished it. On the other hand, a defender who has concentrated his troops to fight as a single force in the right place, and who is determined to attack the enemy in the flank if things come to the worst, is and will be on the right course, backed up by all the advantages defense can offer in his situation. *Sound preparation, composure, confidence, unity, and simplicity* will mark his conduct of the action.

In this connection we cannot help but mention an important historical event, on which the ideas here developed have a close bearing. We do so mainly to prevent erroneous deductions. In October 1806, the Prussian army in Thuringia lay in wait for Bonaparte between the two main roads on which he could advance—the road via Erfurt, and that via Hof toward Leipzig and Berlin. This intermediate position resulted from an earlier intention to advance straight through the Thuringian Forest into Franconia, and later, when this plan had been abandoned, from uncertainty as to which of the two roads the French would use. It should, therefore, have led to a rapid move to block the French advance.

This, in fact, was what the Prussians meant to do if the enemy came by way of Erfurt, for the roads that led there were perfectly accessible. On the other hand, to block the road from Hof was inconceivable, because that road was two or three days' march away, and because the deep valley of the Saale lay between. The Duke of Brunswick had never thought of such a move, nor had any sort of preparations been made for it. It had, however, always been intended by Prince Hohenlohe, or rather by Colonel Massenbach, who tried his utmost to draw the Duke into this scheme. Still less tenable was the idea of leaving the position on the left bank of the Saale to fight an offensive battle against Bonaparte as he advanced—in other words, of falling on his flank in the way described above: if the river was an obstacle to a last-minute interception of the enemy, it would be an even greater obstacle to a sudden attack at a time when he was already in possession, at least partially, of the far bank. The Duke decided, therefore, to remain behind the Saale and await developments—if one can speak of a personal decision at all where such a hydra-headed staff, in a state of chaos and perpetual vacillation, was concerned.

Whatever the truth may be about the decision to wait, the choices that resulted were as follows:

a. The enemy could be attacked as he crossed the Saale to advance on the Prussian army or

b. If he chose to leave the Prussians alone, his line of communication could be harassed, or

c. If it was feasible and advisable, the enemy could still be intercepted at Leipzig by means of a rapid march to the flank.

In the first instance, the depth of the Saale Valley afforded the Prussian army great strategic and tactical superiority. In the second instance, the Prussians' purely strategic superiority was just as great, for the enemy's base between them and the neutral territory of Bohemia was very narrow, whereas the Prussians' was exceptionally broad. Even in the third instance, the Prussians were not at a disadvantage, because they were covered by the Saale. In spite of confusion and uncertainty, all three of these possibilities were actually discussed at headquarters. It is not surprising, however, that while an *idea* might be able to prevail in a state of chaos and indecision, its *execution* was bound to perish in such a maelstrom.

In each of the first two cases, the position on the left bank of the Saale would have amounted to a genuine flank-position, and as such its merits were unquestionably very great; but for an army that was not very sure of itself, pitted against a vastly superior enemy such as *Bonaparte*, a flank position was *a very bold measure* to adopt.

On the thirteenth, after a long period of hesitation, the Duke chose the last of the above three measures, but it was too late. Bonaparte had already begun to cross the Saale and the battles of Jena and Auerstädt were bound to follow. In his vacillation, the Duke had fallen between two stools: he had left the area too late to *intercept* the enemy, and too early to fight a sound battle. Nevertheless, the natural strength of his position was so great that he should have been able to annihilate the right wing of the French at Auerstaedt, while Prince Hohenlohe, by means of a costly rear guard action, should have been able to avoid being trapped at Jena. But at Auerstaedt the Prussians did not dare to hold out for a *certain* victory; at Jena, where it was *completely impossible*, they thought they could count on it.

In any case, Bonaparte's respect for the strategic value of the position on the Saale was such that he did not dare to by-pass it, but preferred to cross the Saale under the enemy's eyes.

The foregoing has, we trust, sufficiently described the relationship of defense to attack in cases calling for decisive action, and defined, according to their position and coherence, the threads that knit individual parts of a defense plan together. We do not intend to explore further details, which would only lead to a boundless maze of individual cases. Once a general has resolved to pursue a specific objective, he will be able to judge how circumstances of geography, statistics and politics, and conditions of matériel and personnel in his own and the enemy's army will fit into it, and he may then adjust his plans accordingly.

The successive stages of defense, which we introduced in the chapter on types of resistance, will now be more clearly defined, and their bearing on the present matter in general more closely examined.

1. The following may be causes for approaching the enemy with the intention of fighting an offensive battle:

a. If one knows that the enemy's forces are widely dispersed and that therefore, even where one's own strength is inferior, there is some prospect of victory.

A dispersed advance is not very likely; such a plan is therefore sound only where one has prior knowledge of the enemy's moves. Merely to make such an *assumption* on insufficient grounds, count on it, and base all one's expectations on it, will usually lead to an unfavorable situation. Circumstances may not turn out as expected, the idea of an offensive battle will have to be abandoned, and no preparation will have been made for a defensive one. An involuntary retreat has to be initiated, and almost everything is left to chance.

That is more or less what happened to the defense undertaken by Dohna's army against the Russians in the campaign of 1759, which came to a disastrous end under General Wedel at the battle of Züllichau.[1]

Because it settles the matter so quickly, planners are only too eager to propose this type of procedure without verifying the underlying assumptions.

b. If, in general, one has enough strength for battle, and

c. If a very clumsy, vacillating enemy invites attack.

In such a case, the effect of surprise may be of greater value than all the benefits of terrain in a favorable position. It is the very essence of good generalship to use the power of psychological forces in this way. Still, a theorist cannot emphasize loudly and strongly enough that there must be *objective reasons* for these assumptions; without such *specific reasons* it is unsuitable and quite unwarranted to talk about surprise, or the merits of an unconventional attack, and to base plans and arguments and criticisms on them.

d. If the composition of the army makes it particularly suited to the offensive.

The army of Frederick the Great was flexible, brave, and confident; it was used to discipline, drilled to perfection, and animated and sustained by a sense of pride. He was certainly not mistaken or unjustified in his belief that this army, trained in his method of oblique attack, was an instrument that in his sure, practiced hand was better suited to attack than for defense. These were the very qualities that his opponents lacked, and it was these qualities that gave him a definite superiority. In most cases, they were of

[1] I.e., Kay. See the note on p. 438 above. Eds.

greater value to him than all the benefits of entrenchments and natural obstacles. Such superiority, however, remains rare: it calls for *more* than a well-trained army used to large-scale movements. One should not give too much credence to Frederick's remark—which has since been incessantly repeated—that the Prussian troops were particularly suited to the offensive: it is normal in a war that spirits and courage are higher in the attacker than in the defender. This is a feeling common to all troops, and there is hardly an army whose generals and officers have not advanced the same claim. One must be careful not to be taken in by this semblance of superiority, while neglecting some solid advantages.

An army's composition may constitute a very natural and weighty argument for fighting an offensive battle—when the army contains abundant cavalry and not much artillery.

To continue the list of reasons for attacking the enemy:

e. If it is impossible to find a favorable position
f. If the need for a decision is urgent
g. Finally, several or all of these reasons may act together.

2. The most natural reasons for awaiting the enemy in an area where one wants to attack him (as at Minden in 1759) can be found in the following:

a. Where the difference between the forces is not greatly to our disadvantage, and we are therefore not obliged to look for a strong entrenched position

b. Where the area is particularly suited to the purpose. The features that determine this belong to the tactical realm; we might just mention the main attributes: easy accessibility for our own approach and numerous obstacles for the enemy's.

3. A position in which one really means to await the enemy should be taken up:

a. If an imbalance of strength forces us to seek shelter behind natural obstacles and entrenchments

b. If the terrain is particularly well suited to such a position.

The second and third types of resistance will deserve greater consideration to the degree to which one does not seek the decision oneself, will be content with a negative success, and can expect the enemy to falter, show indecision, and finally abandon his intentions.

4. An entrenched and unassailable encampment fulfills the purpose only

a. If it is located in an area of particular strategic significance.

The distinctive feature of such a position is that it cannot be overrun; the enemy is thus forced to try every other means available; for instance, pursuing his objective regardless of the position, or surrounding it and starving the garrison. If he is unable to do either, the strategic qualities of such a position are great indeed.

b. If one has reason to expect help from outside.

Such was the case of the Saxon army in its position at Pirna. No matter what has been said about this measure after its unfortunate outcome, the fact remains that 17,000 Saxons could never have neutralized 40,000 Prussians in any other way. If the Austrian army at Lobositz made no better use of the superiority it gained from the Prussians' neutralization, it only goes to show how poor its whole organization and methods were. There is no doubt that Frederick the Great would have chased the Austrians and Saxons beyond Prague in that campaign, and taken the city as well, if the Saxons, instead of occupying the camp at Pirna, had moved in to Bohemia. Whoever denies the value of that feat, and only remembers the capture of the Saxon army, does not know how to calculate matters of this kind, and without calculation there can be no sure result.

But since cases such as a. and b. are very rare, to resort to the measure of entrenched camps requires careful reflection. Only seldom is it properly used. The hope of *impressing* the enemy with such a camp and thereby paralyzing his activities is linked to too great a danger—that of having to fight without a way of retreat. If Frederick gained his objective at Bunzelwitz in that way, one must admire his accurate evaluation of his opponents; but one must also lay more stress than would ordinarily be permissible on the means he would have found to break out with the remnants of his army if things had gone wrong, and—secondly—on the fact that as King, *he was accountable to no one.*

5. If one or more fortresses are located near the frontier, the major question that arises is whether the defender ought to aim at a decision in front of them or behind them. Reasons for the latter are to be found in:

a. The enemy's numerical superiority, which forces us to wear him down before the battle, is joined

b. The proximity of the fortresses, so that the land lost can be kept to an absolute minimum

c. The defensive capacity of the fortresses.

Unquestionably it is, or should be, one of the principal functions of a fortress to brake the enemy's advances, and substantially weaken that part of his forces which we seek to defeat decisively. If fortresses are only rarely used for this purpose, it stems from the fact that only rarely is a decision sought by either party. That, however, is the only case we are considering at the moment. We therefore regard it as a simple but important principle that a defender who has one or more fortresses in the vicinity should keep them in his front, and fight the decisive battle behind them. We will admit that a battle that is lost behind our own fortresses will drive us somewhat further back into our own country than one that is lost with the same tactical results in front of them; but the origins of this difference are more imaginary than real. We also realize that, in a well-chosen position, a battle can be fought on the far side of the fortresses, while a battle on the near side may in many cases turn into an offensive one if the enemy lays siege to a fortress and it is in danger of being taken. But what do these fine points amount to against the advantage of having the enemy's strength in the decisive battle reduced by a quarter or a third—or indeed, if there are several fortresses, by half?

We feel, therefore, that where a *decisive battle is unavoidable*—whether it be desired by the enemy or by our own commander—and where one is not quite sure of victory to begin with, or where the terrain does not dictate the choice of a battlefield farther forward, in such cases a nearby fortress capable of resistance is a powerful argument for retiring behind it in order to seek the decision behind it, and so gaining the benefit of its participation. Besides, if we take up our position so close to the fortress that the enemy can neither besiege nor invest it without driving us off, we will force him to attack us in our position. We therefore think that no defensive measure in a dangerous situation is so simple and as effective as the choice of a good position close to and behind a substantial fortress.

Matters would be different, of course, if the fortress were located far to the rear. One would then have to abandon much of one's theater of operations—a sacrifice that should only be made, as we have seen, if the circumstances demand it. In that event, the measure comes close to being a retreat into the interior of the country.

A fortress's power of resistance is an additional factor. There are fortified localities, large ones in particular, which should never be allowed to come in contact with an enemy, since they are unequal to a strong attack by a substantial force. In such a situation, our positions must at least be close enough behind them to act as a support to the garrison.

6. Finally, withdrawal into the interior of the country is a proper course only under the following circumstances:

a. When our physical and psychological situation vis-à-vis the enemy rules out the possibility of successful resistance at or near the frontier
b. When our main objective is to gain time
c. When the condition of the country is favorable to it, as shown in Chapter Twenty-Five above.

This ends the chapter on defense of a theater of war for cases in which a decision is sought by one side or the other, and is therefore unavoidable. But we must of course remind the reader that in real war things are never so clear-cut. It follows that if one has a mind to apply our statements and arguments to real war, one must also look to Chapter Thirty and note that most generals will have to choose between the two courses, coming *closer* to one or the other, according to circumstances.

CHAPTER TWENTY-NINE

Defense of a Theater of Operations—Continued: *Phased Resistance*

In Chapters Twelve and Thirteen of Book Three we demonstrated that resistance in successive phases goes against the very nature of strategy and that all available forces should be used simultaneously.

So far as mobile forces are concerned, this needs no detailed exposition. But the theater of war considered as a fighting force itself, with all its fortresses, its natural obstacles, and its sheer expanse of surface, is immobile. It may thus either be activated in stages, or else we must immediately retreat so far as to have all the participating components in our front. In that case, every influence that an army's theater of war can exert to weaken the enemy will become effective. The enemy must invest your fortresses, secure the area with garrisons and strongpoints, make lengthy marches, procure supplies from far away, and so forth. He will experience all these effects whether he advances *before the decision or afterwards,* though they will be somewhat more damaging before than after. It follows that if a defender decides early enough to shift the decision farther back in time and space, he will find this to be a means for bringing all immobile forces into play simultaneously.

It is clear, on the other hand, that, strictly speaking, this postponement of the decision will have no effect on the sphere of influence that a victory gives the attacker. This sphere of influence will be examined more closely under the heading of "attack"; but we do want to point out that it will extend to the point at which the superiority (the product of the physical and psychological relation) is exhausted. This superiority is drained by two factors: the demands made on the fighting forces by the theater of war itself, and the losses sustained in battle. Neither of these is changed substantially whether engagements are fought early or late, far forward or to the rear. For example, we believe that in 1812 a victory over the Russians at Vilna would have carried Bonaparte just as far as did the victory at Borodino, provided it had the same dimensions; and that even a victory at Moscow would not have carried him any further. Moscow was the limit of his sphere of victory in any case. Indeed, there is not the slightest doubt that a decisive battle at the frontier would (for different reasons) have produced far greater results, and possibly a broader sphere of victory. This is therefore a consideration that will have no effect on the defender's shifting the point of decision to the rear.

In the chapter on types of resistance we have described, under the label of *withdrawal into the interior of the country,* what can be considered the ultimate form of postponing a decision. This particular form of resistance aims at making the invader wear himself out rather than at defeating him in battle. But a postponement of the decision can be considered as a special *form of resistance* only where that is the primary objective; otherwise, an infinite number of gradations in this method is obviously conceivable and every one of them can be combined with every method of defense. The degree of participation by the theater of war should, therefore, not be considered as a special type of resistance, but merely as an optional admixture of immobile means of resistance, to be used as needed, according to the conditions and circumstances.

If the defender feels that he does not require the aid of these immobile forces, or if the sacrifices involved are too arduous in other respects, these forces remain on hand for a later stage. They can then be used like fresh reinforcements for which the defender had not been able to wait previously, and can become the means by which the mobile forces can follow up the first decision with a second, and possibly even a third. In other words, in this way a gradual application of strength becomes feasible.

If the defender has lost a battle on the frontier, it is quite possible—provided it does not amount to a major defeat—that he may be able to accept another to the rear of his next fortress. Indeed, a serious natural obstacle alone may be enough to bring a not-too-resolute opponent to a halt.

In the utilization of a theater of war, as in everything else, strategy calls for *economy of strength.* The less one can manage with, the better; but manage one must, and here, as in commerce, there is more to it than mere stinginess.

In order to avoid a serious misconception, we want to make it clear that the amount of resistance rendered or attempted after a defeat is not what we are discussing. What matters is the amount of success that can be expected in *advance* from such renewed resistance—how much value the overall plan should place on it. There is hardly more than one way in which a defender should look at this: from the point of view of the enemy, of his character and situation. If his character is weak, if he lacks self-confidence and overriding ambition, and if his freedom of action is closely limited, he will, if he is successful, be content with a modest advantage. Every new chance for decision that the defense dares to offer, will make him hesitate irresolutely. In that case, the defender can count on making the means of resistance that his theater of operations offers felt gradually, in an unremitting series of decisive actions, which, though individually of no great consequence, will always hold out the possibility of turning these decisions in his favor.

Surely it has become obvious by now that all this leads to the subject of campaigns that reach no decision. These are the real arena for the gradual use of strength, and will be discussed in the next chapter.

CHAPTER THIRTY

Defense of a Theater of Operations—Concluded: *Where a Decision Is Not the Objective*

In the last book we shall deal with the question whether and in what way a war can take place if neither side attacks the other—in other words, in which neither side has a positive objective. At this point, this contradiction need not concern us: in the individual theater of operations, we can simply assume the reasons for a defensive stance on both sides to be given by the relations of each of these parts to the whole.

This is not the only type of campaign that lacks the necessary focus of decision. History records numerous cases that do not lack for an aggressor or a positive ambition on one side at least, but where this ambition is not pronounced enough to be relentlessly pursued until it leads to the *inevitable* decision. In such wars, the attacker seeks no advantages beyond those offered by the circumstances at hand. Either he has set himself no objective to pursue, and is merely harvesting the fruit that may ripen in the course of time; or, if he has an objective, he has made the pursuit of it dependent on favorable circumstances.

This type of attack, which ignores the strict logical necessity of pressing on to the goal, is like an idler who strolls through a campaign and takes advantage of the occasional bargain that comes his way. It does not differ greatly from the defense that also allows its general to pick up a bargain, but we shall reserve its more detailed scientific study for the book on attack. At this time we will only state the conclusion, which is that in such a campaign neither the offensive nor the defensive will keep the need for a decision paramount. The decision in that case ceases to be the keystone of the arch upon which all the lines of strategy converge.

The history of war, in every age and country, shows not only that most campaigns are of this type, but that the majority is so overwhelming as to make all other campaigns seem more like exceptions to the rule. Even if this ratio changes in future, it is certain that there will always be a substantial number of campaigns of this kind, and that aspect must have its due in any doctrine of defending a theater of operations. We shall try to indicate those attributes that appear to delimit it. In reality, most wars will probably fall between the two poles, sometimes approaching one, sometimes the other. The practical effect of these attributes becomes evident only as a modification, caused by their contrary action, in the *absolute form* of war.

We have argued in Chapter Three of this book that the state of waiting is one of the greatest advantages the defense enjoys over the attack. It seldom happens in real life, and even less often in war, that everything circumstances would lead one to expect actually takes place. Because of the limits of human insight, the dread that things might go wrong, and accidents that change the course of action, many possible options are never chosen, even though circumstances would have favored them. In war, where imperfect intelligence, the threat of a catastrophe, and the number of accidents are incomparably greater than in any other human endeavor, the amount of missed opportunities, so to speak, is, therefore, also bound to be greater. This is the fertile field in which the defender may glean a harvest he did not sow. Add to this the intrinsic value represented by territory in the conduct of war; what follows is a maxim that is also revered in the battles of civilian life—litigation: *beati sunt possidentes*. This principle here takes the place of the *decision*, which, in all wars directed toward *mutual defeat*, is the focal point of the whole process. It is an uncommonly fertile principle—not, of course, evoked in action, but in disclosures and motives for inaction, as well as for the kind of activity that is meant to result in inaction. Where no decision is intended or expected, there is no reason for abandoning anything; that would be done only in order to purchase advantages when the time for decision arrived. Consequently, the defender's aim is to hold on to—that is, to cover—as much as possible, while the attacker will try to take as much as he can—that is, to spread his forces as widely as possible—without provoking a decision. Only the former concerns us here.

Wherever there is no defending force, the attacker can take possession; then the benefit of waiting is *his*. The defender will, therefore, attempt to cover all territory directly, and then take a chance on the enemy's willingness to attack his covering forces.

Before launching into a more detailed description of the special features of defense, we must anticipate the book on attack by listing the objectives that an offensive usually pursues when no decision is intended. These are the following:

1. Seizure of a considerable amount of territory, if this is possible without a decisive engagement

2. Capture of an important supply depot, on the same conditions

3. Capture of a fortress left without cover (It is true that a siege is a fairly serious business and can cost much effort; but it cannot lead to disaster. If the worst comes to the worst, it can simply be abandoned, without one's suffering any real loss.)

4. Finally, victory in a moderately important engagement, in which not much is risked and consequently not much can be gained; not an engagement fraught with consequence, the climax of a whole strategic plan, but one that is fought for its own sake only—for trophies or for military glory. If that is its purpose, it will not, of course, be entered into at any price: one either waits for any opportunity that may arise, or attempts artificially to provide one.

These four objectives of the offensive now call forth the following efforts on the defender's part:

1. To cover his fortresses by keeping them in his rear
2. To cover the country by spreading out his forces
3. To interpose his forces quickly by means of flank marches wherever his extension is not wide enough
4. To avoid, at the same time, any unfavorable engagements.

The first three aims are obviously intended to result in forcing the initiative on the offensive, and to draw the utmost benefit from waiting. The purpose is so rooted in the nature of the case that it would be foolish to condemn it out of hand. The less a decision can be expected, the more valid this objective will become. It is the governing principle in all such campaigns, even though superficially there may appear to be a lot of brisk activity, in the form of minor skirmishes which do not lead to decisive consequences.

Hannibal as well as Fabius, and Frederick the Great as well as Daun, embraced this principle whenever they did not seek and did not expect a decision. The fourth objective, which serves as a corrective to the other three, is their *conditio sine qua non*.

We shall now discuss these matters in more detail.

To protect a fortress against enemy attack by positioning an army in front of it may at first sight seem absurd and even superfluous when, after all, fortresses are built for the purpose of resisting enemy attacks. Nevertheless, we observe this measure occurring thousands of times. It is typical of the conduct of war that the most common things in it often appear to be the most incomprehensible. Who could muster enough courage, based on this apparent contradiction, to declare that this oft-repeated measure was a mistake every time it was employed? The ever-recurring form proves that there must be some deep-seated reason for it. It is none other than that cited above: sheer psychological inertia.

If we take up a position in front of our fortress, the enemy cannot attack the fortress without first defeating our army. But a battle implies a decision. If the enemy does not want one, he will not fight a battle; and we can hold our fortress without striking a blow. Whenever we doubt that the enemy seeks a decision, we must take a chance: the likelihood is that he will not do so. And in most cases, one still has the possibility of withdrawing behind the fortress if, contrary to expectations, the enemy does decide to attack. This minimizes the dangers of taking a position in front of the fortress; the *practical* certainty that the *status quo* can be maintained without sacrifice will not entail even a *remote* degree of risk.

By taking up position behind the fortress, we would present the attacker with an ideal target. Unless the fortress is very strong, and he is completely unprepared, he will, for better or worse, start to lay siege to it; and in order to prevent its falling into his hands, he will be obliged to relieve it. The positive action, the initiative, is now up to us; the enemy, whose siege can be considered as an advance toward his goal, is in possession. Experience shows that matters always take this turn; they are bound to do so by their very nature. As we have said, a siege does not have to end in disaster. The feeblest, idlest, and least enterprising general who would never bring himself to fight a battle, will cheerfully besiege a fortress just as soon as he can reach it—even with nothing heavier than field-guns. At worst, he can always give up the enterprise without suffering any real losses. To that development must be added the danger to which most fortresses are more or less exposed—that of being taken by storm or in some other unconventional manner. These circumstances must not be overlooked when the defender calculates the possibilities.

Measuring these two chances against one another, the defender, rather than choosing the advantage of fighting on better terms, would naturally choose the virtual certainty of *not* having to fight *at all*. Seen in that light, the practice of taking up position in front of a fortress becomes perfectly natural and understandable. Frederick the Great observed it almost always—at Glogau against the Russians, at Schweidnitz, Neisse, and Dresden against the Austrians. At Breslau, on the other hand, it did not serve the Duke of Bevern well: he could not have been attacked behind Breslau, the Austrians had the upper hand so long as the King was absent, and they knew that their superiority was unlikely to last once he came close. Thus, the juncture at which the battle of Breslau took place *was far from being one at which no decision was to be expected*—which makes the position before Breslau less appropriate. The Duke of Bevern himself would surely have preferred to take up a position on the far side of Breslau; but that would have exposed the city and its supply depots to bombardment, and would have earned him the severe displeasure of the King who, in cases of this kind, could be quite small-minded. One cannot ultimately blame the Duke for *making an attempt* at saving Breslau by taking an entrenched position in front of it: it was entirely possible that it might have stopped Prince Charles of Lorraine from making further advances, since he was quite satisfied with his occupation of Schweidnitz, and was threatened by the King's approach. The best solution would have been to avoid involvement in a battle and to withdraw through Breslau the moment the Austrians advanced. The Duke would then have had all the benefits of waiting without having to pay for them by taking risks.

We have now explained and justified the higher and overriding arguments for a commander's taking up his position in front of a fortress; but we should say there is another, secondary reason for doing so—more obvious, perhaps, but not valid in itself, for it is not of universal application. It is the habit armies have of using the nearest fortress as a depot for their supplies. That is so convenient, and so full of advantages, that a general will not easily be persuaded to draw supplies from far away or to deposit them in unprotected places. Where a fortress serves as a supply depot, it will in many cases become absolutely necessary for an army to position itself before the fortress, and in most cases it will be the natural thing to do. The point is so obvious that those who are disinclined to look further are apt to give it too much weight; but it cannot serve to explain every instance, nor are its implications important enough to account for the final decision.

It is the normal objective of all attacks that do not aim at a major decision to capture one or more fortresses without incurring the risk of a battle; so much so that the prevention of that aim becomes a major item in the efforts of the defense. That is why, in theaters of war where there are plenty of fortresses, almost every movement turns on their possession. The attacker tries to approach them unexpectedly, using various feints, while the defender attempts to forestall this by means of well-planned movements. That was characteristic of almost all campaigns in the Low Countries between the days of Louis XIV and those of Marshal Saxe.

So much for the covering of fortresses.

Covering the country by spreading out our forces is conceivable only in combination with major natural obstacles. The posts of various sizes that have to be established can only acquire a certain capacity of resistance from the strong positions they occupy; since natural obstacles are seldom plentiful enough, the art of entrenchment must come to their assistance. It should, however, be recalled that the strength produced by entrenchments at any given point is only *relative* and should never be regarded as *absolute*. (See the chapter on the significance of the engagement.) It may happen, of course, that a post is able to ward off all attacks and thus achieve an absolute defense; but out of all the numerous posts, each individual one must be regarded as relatively weak in relation to the whole, and vulnerable to possible assault by heavily superior forces. It would be unwise, therefore, to base one's confidence on the resistance that can be rendered by each individual post. With this type of an extended position, one may expect a fairly long resistance at best, but never a proper victory. Even so, the individual post can serve its purpose and contribute to the general objective. In campaigns where one need not be afraid of great decisions, of being relentlessly driven toward disaster, not much risk will be involved if a post becomes engaged in a skirmish, even if it ends in its loss. The stake is seldom more than the post itself, plus a few trophies. It will not be a victory of any consequence; it will not shatter foundations and bring down the walls. At worst, if the whole defensive system was breached by the loss of a single post, there would still be time for the defender to concentrate his forces and, *offer* the enemy the decision that the latter, according to our hypothesis, does not seek. Usually, such a concentration of forces puts an end to the matter, and the attacker's further advance is brought to a halt. A little land, a few men, and some guns constitute the defender's loss and the attacker's adequate gain.

To such a risk, if things go wrong, the defender may safely expose himself if the risk is balanced by the possibility, or better yet the probability, that it will not happen at all: timidity, or prudence—call it what you will—may bring the attacker to a halt before the defender's posts without his battering his head against them. In advancing this argument, we must never forget that it assumes an attacker who will not take great risks, who may well be stopped by a moderate-sized but strong post. Even if he knows that he can take it, he will still wonder at what price, and whether that price might not be too high in relation to the use he can, in his present position, make of his victory.

This demonstrates that from the defender's point of view the strong relative resistance afforded him by a position extending through a long line of posts can make a worthwhile contribution to the sum of his campaign. Reference to military history (to which the reader will turn at this point) will show that such extended positions are most common in the latter part of a campaign. By then the defender has learned to estimate the attacker's aims and potentialities for the current season, while the attacker, for his part, will have lost what little enterprise he started out with.

Defense in an extended position that covers *land, supplies,* and *fortresses,* must, of course, assign a principal role to all major natural obstacles—streams and rivers, mountains, forests and marshes; they are of paramount importance. Their use has been described in earlier chapters.

This paramount importance of topography makes special demands on the type of knowledge and activity that we primarily associate with the General Staff. Since that is the branch of the army that tends to write and publish the most, these aspects of a campaign are the most fully documented. At the same time there is a rather natural tendency to shape these into a system, using the historical interpretation of a single case as the basis of a general principle applicable to all cases. But that is futile and thus wrong. In this more passive and more locally conditioned form of warfare, every case is different and must be treated differently. Even the ablest critical memoirs on these subjects are, therefore, suited only to present the facts, never to serve as a precept. They really revert to being a *military history,* concerned with only one particular aspect of the wars they describe.

The activities of the General Staff that we have, in accordance with the general view of the matter, designated as its special province, are useful and meritorious. However, we must sound a warning against abuses to which they may be put, which frequently harm the enterprise as a whole. The authority acquired by those leading members of the staff who are most competent in military topography often gives them a kind of general dominance over others, especially over the commander himself, which may result in one-sidedness. In the end, the commander-in-chief can see nothing but mountains and passes; and in the place of rational decisions freely arrived at, automatic responses will become second nature to him.

Thus, in the Prussian army of 1793 and 1794, Colonel Grawert, who was the very soul of the General Staff at the time, and a well-known specialist in mountains and passes, managed to persuade two generals with entirely different personal characteristics—the Duke of Brunswick and General Möllendorf—to follow identical courses in their conduct of war.

It is clear that a defensive line that runs parallel to a major natural obstacle may well lead to a cordon war. It would normally be bound to do so if the full extent of the operational theater were really to be directly covered in that manner, for, compared with the breadth of most of such theaters, the natural tactical extension of the forces detailed to defend it is extremely small. External circumstances as well as his own dispositions, however, will limit the attacker to certain general routes and roads. If he were to deviate from these excessively, he would run into too much trouble and inconvenience, no matter how inactive the defender. Therefore, all the latter needs to do is to cover the area for a certain number of miles or marches to either side of the relevant main routes. To effect such cover, one can simply set up defensive posts on the main roads and points of access; observation posts will suffice for the area between them. Obviously an enemy column may then pass between two posts and attack one or the other of them from several directions. The posts, in turn, are somewhat prepared for this eventuality, partly by having flank supports, partly by forming flank defenses (so-called crochets), and partly by being able to obtain aid from reserves in the rear or from units sent by an adjoining post. In this manner, the number of posts is still further reduced, with the result that an army engaged in this type of defense will be grouped into four or five main posts.

Special centers are set up for important points of access, which are distant but may still be in danger. These form small operational zones, so to speak, within the main theater of war. In the Seven Years War, for example, the main Austrian army generally held four or five posts in the mountains of Lower Silesia, while in upper Silesia a similar system of defense was created by a small and partly independent force.

As a system of defense moves away from the method of direct cover, it has to place additional reliance on mobility, active defense, and even offensive measures. Certain units will be held in reserve; one post will send whatever troops it can spare to help another. This support can take the form of coming up from the rear to stiffen a passive resistance and renew it, or of attacking the enemy in flank or even threatening his retreat. If the enemy threatens the flank of a post, not by a direct attack but by taking a position from which he tries to interrupt the post's communications, then either the unit that creates the threat must be attacked, or one can retaliate with a threat to the enemy's communications.

It becomes clear that this kind of defense, though its basic nature is essentially passive, must include a number of active means that will enable it to deal with a wide variety of complex needs. Generally, a defense that makes use of active, even offensive, means is considered superior. However, that depends in part on the nature of the terrain, the composition of the fighting forces, and even the general's ability; also, one may be expecting too much from mobility and other auxiliary measures of an active nature, and too easily discount the local defensive capacity of a major natural obstacle.

We believe we have sufficiently explained what we mean by an extended line of defense, and now turn to the third of the remedies listed: forestalling the enemy by a rapid lateral march.

This method is necessarily part of the mechanism of the type of defense under discussion. In some instances, the defender cannot cover every threatened point of access to the country, no matter how far he extends his position. In many others he must be ready to lead the main strength of his force to those posts that are the enemy's objective, since they would otherwise be too easily overcome. Finally, any general who dislikes tying down his forces to passive defense in an extended position will only achieve his purpose of protecting the country by a greater use of rapid movements, well-conceived and well-executed. The wider the gaps left uncovered, the greater must be the virtuosity of movement in order to forestall the enemy everywhere in time.

The natural consequence of these endeavors will be a search for positions that can be occupied in such eventualities, and are strong enough, once the defender's army, or even just a portion of it, has occupied them, to dispel any thought of attack. These positions can be found everywhere and everything depends on reaching them: they are therefore the keynote, so to speak, of this kind of warfare. That, no doubt, is why it has been called *war of positions*.

Just as extended deployment and limited resistance in a war not aiming at *a great decision* do not run the risks otherwise inherent in them, forestalling an enemy by lateral marches is not so dangerous as it would be if great decisions were involved. A hasty last-minute lateral push into a position, against a determined, able, and purposeful enemy who will not hesitate to expend considerable force, would be meeting disaster half-way: such a rush and scramble into a position could never stand the test of a relentless concentrated blow. It might well succeed, on the other hand, against an enemy who hesitates to get his feet wet, who would not know how to use a great success, or rather its preliminary stages, who seeks merely a limited advantage and that at low cost. Against such an enemy, this type of resistance may certainly be successfully employed.

Consequently this means will also, as a rule, be more common in the second half of a campaign than in the first.

Here, too, the General Staff has an opportunity to use its knowledge of topography in working out a set of interlocking plans concerning the choice and preparation of positions and the roads leading to them.

Where one side's efforts are completely concentrated on reaching a certain point and the other side's efforts are equally bent on preventing it, both will frequently be in the position of having to execute their movements in full view of the enemy. Both will, therefore, move with greater caution and precision than would normally be required. In former times, when the main force was not split into independent parts but was treated as indivisible even when on the march, such caution and precision called for much more complicated movements, and therefore required a great deal of artistry in tactics. Of course, single brigades sometimes had to race ahead of the battle line and seize important points, thereby assuming an independent role in which they were ready to make contact with the enemy even before the main force had arrived. But these were and remained *exceptions*. In general the order of march was always arranged to move the army as a whole with its order undisturbed, and expedients of that kind were avoided as much as possible. Today, the main force is once more divided into independent units, which are permitted to engage the main enemy force, provided the rest of the units are close enough to continue the engagement and fight it to a finish. Such a lateral march is therefore much less difficult today, even in full view of the enemy. What could formerly be accomplished only by the mechanism of the order of march can now be done by dispatching some divisions ahead of time, having others march faster, and by the greater flexibility with which the whole can nowadays be employed.

The defensive measures we have listed are designed to prevent an attacker from taking a fortress, a valuable stretch of country, or a supply depot. He is denied them when, by means of these measures, he encounters engagements at every turn, which either offer too little prospect of success, too great a risk of repercussions in case he loses, or, in general, too great an expenditure of forces in proportion to his aims and situation.

If the defender perseveres, and his skill and dispositions meet with success, the attacker will find even his limited purposes frustrated at every turn by prudent preparations. At that point the offensive principle may often attempt to vent itself by finding satisfaction for honor's sake alone. Victory in any engagement of some consequence will lend a semblance of superiority. It satisfies the vanity of the general, the court, the army, and the people, and thereby in some measure the expectations that are always pinned on an offensive.

Thus the last hope of the attacker will center on a favorable engagement of some consequence, for the sake of the victory and the trophies alone. We are not involving ourselves in a contradiction here, since we are still proceeding on our *own assumption* that the defender, by his foresight, has deprived the enemy of any hope of using a *success* to gain his real objectives. Any such hope would turn on two requirements: first, *a favorable outcome*, and second, *that the victory actually lead on to the further objectives*.

The first of these may well be met without the second; therefore, when the *honors of the field* are the enemy's sole concern, the defender's individual posts and units will more often run the risk of having to fight at a disadvantage than when he is out for additional gains.

If we place ourselves in Daun's position and adopt his way of thinking, it is understandable that, when all he wanted were the trophies of the day, he could take a chance on the assault at Hochkirch without stepping out of character. A victory of consequence, on the other hand, which would have forced the King to abandon Dresden and Neisse, would have been a completely different task which he was not prepared to tackle.

These are not trivial or meaningless distinctions: indeed, we are dealing with one of the most fundamental principles of war. In strategy, the significance of an engagement is what really matters. We cannot repeat often enough that all its essentials always derive from the ultimate intentions of both parties, from the conclusion of the whole sequence of ideas. That is why, strategically speaking, the difference between one battle and another can be so great that the two can no longer be considered as the same instrument.

One can hardly consider that type of victory by the attacker as one that will inflict serious damage on the defender; still, the latter will not be willing to concede even *that much* advantage, especially since one can never tell what else may by chance adhere to it. Therefore he will constantly be concerned with surveying the condition of all his important units and posts. That mainly depends on the sound actions of their own commanders; but inappropriate orders on the part of the general himself can also involve them in inevitable disaster. The fate of Fouqué's corps at Landeshut and of Finck's at Maxen are cases that come to mind.

On both occasions Frederick the Great relied too much on the effectiveness of traditional ways of thinking. He cannot really have believed that 10,000 men in the position at Landeshut could be victorious over 30,000, or that Finck could hold against the overwhelming numbers that poured in on him from all sides. He did assume that the strength of the Landeshut position would go on being taken at face value, and that Daun would find in the demonstration on his flank an adequate excuse for exchanging a poor position in Saxony for a more convenient one in Bohemia. He misjudged Laudon in one case and Daun for once in the other; and in that lay the fault in his dispositions.

That type of error can happen even to commanders who are not too proud, too bold and too obstinate, which Frederick the Great on occasion certainly was. But here the real difficulty lies in the fact that a general cannot always count on his corps commanders all having the sense and good intentions, courage and strength of character that would ideally be desirable. He is, therefore, not able to leave everything to their discretion, but must give them directives, which will restrict their actions and may easily render these inappropriate to the circumstances of the moment. That is a completely unavoidable disadvantage. No army can be properly commanded in the absence of a dominant, authoritarian determination that permeates it down to the last man. Anyone who falls into the habit of thinking and expecting the best of his subordinates at all times is, for that reason alone, unsuited to command an army.

A sharp watch, therefore, must be kept on the condition of each corps and post in order to prevent their becoming involved in unforeseen disaster.

All four of these efforts are intended to preserve the *status quo*. The more successful and fortunate they are, the longer the war will remain static; but the longer the war remains static, the more important the problem of subsistence will become.

Very soon, if not from the start, exactions and requisitions will yield to a system of supply from depots. Farm carts will no longer have to be assembled every time, but a more or less regular transport pool will be set up, either with local carts or military wagons. In short, the practice soon approximates to the highly organized method of supplying food from depots that we outlined in the chapter on maintenance and supply.

But this is not an aspect that exercised great influence on this type of warfare, which by definition and character is confined to very limited areas. It may be partly—and probably is to a large degree—determined by the question of subsistence, but that will not alter its basic characteristics. On the other hand, reciprocal threats to lines of communication will, for two reasons, assume far greater importance: for one thing, since there are no important and decisive means of action in this type of campaign, the general's efforts will have to be directed to such minor ones; for another, there is plenty of time to wait for these measures to take effect. Safeguarding one's lines of communication therefore assumes substantial importance. While their disruption will not be the final purpose of an enemy attack, it can be a very effective means for forcing the defender to retreat, and in the process abandon other points.

Anything done to protect the area occupied by the theater of operations must, of course, also serve to cover the lines of communication. Their safety is partly assured by these measures, and we simply wish to note that concern for their safety will weigh heavily in the choice of position.

One *particular* method of protection is the escorting of individual convoys by small bodies of troops, or even fairly large ones. Even the most extended positions are at times not wide enough to safeguard lines of communication; sometimes such an escort is especially desirable in places where the general has been anxious to avoid an extended position. Hence, Tempelhoff's *History of the Seven Years War* abounds in cases in which Frederick the Great provided his wagon train of bread and flour with escorts—sometimes a single regiment of infantry or cavalry, sometimes a whole brigade. There seems no record of the Austrians doing this, but possibly that is because there was no such thorough historian as Tempelhoff on their side, or because their positions were always a great deal more extended than the Prussians'.

We have now examined all four types of endeavor that, without containing an element of attack, can form the basis of a defense when no decision is intended. At this point we must add a few words about the offensive means with which they may be combined, or, so to speak, spiced. They consist mainly of the following:

1. Action against the enemy's lines of communication, which naturally includes operations against his supply depots
2. Raids and diversions into enemy territory
3. Attacks on enemy posts and units, and even on his main force where conditions are favorable—or simply the threat of such attacks.

In all campaigns of this kind, the first of these means is constantly at work, but silently: it never comes to the fore. Any effective position occupied by the defender derives its value, for the most part, from the fact that it makes the attacker nervous about his own communications. The problem of supply assumes a vital importance in this kind of warfare, as we have already explained in the context of defense; and that holds for the attacker as well. Consequently, the strategic pattern is largely determined by the offensive value latent in the enemy's positions, a topic that will be further dealt with when we come to the subject of the attack.

Such a defense is not limited to the general effect produced by the choice of positions, which, like the effect of pressure in mechanics, works *invisibly*; it may also encompass a true offensive advance by part of the fighting forces. But if it is to be successful, *the location of the lines of communication, the nature of the terrain, or the special qualities of the troops* must be especially favorable.

Raids into enemy territory made for the purpose of retribution or sacking for the sake of booty cannot properly be counted as measures of defense: rather, they are a means of attack. Normally however they can be combined with the objective of an actual diversion aimed at reducing the strength of the opposing side, and can therefore be called a truly defensive measure. But since they can be used offensively just as well, and really are a form of attack, the next book will be a more suitable place to discuss them in detail. We mention raids here only to complete a list of the armory of minor methods of offense, which the defender of an operational theater has at his disposal, and merely wish to note that this is a method that may increase in scope and importance until it gives the whole war the appearance, and along with that the glory of an offensive. That was the nature of Frederick's activities in Poland, Bohemia, and Franconia prior to the campaign of 1759. Though the campaign itself was obviously defensive, these forays into enemy territory gave it the character of an offensive, which may have been particularly important for its psychological impact.

Attacks on enemy units, or even on his main force, should be considered a necessary complement to the defense as a whole, to be used at times when the attacker takes things a little too easily and lays himself wide open at some points. That is the tacit condition for this type of action. But here too, as in his threats to the enemy's lines of communication, the defender can move closer to the attack by always being, like his enemy, on the alert *for a chance to strike a favorable blow*. He can expect a measure of success in this either if he has considerably larger numbers than his opponent—which, though it may happen, is not really consistent with the nature of defense—or if he is skillful and methodical enough to keep his forces better concentrated than the enemy. Then he can use activity and movement to offset the sacrifices that his situation forces on him.

The first was Daun's case in the Seven Years War; the second, Frederick the Great's. Even so, Daun was usually found on the offensive only when Frederick invited it by excessive boldness and contempt of his enemy, as at Hochkirch, Maxen and Landeshut. Frederick, on the other hand, was almost always on the move, aiming to strike at one or another of Daun's corps with his own main force. He succeeded only rarely, or at any rate with only modest results, because Daun combined unusual caution and prudence with his great superiority in numbers. Still, the King's endeavors cannot be considered to have been entirely in vain. They were in fact a most effective form of resistance, for the care and trouble that Daun was driven to take in order to avoid unfavorable engagements neutralized the strength that would otherwise have contributed to a furthering of his offensive. One need only point to the campaign of 1760 in Silesia where Daun and the Russians never managed to advance a single step because they were so apprehensive that the King might attack, now here, now there, and defeat them.

We believe we have now reviewed all the ingredients that in the defense of an operational theater where no decision is intended constitute the dominant ideas, the major efforts, and therefore the backbone of all action. The main reason we brought them all together was to give a coherent view of the articulation of the strategic operation; the particular ways in which each component operates—marches, positions, and so forth—have been previously examined in detail.

When we consider the subject as a whole once more, we are bound to observe that where the offensive principle is so weak, the urge for a decision so faint on both sides, the positive initiative so feeble, and the psychological brakes so numerous as described here, the essential difference between offensive and defensive must gradually disappear. Admittedly, at the start of a campaign one side will invade the other's theater of war and assume the role of attacker; but it may well happen, and often does, that the attacker soon finds himself expending all his energy defending his own country on foreign soil. The two sides then confront each other basically in a state of mutual observation: both are intent on not giving up anything, and perhaps equally intent on achieving a positive gain. Indeed it is possible, as in the case of Frederick the Great, that the actual defender is more aggressive than his opponent.

The more the attacker relinquishes his active advance, the less the defender feels threatened and the less he is narrowly confined to resistance by the urgent need for safety, the more the situation will balance out on both sides. The activity of each will be aimed at gaining an advantage from the other, while avoiding any disadvantage to himself. This is a phase of true strategic *maneuver*, and is certainly more or less characteristic of all campaigns where a major decision is precluded by political motives or the general state of affairs.

The subject of strategic maneuver will have a chapter to itself in the next book. But since theorists have frequently ascribed false importance to this balanced play of forces, particularly in the context of defense, we feel obliged to discuss the subject here in some detail.

We call it a balanced play of forces. Wherever the whole is not in motion, a state of balance exists; and where there is no great purpose to impel it, the whole will not be in motion. That being so, both sides, no matter how unequal, must be deemed to be in balance. Motives for minor actions and lesser purposes are now able to emerge from the balanced state of the whole.

They can develop at this point, for they are no longer under the strain of a great decision or great danger. Thus, whatever may be gained or lost has been converted into smaller tokens, and the conflict as a whole has splintered into minor actions. Given these minor contests for more modest prizes, the two generals now engage in a test of skill. But since chance, or luck, can never be kept completely out of war, this match will never cease to be a *gamble*.

Two further questions now arise. In the course of these maneuvers, will chance play a smaller part in shaping the decision than when everything is concentrated into one great act? And will intelligence play a greater role? The answer to this last question must be positive. The more complex the whole, the more time (with its single moments) and space (with its single points) enter into consideration, then the wider, obviously, will be the field for calculation, and thus the greater the supremacy of the reasoning mind. What reason gains will, in part, be lost to chance, but not necessarily completely. It does not follow, therefore, that the first question need also be answered affirmatively. We should remember in this connection that a reasoning mind is not the general's only mental asset. Courage, energy, determination, prudence, and so forth are attributes that will weigh more heavily where a single great decision is at stake. They will, therefore, count somewhat less in a balanced play of forces, and the primary importance of intelligent calculations increases at the expense of these qualities as well as that of chance. At the hour of a great decision, on the other hand, these brilliant qualities can deprive chance of a great deal of its dominance, and so, in a way, secure some things that the reasoning mind had been forced to release. Obviously, a number of factors are in conflict here, and one cannot flatly say there is more room for chance in a great decision than in the final score of a balanced play of forces. Therefore, when we suggest that the play of forces is mainly a trial of skill, we mean skill in intelligent calculation rather than in the whole range of military virtuosity.

This aspect of strategic maneuver has given it the exaggerated importance that we mentioned above. For one thing, skill in this area has been confused with the sum total of a general's intellectual powers, which is a serious mistake. We must repeat that, at times of great decision, a general's other psychological qualities may control the power of circumstances. Even if this control stems from an impulse prompted by strong emotions and from flashes of almost automatic intuition rather than being the product of a lengthy chain of reasoning, it nonetheless genuinely pertains to the art of war: after all, waging war is not merely an act of reason, nor is reasoning its foremost activity. Second, there has been a feeling that every unsuccessful action in a campaign was the result of skill on the part of one or even both of the generals; actually, its normal and chief basis lay in the prevalent conditions created by war for this kind of gamble.

Since most wars between developed countries have been more a matter of observing the enemy than of defeating him, it follows that strategic maneuver is characteristic of most campaigns. Those in which no famous general was engaged have been ignored; but when some great commander was there to catch the eye, or where indeed there was one on each side as in the case of Turenne and Montecuccoli, their names alone were enough to give the final stamp of approval to the whole art of maneuver. The further consequence has been that this game was rated the highest form of skill, and a product of ultimate perfection. It has, therefore, been treated as the prime source, and outstanding text, for the study of the art of war.

This was the view almost universally held among theorists before the Wars of the French Revolution suddenly opened up a whole new world of military manifestations. At first it was somewhat raw and primitive; later Bonaparte crystallized it into a grandiose system that achieved successes which amazed everyone. At that point, the old models were scrapped; it was assumed that everything was the result of new discoveries, sublime ideas, and so forth, and, indeed, the transformation of the social order. It seemed that the old forms would serve no further use and would never return. But, in such revolutions of thought, factions usually arise; and so the old school found its champions who regarded the new phenomena as outbreaks of the crudest violence, and as a general decay in the art of war. They are convinced that the balanced, sterile, pointless game is the very zenith of development. This is a view so lacking in logic and insight that it must be considered a hopeless confusion of values. The opposite view, however—that nothing of the kind would ever happen again—is also extremely ill-advised. Very few of the new manifestations in war can be ascribed to new inventions or new departures in ideas. They result mainly from the transformation of society and new social conditions. But these, too, while they are in the crisis of fermentation, should not be accepted as permanent. There can therefore be little doubt that many previous ways of fighting will reappear. This is not the time to go more fully into these matters. We are content to have pointed out the place that this balanced play of forces holds in the general conduct of war by means of its importance to, and intrinsic relationship with, the elements concerned; and that it is always the result of constrained circumstances and a greatly reduced military spirit on both sides. It is possible for one general to show greater skill in this game than the other. If he matches him in strength, he may thereby win some advantages; if he is weaker, he may use his greater skill to hold the balance. But it would be a real contradiction to seek a general's highest honor and glory in this area. On the contrary, campaigns of that sort are the surest sign that neither general has great military gifts, or that the one who does is prevented by circumstances from taking the risk of seeking a decision. Where that is the case, however, one will never find the realm of highest military glory.

So far we have discussed strategic maneuver in general. Now we must deal with a particular effect that it has on operations: the fighting forces are often diverted from the important roads and towns to remote or at any rate unimportant areas. Where matters are determined by minor interests of a temporary nature, the influence exerted on the conduct of war by major topographical features will become less important. The fighting forces may be shifted to places where the plain overall needs of the war would never lead them; consequently, the course of war will take much greater twists and turns in its details than in wars leading to a great decision. Take, for instance, the last five campaigns of the Seven Years War: while the overall circumstances remained unchanged, each campaign took a different course, and, strictly speaking, no one operation was repeated; and this despite the fact that the allies showed far more offensive spirit than they had in most earlier campaigns.

The subject of this chapter has been the defense of an operational theater when no great decision is in prospect, and the tendencies to be expected from such an operation—their connection, relationship, and character. Details of the relevant measures have been explored earlier. Now we come to another question: whether a set of all-encompassing principles, rules, and methods may be formulated for these various endeavors. Our reply must be that history has certainly not guided us to any recurrent forms; nevertheless, for a subject of such constantly changing nature one can hardly formulate a theoretical law that is not based on experience. A war in which great decisions are involved is not only simpler but also less inconsistent, more in concert with its own nature, more objective, and more obedient to the law of inherent necessity. In such a case, reason can make rules and laws, but in the type of war we have been describing this seems far more difficult. Two main principles for the conduct of major wars have evolved in our own time: Bülow's "breadth of a base" and Jomini's "interior lines." Even these, when actually applied to the defense of an operational theater, have never proved to be absolute and effective. Yet this is where, as purely formal principles, they should be at their most effective: the more operations expand in time and space, the more rules tend to increase in effectiveness and dominate all other factors in the result. Nevertheless, they turn out to be merely special aspects of the subject, and certainly anything but decisive advantages. It is plain that circumstances exert an influence that cuts across all general principles. The width and careful choice of positions achieved by Daun found its counterpart in the King's practice of closely concentrating his main force, keeping it hard by the enemy and constantly ready to improvise. Both methods are traceable not only to the nature of their armies but also to the circumstances; it is easier for a king to improvise than for a general who must answer for his acts. We must take this opportunity to emphasize once more that a critic has no right to rank the various styles and methods that emerge as if they were stages of excellence, subordinating one to the other. They exist side by side, and their use must be judged on its merits in each individual case.

At this point we do not intend to catalog the various styles that may derive from the characteristics of the army, the country, or the circumstances; their general influence has been pointed out earlier.

We admit, in short, that in this chapter we cannot formulate any principles, rules, or methods: history does not provide a basis for them. On the contrary, at almost every turn one finds peculiar features that are often incomprehensible, and sometimes astonishingly odd. Nevertheless it is useful to study history in connection with this subject, as with others. While there may be no system, and no mechanical way of recognizing the truth, truth does exist. To recognize it one generally needs seasoned judgment and an instinct born of long experience. While history may yield no formula, it does provide an *exercise for judgment* here as everywhere else.

We have but a single comprehensive principle to offer; or rather, we shall express the natural assumption underlying all we have said in the form of an independent principle, so as to increase its impact on the reader's mind.

All means described above have only a *relative* value; all are inhibited by certain limitations on both sides. *Beyond* this sphere, a different set of rules applies, in a totally different universe of phenomena. A general must never forget this; he must never expect to move on the narrow ground of illusory security as if it were *absolute*; he must never permit himself to feel that the means he is using are *absolutely necessary* and *the only ones possible*, and *persist in using them even though he may shudder at the thought of their possible inadequacy*.

The point of view adopted here might well seem to make this type of mistake impossible; but that is not the case in practice, where matters are not so sharply delineated.

Once again we must remind the reader that, in order to lend clarity, distinction, and emphasis to our ideas, only perfect contrasts, the extremes of the spectrum, have been included in our observations. As an actual occurrence, war generally falls somewhere in between, and is influenced by these extremes only to the extent to which it approaches them.

Broadly speaking, then, it is crucial that the general decide from the start whether his opponent is both willing and able to outdo him by using stronger, more decisive measures. If this is what he suspects, he must abandon the minor measures he had employed in order to escape minor disadvantages. Then he may avail himself, through a voluntary sacrifice, of the means of achieving a better position, and will thus be able to cope with a weightier decision. In other words, the first requirement is that the general apply the right standard of measurement in his plan of operations.

This will be made clearer by examples from real life. We shall briefly cite a number of cases of misguided judgments—cases, in our opinion, where a general's operations had been calculated to suit a much less decisive action on the part of his opponent. Let us begin with the opening of the campaign of 1757. The disposition of the Austrian forces proved that they had not expected an offensive as thoroughgoing as the attack that Frederick actually launched. Even the delay of Piccolomini's corps on the Silesian frontier while Duke Charles of Lorraine, with his army, was in danger of having to surrender, shows a similar total misunderstanding of the circumstances.

In 1758, the French were not only completely mistaken about the effects of the Convention of Kloster-Zeven (which has no place in the present discussion), but two months later they were also quite wrong in their estimate of the enemy's capabilities—and therefore lost the whole area between the Weser and the Rhine. We have already mentioned the miscalculations of Frederick the Great in 1759 at Maxen and in 1760 at Landeshut, where he did not expect his enemies to take decisive measures.

We will hardly find a more erroneous standard of measurement in history than that applied in 1792. It was expected that a moderate auxiliary corps would be enough to end a civil war; but the colossal weight of the whole French people, unhinged by political fanaticism, came crashing down on us. We call this a serious error because later on it proved to be one; not because the error would have been easy to avoid at the time. In terms of actual operations it is undeniable that the chief basis for all the subsequent calamitous years can be found in the campaign of 1794. Not only did the allies, in the campaign itself, completely fail to recognize the powerful nature of the enemy offensive, trying to counter it with a paltry system of extended positions and strategic maneuver, but it is evident from the political squabbles between Prussia and Austria, and the foolish abandonment of Belgium and the Netherlands, that the governments involved had no idea of the fury of the oncoming torrent. The Austrians' isolated efforts at resistance in 1796, at Montenotte, Lodi, and so forth, are sufficient proof that they failed to understand what really mattered in a war against Bonaparte.

In 1800 it was not the direct effect of the offensive that brought disaster on Melas, but his erroneous estimate of the results of that assault.

Ulm in 1805, the final knot in a flimsy web of scientific but extremely feeble strategic schemes, would have sufficed to catch a Daun or a Lacy, but it was not strong enough for Bonaparte, Emperor of the Revolution.

The vacillation and confusion of the Prussians in 1806 was brought about by outdated, petty and impracticable views and schemes, which were combined with a few lucid ideas and a sense of the outstanding importance of the moment. If they had been fully conscious and aware of their situation, they would not have kept 30,000 men in Prussia and planned separate operations in Westphalia, nor would they have hoped for any result whatsoever from minor offensives such as those to which Rüchel's and the Weimar corps were detailed. They certainly would not have spent their last moments debating threats to supply depots and the possible forfeiture of a few small tracts of land.

Even the campaign of 1812, the most grandiose of all, was not at first without its share of projects based on miscalculation. At headquarters in Vilna, a group of respected officers were determined to fight a battle on the frontier, in order to show that Russian soil could not be violated with impunity. They were well aware that such a battle might be—indeed would be—lost; even though they did not know that 80,000 Russians would be faced with 300,000 Frenchmen, they realized that they would have to expect a great superiority of numbers. Their chief error lay in their estimate of the battle. They believed it would be a defeat like any other, but one can be practically certain that such an important decision on the frontier would have brought about a completely different series of repercussions. Even the camp at Drissa was a measure that was still based on a totally erroneous estimate of the enemy. If the Russian army had intended to stay there, it would have been cut off and completely isolated, and the French army would not have lacked the means of forcing it to surrender. The man who designed that camp surely had not reckoned with such dimensions of power and determination.

But Bonaparte, too, occasionally applied the wrong standard of measurement. After the armistice of 1813 he believed he could contain the smaller allied armies—Blücher's and the Crown Prince of Sweden's—by using corps that were not able to put up serious resistance. To a cautious opponent, they might have provided an excuse for not incurring risks, which had often happened in earlier wars. He did not take enough account of the deep-seated hatred and the sense of imminent danger that animated Blücher and Bülow.

In general, Napoleon always tended to underestimate old Blücher's enterprising spirit. It was Blücher alone who deprived him of victory at Leipzig. At Laon, Blücher could have destroyed him completely, and the fact that he did not had nothing to do with Bonaparte's own calculations. The penalty for his past mistakes finally caught up with him when lightning struck at Belle Alliance.

BOOK SEVEN

The Attack

CHAPTER ONE

Attack in Relation to Defense

Where two ideas form a true logical antithesis, each complementary to the other, then fundamentally each is implied in the other. If the limitations of our mind do not allow us to comprehend both simultaneously, and discover by antithesis the whole of one in the whole of the other, each will nevertheless shed enough light on the other to clarify many of its details. In consequence we believe that the earlier chapters about defense will have sufficiently illuminated the aspects of attack on which they touch. But this is not always so. No analytical system can ever be explored exhaustively. It is natural that, where the antithesis does not lie so close to the root of the concept as in the previous chapters, what we can say about attack will not follow directly from what was said there about defense. A shift in our viewpoint will bring us nearer the subject, so that we can examine more closely what we previously surveyed from a distance. This will supplement our previous analysis; and what will now be said about attack will frequently also cast more light on defense.

In dealing with attack, we shall largely have to treat topics that have already been discussed. But we do not think we need proceed, as do so many textbooks in engineering, by circumventing or demolishing all the positive values we identified in defense and proving that for every method of defense there is an infallible method of attack. Defense has its strengths and weaknesses. Though the former may not be insurmountable, the cost of surmounting them may be disproportionate. This must hold true whatever way we look at it; otherwise we are contradicting ourselves. Nor do we intend to analyze this interaction exhaustively. Every method of defense leads to a method of attack, but this is often so obvious that we do not need to discuss both in order to perceive it: one follows automatically from the other. We intend to indicate in each case the special features of attack that do not arise directly from the defense. This is bound to call for a number of chapters that have no counterpart in the previous book.

CHAPTER TWO

The Nature of Strategic Attack

As we have seen, defense in general (including of course strategic defense) is not an absolute state of waiting and repulse; it is not total, but only relative passive endurance. Consequently, it is permeated with more or less pronounced elements of the offensive. In the same way, the attack is not a homogeneous whole: it is perpetually combined with defense. The difference between the two is that one cannot think of the defense without that necessary component of the concept, the counterattack. This does not apply to the attack. The offensive thrust or action is complete in itself. It does not have to be complemented by defense; but dominating considerations of time and space do introduce defense as a necessary evil. *In the first place,* an attack cannot be completed in a single steady movement: periods of rest are needed, during which the attack is neutralized, and defense takes over automatically. *Second,* the area left in rear of the advancing forces, an area vital to their existence, is not necessarily covered by the attack, and needs special protection.

The act of attack, particularly in strategy, is thus a constant alternation and combination of attack and defense. The latter, however, should not be regarded as a useful preliminary to the attack or an intensification of it, and so an active principle; rather it is simply a necessary evil, an impeding burden created by the sheer weight of the mass. It is its original sin, its mortal disease.

We call it an *impeding* burden: unless defense contributes to the attack, it will tend to diminish its effect, if only because of the loss of time involved. Is it possible for this defensive component, which is part of every offensive, to be actually disadvantageous? When we assume *attack to be the weaker and defense the stronger form* of war, it seems to follow that the latter cannot be detrimental to the former: if there are enough forces to serve the weaker form, they must surely suffice for the stronger. That is generally so. We shall examine the subject more closely in the chapter on the *culminating point of victory*. However, we must not forget that the superiority of *strategic defense* arises partly from the fact that the attack itself cannot exist without some measure of defense—and defense of a much less effective kind. What was true of defense as a whole no longer holds true for these parts, and it thus becomes clear how these features of defense may positively weaken the attack. It is these very moments of weak defense during an offensive that the positive activity of the offensive principle *in defense* seeks to exploit.

Consider the difference of the situations during the twelve-hour rest period that customarily follows a day's action. The defender holds a well-chosen position which he knows and has prepared with care; the attacker has stumbled into his bivouac like a blind man. A longer halt, such as may be required to obtain supplies, await reinforcements, and so forth, will find the defender close to his fortresses and depots, while the attacker is like a bird perched on a limb. Every attack will anyhow end in a defense whose nature will be decided by the circumstances. These may be very favorable when the enemy forces have been destroyed, but where this is not the case things may be very difficult. Even though this type of defense is no longer part of the offensive, it must affect it and help determine its effectiveness.

It follows that every attack has to take into account the defense that is necessarily inherent in it, in order clearly to understand its disadvantages and to anticipate them.

But in other respects attack remains consistent and unchanged, while defense has its stages, insofar as the principle of waiting is exploited. From these, essentially different forms of action will result, as has been discussed in the chapter on kinds of resistance.

But since attack has but one single active principle (defense in this case being merely a dead weight that clings to it) one will find in it no such differentiations. Admittedly there are tremendous differences in terms of vigor, speed, and striking power, but these are differences of degree, not of kind. It even might be conceivable for the attacker to choose the defensive form to further his aims. He might, for instance, occupy a strong position in the hope that the defender would attack him there. But such cases are so rare that in the light of actual practice they do not require consideration in our listing of concepts and principles. To sum up: there is no growth of intensity in an attack comparable to that of the various types of defense.

Finally, the means of attack available are usually limited to the fighting forces—to which one must of course add any fortresses located close to the theater of war, which may have a substantial influence on the attack. But this influence will weaken as the advance proceeds; clearly, the attacker's fortresses can never play so prominent a part as the defender's, which often become a main feature. Popular support of the attack is conceivable where the inhabitants are more favorably inclined toward the attacker than toward their own army. Finally, the attacker may have allies, but only as a result of special or fortuitous circumstances. Their support is not inherent in the nature of the attack. Thus, while we have included fortresses, popular uprisings and allies among the possible means of defense, we cannot include them among the means of attack. In the first they are inherent, in the second they are rare and then usually accidental.

CHAPTER THREE

The Object of Strategic Attack

In war, the subjugation of the enemy is the end, and the destruction of his fighting forces the means. That applies to attack and defense alike. By means of the destruction of the enemy's forces defense leads to attack, which in turn leads to the conquest of the country. That, then, is the objective, but it need not be the whole country; it may be limited to a part—a province, a strip of territory, a fortress, and so forth. Any one of these may be of political value in negotiations, whether they are retained or exchanged.

The object of strategic attack, therefore, may be thought of in numerous gradations, from the conquest of a whole country to that of an insignificant hamlet. As soon as the objective has been attained the attack ends and the defense takes over. One might therefore think of a strategic attack as an entity with well-defined limits. But practice—seeing things, that is, in the light of actual events—does not bear this out. In practice the stages of the offensive—that is, the intentions and the actions taken—as often turn into defensive action as defensive plans grow into the offensive. It is rare, or at any rate uncommon, for a general to set out with a firm objective in mind; rather, he will make it dependent on the course of events. Frequently his attack may lead him further than he expected; after a more or less brief period of rest he often acquires new strength; but this should not be considered as a second, wholly separate action. At other times he may be stopped earlier than he had anticipated, but without abandoning his plan and moving over to a genuine defensive. So it becomes clear that if a successful defense can imperceptibly turn into attack, the same can happen in reverse. These gradations must be kept in mind if we wish to avoid a misapplication of our general statements on the subject of attack.

CHAPTER FOUR

The Diminishing Force of the Attack

The diminishing force of the attack is one of the strategist's main concerns. His awareness of it will determine the accuracy of his estimate in each case of the options open to him.

Overall strength is depleted:

1. If the object of the attack is to occupy the enemy's country (Occupation normally begins only after the first decisive action, but the attack does not cease with this action.)
2. By the invading armies' need to occupy the area in their rear so as to secure their lines of communication and exploit its resources
3. By losses incurred in action and through sickness
4. By the distance from the source of replacements
5. By sieges and the investment of fortresses
6. By a relaxation of effort
7. By the defection of allies.

But these difficulties may be balanced by other factors that tend to strengthen the attack. Yet it is clear that the overall result will be determined only after these various quantities have been evaluated. For instance, a weakening of the attack may be partially or completely cancelled out or outweighed by a weakening of the defense. This is unusual; in any case one should never compare all the forces in the field, but only those facing each other at the front or at decisive points. Different examples: the French in Austria and Prussia, and in Russia; the allies in France; the French in Spain.

CHAPTER FIVE

The Culminating Point of the Attack

Success in attack results from the availability of superior strength, including of course both physical and moral. In the preceding chapter we pointed out how the force of an attack gradually diminishes; it is possible in the course of the attack for superiority to increase, but usually it will be reduced. The attacker is purchasing advantages that may become valuable at the peace table, but he must pay for them on the spot with his fighting forces. If the superior strength of the attack—which diminishes day by day—leads to peace, the object will have been attained. There are strategic attacks that have led directly to peace, but these are the minority. Most of them only lead up to the point where their remaining strength is just enough to maintain a defense and wait for peace. Beyond that point the scale turns and the reaction follows with a force that is usually much stronger than that of the original attack. This is what we mean by the culminating point of the attack. Since the object of the attack is the possession of the enemy's territory, it follows that the advance will continue until the attacker's superiority is exhausted; it is this that drives the offensive on toward its goal and can easily drive it further. If we remember how many factors contribute to an equation of forces, we will understand how difficult it is in some cases to determine which side has the upper hand. Often it is entirely a matter of the imagination.

What matters therefore is to detect the culminating point with discriminative judgment. We here come up against an apparent contradiction. If defense is more effective than attack, one would think that the latter could never lead too far; if the less effective form is strong enough the more effective form should be even stronger.[1]

[1] The manuscript concludes with the passage: "Development of this subject after Book Three, in the essay on the culminating point of victory."

An essay by that title has been found in a folder marked "Various Essays: Materials [for a revision of the manuscript]." It appears to be an expansion of the chapter that is merely outlined here, and is printed at the end of Book Seven. Marie von Clausewitz.

CHAPTER SIX

Destruction of the Enemy's Forces

Destruction of the enemy's forces is the means to the end. What does this mean? At what price?

Different points of view that are possible:

1. To destroy only what is needed to achieve the object of the attack
2. To destroy as much as possible
3. The preservation of one's own fighting forces as the dominant consideration
4. This can go so far that the attacker will attempt destructive action only under favorable circumstances, which may also apply to the achievement of the objective, as has been mentioned in Chapter Three.

The engagement is the only means of destroying the enemy's forces, but it may act in two different ways, either directly or indirectly, by a combination of engagements. Thus while a battle is the principal means, it is not the only one. The capture of a fortress or a strip of territory also amounts to a destruction of enemy forces. It may lead to further destruction, and thereby become an indirect means as well.

So the occupation of an undefended strip of territory may, aside from its direct value in achieving an aim, also have value in terms of destruction of enemy forces. Maneuvering the enemy out of an area he has occupied is not very different from this, and should be considered in the same light, rather than as a true success of arms. These means are generally overrated; they seldom achieve so much as a battle, and involve the risk of drawbacks that may have been overlooked. They are tempting because they cost so little.

They should always be looked upon as minor investments that can only yield minor dividends, appropriate to limited circumstances and weaker motives. But they are obviously preferable to pointless battles—victories that cannot be fully exploited.

CHAPTER SEVEN

The Offensive Battle

What we have said about the defensive battle will have already cast considerable light on the offensive battle.

We were thinking of the kind of battle in which the defensive is most prominent, in order to clarify the nature of the defensive. But very few battles are of that type; most of them are in part encounters (*demi-rencontres*) in which the defensive element tends to get lost. This is not so with the offensive battle, which retains its character under all circumstances, and can assert it all the more since the defender is not in his proper element. So there remains a certain difference in the character of the battle—the way in which it is conducted by one side or the other—between those battles that are not really defensive and those that are true encounters (*rencontres*). The main feature of an offensive battle is the outflanking or by-passing of the defender—that is, taking the initiative.

Enveloping actions obviously possess great advantages; they are, however, a matter of tactics. The attacker should not forego these advantages simply because the defender has a means of countering them; it is a means the attacker cannot use, for it is too much bound up with the rest of the defender's situation. A defender, in order to outflank an enemy who is trying to outflank him, must be operating from a well-chosen, well-prepared position. Even more important is the fact that the defender cannot actually use the full potential offered by his situation. In most cases, defense is a sorry, makeshift affair; the defender is usually in a tight and dangerous spot in which, because he expects the worst, he meets the attack half-way. Consequently, battles that make use of enveloping lines or reversed fronts—which ought to be the result of advantageous lines of communication—tend in reality to be the result of moral and physical superiority. For examples, see Marengo, Austerlitz, and Jena. And in the opening battle of a campaign the attacker's base-line, even if it is not superior to the defender's, will usually be very wide, because the frontier is so close, and he can thus afford to take risks. Incidentally a flank-attack—that is, a battle in which the front has been shifted—is more effective than an enveloping one. It is a mistake to assume that an enveloping strategic advance must be linked with it from the start, as it was at Prague. They seldom have anything in common, and the latter is a very precarious business about which we shall have more to say when we discuss the attack on a theater of operations. Just as the commander's aim in a defensive battle is to postpone the decision as long as possible in order to gain time (because a defensive battle that remains undecided at sunset can usually be considered a battle won), the aim of the commander in an offensive battle is to expedite the decision. Too much haste, on the other hand, leads to the risk of wasting one's forces. A peculiarity in most offensive battles is doubt about the enemy's position; they are characterized by groping in the dark—as, for example, at Austerlitz, Wagram, Hohenlinden, Jena, and Katzbach. The more this is so, the more it becomes necessary to concentrate one's forces, and to outflank rather than envelop the enemy. In Chapter Twelve of Book Four it has been demonstrated that the real fruits of victory are won only in pursuit. By its very nature, pursuit tends to be a more integral part of the action in an offensive battle than in a defensive one.

CHAPTER EIGHT

River Crossings

1. A major river that cuts across the line of attack is a great inconvenience to the attacker. Having crossed it he is usually limited to a single bridge, so unless he stays close to the river his actions will be severely hampered. Worse, if he intends to offer a decisive battle on the far side, or if he expects the enemy to attack him, he will expose himself to grave danger. So no general will place himself in such a position unless he can count on substantial moral and material superiority.

2. The very difficulty involved in taking up a position beyond a river also greatly increases the possibility of its effective defense. Assuming that this defense is not regarded as the only resource available, but is planned so that even if it fails a stand along the river remains possible, the resistance that the attacker will meet from the defender of the river must be added to all the advantages listed under no. 1. Taken together, all this explains the respect in which an attack on a defended river is held by most generals.

3. But in the previous book we have seen that under certain conditions the defense of the river itself promises good results. We must admit that in practice these successes occur even more frequently than theory would lead one to expect. Theory takes into account only a set of known circumstances, but in practice these will appear more difficult to the attacker than they really are, and so will serve as a powerful brake on his actions.

If the attack under discussion is not aimed at an important decision and is made without dash or determination, the attacker is bound in carrying it out to meet a number of awkward little obstacles and accidents, things no theorist ever took into account, which will be to his disadvantage simply because he is taking the initiative and they therefore happen to him first. Let us only think how often the streams of Lombardy, insignificant in themselves, have been successfully defended. If military history also shows cases in which the defense of rivers failed to fulfill expectations, this is merely evidence that sometimes too much has been expected that was based not on tactical possibilities but on the lessons of past experience stretched beyond their limits.

4. A defended river can be considered as a form of resistance that favors the attacker only if the defender makes the mistake of staking his whole future on this defense. If the crossing is forced after all, he runs into grave difficulties and risks catastrophe; it is easier to force a river-crossing than to win an ordinary battle.

5. It follows that the defense of a river can be really useful where a great decision is not sought; but if, because of the superiority and vigor of the enemy, such a decision can reasonably be expected, a miscalculated river defense can be of positive value to the attacker.

6. There are very few defensive river lines that cannot be turned, either along the whole of their length or at some particular point. If the attacker is stronger and eager to strike a major blow, he can make a diversion at one point while he crosses at another. He can then make up for any setbacks in the early stages of the engagement by using his superior strength to press ahead ruthlessly. Rarely if ever is a river crossing actually forced by tactical means—by driving off one of the main defensive posts by superior firepower and superior valor. "Forcing a passage" is a phrase to be interpreted in the strategic sense alone: the attacker, by crossing at an undefended or only lightly defended point, still faces all the dangers that the crossing should, from the defender's point of view, bring down on him. But the worst thing an attacker can do is to cross simultaneously at several points, unless these are so close together as to afford mutual support. The defender has to disperse his strength, but if the attacker does the same he throws away a natural advantage. That is how Bellegarde lost the battle on the Mincio in 1814: both armies happened to be crossing simultaneously at different points, but the Austrians were more widely separated than the French.

7. If the defender remains on the same side of the river as the attacker, there are obviously two ways of obtaining a strategic advantage. The first is to cross anywhere regardless of his presence, and thus turn the tables on him. The second is to fight a battle. The main determinant in the first case should be the relationship between the base and the lines of communication, though the particular circumstances are often more decisive than the general situation. This can to some extent be ignored by the army that has chosen better sites for its posts, made better tactical dispositions, commands greater discipline, or can march faster. As for the other course, one must assume that the attacker has the means, the right conditions and the determination to fight a battle. If that is so, the defender will not lightly run the risk of this kind of river defense.

8. Let us sum up, then, by saying that river crossings as such seldom present major difficulties; but, unless a great decision is involved, there will be enough misgivings about their consequences and their further implications for them easily to bring the attacker to a halt. He will either leave the defender on the near side of the river or, at most, cross the river but stay close to it. It is rare for two armies to face one another across a river for any length of time.

But even when a great decision is involved, a river is a substantial factor, for it always weakens and dislocates the offensive. In such a case, one can only hope that the defender will make the mistake of treating the river as a tactical obstacle and making its direct defense the central point of his resistance, thus handing the attacker the advantage of dealing him the decisive blow with a minimum of effort. It must be understood that this blow will not bring about the enemy's complete collapse immediately, but will lead to a series of successful engagements that will gradually create severely unfavorable general conditions, as was the case with the Austrians on the lower Rhine in 1796.

CHAPTER NINE

Attack on Defensive Positions

In the book on defense there is a detailed discussion on the extent to which defensive positions compel an enemy either to attack them or to abandon his advance. Only those that achieve these aims are appropriate: wearing the enemy forces down, whether totally or partially, or neutralizing them. The attack cannot prevail against them; it has no means at its disposal to counteract their advantage. In practice, not all defensive positions are like this. If the attacker sees that he can get his way without assaulting them, it would be stupid of him to attempt it. If he cannot, the question is whether he can maneuver the defender out by threatening his flank. He will decide to attack a good position only where these means are ineffective; in that case a flank attack will always pose somewhat fewer problems. The choice between the two flanks will then be determined by the location and direction of each side's lines of communication—in other words, the threat to the enemy's retreat and the security of one's own. These two factors may easily conflict, in which case threatening the enemy's line should receive preference. Its nature is offensive and therefore of the same type as the attack, whereas the nature of the other is defensive. One thing, however, is sure, and fundamental to the issue: *it is a risky business to attack an able opponent in a good position.* True, there is no lack of examples of such battles succeeding, as at Torgau and Wagram (I am not including Dresden, because the enemy there could not be described as able). On the whole, however, the number[1] is small and insignificant when compared with the immense number of cases in which the most resolute of generals did not attack such positions.

But our topic should not be confused with ordinary battles. Most battles are true clashes (*rencontres*) in which one side is admittedly on the defensive but not in entrenched positions.

[1] The first edition has *Gefahr,* which makes no sense. The second edition tries to make sense by an inappropriate insertion. For *Gefahr* we read *Zahl.* Eds.

CHAPTER TEN

Attack on Entrenched Camps

It was once fashionable to belittle entrenchments and their effectiveness. The cordons on the French frontier which were pierced so often, the entrenched camp at Breslau where the Duke of Bevern was defeated, the battle of Torgau, and a number of other examples caused this prejudice. Moreover the victories that Frederick the Great won by mobility and aggressiveness had cast a shadow over defense as such, over all fixed positions and especially all entrenchments, which further increased this disdain. Certainly if a few thousand men are expected to defend several miles, or if entrenchments are nothing more than lateral communication trenches, entrenchments will not be worth anything. Any confidence placed in them is dangerously misleading. But it must surely be a contradiction or even nonsense when this opinion is extended to the very concept of entrenchment—as Tempelhoff does in his blustering fashion. What use would entrenchments be anyhow, if they did not help the defender? No, not only reason, but hundreds and thousands of examples show that a well-prepared, well-manned, and well-defended entrenchment *must generally be considered as an impregnable point*, and is indeed regarded as such by the attacker. If we proceed from this factor of the effectiveness of a single trench, we cannot really doubt that the assault on an entrenched camp is a very difficult and usually an impossible task for the attacker.

By their very nature, entrenched camps are thinly manned; but if the natural obstacles are favorable and the entrenchments well constructed, they can be held against heavily superior numbers. Frederick the Great thought that an attack on the camp at Pirna would not be feasible, even though his strength was twice the garrison's. Since then it has sometimes been alleged that the camp would not have been too hard to take; but the only proof for this view is based on the extremely poor condition of the Saxon garrison, which is of course no argument against the value of the entrenchments. But it is questionable whether those who afterward claimed that an attack would not only have been feasible but even easy would themselves have opted for it at the crucial moment.

In our opinion, then, the offensive should only very rarely resort to an attack on an entrenched camp. Such an attack is advisable only if the defenses have been executed hurriedly, left incomplete and lack obstacles to access; or in general if, as often happens, the camp is a mere sketch of what it ought to be—a half-completed ruin. Then an attack may be advisable and an easy way to vanquish the enemy.

CHAPTER ELEVEN

Attack on a Mountainous Area

The general strategic significance of a mountainous area, both in attack and in defense, was fully explained in Chapter Five and the following chapters of Book Six. There we also attempted to indicate the part that mountains play as an actual line of defense; and from this their significance from the point of view of the attack can be developed. Consequently little remains to be said here on this important subject. The main conclusion reached was that the defender has to accept one of two widely differing situations: a subordinate engagement or a major battle. In the former case, an attack on a mountain range is at best a necessary evil, for every factor will be adverse; but in the case of a major battle, the advantages will all be on the side of the attacker.

An offensive endowed with the means and the resolve to fight a battle will meet the enemy in the mountains and will certainly profit by doing so.

But here again we have to repeat that it will be difficult to gain acceptance for this conclusion because it goes against all appearances and, at first sight, against all experience. In most cases it still holds true that an army on the offensive, whether or not it is bent on a decisive battle, would consider itself extraordinarily lucky to find that the enemy had not occupied a mountain range between them. It would hurry to get there first, and no one would regard this as contradictory to the nature of the offensive. We agree; but it is necessary to make more precise distinctions.

An army advancing upon an enemy with the intention of fighting a decisive battle will, if it has to cross an unoccupied mountan range, naturally be concerned lest the enemy might at the last moment block the very passes it intended to use. In that case the attacker could not enjoy the advantages that would have been his if the enemy had occupied an ordinary mountain position. The enemy is now no longer overextended; he is no longer in doubt about the route the attacker is taking; and the latter has not been able to choose his roads with the enemy's position in mind. So this battle in the mountains does not confer all the advantages on the attacker we have described in Book Six. Under these circumstances it is possible to regard the defender as being in an unassailable position. So the defender may after all command the means of turning the mountains to his own advantage in a decisive battle. This may indeed be possible; but when one considers the difficulties of taking up a favorable mountain position at the last moment, especially if it has previously been left completely unoccupied, one will realize that this is a totally unreliable method of defense. So the circumstances that the attacker has reason to dread are most *unlikely* to occur. Yet even though they are unlikely, it remains natural to fear them: in war it is often the case that a particular concern is entirely natural even if somewhat unnecessary.

To the attacker, another cause for worry is a preliminary defense of the mountains by an advance guard or a chain of posts. While this measure too will very seldom be to the defender's interest, the attacker is not in a position to distinguish whether or not this is likely to happen, and he will therefore fear the worst.

Our view of the matter, moreover, does not exclude the possibility that the mountainous nature of the ground may make a position really unassailable. Such positions do exist, though they do not have to be in mountain country: Pirna, Schmottseifen, Meissen and Feldkirch are examples. These are all the more useful because they are not in mountains. Still, it is conceivable that they may be found in mountains—on high plateaus, for instance, where the defender is able to avoid the ordinary drawbacks of a mountainous position. But they are exceptions, and we must address ourselves to the majority of cases.

Military history shows clearly how unsuited mountains are for decisive defensive battles. Great commanders bent on such a battle usually preferred to take up a position on open ground. There is not a single example in the history of war of a decisive engagement in the mountains, with the exception of the Revolutionary Wars. There, apparently a mistaken application and analogy led to the use of mountain positions even where decisive engagements could be expected, as in the Vosges in 1793 and 1794, and in Italy in 1795, 1796, and 1797. Mélas has been widely criticized for not having occupied the Alpine passes in 1800, but that is a rash criticism—one could call it a rather immature and superficial judgment. Bonaparte, in Mélas' place, would not have occupied them either.

Preparations for attacking a mountain position are mostly tactical. However, we ought to list the following as a preliminary outline, applicable to the parts closest to strategy and coinciding with it.

1. Mountains, as distinguished from other terrain, do not allow one to diverge from the road and split one column into two or three as the needs of the moment may require. Usually everything comes to a halt in long defiles. The advance, therefore, should proceed on several roads from the start or, better still, be made on a somewhat broader front.

2. When attacking a widely extended line of defense in mountains, one will of course do so with a concentrated force; the whole position cannot possibly be outflanked. If one is aiming at a major victory it will have to be accomplished by piercing the enemy's lines and forcing his wings apart, rather than by surrounding the force and cutting it off. The attacker's natural intention will then be a rapid irresistible advance along the enemy's main line of retreat.

3. If, however, the enemy has to be attacked in a more concentrated

mountain position,[1] flanking operations will form a major part of the plan, since frontal attacks will meet with maximum opposition; but they must be aimed more at actually cutting off the enemy forces than at tactical assaults in flank or rear, for even the rear of a mountain position can offer strong resistance if the forces are available. The fastest way of getting results is always to give the enemy reason to fear having his line of retreat cut. That fear is aroused more quickly and effectively in mountain warfare, for there it is not so easy to cut one's way out if worst comes to worst. But for the offensive a simple demonstration will not be enough: at best it might maneuver the enemy out of his position, but it would not yield a definite result. One must therefore aim at really cutting him off.

[1] Although all texts consulted give *weniger gesammelten*, this paragraph makes sense only on the assumption that *weniger* should read *mehr*, and we translate accordingly. Eds.

CHAPTER TWELVE

Attack on Cordons

If a major decision is intended by a defense of or an attack on cordons, it is the attacker who has the greater advantage, for their vast extension makes them even less suited to the needs of a decisive battle than is the direct defense of a river or a mountain range. Prince Eugene's lines at Denain in 1712 are an illustration: their loss would have been comparable to a lost battle, but Villars could hardly have been victorious against Eugene in a concentrated position. If a decisive victory is beyond the means of the attacker, he will defer even to lines, especially when they are occupied by the enemy's main force. For instance, in 1703 the lines at Stollhofen held by Louis of Baden were respected even by Villars. But if they are held only by a secondary force, everything will of course depend on the strength of the corps that can be used in the attack. Resistance in that case is not likely to amount to much, but neither, of course, is the ensuing victory.

The lines of circumvallation of a besieging force will have their own peculiar character, which I intend to discuss in the chapter on the attack on a theater of operations.

It is a peculiarity of all cordon-like positions, such as reinforced outpost-lines and so forth,[1] that they are easily pierced; but where this is done without the intention of pressing on and obtaining a decision, it will result only in a minor success, which usually is not worth the effort expended on it.

[1] See above, Book Five, Chapter Seven. Eds.

CHAPTER THIRTEEN

Maneuver

1. The subject of maneuver has already been touched on in Chapter Thirty of Book Six. While the device is common to attacker and defender, its nature is more closely related to attack than to defense, and we shall now therefore define it more closely.

2. Maneuver must be distinguished, not only from aggressive conduct of the attack by means of major engagements, but from every operation that arises immediately out of such an attack; whether it be a diversion, pressure on the enemy's lines of communication or on his retreat, and so forth.

3. In its ordinary meaning the term maneuver carries the idea of an effect created out of nothing, so to speak—that is to say, out of a state of *equilibrium*—by using the mistakes into which the enemy can be lured. It can be compared to the opening gambits in a game of chess. It is, in fact, a play of balanced forces whose aim is to bring about favorable conditions for success and then to use them to gain an advantage over the enemy.

4. Considerations to be borne in mind, partly as goals and partly as a frame of reference for our actions, are the following:

a. The enemy's food supplies, which one aims to cut off or reduce

b. A combination with other units

c. A threat to other communications with the interior of the country or with other armies or detachments

d. A threat to the retreat

e. An attack on individual points with superior forces.

These five factors can be found in the smallest detail of the particular situation, which then becomes the object around which, for a time, everything revolves. It may be a bridge, a highway or an entrenchment; but in every case it would be easy to show that its importance is derived entirely from its relation to one of the factors named above.

f. For the attacker, or rather for the active party (who admittedly may be the defender) the outcome of a successful maneuver will consist of a strip of land, supply depot, or the like.

g. A strategic maneuver comprises two pairs of opposites, which appear distinct and may well have been used for formulating misleading rules and maxims. In fact their four elements are all basically essential parts of one whole and must be considered as such. The first pair of opposites consists of outflanking the enemy or of operating on interior lines; the second, of concentrating one's forces or of extending them over numerous posts.

h. Concerning the first pair of opposites, one cannot possibly say that one of the two elements is generally superior to the other. The reason is partly that an attempt at one will naturally evoke the other as the obvious countermove, the proper antidote; and partly because envelopment is associated with attack, whereas the use of interior lines is associated with defense: so that, broadly speaking, the former suits the attack, and the latter the defense. The superior form is that which is best carried out.

i. It is just as impossible to rank the elements of the other pair of opposites. A stronger force can afford to extend itself. It will thus, in many respects, establish a convenient strategic posture and spare the troops unnecessary effort. The weaker side must remain more closely concentrated and make up for the resulting disadvantages by mobility. This greater mobility assumes a greater degree of competence in marching. The weaker side, therefore, must exert itself more, both physically and morally. That is the inevitable conclusion, of course, if our argument has been consistent; in fact, it can virtually be regarded as its appropriate test. The campaigns of Frederick the Great against Daun in 1759 and 1760, and Laudon in 1761, and of Montecuccoli against Turenne in 1673 and 1675 have always been considered as the most brilliant examples of this form, and our opinions are largely based on them.

j. Just as one must avoid misusing the four factors of the two postulated opposites for formulating misleading rules and maxims, one should be on the alert against attributing more importance and decisive influence than they possess to such other general circumstances as baseline, terrain, and so forth. The smaller the importance of the stake, the greater the importance of the details of the momentary situation; wider and more general factors fade into the background, being in a way too large in scale for the matter in hand. Could there, broadly speaking, be a more preposterous situation than that of Turenne in 1675? He stood with his back close to the Rhine, extended over fifteen miles, the bridge for his retreat being at his extreme right wing. All the same, his dispositions worked, and they have rightly been held to show a high degree of skill and judgment. But this degree of success and skill can only be fully understood when one pays attention to the details, and appreciates them according to their value in each individual case.

We are therefore certain that no rules of any kind exist for maneuver, and no method or general principle can determine the value of the action; rather, superior application, precision, order, discipline, and fear will find the means to achieve palpable advantage in the most singular and minute circumstances. It is on these qualities that victory in this type of contest largely depends.

CHAPTER FOURTEEN

Attacks on Swamps, Flooded Areas, and Forests

Swamps—impassable marshes crossed only by a few causeways—present the attacker with several problems, as has already been explained under the heading of defense. They are too wide to enable us to drive the enemy off the opposite bank by cannon fire and permit us to construct our own means of crossing. Strategically, the consequence is that one avoids an attack on swamps and tries to by-pass them. Where the country is so densely cultivated—as it is in many low-lying areas—that the means of passing are innumerable, the defenders' resistance may still be relatively strong; but for an absolute decision it will be that much weaker, and therefore inappropriate. If, on the other hand, the low-lying ground can be fortified by flooding, as in Holland, resistance can grow to be absolute, and then any attack is bound to fail. This was demonstrated in Holland in 1672. After every fortress outside the flooded area was captured and occupied by the French, they still had 50,000 men to spare; but, first under Condé and then under Luxembourg, they still failed to penetrate the flooded area, though only 20,000 men defended it. True, the Prussian campaign against the Dutch in 1787 under the Duke of Brunswick shows the opposite result, the line being forced by only slightly greater numbers and with insignificant losses. The reason for this, however, can be laid to the fact that the defenders were divided by political dissension and lacked unified command. Yet nothing is more certain than that the success of the campaign—an advance through the final line of inundation to the very gates of Amsterdam—depended on such a fine point that one cannot possibly draw a conclusion from it. This fine point was the Haarlemer Meer, which had been left unguarded. There the Duke could by-pass the line of defense and approach the post at Amselvoen from the rear. If the Dutch had had a few ships on the Haarlemer Meer, the Duke would never have reached Amsterdam, for he was at the end of his tether. What effect this might have had on the conclusion of the peace does not concern us here, but certainly there could have been no further question of forcing the final line of inundation.

Winter is, of course, the natural enemy of this means of defense, as the French made clear in 1794–1795; but it has to be a *hard* winter.

Forests that are scarcely penetrable are another powerful asset to defense, as we have said. If they are not too thick the attacker can traverse them by several neighboring roads and reach a more favorable area. The tactical strength of the individual positions will not be great, because a forest can never be considered impassable in the same way as a swamp or a river. In

Russia and Poland, on the other hand, vast tracts of land are almost completely covered by forests; if an attacker is not strong enough to get to the far side, he will be in a most difficult situation. One only has to remember the problems of supply with which he has to contend. What is more, in the depth of the forest he will hardly be in a position to impress the omnipresent enemy with the superior weight of his numbers. This is without doubt one of the worst situations in which an attacker can find himself.

CHAPTER FIFTEEN

Attack on a Theater of War: Seeking a Decision

Most aspects of this question have already been touched upon in Book Six, "On Defense," which will have reflected sufficient light on the subject of attack.

The concept of a self-contained theater of operations is in any case more closely associated with defense than with attack. A number of salient points, such as *the object of the attack* and *the sphere of effectiveness of the victory*, have already been dealt with in Book Six, and the really basic and essential features of attack can be expounded only in connection with the subject of war plans. Still, enough remains to be set forth here, and we shall once again begin by discussing a campaign intended to force a major decision.

1. The immediate object of an attack is victory. Only by means of his superior strength can the attacker make up for all the advantages that accrue to the defender by virtue of his position, and possibly by the modest advantage that his army derives from the knowledge that it is on the attacking, the advancing side. Usually this latter is much overrated: it is short-lived and will not stand the test of serious trouble. Naturally we assume that the defender will act as sensibly and correctly as the attacker. We say this in order to exclude certain vague notions about sudden assaults and surprise attacks, which are commonly thought of as bountiful sources of victory. They will only be that under exceptional circumstances. We have already discussed elsewhere the nature of a genuine strategic surprise.

If an attack lacks material superiority, it must have moral superiority to make up for its inherent weakness. Where even moral superiority is lacking, there is no point in attacking at all, for one cannot expect to succeed.

2. Prudence is the true spirit of defense, courage and confidence the true spirit of attack. Not that either form can do without both qualities, but each has a stronger affinity with one of them. After all, these qualities are necessary only because action is no mathematical construction, but has to operate in the dark, or at best in twilight. Trust must be placed in the guide whose qualifications are best suited to our purposes. The lower the defender's morale, the more daring the attacker should be.

3. Victory presupposes a clash of the two main forces. This presents less uncertainty to the attacker. His role is to confront the defender, whose positions are usually already known. In our discussion of the defense, on the other hand, we argued that if the defender has chosen a poor position the attacker should not seek him out, because the defender would have in that

case to seek *him* out instead, and he would then have the advantage of catching the defender unprepared. In that case, everything would depend on the most important road and its general direction. This point was not discussed in the previous book, but was left until this chapter. We must therefore examine it now.

4. The possible objectives of an attack, and, consequently, the *aims* of victory, have already been discussed. If these lie within the theater of war that we intend to attack, and within the probable sphere of victory, the natural direction of the blow will be determined by the roads leading to them. But one should not forget that the object of the attack usually gains significance only with victory; victory must always be conceived in conjunction with it. So the attacker is not interested simply in reaching the objective: he must get there as victor. Consequently, his blow must be aimed not just at the objective but at the road that the enemy will have to take to reach it. The road then becomes the first objective. Victory can be made more complete if we encounter the enemy before he has reached that objective, cutting him off from it and getting there first. If for instance the main objective of the attack is the enemy's capital and the defender has not taken up a position between it and the attacker, the latter would be making a mistake if he advanced straight on the city. He would do better to strike at the communications between the enemy army and its capital and there seek the victory which will bring him to the city.

If there is no major objective within the area affected by the victory, the point of paramount importance is the enemy's line of communication with the nearest significant objective. Every attacker, therefore, has to ask himself how he will exploit his victory after the battle. The next objective to be won will then indicate the natural direction of his blow. If the defender has taken up his new position in that area he has made the correct choice, and the attacker has got to seek him out there. If that position is too strong, the attacker must try to by-pass it, making a virtue of necessity. But if the defender is not where he ought to be, the attacker must move in that direction himself. As soon as he is level with the defender—assuming the latter has made no lateral movement in the meantime—he should wheel toward the enemy's lines of communication with the proper objective of seeking out his enemy there. If the latter has not moved at all, the attacker will have to turn and take him in the rear.

Among the roads from which the attacker may choose, the great commercial highways are the most obvious and suitable. But wherever they form too large a detour, one should take a more direct, even if a narrower road. A line of retreat that deviates considerably from a straight line always involves a serious risk.

5. An attacker bent on a major decision has no reason whatever to divide his forces. If in fact he does so, it may usually be ascribed to a state of confusion. His columns should advance on no wider a front than will allow them to be brought into action simultaneously. If the enemy force is divided, so much the better; in that case, minor diversions are in order—strategic

feints, made with the object of maintaining one's advantage. Should the attacker choose to divide his forces for that purpose he would be quite justified in doing so.

The division of the army into several columns, which in any case is indispensable, must be the basis for envelopment in the tactical attack; for envelopment is the most natural form of attack, and should not be disregarded without good cause. But the envelopment must be tactical; a strategic envelopment concurrent with a major blow is a complete waste of strength. It can only be justified if the attacker is strong enough not to have any doubts about the outcome.

6. But attack also requires caution: the attacker himself has a rear and communications to protect. This protection should, if possible, consist in the direction of advance—that is, it should be automatically provided by the army itself. If forces have to be detached for this purpose, thus causing a diversion of strength, it can only lessen the impact of the blow. A large army always advances on a front at least a day's march in width; so if the lines of communication and retreat do not deviate too much from the perpendicular, the front itself usually provides all the cover necessary.

Dangers of this sort to which the attacker is exposed can be gauged chiefly by the enemy's character and situation. If everything is subordinated to the pressure of an imminent major decision, the defender will have little scope for auxiliary operations, and the attacker, therefore, will not ordinarily be in great danger. But once the advance is over and the attacker gradually goes over to a state of defense, the protection of the rear assumes increasing urgency and importance. The attacker's rear is inherently more vulnerable than the defender's; so the latter may have started operations against the attacker's lines of communication long before he goes over to an actual offensive, and even while he is still on the retreat.

CHAPTER SIXTEEN

Attack on a Theater of War: Not Seeking a Decision

1. Even where determination and strength will not suffice to bring about a great decision, one may still want to mount a strategic attack against a minor objective. If that attack succeeds and the objective is attained, the situation reverts to a state of rest and balance. If difficulties are encountered to any serious extent, the advance is halted at an earlier stage. It will then be replaced either by offensives of opportunity or by mere strategic maneuver. That is the nature of most campaigns.

2. Objectives of such an offensive may be:

a. *A stretch of territory.* This may yield food-supplies; possibly contributions; protection of one's own territory; or a bargaining counter in peace negotiations. Sometimes the concept of military glory may play a part, as it constantly did in the campaigns fought by the French marshals under Louis XIV. The essential distinction lies in whether the territory can be held. As a rule, that is possible only if it borders on one's own theater of operations and forms a natural extension to it. Only this type can constitute a bargaining counter at the peace table; all others are usually held temporarily, for the duration of the campaign, to be abandoned in the winter.

b. *An important depot.* If it were not important, it would hardly be considered an objective for an offensive taking up a whole campaign. It may in itself constitute a loss to the defender and a gain to the attacker; but the chief advantage to the latter lies in the fact that it will force the defender to withdraw and abandon territory which he would otherwise have held. Thus the capture of the depot is actually more of a means, and is listed here as an end only because it is the nearest immediate objective of action.

c. *The capture of a fortress.* We refer the reader to the separate chapter devoted to the capture of fortresses. It is clear from the arguments developed there why fortresses have always been the preferred and most desirable objectives in offensives or campaigns that could aim *neither* at the enemy's total defeat *nor* at the seizure of an important part of his country. So it is easily explained why in a country like the Netherlands, which is full of fortresses, the aim of operations has always been the capture of one fortress or another, the eventual seizure of the whole area *rarely emerging as the objective of the campaign.* Each fortress was deemed a discrete unit, and prized for its own sake. Apparently more attention was paid to the convenience and ease of the enterprise than to the actual value of the place.

Still, the siege of a fortress of any size is always an important operation because it is very expensive—an important consideration in wars that are not fought for major issues. That is why such a siege must be included among the significant elements of a strategic attack. The less important the place, the less determined the siege, the fewer the preparations made for it, the greater the likelihood of an air of improvisation, then the more the strategic objective will shrink in significance, and the weaker the forces and intentions to which it is suited. Such cases often end up as shadow-boxing, simply aimed at terminating the campaign honorably: as the attacker, one is after all bound to do something.

d. A *successful engagement*, *encounter*, or even *battle*, whether for the sake of trophies, or possibly simply of honor, and at times merely to satisfy a general's ambition. Anyone who doubts that this occurs does not know military history. Most of the offensive battles in the French campaigns during the age of Louis XIV were of this type. It is more important to note, however, that these considerations are not without weight, mere quirks of vanity: they have a very definite bearing on the peace and hence they lead fairly straight to the goal. Military honor and the renown of an army and its generals are factors that operate invisibly, but they constantly permeate all military activity.

Such engagements, to be sure, are based on the following assumptions that: (a) there is a fair prospect of victory; and (b) if they end in defeat, not too much is lost. One must be careful not to confuse this type of battle, fought under restricted conditions for limited objectives, with victories that were not followed up for want of moral fiber.

3. With the exception of the last of these categories, (d), all can be achieved without major engagements. The means that the offensive can use for this purpose derive from the interests that the defender has to protect in his theater of war. They will, therefore, consist in threatening his lines of communication, with its depots, rich provinces, important towns, or key points such as bridges, passes, etc.; or in the occupation of strong positions uncomfortably located for the defender;[1] or in the occupation of important towns, fertile agricultural areas, or disaffected districts which can be seduced into revolt; or in threatening his weaker allies, and so on. If the attacker manages to disrupt communications to the point where the enemy cannot restore them without serious loss, if he sets out to seize these points, he will force the defender to take up another position to the rear or to the flank so as to cover them, even if it means giving up lesser ones. Thus an area is left uncovered, or a depot or fortress exposed—the former open to conquest, the latter to siege. Major or minor engagements may result, but they will neither be sought, nor will they be treated as objectives in them-

[1] The first edition is so obscure that we here follow the text of the second. Eds.

selves, but rather as necessary evils. They cannot rise above a certain level of magnitude and importance.

4. An operation on the part of a defender against the attacker's lines of communication is a type of reaction which, in a war aiming at major decisions, can take place only if those lines become very long. But in wars not seeking great decisions this type of reaction is more appropriate. Admittedly the enemy's lines will rarely be very extended, but the point here is not to inflict severe damage on him. It will often be enough to harass him and keep him short of supplies; and what the lines lack in length is to some extent made up by the length of time that can be spent on this kind of fighting. That is why the cover of his strategic flanks is of great importance to the attacker. If this kind of contest or rivalry develops between the attacker and the defender, the former will have to make up for his natural disadvantages by means of his superior numbers. If his strength and determination are still enough for him to risk a decisive blow at an enemy unit or even at the main enemy force, this threat, held over the defender's head, remains his best way of covering himself.

5. In conclusion, we must mention one other important advantage which the attacker enjoys in this sort of war: he is better placed to gauge the enemy's intentions and resources than the defender is to gauge his. It is a great deal harder to predict the degree of vigor and daring with which the attacker will act than it is to predict whether the defender is contemplating a major stroke. In practice, the mere choice of the defensive form of warfare generally assures a lack of positive intentions. Besides, the difference between preparations for a major counterstroke and ordinary means of defense is much more marked than that between the preparations for a major attack and for a minor one. Finally, the defender is forced to make his dispositions earlier, thus giving the attacker the advantage of a counter-riposte.

CHAPTER SEVENTEEN

Attack on Fortresses

We shall naturally not discuss this topic from the technical point of view. Rather we will consider, first its strategic object; second, deciding which fortress to attack; and third, the way in which the siege is to be protected.

The loss of a fortress weakens the enemy's defense, especially where it constituted a vital part of it. Its occupation provides various benefits for the attacker: he may use it as a storehouse and depot, to cover the countryside and his billets, and so forth. And if the attack ends up as a defense, such fortresses will be its strongest support. These relations between fortresses and theaters of war during the progress of hostilities have all been sufficiently treated in our discussion of fortresses in the book on defense; the reflection of what was said there will shed the necessary light on them in the context of attack.

Attacks on fortresses are another case in which campaigns that aim at a great decision differ substantially from other kinds. In the former instance, the capture of a fortress must always be considered as a necessary evil. While a decision is still in the balance a siege will be undertaken only where it cannot be avoided. Once the decision has been made, the crisis is past, the tension relaxed for the moment, and a state of rest has set in, then the occupation of fortresses will serve as a consolidation of the conquest. At that point they can usually be taken without risk, if not without some effort and expenditure of strength. During the crisis itself, besieging a fortress increases the problems of the attacker. Clearly, there is nothing that will diminish his strength so much and is therefore so likely temporarily to rob him of his superiority. Still, there are times when a siege is unavoidable if the attack is to progress at all. In such cases a siege should be considered as an intensification of the attack. The fewer the decisions already reached, the deeper the crisis will be. Further discussion of the subject belongs to the book on war plans.

When the aim of a campaign is limited, a fortress is usually not the means, but the end. It will rank as a small, independent conquest, and as such will have the following advantages:

1. A fortress is a minor conquest with precisely defined boundaries. Taking it does not call for a major effort and one need therefore not be worried about a setback.
2. It is a useful bargaining counter at the peace table.
3. A siege is, or at least appears to be, an intensification of the attack,

often without the decrease in strength brought about by other forms of offensive advance.

4. A siege is an operation that cannot result in disaster.

All these factors combine to make the capture of one or several fortresses a frequent objective for the type of strategic attack that cannot aim at a higher objective.

If there is any doubt concerning which of several fortresses is to be besieged, the choice should be based on the following principles:

a. The fortress should be easy to hold, and therefore constitute an important bargaining counter at the peace table.

b. If the means available for its conquest are limited, only minor fortresses can be conquered; actual occupation of a minor fortress is preferable to an unsuccessful attack on a major one.

c. Obviously the strength of the fortifications often bears no relation to the importance of the place. Nothing could be more foolish than to waste one's efforts on a place that is very strong but relatively insignificant when one could be attacking a weaker one.

d. The strength of its armament—which includes, of course, the garrison. If the fortress is lightly armed and held, it will naturally be easier to take. But we must point out that the strength of the armament and garrison are necessarily factors in the *importance* of a fortress: they are direct components of the enemy's armed strength. This is not true to the same extent of the fortifications, so that the capture of a strongly garrisoned fortress is more likely to be worth the sacrifices it entails than the capture of a strongly fortified one.

e. The ease of providing and maintaining a siege-train. Most sieges fail because of lack of equipment. The best known examples are Prince Eugene's siege of Landrecies in 1712 and Frederick's siege of Olmütz in 1758.

f. Finally, the ease of protecting the siege is a point to bear in mind.

There are two basically different ways of covering a siege: one is by entrenching the besieging force—that is, by the construction of a line of circumvallation; the other is by a so-called line of observation. The first has gone completely out of fashion in spite of an important point in its favor: it permits the attacker to avoid being weakened by a dispersion of strength, which is generally to his decided disadvantage. Still, his strength will be markedly impaired in other ways:

1. As a rule, a position encircling the fortress requires too great an extension in relation to the strength of the army.

2. The garrison, together with the enemy's relieving forces, would normally constitute simply the original enemy strength confronting us; but *now* it has to be seen as an enemy unit inside our own camp, which, behind the protection of its ramparts, is *invulnerable* or, at any rate, cannot be overrun; which greatly increases its potential.

3. The defense of a line of circumvallation will only permit the use of the absolute form. A circular position, facing outward, is the weakest, most awkward order of battle one can imagine, and makes an advantageous sortie virtually impossible. There is no choice but to defend oneself to the last within one's entrenchments. It is quite conceivable that such a situation may lead to a reduction of defensive strength far greater than the one-third that could probably be expected if a corps of observation were detached. If one also bears in mind that since the days of Frederick the Great there has been a general preference for so-called offensives (though they are not always that in practice), for mobility and maneuver, and a general aversion to entrenchments, one will not be surprised that a line of circumvallation is no longer in vogue. Yet its weakening of the tactical defense is by no means its only drawback; the bias that intrudes has merely been listed along with each of the drawbacks, since they are closely related. Basically, a line of circumvallation only covers that area of the theater of war which it encloses: the rest is more or less abandoned to the enemy, except where special detachments are deployed to cover it. But that would amount to a division of forces, which is just what one tries to avoid. The besieger will be constantly worried and alarmed about his supplies. In any case, the use of a line of circumvallation to protect one's supply lines when the army and the supplies it needs are substantial and the enemy is in the field in considerable strength, is possible only under conditions resembling those of the Netherlands. There a whole system of fortresses, close to each other and linked by a network of entrenchments, covers the rest of the operational theater and greatly shortens supply lines. In the days before Louis XIV, the movements of an army were not yet connected with the concept of a theater of war. Armies, particularly in the Thirty Years War, moved about sporadically, confronting any fortress when no hostile force was close by and besieging it for as long as their supplies held out or until an enemy army approached to relieve it. Then lines of circumvallation were natural enough.

In the future they will probably not often be employed except where conditions come close to those described above: when the enemy's army in the field is fairly weak, and the concept of an operational theater is rated lower than that of a siege. Then it will be natural to concentrate one's forces on the siege itself, which will thus undoubtedly considerably gain in energy.

The lines of circumvallation under Louis XIV at Cambrai and Valenciennes were of little use: Turenne took the former by force from Condé, and Condé the latter from Turenne. Still, one should not forget the countless occasions when they have been treated with respect, even when there was a most urgent demand for relief and the defending commander was a man of great enterprise. At Lille in 1708 Villars did not dare attack the allies in their lines. At Olmütz in 1758, and at Dresden in 1760, Frederick the Great, while not using a true line of circumvallation, employed a system that was essentially identical: he used the same army to conduct the siege and to cover it. He was led to do this by the distance of the Austrian army from Olmütz, but he had cause to regret it when he lost his convoys at

Domstadtl. His reason for adopting the system at Dresden in 1760 lay in his low opinion of the Imperial army and his eagerness to capture the city.

A final drawback to such lines is that it is harder to save the siege-train if things go wrong. Where the decision is reached as far as a day's march or two away, the siege can be raised before the enemy arrives, and then the main transport may even gain a day's march on the enemy.

When it comes to deploying an army of observation, the main problem is how far from the siege it should be stationed. The question is usually decided by the terrain or by the position of other armies or units with which the besieging army wishes to remain in contact. In most respects, it is quite obvious that the siege is more securely covered where the distance is greater; on the other hand, a smaller distance, of no more than a few miles, will permit both armies to come to each other's help.

CHAPTER EIGHTEEN

Attack on Convoys

The attack and defense of a convoy is a tactical question, and we should have nothing to say about it here if it were not in some sense necessary to demonstrate that it is feasible at all, which can only be shown by reference to strategic needs and circumstances. We could have treated the subject earlier, in connection with defense; but it is more important offensively, and what little can be said about it may here be summarized for both attack and defense.

An average convoy of three to four hundred wagons, whatever its contents, will be two miles long; a major convoy will be considerably longer. How can one hope to cover this length with the handful of men that are normally assigned as escort? Added to this difficulty is the ponderousness of the whole, which crawls slowly along and is always in danger of ending up in confusion. Moreover, every part requires the same degree of cover, otherwise the whole train would stop and fall into disarray if any part were attacked. One may well ask how the protection and defense of such a convoy is possible at all. In other words, why is not every convoy taken once it has been attacked, and why is not every convoy attacked if it is worth an escort at all—that is to say, if it comes within the enemy's reach? Various tactical remedies have been proposed, such as Tempelhoff's most impractical idea of shortening convoys by constantly stopping to reassemble and starting up again; and Scharnhorst's far more reasonable scheme of splitting them into several columns. But these are only modest palliatives for a deep-seated problem.

The explanation lies in the fact that most convoys are better protected by their general strategic situation than is any other part of the army that the enemy may attack, and hence their limited means of defense are decidedly more effective. That is because convoys, as a rule, move in the rear of their own army, or at least at a considerable distance from the enemy's. Consequently, only minor forces can be detached to attack them, which must protect themselves by strong reserves in order to cover their own flank or rear against another enemy force which might suddenly materialize. Add to this that the very unwieldiness of the wagons makes it difficult to take them away: the attacker must usually be satisfied with cutting the traces, taking out the horses, blowing up the ammunition carts, and so on. This will halt the convoy and throw it into confusion, but it will not actually be lost. It thus becomes even clearer that the convoy's safety lies much more in the general situation than its escort's capacity to defend it. But if the escort actually does so with determination—not by trying to protect the wagons themselves, but by disrupting the enemy's system of attack—it appears that in the end an attack on convoys, far from being easy and reliable, is quite difficult and uncertain.

One more main point remains to be considered: the danger that the enemy army, or part of it, may take revenge on the attacker by inflicting a defeat on him later, as punishment for the operation. This possibility prevents a great many such raids without the reason ever being admitted. We credit the escort for the convoy's safety, and are amazed that such a meager device should be accorded so much respect. To grasp this truth, one should remember the famous retreat that Frederick the Great made through Bohemia in 1758, after the siege of Olmütz: half his army was broken up into sections in order to escort a convoy of four thousand wagons. What prevented Daun from attacking this monstrosity? The fear that Frederick would fall on him with the rest of his army, and engage him in a battle he wanted to avoid. And what prevented Laudon, who was on the convoy's flank throughout, from attacking it sooner and with greater determination than he did at Zischbowitz? The fear of having his knuckles rapped. He was fifty miles away from his main force, and separated from it by the Prussian army. He therefore believed himself to be in danger of serious defeat if the King, who was not being kept busy by Daun, were to turn on him with the bulk of his forces.

Convoys will be in real danger only where an army is forced by its strategic situation to make the unnatural move of drawing its supplies from the flank or even from the front. Then the transports do become a worthwhile target for attack, assuming the enemy can spare the necessary forces. One can point to the complete success of this kind of operation in the same campaign of 1758: the capture of the Domstadtl convoy. The road to Neisse ran to the left of the Prussian positions; Frederick's forces were neutralized by the siege and by having deployed a corps against Daun; the raiders did not have to worry about their own safety, and could attack the convoy at their leisure.

When Prince Eugene was besieging Landrecies in 1712, he procured his supplies from Bouchain, via Denain—that is to say, from the front of his strategic position. Everyone knows the measures he adopted to provide cover under these difficult conditions, and the difficulties in which he got involved; the outcome being a complete transformation of the situation.

We may therefore conclude that, while tactically it may look easy, attacking a convoy is strategically not very advantageous. It promises worthwhile results only in the unusual event of seriously exposed lines of communication.

CHAPTER NINETEEN

Attack on an Enemy Army in Billets

In the book on defense we did not treat this subject, since a line of billets cannot be considered as a means of defense. It is merely the army in a certain state of existence, and one which implies a low capacity for action at that. Our discussion in Chapter Thirteen of Book Five limited itself to discussing this capacity.

In the context of the attack, however, an enemy army in billets must be treated as a separate topic. For one thing, such an attack is a highly specialized operation; for another, it can be considered as a very effective strategic move. We are not talking about an attack on a single billet or on a small unit quartered in a few villages: that would be a purely tactical affair. We are concerned with an attack on a substantial force in billets covering a fairly wide area. The aim is no longer an assault on an individual billet, but the prevention of the enemy's ability to concentrate.

An attack on an army in billets is therefore an attack on a dispersed army. The assault can be considered a success if the enemy is unable to reach his prearranged assembly point and has to find another further to the rear. Such a redeployment in a state of emergency will seldom be made at a cost of less than a day's march, and usually takes longer. The loss of ground involved is no small matter, and that is the first advantage which falls to the attacker.

An assault of this sort may be designed to affect the general situation, but initially it may bear upon individual billets at the same time. Certainly it will not affect them all or even very many of them, since that would entail far greater extension and dispersal of the attacking army than would be advisable. Therefore, only the nearest billets—those that lie in the attacker's path—can be taken by surprise; and even this will seldom be completely successful, for a substantial force is unlikely to be able to approach unnoticed. Still, this is an aspect of the attack that should by no means be disregarded; its results can be considered as the second advantage of such an assault.

A third advantage lies in the separate actions that the enemy is forced to fight, and in which he may suffer serious losses. After all, a substantial force does not assemble by battalions at the main point of concentration. It usually first forms up in brigades, divisions, and even army corps. Units of that size cannot simply flee to the rendezvous; on the contrary, if they make contact with an enemy column, they must accept battle. They may win of course—especially if the attacking column is not strong enough—but even then they will lose time, and, we need hardly add, the general rearward movement means that they are unlikely, as a rule, to make good use of their success. On the other hand, they may be beaten, which is inherently more likely because they have not had the time to organize an effective resistance. It is therefore quite probable that, if an assault has been well-planned and executed, these separate actions may yield the attacker important trophies, which in turn may become major parts of the general result.

A fourth and last advantage, and the keystone for the whole operation, is the fact that the enemy is temporarily thrown off balance and demoralized, so that he can seldom make use of his force when it is finally assembled. Usually, he will have to yield even more ground, and in general completely change his plan of operations.

Such are the typical rewards that may be gained by a successful surprise on enemy billets—that is to say, one that prevents the enemy from assembling his force without losses at a previously chosen rendezvous. But depending on the circumstances, there are many degrees of success: the results may be worth a great deal in one case, and next to nothing in another. Yet even when the operation is so successful that it has significant results, these can seldom compare with those of a major victory; partly because the trophies are rarely so impressive, and partly because the psychological effect is not comparable.

Let us remember that that is all there is to the result; otherwise one may expect more from such an operation than it can deliver. There are people who consider it the paragon of offensive effectiveness, but, as we see from this analysis as well as from military history, that is by no means the case.

One of the most brilliant of surprises was that which the Duke of Lorraine sprang on the French billets under General Rantzau at Tuttlingen in 1643. The corps was 16,000 strong, and lost its commanding general and 7,000 men. It was a total defeat, caused by the complete absence of outposts.

In 1644, when Turenne was surprised at Mergentheim (or, as the French call it, Mariendal) the effect was also tantamount to a defeat. He lost 3,000 of his 8,000 men, mainly because, once his troops were assembled, he badly mistimed his resistance. One should thus not count on frequent results of this sort. In this case they were caused not so much by the surprise itself as by a badly handled fight. Turenne could quite well have avoided the action and joined up somewhere else with the troops stationed in more distant billets.

A third surprise that has become famous is Turenne's operation of 1674 against the allied positions in Alsace under the Great Elector, the Imperial General Bournonville, and the Duke of Lorraine. The trophies were modest, and the allies only lost 2,000 or 3,000 men, which, from a force of 50,000, was far from being decisive. Still, they no longer felt able to put up any resistance in Alsace, and withdrew across the Rhine. This strategic success was all that Turenne wanted, but it was not caused by the assault itself. Turenne surprised the enemy's plans rather than his troops; disagreements among the allied generals and the proximity of the Rhine did the rest. This whole affair actually deserves much closer study because it is usually misinterpreted.

When Neipperg surprised the King in billets in 1741, the only effect was to make Frederick change front and fight the battle of Mollwitz before his troops were completely assembled.

In 1745, when Frederick surprised the Duke of Lorraine in billets in Lusatia, success was mainly due to the actual assault on one of the most important billets, namely Hennersdorf, where the Austrians lost 2,000 men. As a general consequence, the Duke of Lorraine withdrew to Bohemia by way of upper Lusatia. This did not prevent him, however, from reentering Saxony on the left bank of the Elbe, so no important results would have been achieved at all but for the battle of Kesselsdorf.

In 1758 Duke Ferdinand of Brunswick surprised the French in billets. As a direct result, they lost several thousand men and had to take up position behind the river Aller. The psychological effect may well have been more far-reaching, and may have had some influence on the subsequent evacuation of the whole of Westphalia.

Should one wish to draw a general conclusion from these examples as to the value of such attacks, only the first two may be equated with victories in battle. In these cases, however, the corps were small, and the absence of outposts, characteristic of war at the time, was a very favorable circumstance. The other four cases, even though they must be counted among the greatest successes of their kind, were, judging by their results, obviously not in the same class as victory in battle. General success can only be achieved against a weak-willed and characterless adversary; so it was not obtained against Frederick in 1741.

In 1806 the Prussian army intended to surprise the French in Franconia. The chances seemed good: Bonaparte had not yet arrived, and the French were widely scattered in billets. Under these circumstances, the Prussians, had they attacked with speed and determination, might well have expected to drive the French back over the Rhine with appreciable losses. But that was all. Had they aimed at anything more—at pursuing their advantage across the Rhine, for instance, or at gaining a psychological superiority great enough to keep the French from returning to the right bank of the Rhine for the rest of the campaign—their calculations would have lacked any real basis.

Early in August of 1812, the Russians intended a surprise attack from Smolensk on the billets when Napoleon's army had halted near Vitebsk. But their courage failed them when it came to executing the plan—which was just as well for them. Not only was the French center twice as strong as their own, but the French commander was the most determined general the world has ever seen. Anyhow, the loss of a few miles would have settled nothing; there was no natural obstacle close enough on which they could have driven an advantage home and established a reasonably secure position. This was not the kind of campaign that drags feebly on to its conclusion, but the first plan ever made by an attacker bent on the complete destruction of his adversary.[1] The minor advantages to be gained by a surprise attack on billets could, therefore, never have been anything but grossly disproportionate to the needs of the situation; nor could they possibly offset so great a disparity of strength and resources. The attempt, however, showed how easily a confused idea of what the operation might achieve could lead to a completely incorrect application.

Up to this point, the subject has been treated as a strategic device. Its execution is not merely a tactical matter, however, but in part also belongs to strategy. The reason is that such an assault is normally made on a front of considerable breadth and the attacking army may not, and generally will not, have enough time to concentrate before going into action. The whole affair, therefore, consists of a number of separate engagements. Accordingly we now propose to say a few words on how an attack of this sort can best be organized.

The first condition is that the attack on a line of billets be made on a certain width of front. This is the only way in which one can actually assault some quarters, cut off others, and in general throw the enemy force into the desired state of chaos. Conditions will determine how many columns should be used, and how far apart they ought to be.

Second, the columns must converge upon some chosen meeting point. The enemy will end up concentrated to some extent, so our troops must do the same. If possible, that point should be in the same place as the enemy's or located along his line of retreat—preferably where the latter crosses a natural obstacle.

Third, as each column comes into contact with the enemy, it must engage him with great determination, daring and bravery. Circumstances are by and large on its side, and where that is so audacity is in order. For this reason the commander of each column should be given a wide degree of latitude and authority.

Fourth, the tactical plan for attacking the first enemy units that make a stand must be aimed at turning their flanks, since the key to success is always to split the enemy and isolate each part.

Fifth, individual columns must be composed of all arms and should not be short of cavalry. In fact, it may be advantageous to distribute the reserve cavalry among them: it would be a serious mistake to assume that a cavalry reserve on its own could play a major part in this kind of operation. It would be held up at the first village it came to, the smallest bridge, or the most insignificant thicket.

Sixth, the very nature of surprises, of course, forbids one sending an advance guard far ahead—though this holds true only until the first approach has been made. Once fighting has started in the line of enemy billets, the actual surprise has been gained. Then every column must dispatch advance

[1] This translates the word *erste*, as given in the first edition. Subsequent editions, however, give *ernste*: "the determined plan of an attacker, etc." Eds.

guards of all arms as far ahead as possible. These, moving fast, will be able to add substantially to the enemy's confusion. This is the only means of occasionally capturing the welter of baggage and guns, supply units, and men on special duty that usually trail in the wake of troops that have suddenly abandoned their billets; and the advance guards will become the most effective means of turning the enemy and cutting off isolated units.

Finally, seventh, in case the operation should end in failure, one must arrange for a withdrawal, and assign a rallying point for the army.

CHAPTER TWENTY

Diversions

The term "diversion" in ordinary usage means an attack on enemy territory that draws off the enemy's forces from the main objective. Only where this, rather than the capture of the point attacked, is the chief intention, is a diversion a distinct operation. Otherwise it remains an ordinary attack.

In such a diversion there must, of course, be an objective to attack. Only the value of this objective can induce the enemy to dispatch troops for its protection. Besides, if the operation fails as a diversion, the objective will serve as a compensation for the effort expended on capturing it.

These objectives may be fortresses, important depots, large and wealthy towns—especially capital cities—contributions of all sorts, and, finally, cooperation with disaffected subjects of the enemy.

Diversions can obviously be useful but this is not by any means invariably so. Sometimes they can actually do harm. The main requirement is that the enemy should withdraw more men from the main scene of operations than are used for the diversion. If the numbers are even, the effectiveness of the diversion as such ceases, and the operation becomes merely a subordinate attack. Even where a subordinate attack is called for because a major objective might be achieved by a very small expenditure of strength—the easy capture of an important fortress, for example—one should not call it a diversion. Another kind of action is commonly called a diversion: when a state, while defending itself against another, is attacked by a third; but the only difference between that attack and an ordinary one is its direction. There is no reason for giving it a special name: in theoretical discussion, particular terms should be reserved for particular qualities.

If small forces are to draw off larger ones, there must obviously be special circumstances at the root of it. For a diversion to be effective it is not enough arbitrarily to dispatch troops to a previously unoccupied place.

Suppose the attacker decides to raid an enemy district outside the theater of operations with a small force, say a thousand men—to levy contributions, and so forth, the enemy cannot, of course, expect to stop it by sending out a thousand men of his own: it will certainly take more than that to keep the area free of raiders. But, one may ask, could not the defender, instead of protecting his own ground, restore the balance by sending a force of equal strength to raid an equivalent area of the enemy's? Indeed, if the attacker is to reap an advantage, he must start by making sure that there is more to be carried off or threatened in the defender's area than in his own. Where that is so, even a fairly weak diversion cannot fail to keep a much larger enemy force occupied. On the other hand, the advantage will in the nature of the case diminish as the numbers grow: 50,000 men can defend a fairly large area not only against an equal number, but even against a slightly larger one. So the value of large-scale diversion is very doubtful; and the larger it is the more the remaining circumstances must favor the diversion if it is to be successful at all.

The following factors may be favorable:

a. Forces that the attacker can make available for the diversion without detracting from his main offensive
b. Vulnerable objectives of great importance to the enemy
c. Disaffected enemy subjects
d. An area rich enough to yield substantial war material.

If diversions are attempted only if they promise success after having passed these various tests, we will find that favorable opportunities do not arise very often.

Another important point remains to be considered. Diversions always bring the war into an area that would otherwise have been left untouched. Enemy forces that would otherwise be dormant are consequently in some degree brought to life. This will be very marked if the enemy's war plans have included a militia and arms are available for distribution to the populace.

It is quite natural, and experience has frequently illustrated this, that when an area is suddenly threatened and no preparations have been made to defend it, such capable officials as there may be on the spot will mobilize all available extraordinary means to deal with the danger. New means of resistance are created—means that border on guerrilla warfare and can easily bring it about.

This point should be kept very much in mind when a diversion is considered; otherwise one may be digging one's own grave.

Take, for instance, the landings in North Holland in 1799 and on Walcheren in 1809. As diversions they can only be justified by the fact that the British troops could not be used in any other way; yet they undeniably left the French defenses stronger than before, just as a landing in France itself would have done. Much, of course, was to be gained by threatening the coast of France, for the threat alone would neutralize a large force that had to be detailed to guard against it; but one can only justify a landing in force if one can count on the support of the area against its government.

The more remote the likelihood of a great decision in a war, the more legitimate it is to make diversions—but, of course, the smaller are the gains one can expect. Such diversions are simply a means of stirring up a situation.

Execution

1. A diversion may include a real attack. In that event its execution calls for no special characteristics apart from speed and daring.

2. It may, however, be calculated to look more important than it is, thus being at the same time a feint. The exact means that should be used to achieve this can only be determined by an acute mind, with close knowledge of the circumstances and forces involved. It will inevitably involve considerable dispersal of forces.

3. If the forces involved are not inconsiderable, and the retreat is restricted to certain points, it is essential to maintain a reserve on which the rest can fall back.

CHAPTER TWENTY-ONE

Invasion

Almost all we wish to say about invasion consists in a definition of the term. It is often used by modern writers—indeed, even with the air of designating a special quality. The French are always writing about *guerre d'invasion*. What they understand by it is any attack that penetrates deep into enemy territory, and they would like if possible to establish its meaning as the opposite of a routine attack—that is, one that merely nibbles at a frontier. That, however, is unscientific linguistic confusion. Whether an attack will halt at the frontier or penetrate into the heart of the enemy's territory, whether its main concern is to seize the enemy's fortresses or to seek out the core of enemy resistance and pursue it relentlessly, is not a matter that depends on form: it depends on circumstances. Theory, at least, permits no other answer. In some cases it may be more methodical and even more prudent to penetrate some distance rather than stay close to the frontier, but usually this is nothing but the successful outcome of a vigorous *attack*, and so cannot be distinguished from it in any way.

CHAPTER TWENTY-TWO

The Culminating Point of Victory[1]

It is not possible in every war for the victor to overthrow his enemy completely. Often even victory has a culminating point. This has been amply demonstrated by experience. Because the matter is particularly important in military theory and forms the keystone for most plans of campaign, and because its surface is distorted by apparent contradictions, like the dazzling effect of brilliant colors, we shall examine it more closely and seek out its inner logic.

Victory normally results from the superiority of one side; from a greater aggregate of physical and psychological strength. This superiority is certainly augmented by the victory, otherwise it would not be so coveted or command so high a price. That is an automatic consequence of victory *itself*. Its effects exert a similar influence, but only up to a point. That point may be reached quickly—at times so quickly that the total consequences of a victorious battle may be limited to an increase in psychological superiority alone. We now propose to examine how that comes about.

As a war unfolds, armies are constantly faced with some factors that increase their strength and with others that reduce it. The question therefore is one of superiority. Every reduction in strength on one side can be considered as an increase on the other. It follows that this two-way process is to be found in attack as well as in defense.

What we have to do is examine the principal cause of this change in one of these instances, and so at the same time determine the other.

In an advance, the principal causes of additional strength are:

1. The losses suffered by the defending forces are usually heavier than those of the attacker.
2. The defender's loss of fixed assets such as magazines, depots, bridges, and the like, is not experienced by the attacker.
3. The defender's loss of ground, and therefore of resources, from the time we enter his territory.
4. The attacker benefits from the use of some of these resources; in other words, he can live at the enemy's expense.
5. The enemy loses his inner cohesion and the smooth functioning of all components of his force.
6. Some allies are lost to the defender, others turn to the invader.
7. Finally, the defender is discouraged, and so to some extent disarmed.

[1] Compare Chapters Four and Five above, and note on p. 528. Eds.

The causes of loss in strength for an invading army are:

1. The invader has to besiege, assault or observe the enemy's fortresses; while the defender, if he has previously been doing the same, will now add the units so employed to his main force.
2. The moment an invader enters enemy territory, the nature of the operational theater changes. It becomes hostile. It must be garrisoned, for the invader can control it only to the extent that he has done so; but this creates difficulties for the entire machine, which will inevitably weaken its effectiveness.
3. The invader moves away from his sources of supply, while the defender moves closer to his own. This causes delay in the replacement of his forces.
4. The danger threatening the defender will bring allies to his aid.
5. Finally, the defender, being in real danger, makes the greater effort, whereas the efforts of the victor slacken off.

All these advantages and disadvantages may coexist; they can meet, so to speak, and pursue their ways in opposite directions. Only the last meet as true opposites: they cannot by-pass one another, so they are mutually exclusive. That alone is enough to show the infinite range of effects a victory can have—depending on whether they stun the loser or rouse him to greater efforts.

We shall try to qualify each of the above points in a few brief comments.

1. The enemy's losses may be at their maximum directly after his defeat, and then diminish daily until the point is reached where his strength equals ours. On the other hand, his losses may grow progressively day by day. All depends on differences in the overall situation and circumstances. Generally speaking, one can only say that the former is more likely to occur with a good army, the latter with a bad one. The most important factor besides the spirit of the troops is the spirit of the government. It is vital in war to distinguish between the two, or one may stop at the very point where one should really start, and vice versa.
2. The enemy's loss of fixed assets may decrease or increase in the same way, depending on the location and nature of his supply depots. Nowadays, incidentally, this point is no longer so important as the others.
3. The third advantage cannot fail to grow with the progress of the advance. Indeed one can say that it only begins to count when the attack has penetrated deep into enemy territory—when a third or at least a quarter has been taken. A further factor is an area's intrinsic value in relation to the war effort.
4. The fourth advantage is also bound to increase as the advance proceeds.

In connection with these two last points, it should be noted that they seldom have an immediate effect on troops in action. Their work is slow and indirect. Therefore one should not on their account make too great an effort and so place oneself in too dangerous a situation.

5. The fifth advantage also only begins to tell after an army has advanced some distance, and when the configuration of the enemy's country provides an opportunity to isolate certain areas from the rest. Like tightly constricted branches, these will then tend to wither away.

6. and 7. It is probable, at all events, that the sixth and seventh advantages will increase with the advance. We shall return to them later on.

Now let us turn to the causes for loss in strength.

1. In most cases as an advance proceeds, there will be more sieges, assaults, and investments of fortresses. This on its own is so debilitating to *the available fighting forces* that it may easily cancel out all other advantages. True, in modern times one has begun to assault fortresses with very few troops and to observe them with still smaller numbers, and the enemy has, of course, to find garrisons for them. Nevertheless, fortresses remain an important element of security. Half the garrisons may usually consist of men who have not so far taken part in the war; yet one must still leave twice their strength in front of the fortresses on one's line of communication; and if even a single important place has to be formally besieged or starved out, it will call for a small army.

2. The second cause of weakening, the establishment of a theater of operations in enemy territory, grows, of course, with the advance. It may not immediately deplete the strength of the forces, but in the long run it will be even more effective than the first factor.

The only parts of enemy territory one can treat as being within one's theater of operations are those one has actually occupied—by leaving small units in the field, by intermittent garrisons stationed in the major towns, by units established at the relay stations, and so on. Small as each of these garrisons may be, they all deplete the army's fighting strength. But that is the least important part.

Every army has strategic flanks—that is, the areas along both sides of its lines of communication; but since the enemy's army has the same, these are hardly considered a source of weakness. That, however, only holds true in one's own country. Once on enemy soil the weakness becomes palpable. If a long line of communications is covered poorly or not at all, the smallest operation against it holds out promise of success; and in enemy territory raiders may appear from any quarter.

The further the advance, the longer these flanks become, and the risks they represent will progressively increase. Not only are they hard to cover, but the very length of unprotected lines of communication tends to challenge the enemy's spirit of enterprise; and the consequences their loss can have in the event of a retreat are very grave indeed.

All this contributes to place a new burden on an advancing army with every step it takes; so unless it started with exceptional superiority, it will find its freedom of action dwindling and its offensive power progressively reduced. In the end, it will feel unsure of itself and nervous about its situation.

3. The third factor, the distance from the sources that must send continual replacements for this steadily weakening army, will increase proportionately with the advance. In this respect a conquering army is like the light of a lamp; as the oil that feeds it sinks and draws away from the focus, the light diminishes until at last it goes out altogether.

It is true that the wealth of the conquered areas may mitigate this problem, but it can never eliminate it altogether. There are always things that must be supplied from home—especially men. *In general,* deliveries from enemy resources are neither so prompt nor so reliable as those from one's own. In an emergency, help takes longer to arrive, while misunderstandings and mistakes of all kinds cannot be brought to light and rectified so promptly.

If a monarch does not command his troops in person, as has become customary in recent wars, if he is no longer easily available, a new and very serious handicap arises from the loss of time involved in the transmission of messages. Even the widest powers conferred on a commander will not suffice to meet every contingency that may arise in his sphere of action.

4. The change in political alignments. If these changes, resulting from his victories, are likely to be to the disadvantage of the victor, they will probably be so in direct proportion to his advance—which is also the case if they are to his advantage. All depends on the existing political affiliations, interests, traditions, lines of policy, and the personalities of princes, ministers, favorites, mistresses, and so forth. The only general comment one can make is that after the defeat of a major power with lesser allies, these will quickly desert their leader. In this respect, the victor will then gain strength with every blow. If, on the other hand, the defeated state is smaller, protectors will appear much sooner if its very existence is threatened. Others who may have helped to endanger it will detach themselves if they believe that the success is becoming too great.

5. The increased resistance aroused in the enemy. Sometimes, stunned and panic-stricken, the enemy may lay down his arms, at other times he may be seized by a fit of enthusiasm: there is a general rush to arms, and resistance is much stronger after the first defeat than it was before. The information from which one must guess at the probable reaction include the character of the people and the government, the nature of the country, and its political affiliations.

The last two points alone can make an infinite difference to the plans that one can and must make in war to take account of either possibility.

While one man may lose his best chance through timidity and following so-called orthodox procedures, another will plunge in head first and end up looking as dazed and surprised as if he had just been fished out of the water.

Further, one should be conscious of the slackening of effort that not infrequently occurs on the part of the victor after the danger has been overcome, and when, on the contrary, fresh efforts are called for to follow up the victory. If we take an overall view of these differing and opposing principles, we will doubtless conclude that the utilization of the victory, a continued advance in an offensive campaign, will usually swallow up the superiority with which one began or which was gained by the victory.

At this point we are bound to ask: if all this is true, why does the winner persist in pursuing his victorious course, in advancing his offensive? Can one really still call this a "utilization of victory?" Would he not do better to stop before he begins to lose the upper hand?

The obvious answer is that superior strength is not the end but only the means. The end is either to bring the enemy to his knees or at least to deprive him of some of his territory—the point in that case being *not to improve the current military position* but to improve one's general prospects in the war and in the peace negotiations. Even if one tries to destroy the enemy completely, one must accept the fact that every step gained may weaken one's superiority—though it does not necessarily follow that it must fall to zero before the enemy capitulates. He may do so at an earlier point, and if this can be accomplished with one's last ounce of superiority, it would be a mistake not to have used it.

Thus the superiority one has or gains in war is only the means and not the end; it must be risked for the sake of the end. But one must know the point to which it can be carried in order not to overshoot the target; otherwise instead of gaining new advantages, one will disgrace oneself.

There is no need to cite historical examples in order to prove that this is how loss of superiority affects a strategic attack. Indeed, such instances occur so frequently that we have felt it necessary to investigate their underlying causes. Only with the rise of Bonaparte have there been campaigns between civilized states where superiority has consistently led to the enemy's collapse. Before his time, every campaign had ended with the winning side attempting to reach a state of balance in which it could maintain itself. At that point, the progress of victory stopped, and a retreat might even be called for. This culminating point in victory is bound to recur in every future war in which the destruction of the enemy cannot be the military aim, and this will presumably be true of most wars. The natural goal of all campaign plans, therefore, is the turning point at which attack becomes defense.

If one were to go beyond that point, it would not merely be a *useless* effort which could not add to success. It would in fact be a *damaging* one, which would lead to a reaction; and experience goes to show that such reactions usually have completely disproportionate effects. This is such a universal experience, and appears so natural and easy to understand, that there is no need for a laborious investigation of its causes. The main causes are always the lack of organization in newly occupied territory, and the psychological effect of the stark contrast between the serious losses sustained and the successes that had been hoped for. There is an unusually active interplay between the extremes of morale—on the one hand, encouragement often verging on bravado, and on the other, depression. As a result, losses will be heavier during a retreat, and one can usually be grateful if one has to sacrifice only conquered territory, and not one's native soil.

At this point we must eliminate an apparent inconsistency.

This rests on the assumption that so long as an attack progresses there must still be some superiority on its side; further, that since defense (the more effective form of war) must start when the advance ends, one may not really be in much danger of imperceptibly becoming the weaker side. Yet that is what happens; history forces us to admit that the risk of a setback often does not reach its peak until the moment when the attack has lost its impetus and is turning into defense. We must look for the reason.

The superiority that I have attributed to the defensive form of warfare rests on the following:

1. The utilization of terrain
2. The possession of an organized theater of operations
3. The support of the population
4. The advantage of being on the waiting side.

It is obvious that these factors will not everywhere be found in equal strength, or always be equally effective. One defense is therefore not exactly like another, nor will defense always enjoy the same degree of superiority over attack. In particular this will be the case in a defense that follows directly the exhaustion of an offensive—a defense whose theater of operations is located at the apex of an offensive wedge thrust forward deep into hostile territory. Only the first of the four factors listed above, the utilization of terrain, will remain unchanged in such a defense; the second is usually eliminated, the third works in reverse, and the fourth is much reduced in strength. A word or two in explanation of this last point may be useful.

In an imaginary equilibrium, whole campaigns might often end without result because the side that should take the initiative lacks determination. That, in our view, is exactly why it is an advantage to be able to await the enemy. But if an offensive act upsets this equilibrium, damages the enemy's interests and impels him into action, he is far less likely to remain inactive and irresolute. A defense is far more provocative in character when it is undertaken on occupied territory than it is on one's own; it is, so to speak, infected with the virus of attack, and this weakens its basic character. In Silesia and Saxony Daun granted Frederick a period of calm that he would never have allowed in Bohemia.

It is clear, therefore, that a defense that is undertaken in the framework of an offensive is weakened in all its key elements. It will thus no longer possess the superiority which basically belongs to it.

Just as no defensive campaign consists simply of defensive elements, so no offensive campaign consists purely of offensive ones. Apart from the short intervals in every campaign during which both sides are on the defensive, every attack which does not lead to peace must necessarily end up as a defense.

It is thus defense itself that weakens the attack. Far from this being idle sophistry, we consider it to be the greatest disadvantage of the attack that one is eventually left in a most awkward defensive position.

This will explain why there is a gradual reduction in the difference between the original effectiveness of attack and defense as forms of warfare. We now propose to show how this difference can for a time vanish altogether and reverse itself completely.

We can be more succinct if we may use an analogy from nature. Every physical force requires time to become effective. A force that, if gently and gradually applied, would suffice to arrest a body in motion, will be overcome by it if there is not enough time for it to operate. This law of physics provides a pertinent image of many features of our own psychology. Once our train of thought is set in a certain direction, many reasons which would otherwise be basically adequate to do so will not be able to deflect or arrest it. Time, repose, and a sustained impact on one's consciousness are needed. It is the same in war. Once the mind is set on a certain course toward its goal, or once it has turned back toward a refuge, it may easily happen that arguments which would compel one man to stop, and justify another in acting, will not easily be fully appreciated. Meanwhile the action continues, and in the sweep of motion one crosses the threshold of equilibrium, the line of culmination, without knowing it. It is even possible that the attacker, reinforced by the psychological forces peculiar to attack, will in spite of his exhaustion find it less difficult to go on than to stop—like a horse pulling a load uphill. We believe that this demonstrates without inconsistency how an attacker can overshoot the point at which, if he stopped and assumed the defensive, there would still be a chance of success—that is, of equilibrium. It is therefore important to calculate this point correctly when planning the campaign. An attacker may otherwise take on more than he can manage and, as it were, get into debt; a defender must be able to recognize this error if the enemy commits it, and exploit it to the full.

In reviewing the whole array of factors a general must weigh before making his decision, we must remember that he can gauge the direction and value of the most important ones only by considering numerous other possibilities—some immediate, some remote. He must *guess*, so to speak: guess whether the first shock of battle will steel the enemy's resolve and stiffen his resistance, or whether, like a Bologna flask, it will shatter as soon as its surface is scratched; guess the extent of debilitation and paralysis that the drying up of particular sources of supply and the severing of certain lines of communication will cause in the enemy; guess whether the burning pain of the injury he has been dealt will make the enemy collapse with exhaustion or, like a wounded bull, arouse his rage; guess whether the other powers will be frightened or indignant, and whether and which political alliances will be dissolved or formed. When we realize that he must hit upon all this and much more by means of his discreet judgment, as a marksman hits a target, we must admit that such an accomplishment of the human mind is no small achievement. Thousands of wrong turns running in all directions tempt his perception; and if the range, confusion and complexity of the issues are not enough to overwhelm him, the dangers and responsibilities may.

This is why the great majority of generals will prefer to stop well short of their objective rather than risk approaching it too closely, and why those with high courage and an enterprising spirit will often overshoot it and so fail to attain their purpose. Only the man who can achieve great results with limited means has really hit the mark.

BOOK EIGHT

War Plans

CHAPTER ONE

Introduction

In the chapter on the nature and purpose of war we roughly sketched the general concept of war and alluded to the connections between war and other physical and social phenomena, in order to give our discussion a sound theoretical starting point. We indicated what a variety of intellectual obstacles besets the subject, while reserving detailed study of them until later; and we concluded that the grand objective of all military action is to overthrow the enemy—which means destroying his armed forces. It was therefore possible to show in the following chapter that battle is the one and only means that warfare can employ. With that, we hoped, a sound working hypothesis had been established.

Then we examined, one by one, the salient patterns and situations (apart from battle itself) that occur in warfare, trying to gauge the value of each with greater precision, both according to its inherent characteristics and in the light of military experience. We also sought to strip away the vague, ambiguous notions commonly attached to them, and tried to make it absolutely clear that the destruction of the enemy is what always matters most.

We now revert to warfare as a whole, to the discussion of the planning of a war and of a campaign, which means returning to the ideas put forward in Book One.

The chapters that follow will deal with the problem of war as a whole. They cover its dominant, its most important aspect: pure strategy. We enter this crucial area—the central point on which all other threads converge—not without some diffidence. Indeed, this diffidence is amply justified.

On the one hand, military operations appear extremely simple. The greatest generals discuss them in the plainest and most forthright language; and to hear them tell how they control and manage that enormous, complex apparatus one would think the only thing that mattered was the speaker, and that the whole monstrosity called war came down, in fact, to a contest between individuals, a sort of duel. A few uncomplicated thoughts seem to account for their decisions—either that, or the explanation lies in various emotional states; and one is left with the impression that great commanders manage matters in an easy, confident and, one would almost think, off-hand sort of way. At the same time we can see how many factors are involved and have to be weighed against each other; the vast, the almost infinite distance there can be between a cause and its effect, and the countless ways in which these elements can be combined. The function of theory is to put all this in systematic order, clearly and comprehensively, and to trace each action to an adequate, compelling cause. When we contemplate all this, we are overcome by the fear that we shall be irresistibly dragged down to a state of dreary pedantry, and grub around in the underworld of ponderous concepts where no great commander, with his effortless *coup d'oeil,* was ever seen. If that were the best that theoretical studies could produce it would be better never to have attempted them in the first place. Men of genuine talent would despise them and they would quickly be forgotten. When all is said and done, it really is the commander's *coup d'oeil,* his ability to see things simply, to identify the whole business of war completely with himself, that is the essence of good generalship. Only if the mind works in this comprehensive fashion can it achieve the freedom it needs to dominate events and not be dominated by them.

We resume our task then, with some diffidence; and we shall fail unless we keep to the path we set ourselves at the beginning. Theory should cast a steady light on all phenomena so that we can more easily recognize and eliminate the weeds that always spring from ignorance; it should show how one thing is related to another, and keep the important and the unimportant separate. If concepts combine of their own accord to form that nucleus of truth we call a principle, if they spontaneously compose a pattern that becomes a rule, it is the task of the theorist to make this clear.

The insights gained and garnered by the mind in its wanderings among basic concepts are benefits that theory can provide. Theory cannot equip the mind with formulas for solving problems, nor can it mark the narrow path on which the sole solution is supposed to lie by planting a hedge of principles on either side. But it can give the mind insight into the great mass of phenomena and of their relationships, then leave it free to rise into the higher realms of action. There the mind can use its innate talents to capacity, combining them all so as to seize on what is *right* and *true* as though this were a single idea formed by their concentrated pressure—as though it were a response to the immediate challenge rather than a product of thought.

CHAPTER TWO

Absolute War and Real War

War plans cover every aspect of a war, and weave them all into a single operation that must have a single, ultimate objective in which all particular aims are reconciled. No one starts a war—or rather, no one in his senses ought to do so—without first being clear in his mind what he intends to achieve by that war and how he intends to conduct it. The former is its political purpose; the latter its operational objective. This is the governing principle which will set its course, prescribe the scale of means and effort which is required, and make its influence felt throughout down to the smallest operational detail.

We said in the opening chapter that the natural aim of military operations is the enemy's overthrow, and that strict adherence to the logic of the concept can, in the last analysis, admit of no other. Since both belligerents must hold that view it would follow that military operations could not be suspended, that hostilities could not end until one or other side were finally defeated.

In the chapter on the suspension of military activity[1] we showed how factors inherent in the war-machine itself can interrupt and modify the principle of enmity as embodied in its agent, man, and in all that goes to make up warfare. Still, that process of modification is by no means adequate to span the gap between the pure concept of war and the concrete form that, as a general rule, war assumes. Most wars are like a flaring-up of mutual rage, when each party takes up arms in order to defend itself, to overawe its opponent, and occasionally to deal him an actual blow. Generally it is not a case in which two mutually destructive elements collide, but one of tension between two elements, separate for the time being, which discharge energy in discontinuous, minor shocks.

But what exactly is this nonconducting medium, this barrier that prevents a full discharge? Why is it that the theoretical concept is not fulfilled in practice? The barrier in question is the vast array of factors, forces and conditions in national affairs that are affected by war. No logical sequence could progress through their innumerable twists and turns as though it were a simple thread that linked two deductions. Logic comes to a stop in this labyrinth; and those men who habitually act, both in great and minor affairs, on particular dominating impressions or feelings rather than according to strict logic, are hardly aware of the confused, inconsistent, and ambiguous situation in which they find themselves.

[1] Book Three, Chapter Sixteen. Eds.

The man in overall command may actually have examined all these matters without losing sight of his objective for an instant; but the many others concerned cannot all have achieved the same insight. Opposition results, and in consequence something is required to overcome the vast inertia of the mass. But there is not usually enough energy available for this.

This inconsistency can appear in either belligerent party or in both, and it is the reason why war turns into something quite different from what it should be according to theory—turns into something incoherent and incomplete.

This is its usual appearance, and one might wonder whether there is any truth at all in our concept of the absolute character of war were it not for the fact that with our own eyes we have seen warfare achieve this state of absolute perfection. After the short prelude of the French Revolution, Bonaparte brought it swiftly and ruthlessly to that point. War, in his hands, was waged without respite until the enemy succumbed, and the counterblows were struck with almost equal energy. Surely it is both natural and inescapable that this phenomenon should cause us to turn again to the pure concept of war with all its rigorous implications.

Are we then to take this as the standard, and judge all wars by it, however much they may diverge? Should we deduce our entire theory from it? The question is whether that should be the only kind of war or whether there can be other valid forms. We must make up our minds before we can say anything intelligent about war plans.

If the first view is right, our theory will everywhere approximate to logical necessity, and will tend to be clear and unambiguous. But in that case, what are we to say about all the wars that have been fought since the days of Alexander—excepting certain Roman campaigns—down to Bonaparte? We should have to condemn them outright, but might be appalled at our presumption if we did so. Worse still, we should be bound to say that in spite of our theory there may even be other wars of this kind in the next ten years, and that our theory, though strictly logical, would not apply to reality. We must, therefore, be prepared to develop our concept of war as it ought to be fought, not on the basis of its pure definition, but by leaving room for every sort of extraneous matter. We must allow for natural inertia, for all the friction of its parts, for all the inconsistency, imprecision, and timidity of man; and finally we must face the fact that war and its forms result from ideas, emotions, and conditions prevailing at the time—and to be quite honest we must admit that this was the case even when war assumed its absolute state under Bonaparte.

If this is the case, if we must admit that the origin and the form taken by a war are not the result of any ultimate resolution of the vast array of circumstances involved, but only of those features that happen to be dominant. It follows that war is dependent on the interplay of possibilities and probabilities, of good and bad luck, conditions in which strictly logical reasoning often plays no part at all and is always apt to be a most unsuitable and awkward intellectual tool. It follows, too, that war can be a matter of degree.

Theory must concede all this; but it has the duty to give priority to the absolute form of war and to make that form a general point of reference, so that he who wants to learn from theory becomes accustomed to keeping that point in view constantly, to measuring all his hopes and fears by it, and to approximating it *when he can* or *when he must*.

A principle that underlies our thoughts and actions will undoubtedly lend them a certain tone and character, though the immediate causes of our action may have different origins, just as the tone a painter gives to his canvas is determined by the color of the underpainting.

If theory can effectively do this today, it is because of our recent wars. Without the cautionary examples of the destructive power of war unleashed, theory would preach to deaf ears. No one would have believed possible what has now been experienced by all.

Would Prussia in 1792[2] have dared to invade France with 70,000 men if she had had an inkling that the repercussions in case of failure would be strong enough to overthrow the old European balance of power? Would she, in 1806, have risked war with France with 100,000 men, if she had suspected that the first shot would set off a mine that was to blow her to the skies?

[2] The German has 1798, which obviously is a misprint. Eds.

CHAPTER THREE

A. Interdependence of the Elements of War

Since war can be thought of in two different ways—its absolute form or one of the variant forms that it actually takes—two different concepts of success arise.

In the absolute form of war, where everything results from necessary causes and one action rapidly affects another, there is, if we may use the phrase, no intervening neutral void. Since war contains a host of interactions[1] since the whole series of engagements is, strictly speaking, linked together,[2] since in every victory there is a culminating point beyond which lies the realm of losses and defeats[3]—in view of all these intrinsic characteristics of war, we say there is only one result that counts: *final victory*. Until then, nothing is decided, nothing won, and nothing lost. In this form of war we must always keep in mind that it is the end that crowns the work. Within the concept of absolute war, then, war is indivisible, and its component parts (the individual victories) are of value only in their relation to the whole. Conquering Moscow and half of Russia in 1812 was of no avail to Bonaparte unless it brought him the peace he had in view. But these successes were only a part of his plan of campaign: what was still missing was the destruction of the Russian army. If that achievement had been added to the rest, peace would have been as sure as things of that sort ever can be. But it was too late to achieve the second part of his plan; his chance had gone. Thus the successful stage was not only wasted but led to disaster.

Contrasting with this extreme view of the connection between successes in war, is another view, no less extreme; which holds that war consists of separate successes each unrelated to the next, as in a match consisting of several games. The earlier games have no effect upon the later. All that counts is the total score, and each separate result makes its contribution toward this total.

The first of these two views of war derives its validity from the nature of the subject; the second, from its actual history. Countless cases have occurred where a small advantage could be gained without an onerous condition being attached to it. The more the element of violence is moderated, the commoner these cases will be; but just as absolute war has never in fact been achieved, so we will never find a war in which the second concept is so prevalent that the first can be disregarded altogether. If we postulate the first of the two concepts, it necessarily follows from the start that every

[1] See Chapter One, Book One. Cl. [2] See Chapter Two, Book One. Cl.
[3] See Chapters Four and Five, Book Seven. Cl.

war must be conceived of as a single whole, and that with his first move the general must already have a clear idea of the goal on which all lines are to converge.

If we postulate the second concept, we will find it legitimate to pursue minor advantages for their own sake and leave the future to itself.

Since both these concepts lead to results, theory cannot dispense with either. Theory makes this distinction in the application of the two concepts: all action must be based on the former, since it is the fundamental concept; the latter can be used only as a modification justified by circumstances.

In 1742, 1744, 1757, and 1758, when Frederick, operating from Silesia and Saxony, thrust new spearheads into Austria, he was well aware that they could not lead to another permanent acquisition such as Silesia and Saxony. His aim was not to overthrow the Austrian Empire but a secondary one, namely to gain time and strength. And he could pursue this secondary aim without any fear of risking his own existence.[4]

However, when Prussia in 1806, and Austria in 1805 and 1809, adopted a still more modest aim—to drive the French across the Rhine—it would have been foolish if they had not begun by carefully reviewing the whole chain of events that success or failure would be likely to bring in consequence of the initial step, and which would lead to peace. Such a review was indispensable, both in order to decide how far they could safely exploit their successes and also how and where any enemy successes could be arrested.

Careful study of history shows where the difference between these cases lies. In the eighteenth century, in the days of the Silesian campaigns, war was still an affair for governments alone, and the people's role was simply that of an instrument. At the onset of the nineteenth century, peoples themselves were in the scale on either side. The generals opposing Frederick the Great were acting on instructions—which implied that caution was one of their distinguishing characteristics. But now the opponent of the Austrians and Prussians was—to put it bluntly—the God of War himself.

Such a transformation of war might have led to new ways of thinking about it. In 1805, 1806, and 1809 men might have recognized that total ruin was a possibility—indeed it stared them in the face. It might have

[4] If Frederick had won the battle of Kolin and in consequence had captured the main Austrian army in Prague with both its senior commanders, it would indeed have been such a shattering blow that he might well have thought of pressing on to Vienna, shaking the foundations of the monarchy and imposing peace. That would have been an unparalleled success for those days, as great as the triumphs of the Napoleonic wars, but still more wonderful and brilliant for the disparity in size between the Prussian David and the Austrian Goliath. Victory at Kolin would almost certainly have made this success possible. But that does not invalidate the assertion made above, which only concerned the original purpose of the King's offensive. To surround and capture the enemy's main army, on the other hand, was something wholly unprovided for and the King had never given it a thought—at least until the Austrians invited it by the inadequate position they took up at Prague. Cl.

stimulated them to different efforts that were directed toward greater objectives than a couple of fortresses and a medium-sized province.

They did not, however, change their attitude sufficiently, although the degree of Austrian and Prussian rearmament shows that the storm clouds massing in the political world had been observed. They failed because the transformations of war had not yet been sufficiently revealed by history. In fact the very campaigns of 1805, 1806, and 1809, and those that followed are the ones that make it easier for us to grasp the concept of modern, absolute war in all its devastating power.

Theory, therefore, demands that at the outset of a war its character and scope should be determined on the basis of the political probabilities. The closer these political probabilities drive war toward the absolute, the more the belligerent states are involved and drawn in to its vortex, the clearer appear the connections between its separate actions, and the more imperative the need not to take the first step without considering the last.

B. Scale of the Military Objective and of the Effort To Be Made

The degree of force that must be used against the enemy depends on the scale of political demands on either side. These demands, so far as they are known, would show what efforts each must make; but they seldom are fully known—which may be one reason why both sides do not exert themselves to the same degree.

Nor are the situation and conditions of the belligerents alike. This can be a second factor.

Just as disparate are the governments' strength of will, their character and abilities.

These three considerations introduce uncertainties that make it difficult to gauge the amount of resistance to be faced and, in consequence, the means required and the objectives to be set.

Since in war too small an effort can result not just in failure but in positive harm, each side is driven to outdo the other, which sets up an interaction.

Such an interaction could lead to a maximum effort if a maximum could be defined. But in that case all proportion between action and political demands would be lost: means would cease to be commensurate with ends, and in most cases a policy of maximum exertion would fail because of the domestic problems it would raise.

In this way the belligerent is again driven to adopt a middle course. He would act on the principle of using no greater force, and setting himself no greater military aim, than would be sufficient for the achievement of his political purpose. To turn this principle into practice he must renounce the need for absolute success in each given case, and he must dismiss remoter possibilities from his calculations.

At this point, then, intellectual activity leaves the field of the exact sciences of logic and mathematics. It then becomes an art in the broadest meaning of the term—the faculty of using judgment to detect the most important and decisive elements in the vast array of facts and situations. Undoubtedly this power of judgment consists to a greater or lesser degree in the intuitive comparison of all the factors and attendant circumstances; what is remote and secondary is at once dismissed while the most pressing and important points are identified with greater speed than could be done by strictly logical deduction.

To discover how much of our resources must be mobilized for war, we must first examine our own political aim and that of the enemy. We must gauge the strength and situation of the opposing state. We must gauge the character and abilities of its government and people and do the same in regard to our own. Finally, we must evaluate the political sympathies of other states and the effect the war may have on them. To assess these things in all their ramifications and diversity is plainly a colossal task. Rapid and correct appraisal of them clearly calls for the intuition of a genius; to master all this complex mass by sheer methodical examination is obviously impossible. Bonaparte was quite right when he said that Newton himself would quail before the algebraic problems it could pose.

The size and variety of factors to be weighed, and the uncertainty about the proper scale to use, are bound to make it far more difficult to reach the right conclusion. We should also bear in mind that the vast, unique importance of war, while not increasing the complexity and difficulty of the problem, does increase the value of the correct solution. Responsibility and danger do not tend to free or stimulate the average person's mind—rather the contrary; but wherever they do liberate an individual's judgment and confidence we can be sure that we are in the presence of exceptional ability.

At the outset, then, we must admit that an imminent war, its possible aims, and the resources it will require, are matters that can only be assessed when every circumstance has been examined in the context of the whole, which of course includes the most ephemeral factors as well. We must also recognize that the conclusion reached can be no more wholly objective than any other in war, but will be shaped by the qualities of mind and character of the men making the decision—of the rulers, statesmen, and commanders, whether these roles are united in a single individual or not.

A more general and theoretical treatment of the subject may become feasible if we consider the nature of states and societies as they are determined by their times and prevailing conditions. Let us take a brief look at history.

The semibarbarous Tartars, the republics of antiquity, the feudal lords and trading cities of the Middle Ages, eighteenth-century kings and the rulers and peoples of the nineteenth century—all conducted war in their own particular way, using different methods and pursuing different aims.

The Tartar hordes searched for new land. Setting forth as a nation, with women and children, they outnumbered any other army. Their aim was to subdue their enemies or expel them. If a high degree of civilization could have been combined with such methods, they would have carried all before them.

The republics of antiquity, Rome excepted, were small and their armies smaller still, for the *plebs*, the mass of the people, was excluded. Being so many and so close together these republics found that the balance that some law of nature will always establish among small and unconnected units formed an obstacle to major enterprises. They therefore limited their wars to plundering the countryside and seizing a few towns, in order to gain a degree of influence over them.

Rome was the one exception to this rule, and then only in her later days. With little bands of men, she had for centuries carried on the usual struggle with her neighbors for booty or alliance. She grew not so much by conquest as by the alliances she made, for the neighboring peoples gradually merged with her and were assimilated into a greater Rome. Only when this process had spread the rule of Rome through Southern Italy did she begin to expand by way of actual conquest. Carthage fell; Spain and Gaul were taken; Greece was subjugated; and Roman rule was carried into Asia and Egypt. Rome's military strength at that period was immense, without her efforts being equally great. Her armies were kept up by her wealth. Rome was no longer like the Greek republics, nor was she even faithful to her own past. Her case is unique.

As singular in their own way were the wars of Alexander. With his small but excellently trained and organized army, Alexander shattered the brittle states of Asia. Ruthlessly, without pause, he advanced through the vast expanse of Asia until he reached India. That was something no republic could have achieved; only a king who in a sense was his own *condottiere* could have accomplished it so rapidly.

Mediaeval monarchs, great and small, waged war with feudal levies, which limited operations. If a thing could not be finished quickly it was impossible. The feudal army itself was an assemblage of vassals and their servants, brought and held together in part by legal obligation, in part by voluntary alliance—the whole amounting to true confederation. Weapons and tactics were based on individual combat, and thus unsuited to the organized action of large numbers. And indeed, cohesion in the state was never weaker or the individual so independent. It was the combination of these factors that gave mediaeval wars their special character. They were waged relatively quickly; not much time was wasted in the field; but their aim was usually to punish the enemy, not subdue him. When his cattle had been driven off and his castles burned, one could go home.

The great commercial cities and the small republics created *condottieri*. They were an expensive and therefore small military force. Even smaller was their fighting value: extremes of energy or exertion were conspicuous by their absence and fighting was generally a sham. In brief, then, hatred and enmity no longer drove the state to take matters into its own hands; they became an element in negotiation. War lost many of its risks; its character was wholly changed, and no deduction from its proper nature was still applicable.

Gradually the feudal system hardened into clearly delimited territorial sovereignty. States were closer knit; personal service was commuted into dues in kind, mostly in the form of money, and feudal levies were replaced by mercenaries. The transition was bridged by the *condottieri*. For a period they were also the instrument of the larger states. But soon the soldier hired on short-term contract evolved into the *permanent mercenary*, and the armed force of the state had become a standing army, paid for by the treasury.

The slow evolution toward this goal naturally brought with it numerous overlappings of these three types of military institutions. Under Henry IV of France feudal levies, *condottieri* and a standing army were used side by side. The *condottieri* survived into the Thirty Years War, and indeed faint traces of them can be found in the eighteenth century.

Just as the character of the military institutions of the European states differed in the various periods, so did all their other conditions. Europe, essentially, had broken down into a mass of minor states. Some were turbulent republics, other precarious small monarchies with very limited central power. A state of that type could not be said to be genuinely united; it was rather an agglomeration of loosely associated forces. Therefore we should not think of such a state as a personified intelligence acting according to simple and logical rules.

This is the point of view from which the policies and wars of the Middle Ages should be considered. One need only think of the German emperors with their constant descents into Italy over a period of five hundred years. These expeditions never resulted in any complete conquest of the country; nor were they ever meant to do so. It would be easy to regard them as a chronic error, a delusion born of the spirit of the times; but there would be more sense in attributing them to a host of major causes, which we may possibly assimilate intellectually, but whose dynamic we will never comprehend as clearly as did the men who were actually obliged to contend with them. So long as the great powers that eventually grew out of this chaos needed time to consolidate and organize themselves, most of their strength and energies went into that process. Foreign wars were fewer, and those that did take place betrayed the marks of immature political cohesion.

The wars of the English against the French are the first to stand out. But France could not yet be considered as a genuine monarchy—she was rather an agglomeration of duchies and counties; while England, though displaying greater unity, still fought with feudal levies amid much domestic strife.

Under Louis XI France took the greatest step toward internal unity. She became a conquering power in Italy under Charles VIII, and her state and her army reached a peak under Louis XIV.

Spanish unity began to form under Ferdinand of Aragon. Under Charles V, as a result of favorable marriages, a mighty Spanish monarchy suddenly emerged, composed of Spain and Burgundy, Germany and Italy. What this colossus lacked in cohesion and domestic stability was made up for by its wealth. Its standing army first encountered that of France. On the abdication of Charles V the colossus broke into two parts—Spain and Austria. The latter, strengthened by Hungary and Bohemia, now emerged as a major power, dragging behind her the German confederation like a dinghy.

The end of the seventeenth century, the age of Louis XIV, may be regarded as that point in history when the standing army in the shape familiar to the eighteenth century reached maturity. This military organization was based on money and recruitment. The states of Europe had achieved complete internal unity. With their subjects' services converted into money payments, the strength of governments now lay entirely in their treasuries. Thanks to cultural developments and to a progressively more sophisticated administration, their power was very great compared with earlier days. France put several hundred thousand regular troops in the field, and other states could do likewise in proportion to their populations.

International relations had changed in other ways as well. Europe was now split between a dozen monarchies and a handful of republics. It was conceivable that two states could fight a major war without, as in former times, involving twenty others. The possible political alignments were still many and various; but they could be surveyed, and their probability at each given instant could be evaluated.

Domestically almost every state had been reduced to an absolute monarchy; the privileges and influence of the estates had gradually disappeared. The executive had become completely unified and represented the state in its foreign relations. Political and military institutions had developed into an effective instrument, with which an independent will at the center could now wage war in a form that matched its theoretical concept.

During this period, moreover, three new Alexanders appeared—Gustavus Adolphus, Charles XII, and Frederick the Great. With relatively limited but highly efficient forces each sought to turn his small state into a large monarchy, and crush all opposition. Had they been dealing only with Asiatic empires they might have resembled Alexander more closely. But in terms of risks that they ran, they undeniably foreshadowed Bonaparte.

But, if war gained in power and effectiveness, it lost in other respects. Armies were paid for from the treasury, which rulers treated almost as their privy purse or at least as the property of the government, not of the people. Apart from a few commercial matters, relations with other states did not concern the people but only the treasury or the government. That at least was the general attitude. A government behaved as though it owned and managed a great estate that it constantly endeavored to enlarge—an effort in which the inhabitants were not expected to show any particular interest. The Tartar people and army had been one; in the republics of antiquity and during the Middle Ages the people (if we confine the concept to those who had the rights of citizens) had still played a prominent part; but in the circumstances of the eighteenth century the people's part had been extinguished. The only influence the people continued to exert on war was an indirect one—through its general virtues or shortcomings.

War thus became solely the concern of the government to the extent that governments parted company with their peoples and behaved as if they were themselves the state. Their means of waging war came to consist of the money in their coffers and of such idle vagabonds as they could lay their hands on either at home or abroad. In consequence the means they had available were fairly well defined, and each could gauge the other side's potential in terms both of numbers and of time. War was thus deprived of its most dangerous feature—its tendency toward the extreme, and of the whole chain of unknown possibilities which would follow.

The enemy's cash resources, his treasury and his credit, were all approximately known; so was the size of his fighting forces. No great expansion was feasible at the outbreak of war. Knowing the limits of the enemy's strength, men knew they were reasonably safe from total ruin; and being aware of their own limitations, they were compelled to restrict their own aims in turn. Safe from the threat of extremes, it was no longer necessary to go to extremes. Necessity was no longer an incentive to do so, and the only impulse could come from courage and ambition. These, on the other hand, were strongly curbed by the prevailing conditions of the state. Even a royal commander had to use his army with a minimum of risk. If the army was pulverized, he could not raise another, and behind the army there was nothing. That enjoined the greatest prudence in all operations. Only if a decisive advantage seemed possible could the precious instrument be used, and to bring things to that point was a feat of the highest generalship. But so long as that was not achieved, operations drifted in a kind of vacuum; there was no reason to act, and every motivating force seemed inert. The original motive of the aggressor faded away in prudence and hesitation.

The conduct of war thus became a true game, in which the cards were dealt by time and by accident. In its effect it was a somewhat stronger form of diplomacy, a more forceful method of negotiation, in which battles and sieges were the principal notes exchanged. Even the most ambitious ruler had no greater aims than to gain a number of advantages that could be exploited at the peace conference.

This limited, constricted form of war was due, as we said, to the narrow base on which it rested. But the explanation why even gifted commanders and monarchs such as Gustavus Adolphus, Charles XII, and Frederick the Great, with armies of exceptional quality, should have risen so little above the common level of the times, why even they had to be content with moderate success, lies with the balance of power in Europe. With the multitude of minor states in earlier times, any one of them was prevented from rapidly expanding by such immediate and concrete factors as their proximity and contiguity, their family ties and personal acquaintances. But now that states were larger and their centers farther apart, the wide spread of interests they had developed became the factor limiting their growth. Political relations, with their affinities and antipathies, had become so sensitive a nexus that no cannon could be fired in Europe without every government feeling its interest affected. Hence a new Alexander needed more than his own sharp sword: he required a ready pen as well. Even so, his conquests rarely amounted to very much.

Even Louis XIV, though bent on destroying the balance of power in Europe and little troubled by the general hostility he faced by the end of the seventeenth century, continued waging war along traditional lines. While his military instrument was that of the greatest and richest monarch of all, its character was no different from that of his opponents'.

It had ceased to be in harmony with the spirit of the times to plunder and lay waste the enemy's land, which had played such an important role in antiquity, in Tartar days and indeed in mediaeval times. It was rightly held to be unnecessarily barbarous, an invitation to reprisals, and a practice that hurt the enemy's subjects rather than their government—one therefore that was ineffective and only served permanently to impede the advance of general civilization. Not only in its means, therefore, but also in its aims, war increasingly became limited to the fighting force itself. Armies, with their fortresses and prepared positions, came to form a state within a state, in which violence gradually faded away.

All Europe rejoiced at this development. It was seen as a logical outcome of enlightenment. This was a misconception. Enlightenment can never lead to inconsistency: as we have said before and shall have to say again, it can never make two and two equal five. Nevertheless this development benefited the peoples of Europe, although there is no denying that it turned war even more into the exclusive concern of governments and estranged it still further from the interests of the people. In those days, an aggressor's usual plan of war was to seize an enemy province or two. The defender's plan was simply to prevent him doing so. The plan for a given campaign was to take an enemy fortress or prevent the capture of one's own. No battle was ever sought, or fought, unless it were indispensable for that purpose. Anyone who fought a battle that was not strictly necessity, simply out of innate desire for victory, was considered reckless. A campaign was usually spent on a single siege, or two at the most. Winter quarters were assumed to be necessary for everyone. The poor condition of one side did not constitute an advantage to the other, and contact almost ceased between both. Winter quarters set strict limits to the operations of a campaign.

If forces were too closely balanced, or if the more enterprising side was also clearly the weaker of the two, no battle was fought and no town was besieged. The whole campaign turned on the retention of certain positions and depots and the systematic exploitation of certain areas.

So long as this was the general style of warfare, with its violence limited in such strict and obvious ways, no one saw any inconsistency in it. On the contrary, it all seemed absolutely right; and when in the eighteenth century critics began to analyze the art of war, they dealt with points of detail, without bothering much about fundamentals. Greatness, indeed perfection, was discerned in many guises, and even the Austrian Field-Marshal Daun—to whom it was mainly due that Frederick the Great completely attained his object and Maria Theresa completely failed in hers—could be considered a great commander. Only from time to time someone of penetrating judgment—of real common sense—might suggest that with superior forces one should achieve positive results; otherwise the war, with all its artistry, was being mismanaged.

This was the state of affairs at the outbreak of the French Revolution. Austria and Prussia tried to meet this with the diplomatic type of war that we have described. They soon discovered its inadequacy. Looking at the situation in this conventional manner, people at first expected to have to deal only with a seriously weakened French army; but in 1793 a force appeared that beggared all imagination. Suddenly war again became the business of the people—a people of thirty millions, all of whom considered themselves to be citizens. We need not study in detail the circumstances that accompanied this tremendous development; we need only note the effects that are pertinent to our discussion. The people became a participant in war; instead of governments and armies as heretofore, the full weight of the nation was thrown into the balance. The resources and efforts now available for use surpassed all conventional limits; nothing now impeded the vigor with which war could be waged, and consequently the opponents of France faced the utmost peril.

The effects of this innovation did not become evident or fully felt until the end of the revolutionary wars. The revolutionary quarrels did not yet advance inevitably toward the ultimate conclusion: the destruction of the European monarchies. Here and there the German armies were still able to resist them and stem the tide of victory. But all this was really due only to technical imperfections that hampered the French, and which became evident first in the rank and file, then in their generals, and under the Directory in the government itself.

Once these imperfections were corrected by Bonaparte, this juggernaut of war, based on the strength of the entire people, began its pulverizing course through Europe. It moved with such confidence and certainty that whenever it was opposed by armies of the traditional type there could never be a moment's doubt as to the result. Just in time, the reaction set in. The Spanish War spontaneously became the concern of the people. In 1809 the Austrian government made an unprecedented effort with reserves and militia; it came within sight of success and far surpassed everything Austria had earlier considered possible. In 1812 Russia took Spain and Austria as models: her immense spaces permitted her measures—belated though they were—to take effect, and even increased their effectiveness. The result was brilliant. In Germany, Prussia was first to rise. She made the war a concern of the people, and with half her former population, without money or credit, she mobilized a force twice as large as she had in 1806. Little by little the rest of Germany followed her example, and Austria too—though her effort did not equal that of 1809—exerted an exceptional degree of energy. The result was that in 1813 and 1814 Germany and Russia put about a million men into the field against France—counting all who fought and fell in the two campaigns.

Under these conditions the war was waged with a very different degree of vigor. Although it did not always match the intensity of the French, and was at times even marked by timidity, campaigns were on the whole conducted in the new manner, not in that of the past. In the space of only eight months the theater of operations changed from the Oder to the Seine. Proud Paris had for the first time to bow her head, and the terrible Bonaparte lay bound and chained.

Since Bonaparte, then, war, first among the French and subsequently among their enemies, again became the concern of the people as a whole, took on an entirely different character, or rather closely approached its true character, its absolute perfection. There seemed no end to the resources mobilized; all limits disappeared in the vigor and enthusiasm shown by governments and their subjects. Various factors powerfully increased that vigor: the vastness of available resources, the ample field of opportunity, and the depth of feeling generally aroused. The sole aim of war was to overthrow the opponent. Not until he was prostrate was it considered possible to pause and try to reconcile the opposing interests.

War, untrammeled by any conventional restraints, had broken loose in all its elemental fury. This was due to the peoples' new share in these great affairs of state; and their participation, in turn, resulted partly from the impact that the Revolution had on the internal conditions of every state and partly from the danger that France posed to everyone.

Will this always be the case in future? From now on will every war in Europe be waged with the full resources of the state, and therefore have to be fought only over major issues that affect the people? Or shall we again see a gradual separation taking place between government and people? Such questions are difficult to answer, and we are the last to dare to do so. But the reader will agree with us when we say that once barriers—which in a sense consist only in man's ignorance of what is possible—are torn down, they are not so easily set up again. At least when major interests are at stake, mutual hostility will express itself in the same manner as it has in our own day.

At this point our historical survey can end. Our purpose was not to assign, in passing, a handful of principles of warfare to each period. We wanted to show how every age had its own kind of war, its own limiting conditions, and its own peculiar preconceptions. Each period, therefore, would have held to its own theory of war, even if the urge had always and universally existed to work things out on scientific principles. It follows that the events of every age must be judged in the light of its own peculiarities. One cannot, therefore, understand and appreciate the commanders of the past until one has placed oneself in the situation of their times, not so much by a painstaking study of all its details as by an accurate appreciation of its major determining features.

But war, though conditioned by the particular characteristics of states and their armed forces, must contain some more general—indeed, a universal—element with which every theorist ought above all to be concerned.

The age in which this postulate, this universally valid element, was at its strongest was the most recent one, when war attained the absolute in violence. But it is no more likely that war will always be so monumental in character than that the ample scope it has come to enjoy will again be severely restricted. A theory, then, that dealt exclusively with absolute war would either have to ignore any case in which the nature of war had been deformed by outside influence, or else it would have to dismiss them all as misconstrued. That cannot be what theory is for. Its purpose is to demonstrate what war is in practice, not what its ideal nature ought to be. So the theorist must scrutinize all data with an inquiring, a discriminating, and a classifying eye. He must always bear in mind the wide variety of situations that can lead to war. If he does, he will draw the outline of its salient features in such a way that it can accommodate both the dictates of the age, and those of the immediate situation.

We can thus only say that the aims a belligerent adopts, and the resources he employs, must be governed by the particular characteristics of his own position; but they will also conform to the spirit of the age and to its general character. Finally, they must always be governed by the general conclusions to be drawn from the nature of war itself.

CHAPTER FOUR

Closer Definition of the Military Objective: The Defeat of the Enemy

The aim of war should be what its very concept implies—to defeat the enemy. We take that basic proposition as our starting point.

But what exactly does "defeat" signify? The conquest of the whole of the enemy's territory is not always necessary. If Paris had been taken in 1792 the war against the Revolution would almost certainly for the time being have been brought to an end. There was no need even for the French armies to have been defeated first, for they were not in those days particularly powerful. In 1814, on the other hand, even the capture of Paris would not have ended matters if Bonaparte had still had a sizable army behind him. But as in fact his army had been largely eliminated, the capture of Paris settled everything in 1814 and again in 1815. Again, if in 1812 Bonaparte had managed, before or after taking Moscow, to smash the Russian army, 120,000 strong, on the Kaluga road, just as he smashed the Austrians in 1805 and the Prussians the following year, the fact that he held the capital would probably have meant that he could make peace in spite of the enormous area still unoccupied. In 1805 Austerlitz was decisive. The possession of Vienna and two-thirds of the Austrian territory had not sufficed to bring about a peace. On the other hand, after Austerlitz the fact that Hungary was still intact did nothing to prevent peace being made. The final blow required was to defeat the Russian army; the Czar had no other near at hand and this victory would certainly have led to peace. Had the Russian army been with the Austrians on the Danube in 1805 and shared in their defeat, it would hardly have been necessary to take Vienna; peace could have been imposed at Linz. Equally, a country's total occupation may not be enough. Prussia in 1807 is a case in point. When the blow against the Russian ally in the uncertain victory of Eylau was not sufficiently decisive, the decisive victory of Friedland had to be gained in order to achieve what Austerlitz had accomplished the year before.

These events are proof that success is not due simply to general causes. Particular factors can often be decisive—details only known to those who were on the spot. There can also be moral factors which never come to light; while issues can be decided by chances and incidents so minute as to figure in histories simply as anecdotes.

What the theorist has to say here is this: one must keep the dominant characteristics of both belligerents in mind. Out of these characteristics a certain center of gravity develops, the hub of all power and movement, on which everything depends. That is the point against which all our energies should be directed.

Small things always depend on great ones, unimportant on important, accidentals on essentials. This must guide our approach.

For Alexander, Gustavus Adolphus, Charles XII, and Frederick the Great, the center of gravity was their army. If the army had been destroyed, they would all have gone down in history as failures. In countries subject to domestic strife, the center of gravity is generally the capital. In small countries that rely on large ones, it is usually the army of their protector. Among alliances, it lies in the community of interest, and in popular uprisings it is the personalities of the leaders and public opinion. It is against these that our energies should be directed. If the enemy is thrown off balance, he must not be given time to recover. Blow after blow must be aimed in the same direction: the victor, in other words, must strike with all his strength and not just against a fraction of the enemy's. Not by taking things the easy way—using superior strength to filch some province, preferring the security of this minor conquest to great success—but by constantly seeking out the center of his power, by daring all to win all, will one really defeat the enemy.

Still, no matter what the central feature of the enemy's power may be—the point on which your efforts must converge—the defeat and destruction of his fighting force remains the best way to begin, and in every case will be a very significant feature of the campaign.

Basing our comments on general experience, the acts we consider most important for the defeat of the enemy are the following:

1. Destruction of his army, if it is at all significant
2. Seizure of his capital if it is not only the center of administration but also that of social, professional, and political activity
3. Delivery of an effective blow against his principal ally if that ally is more powerful than he.

Up till now we have assumed—as is generally permissible—that the enemy is a single power. But having made the point that the defeat of the enemy consists in overcoming the resistance concentrated in his center of gravity, we must abandon this assumption and examine the case when there is more than one enemy to defeat.

If two or more states combine against another, the result is still politically speaking a *single* war. But this political unity is a matter of degree. The question is then whether each state is pursuing an independent interest and has its own independent means of doing so, or whether the interests and forces of most of the allies are subordinate to those of the leader. The more this is the case, the easier will it be to regard all our opponents as a single entity, hence all the easier to concentrate our principal enterprise into one great blow. If this is at all feasible it will be much the most effective means to victory.

I would, therefore, state it as a principle that if you can vanquish all your enemies by defeating one of them, that defeat must be the main objective in the war. In this one enemy we strike at the center of gravity of the entire conflict.

There are very few cases where this conception is not applicable—where it would not be realistic to reduce several centers of gravity to one. Where this is not so, there is admittedly no alternative but to act as if there were two wars or even more, each with its own object. This assumes the existence of several independent opponents, and consequently great superiority on their part. When this is the case, to defeat the enemy is out of the question.

We must now address ourselves more closely to the question: when is this objective both feasible and sound?

To begin with, our forces must be adequate:

1. To score a decisive victory over the enemy's
2. To make the effort necessary to pursue our victory to the point where the balance is beyond all possible redress.

Next, we must be certain our political position is so secure that this success will not bring further enemies against us who could force us immediately to abandon our efforts against our first opponent.

France could annihilate Prussia in 1806 even if this brought down Russia on her in full force, since she could defend herself against the Russians on Prussian soil. In 1808 she could do the same in Spain against England; but in respect of Austria she could not. By 1809, France had to reduce her forces in Spain considerably, and would have had to relinquish Spain altogether if she had not already enjoyed a great moral and material advantage over the Austrians.

These three examples call for careful study. One can win the first decision in a case but lose it on appeal and end by having to pay costs as well.

When the strength and capability of armed forces are being calculated, time is apt to be treated as a factor in total strength on the analogy of dynamics. It is assumed in consequence that half the effort or half the total forces could achieve as much in two years as the whole could do in one. This assumption, which rests, sometimes explicitly, sometimes implicitly, at the basis of military planning, is entirely false.

Like everything else in life, a military operation takes time. No one, obviously, can march from Vilna to Moscow in a week; but here there is no trace of that reciprocal relationship between time and energy that we would find in dynamics.

Both belligerents need time; the question is only which of the two can expect to derive *special advantages* from it in the light of his own situation. If the position on each side is carefully considered, the answer will be obvious: it is the weaker side—but thanks to the laws of psychology rather than those of dynamics. Envy, jealousy, anxiety, and sometimes perhaps even generosity are the natural advocates of the unsuccessful. They will win new friends for him as well as weaken and divide his enemies. Time, then, is less likely to bring favor to the victor than to the vanquished. There is a further point to bear in mind. As we have shown elsewhere, the exploitation of an initial victory requires a major effort. This effort must not only be made but be sustained like the upkeep of a great household. Conquered enemy provinces can, of course, bring additional wealth, but they may not always be enough to meet the additional outlay. If they do not, the strain will gradually increase and in the end resources may be exhausted. Time is thus enough to bring about a change unaided.

Could the money and resources that Bonaparte drew from Russia and Poland in 1812 furnish the men by the hundred thousand whom he needed in Moscow to maintain his position there?

But if the conquered areas are important enough, and if there are places in them vital to the areas still in enemy hands, the rot will spread, like a cancer, by itself; and if only that and nothing else happens, the conqueror may well enjoy the net advantage. Time alone will then complete the work, provided that no help comes from outside, and the area that is still unconquered may well fall without more ado. Thus time can become a factor in the conqueror's strength as well; but only on condition that a counterattack on him is no longer possible, that no reversal is conceivable—when indeed this factor is no longer of value since his main objective has been achieved, the culminating crisis is past, and the enemy, in short, laid low.

That chain of argument was designed to show that no conquest can be carried out too quickly, and that to spread it over a *longer period* than the minimum needed to complete it *makes it not less difficult, but more.* If that assertion is correct, it follows equally that if one's strength in general is great enough to make a certain conquest one must also have the strength to do so in a single operation, not by stages. By "stages" naturally, we do not mean to exclude the minor halts that are needed for reassembling one's forces or for administrative reasons.

We hope to have made it clear that in our view an offensive war requires above all a quick, irresistible decision. If so, we shall have cut the ground from under the *alternative* idea that a slow, allegedly systematic occupation is safer and wiser than conquest by continuous advance. Nonetheless, even those who have followed us thus far may very likely feel that our views have an air of paradox, of contradicting first impressions and of contradicting views that are as deeply rooted as ancient prejudice and that constantly appear in print. This makes it desirable to examine the alleged objections in some detail.

It is of course easier to reach a nearby object than a more distant one. But if the first does not suit our purpose, a pause, a suspension of activity, will not necessarily make the second half of the journey any easier to complete. A short jump is certainly easier than a long one: but no one wanting to get across a wide ditch would begin by jumping half-way.

If the ideas that underlie the concept of so-called methodical offensive operations are examined, we will usually find the following:

1. Capture the enemy fortresses in your path.
2. Accumulate the stores you need.
3. Fortify important points like *depots, bridges, positions,* and so forth.
4. Rest your troops in winter quarters and rest-camps.
5. Wait for next year's reinforcements.

If you halt an offensive altogether and stop the forward movement in order to make sure of all the above, you allegedly acquire a new base, and in theory revive your strength as though the whole of your country were immediately to your rear and the army's vigor were renewed with each campaign.

All these are admirable aims, and no doubt they could make offensive war easier; but they cannot make its results more certain. They usually camouflage misgivings on the part of the general or vacillation on the part of the government. We shall now try to roll them up from the left flank.

1. Waiting for reinforcements is just as useful to the other side—if not in our opinion more. Besides, a country can naturally raise almost as many troops in one year as in two, for the net increase in the second year will be very small in relation to the whole.

2. The enemy will rest his troops while we are resting ours.

3. Fortifying towns and positions is no business for the army and therefore no excuse for suspending operations.

4. Given the way in which armies are supplied today they need depots more when they are halted than when on the move. So long as the advance goes properly, enemy supplies will fall into our hands and make up for any shortage in barren districts.

5. Reducing an enemy fortress does not amount to halting the offensive. It is a means of strengthening the advance, and though it causes an apparent interruption it is not the sort of case we have in mind: it does not involve a suspension or a reduction of effort. Only circumstances can decide whether the right procedure is a regular siege, a mere investment, or simply to keep some fortress or other under observation. But we can make the general comment that the answer to this question turns on the answer to another; namely whether it would be too risky to press on and leave no more than an investing force behind. If it is not, and if you still have room to deploy your forces, the right course is to delay a regular siege until all offensive movement is complete. It is important, therefore, not to give way to the idea of quickly securing everything you have taken, for fear you end by missing something more important.

Such a further advance, admittedly, does seem to place in jeopardy the gains already made.

Our belief then is that any kind of interruption, pause, or suspension of activity is inconsistent with the nature of offensive war. When they are unavoidable, they must be regarded as necessary evils, which make success not more but less certain. Indeed, if we are to keep strictly to the truth, when weakness does compel us to halt, a second run at the objective normally becomes impossible; and if it does turn out to be possible it shows that there was no need for a halt at all. When an objective was beyond one's strength in the first place, it will always remain so.

This seems to us to be generally the case. In drawing attention to it we desire only to dispose of the idea that time, in itself, can work for the attacker. But the political situation can change from year to year, and on that account alone there will often be cases to which this generalization does not apply.

We may perhaps appear to have forgotten our initial thesis and only considered offensive war; but this is not so. Certainly a man who can afford to aim at the enemy's total defeat will rarely have recourse to the defensive, the immediate aim of which is the retention of what one has. But we must insist that defense without an active purpose is self-contradictory both in strategy and in tactics, and in consequence we must repeat that within the limits of his strength a defender must always seek to change over to the attack as soon as he has gained the benefit of the defense. So it follows that among the aims of such an attack, which is to be regarded as the real aim of the defense, however significant or insignificant this may be, the defeat of the enemy could be included. There are situations when the general, even though he had that grand objective well in mind, yet preferred to start on the defensive. That this is no mere abstraction is shown by the campaign of 1812. When Emperor Alexander took up arms he may not have dreamed he would ever completely destroy his enemy—as in the end he did. But would the idea have been absurd? And would it not have been natural in any case for the Russians to adopt the defensive at the outset of the war?

CHAPTER FIVE

Closer Definition of the Military Objective—Continued: *Limited Aims*

In the last chapter we stated the defeat of the enemy, assuming it to be at all possible, to be the true, the essential aim of military activity. We now propose to consider what can be done if circumstances rule that out.

The conditions for defeating an enemy presuppose great physical or moral superiority or else an extremely enterprising spirit, an inclination for serious risks. When neither of these is present, the object of military activity can only be one of two kinds: seizing a small or larger piece of enemy territory, or holding one's own until things take a better turn. This latter is normally the aim of a defensive war.

In considering which is the right course, it is well to remember the phrase used about the latter, *waiting until things take a better turn,* which assumes that there is ground for expecting this to happen. That prospect always underlies a "waiting" war—that is, a defensive war. The offensive—that is exploiting the advantages of the moment—is advisable whenever the future affords better prospects to the enemy than it does to us. A third possibility, perhaps the most usual, arises when the future seems to promise nothing definite to either side and hence affords no grounds for a decision. Obviously, in that case, the offensive should be taken by the side that possesses the political initiative—that is, the side that has an active purpose, the aim for which it went to war. If any time is lost without good reason, the initiator bears the loss.

The grounds we have just defined for choosing offensive or defensive war have nothing to do with the relative strength of the two sides, although one might suppose that to be the main consideration. But we believe that if it were, the wrong decision would result. No one can say the logic of our simple argument is weak; but does it in practice lead to absurd conclusions? Supposing that a minor state is in conflict with a much more powerful one and expects its position to grow weaker every year. If war is unavoidable, should it not make the most of its opportunities before its position gets still worse? In short, it should attack—but not because attack in itself is advantageous (it will on the contrary increase the disparity of strength) but because the smaller party's interest is either to settle the quarrel before conditions deteriorate or at least to acquire some advantages so as to keep its efforts going. No one could consider this a ludicrous argument. But if the smaller state is quite certain its enemy will attack, it can and should stand on the defensive, so as to win the first advantage. By doing so, it will not be placed at any disadvantage because of the passage of time.

Again, suppose a small power is at war with a greater one, and that the future promises nothing that will influence either side's decisions. If the political initiative lies with the smaller power, it should take the military offensive. Having had the nerve to assume an active role against a stronger adversary, it must do something definite—in other words, attack the enemy unless he obliges it by attacking first. Waiting would be absurd, unless the smaller state had changed its political decision at the moment of executing its policy. That is what often happens, and partly explains why the indeterminate character of some wars leaves a student very much perplexed.

Our discussion of the limited aim suggests that two kinds of limited war are possible: offensive war with a limited aim, and defensive war. We propose to discuss them in separate chapters. But first there is a further point to consider.

The possibility that a military objective can be modified is one we have treated hitherto as deriving only from domestic arguments, and we have considered the nature of the political aim only to the extent that it has or does not have an active content. From the point of view of war itself, no other ingredient of policy is relevant at all. Still, as we argued in the second chapter of Book One (purpose and means in war), the nature of the political aim, the scale of demands put forward by either side, and the total political situation of one's own side, are all factors that in practice must decisively influence the conduct of war. We therefore intend to give them special attention in the following chapter.

CHAPTER SIX

A. The Effect of the Political Aim on the Military Objective

One country may support another's cause, but will never take it so seriously as it takes its own. A moderately-sized force will be sent to its help; but if things go wrong the operation is pretty well written off, and one tries to withdraw at the smallest possible cost.

It is traditional in European politics for states to make offensive and defensive pacts for mutual support—though not to the point of fully espousing one another's interests and quarrels. Regardless of the purpose of the war or the scale of the enemy's exertions, they pledge each other in advance to contribute a fixed and usually modest force. A country that makes this sort of alliance does not consider itself thereby involved in actual war with anyone, for that would require a formal declaration and would need a treaty of peace to end it. But even that has never been clearly settled, and practice in the matter varies.

It would all be tidier, less of a theoretical problem, if the contingent promised—ten, twenty, or thirty thousand men—were placed entirely at the ally's disposal and he were free to use it as he wished. It would then in effect be a hired force. But that is far from what really happens. The auxiliary force usually operates under its own commander; he is dependent only on his own government, and the objective the latter sets him will be as ambiguous as its aims.

But even when both states are in earnest about making war upon the third, they do not always say, "we must treat this country as our common enemy and destroy it, or we shall be destroyed ourselves." Far from it: the affair is more often like a business deal. In the light of the risks he expects and the dividend he hopes for, each will invest about 30,000 to 40,000 men, and behave as if that were all he stood to lose.

Nor is that attitude peculiar to the case where one state gives another support in a matter of no great moment to itself. Even when both share a major interest, action is clogged with diplomatic reservations, and as a rule the negotiators only pledge a small and limited contingent, so that the rest can be kept in hand for any special ends the shifts of policy may require.

This used to be the universal way in which an alliance operated. Only in recent times did the extreme danger emanating from Bonaparte, or his own unlimited driving power, force people to act in a natural manner. The old way was a half-and-half affair; it was an anomaly, since in essence war and peace admit of no gradations. Nevertheless, the old way was no mere diplomatic archaism that reason could ignore, but a practice deeply rooted in the frailties and shortcomings of the human race.

Finally, some wars are fought without allies; and, political considerations will powerfully affect their conduct as well.

Suppose one merely wants a small concession from the enemy. One will only fight until some modest *quid pro quo* has been acquired, and a moderate effort should suffice for that. The enemy's reasoning will be much the same. But suppose one party or the other finds he has miscalculated, that he is not, as he had thought, slightly stronger than the enemy, but weaker. Money and other resources are usually running short and his moral impulse is not sufficient for a greater effort. In such a case he does the best he can; he hopes that the outlook will improve although he may have no ground for such hopes. Meanwhile, the war drags slowly on, like a faint and starving man.

Thus interaction, the effort to outdo the enemy, the violent and compulsive course of war, all stagnate for lack of real incentive. Neither side makes more than minimal moves, and neither feels itself seriously threatened.

Once this influence of the political objective on war is admitted, as it must be, there is no stopping it; consequently we must also be willing to wage such minimal wars, which consist in *merely threatening the enemy*, with *negotiations held in reserve*.

This poses an obvious problem for any theory of war that aims at being thoroughly scientific. All imperatives inherent in the concept of a war seem to dissolve, and its foundations are threatened. But the natural solution soon emerges. As the modifying principle gains a hold on military operations, or rather, as the incentive fades away, the active element gradually becomes passive. Less and less happens, and guiding principles will not be needed. The art of war will shrivel into prudence, and its main concern will be to make sure the delicate balance is not suddenly upset in the enemy's favor and the half-hearted war does not become a real war after all.

B. War Is an Instrument of Policy

Up to now we have considered the incompatibility between war and every other human interest, individual or social—a difference that derives from human nature, and that therefore no philosophy can resolve. We have examined this incompatibility from various angles so that none of its conflicting elements should be missed. Now we must seek out the unity into which these contradictory elements combine in real life, which they do by partly neutralizing one another. We might have posited that unity to begin with, if it had not been necessary to emphasize the contradictions with all possible clarity and to consider the different elements separately. This unity lies in *the concept that war is only a branch of political activity; that it is in no sense autonomous.*

It is, of course, well-known that the only source of war is politics—the intercourse of governments and peoples; but it is apt to be assumed that war suspends that intercourse and replaces it by a wholly different condition, ruled by no law but its own.

We maintain, on the contrary, that war is simply a continuation of political intercourse, with the addition of other means. We deliberately use the phrase "with the addition of other means" because we also want to make it clear that war in itself does not suspend political intercourse or change it into something entirely different. In essentials that intercourse continues, irrespective of the means it employs. The main lines along which military events progress, and to which they are restricted, are political lines that continue throughout the war into the subsequent peace. How could it be otherwise? Do political relations between peoples and between their governments stop when diplomatic notes are no longer exchanged? Is war not just another expression of their thoughts, another form of speech or writing? Its grammar, indeed, may be its own, but not its logic.

If that is so, then war cannot be divorced from political life; and whenever this occurs in our thinking about war, the many links that connect the two elements are destroyed and we are left with something pointless and devoid of sense.

This conception would be ineluctable even if war were total war, the pure element of enmity unleashed. All the factors that go to make up war and determine its salient features—the strength and allies of each antagonist, the character of the peoples and their governments, and so forth, all the elements listed in the first chapter of Book I—are these not all political, so closely connected with political activity that it is impossible to separate the two? But it is yet more vital to bear all this in mind when studying actual practice. We will then find that war does not advance relentlessly toward the absolute, as theory would demand. Being incomplete and self-contradictory, it cannot follow its own laws, but has to be treated as a part of some other whole; the name of which is policy.

In making use of war, policy evades all rigorous conclusions proceeding from the nature of war, bothers little about ultimate possibilities, and concerns itself only with immediate probabilities. Although this introduces a high degree of uncertainty into the whole business, turning it into a kind of game, each government is confident that it can outdo its opponent in skill and acumen.

So policy converts the overwhelmingly destructive element of war into a mere instrument. It changes the terrible battle-sword that a man needs both hands and his entire strength to wield, and with which he strikes home once and no more, into a light, handy rapier—sometimes just a foil for the exchange of thrusts, feints and parries.

Thus the contradictions in which war involves that naturally timid creature, man, are resolved; if this is the solution we choose to accept.

If war is part of policy, policy will determine its character. As policy becomes more ambitious and vigorous, so will war, and this may reach the point where war attains its absolute form. If we look at war in this light, we do not need to lose sight of this absolute: on the contrary, we must constantly bear it in mind.

Only if war is looked at in this way does its unity reappear; only then can we see that all wars are things of the *same* nature; and this alone will provide the right criteria for conceiving and judging great designs.

Policy, of course, will not extend its influence to operational details. Political considerations do not determine the posting of guards or the employment of patrols. But they are the more influential in the planning of war, of the campaign, and often even of the battle.

That is why we felt no urge to introduce this point of view at the start. At the stage of detailed study it would not have been much help and might have been distracting. But when plans for a war or a campaign are under study, this point of view is indispensable.

Nothing is more important in life than finding the right standpoint for seeing and judging events, and then adhering to it. One point and *one only* yields an integrated view of all phenomena; and only by holding to that point of view can one avoid inconsistency.

If planning a war precludes adopting a dual or multiple point of view—that is, applying first a military, then an administrative eye, then a political, and so on—the question arises whether *policy* is bound to be given precedence over everything.

It can be taken as agreed that the aim of policy is to unify and reconcile all aspects of internal administration as well as of spiritual values, and whatever else the moral philosopher may care to add. Policy, of course, is nothing in itself; it is simply the trustee for all these interests against other states. That it can err, subserve the ambitions, private interests, and vanity of those in power, is neither here nor there. In no sense can the art of war ever be regarded as the preceptor of policy, and here we can only treat policy as representative of all interests of the community.

The only question, therefore, is whether, when war is being planned, the political point of view should give way to the purely military (if a purely military point of view is conceivable at all): that is, should it disappear completely or subordinate itself, or should the political point of view remain dominant and the military be subordinated to it?

That the political view should wholly cease to count on the outbreak of war is hardly conceivable unless pure hatred made all wars a struggle for life and death. In fact, as we have said, they are nothing but expressions of policy itself. Subordinating the political point of view to the military would be absurd, for it is policy that has created war. Policy is the guiding intelligence and war only the instrument, not vice versa. No other possibility exists, then, than to subordinate the military point of view to the political.

If we recall the nature of actual war, if we remember the argument in Chapter 3 above—that *the probable character and general shape of any war should mainly be assessed in the light of political factors and conditions*—and that war should often (indeed today one might say *normally*) be conceived as an organic whole whose parts cannot be separated, so that each individual act contributes to the whole and itself originates in the central concept, then it will be perfectly clear and certain that the supreme standpoint for the conduct of war, the point of view that determines its main lines of action, can only be that of policy.

It is from this point of view, then, that plans are cast, as it were, from a mold. Judgment and understanding are easier and more natural; convictions gain in strength, motives in conviction, and history in sense.

From this point of view again, no conflict need arise any longer between political and military interests—not from the nature of the case at any rate—and should it arise it will show no more than lack of understanding. It might be thought that policy could make demands on war which war could not fulfill; but that hypothesis would challenge the natural and unavoidable assumption that policy knows the instrument it means to use. If policy reads the course of military events correctly, it is wholly and exclusively entitled to decide which events and trends are best for the objectives of the war.

In short, at the highest level the art of war turns into policy—but a policy conducted by fighting battles rather than by sending diplomatic notes.

We can now see that the assertion that a major military development, or the plan for one, should be a matter for *purely military* opinion is unacceptable and can be damaging. Nor indeed is it sensible to summon soldiers, as many governments do when they are planning a war, and ask them for *purely military advice*. But it makes even less sense for theoreticians to assert that all available military resources should be put at the disposal of the commander so that on their basis he can draw up purely military plans for a war or a campaign. It is in any case a matter of common experience that despite the great variety and development of modern war its major lines are still laid down by governments; in other words, if we are to be technical about it, by a purely political and not a military body.

This is as it should be. No major proposal required for war can be worked out in ignorance of political factors; and when people talk, as they often do, about harmful political influence on the management of war, they are not really saying what they mean. Their quarrel should be with the policy itself, not with its influence. If the policy is right—that is, successful—any intentional effect it has on the conduct of the war can only be to the good. If it has the opposite effect the policy itself is wrong.

Only if statesmen look to certain military moves and actions to produce effects that are foreign to their nature do political decisions influence operations for the worse. In the same way as a man who has not fully mastered a foreign language sometimes fails to express himself correctly, so statesmen often issue orders that defeat the purpose they are meant to serve. Time and again that has happened, which demonstrates that a certain grasp of military affairs is vital for those in charge of general policy.

Before continuing, we must guard against a likely misinterpretation. We are far from believing that a minister of war immersed in his files, an erudite engineer or even an experienced soldier would, simply on the basis of their particular experience, make the best director of policy—always assuming that the prince himself is not in control. Far from it. What is needed in the post is distinguished intellect and strength of character. He can always get the necessary military information somehow or other. The military and political affairs of France were never in worse hands than when the brothers Belle-Isle and the Duc de Choiseul were responsible—good soldiers though they all were.

If war is to be fully consonant with political objectives, and policy suited to the means available for war, then unless statesman and soldier are combined in one person, the only sound expedient is to make the commander-in-chief a member of the cabinet, so that the cabinet can share in the major aspects of his activities.[1] But that, in turn, is only feasible if the cabinet—that is, the government—is near the theater of operations, so that decisions

[1] The first edition has: "so bleibt . . . nur ein gutes Mittel übrig, nämlich den obersten Feldherrn zum Mitglied des Kabinets zu machen, damit dasselbe Theil an den Hauptmomenten seines Handelns nehme." In the second edition, which appeared in 1853, the last part of the sentence was changed to: "damit er in den wichtigsten Momenten an dessen Beratungen und Beschlüssen teilnehme." In his 1943 translation, based on the second or on a still later edition, O.J.M. Jolles rendered this alteration correctly as: "that he may take part in its councils and decisions on important occasions." That, of course, is a reversal of Clausewitz's original sense. By writing that the commander-in-chief must become a member of the cabinet so that the cabinet can share in the major aspects of his activities, Clausewitz emphasizes the cabinet's participation in military decisions, not the soldier's participation in political decisions.

Of the several hundred alterations of the text that were introduced in the second edition of *On War*, and became generally accepted, this is probably the most significant change. Eds.

can be taken without serious loss of time. That is what the Austrian Emperor did in 1809, and the allied sovereigns in 1813–1815. The practice justified itself perfectly.

What is highly dangerous is to let any soldier but the commander-in-chief exert an influence in cabinet. It very seldom leads to sound vigorous action. The example of France between 1793 and 1795, when Carnot ran the war from Paris, is entirely inapplicable, for terror can be used as a weapon only by a revolutionary government.

Let us conclude with some historical observations.

In the last decade of the eighteenth century, when that remarkable change in the art of war took place, when the best armies saw part of their doctrine become ineffective and military victories occurred on a scale that up to then had been inconceivable, it seemed that all mistakes had been military mistakes. It became evident that the art of war, long accustomed to a narrow range of possibilities, had been surprised by options that lay beyond this range, but that certainly did not go against the nature of war itself.

Those observers who took the broadest view ascribed the situation to the general influence that policy had for centuries exerted, to its serious detriment, on the art of war, turning it into a half-and-half affair and often into downright make-believe. The facts were indeed as they saw them; but they were wrong to regard them as a chance development that could have been avoided. Others thought the key to everything was in the influence of the policies that Austria, Prussia, England and the rest were currently pursuing.

But is it true that the real shock was military rather than political? To put it in the terms of our argument, was the disaster due to the effect of policy on war, or was policy itself at fault?

Clearly the tremendous effects of the French Revolution abroad were caused not so much by new military methods and concepts as by radical changes in policies and administration, by the new character of government, altered conditions of the French people, and the like. That other governments did not understand these changes, that they wished to oppose new and overwhelming forces with customary means: all these were political errors. Would a purely military view of war have enabled anyone to detect these faults and cure them? It would not. Even if there really had existed a thoughtful strategist capable of deducing the whole range of consequences simply from the nature of the hostile elements, and on the strength of these of prophesying their ultimate effects, it would have been quite impossible to act on his speculations.

Not until statesmen had at last perceived the nature of the forces that had emerged in France, and had grasped that new political conditions now obtained in Europe, could they foresee the broad effect all this would have on war; and only in that way could they appreciate the scale of the means that would have to be employed, and how best to apply them.

In short, we can say that twenty years of revolutionary triumph were mainly due to the mistaken policies of France's enemies.

It is true that these mistakes became apparent only in the course of the wars, which thoroughly disappointed all political expectations that had been placed on them. But the trouble was not that the statesmen had ignored the soldiers' views. The military art on which the politicians relied was part of a world they thought was real—a branch of current statecraft, a familiar tool that had been in use for many years. But *that* form of war naturally shared in the errors of policy, and therefore could provide no corrective. It is true that war itself has undergone significant changes in character and methods, changes that have brought it closer to its absolute form. But these changes did not come about because the French government freed itself, so to speak, from the harness of policy; they were caused by the new political conditions which the French Revolution created both in France and in Europe as a whole, conditions that set in motion new means and new forces, and have thus made possible a degree of energy in war that otherwise would have been inconceivable.

It follows that the transformation of the art of war resulted from the transformation of politics. So far from suggesting that the two could be dissociated from each other, these changes are a strong proof of their indissoluble connection.

Once again: war is an instrument of policy. It must necessarily bear the character of policy and measure by its standards. The conduct of war, in its great outlines, is therefore policy itself, which takes up the sword in place of the pen, but does not on that account cease to think according to its own laws.

CHAPTER SEVEN

The Limited Aim: Offensive War

Even when we cannot hope to defeat the enemy totally, a direct and positive aim still is possible: the occupation of part of his territory.

The point of such a conquest is to reduce his national resources. We thus reduce his fighting strength and increase our own. As a result we fight the war partly at his expense. At the peace negotiations, moreover, we will have a concrete asset in hand, which we can either keep or trade for other advantages.

This is a very natural view to take of conquered territory, the only drawback being the necessity of defending that territory once we have occupied it, which might be a source of some anxiety.

In the chapter on the culminating point of victory[1] we dealt at some length with the way in which an offensive weakens the attacking force, and showed how a situation might develop that could give rise to serious consequences.

Capturing enemy territory will reduce the strength of our forces in varying degrees, which are determined by the location of the occupied territory. If it adjoins our own—either as an enclave within our territory or adjoining it—the more directly it lies on the line of our main advance, the less our strength will suffer. Saxony in the Seven Years War was a natural extension of the Prussian theater, and its occupation by Frederick the Great made his forces stronger instead of weaker; for Saxony is nearer Silesia than it is to the Mark, and covers both of them.

Even the conquest of Silesia in 1740 and 1741, once completed, was no strain on Frederick's strength on account of its shape and location and the contour of its frontiers. So long as Saxony was not in Austrian hands, Silesia offered Austria only a narrow frontier, which in any case lay on the route that either side would have to take in advancing.

If, on the other hand, the territory taken is a strip flanked by enemy ground on either side, if its position is not central and its configuration awkward, its occupation will become so plain a burden as to make an enemy victory not just easier but perhaps superfluous. Every time the Austrians invaded Provence from Italy they were forced to give it up without any fighting. In 1744 the French thanked God for allowing them to leave Bohemia without having suffered a defeat. Frederick in 1758 found it impossible to hold his ground in Bohemia and Moravia with the same force that had fought so brilliantly the previous year in Silesia and Saxony. Of armies

[1] Book Seven, Chapter Five. Eds.

that had to give up some captured territory just because its conquest had so weakened them, examples are so common that we need not trouble to quote any more of them.

The question whether one should aim at such a conquest, then, turns on whether one can be sure of holding it or, if not, whether a temporary occupation (by way of invasion or diversion) will really be worth the cost of the operation and, especially, whether there is any risk of being strongly counterattacked and thrown off balance. In the chapter on the culminating point, we emphasized how many factors need to be considered in each particular case.

Only one thing remains to be said. An offensive of this type is not always appropriate to make up for losses elsewhere. While we are busy occupying one area, the enemy may be doing the same somewhere else. If our project is not of overwhelming significance, it will not compel the enemy to give up his own conquest. Thorough consideration is therefore necessary in order to decide whether on balance we will gain or lose.

In general one tends to lose more from occupation by the enemy than one gains from conquering his territory, even if the value of both areas should be identical. The reason is that a whole range of resources are denied to us. But since this is also the case with the enemy, it ought not to be a reason for thinking that retention is more important than conquest. Yet this is so. The retention of one's own territory is always a matter of more direct concern, and the damage that our state suffers may be balanced and so to speak neutralized only if retaliation promises sufficient advantage—that is to say the gains are substantially greater.

It follows from all this that a strategic attack with a limited objective is burdened with the defense of other points that the attack itself will not directly cover—far more burdened than it would be if aimed at the heart of the enemy's power. The effect is to limit the scale on which forces can be concentrated, both in time and in space.

If this concentration is to be achieved, at least in terms of time, the offensive must be launched from every practicable point at once. Then, however, the attack loses the other advantage of being able to stay on the defensive here and there and thus make do with a much smaller force. The net result of having such a limited objective is that everything tends to cancel out. We cannot then put all our strength into a single massive blow, aimed in accordance with our major interest. Effort is increasingly dispersed; friction everywhere increases and greater scope is left for chance.

That is how events tend to develop, dragging the commander down, frustrating him more and more. The more conscious he is of his own powers, the greater his self-confidence, the larger the forces he commands, then the more he will seek to break loose from this tendency, in order to give some one point a preponderant importance, even if this should be possible only by running greater risks.

CHAPTER EIGHT

The Limited Aim: Defensive War

The ultimate aim of a defensive war, as we have seen, can never be an absolute negation. Even the weakest party must possess some way of making the enemy conscious of its presence, some means of threatening him.

No doubt that end could in theory be pursued by wearing the enemy down. He has the positive aim, and any unsuccessful operation, even though it only costs the forces that take part in it, has the same effect as a retreat. But the defender's loss is not incurred in vain: he has held his ground, which is all he meant to do. For the defender then, it might be said, his positive aim is to hold what he has. That might be sound if it were sure that a certain number of attacks would actually wear the enemy down and make him desist. But this is not necessarily so. If we consider the relative exhaustion of forces on both sides, the defender is at a *disadvantage*. The attack may weaken, but only in the sense that a turning point may occur. Once that possibility is gone, the defender weakens more than the attacker, for two reasons. For one thing, he is weaker anyway, and if losses are the same on both sides, it is he who is harder hit. Second, the enemy will usually deprive him of part of his territory and resources. In all this we can find no reason for the attacker to desist. We are left with the conclusion that if the attacker sustains his efforts while his opponent does nothing but ward them off, the latter can do nothing to neutralize the danger that sooner or later an offensive thrust will succeed.

Certainly the exhaustion or, to be accurate, the fatigue of the stronger has often brought about peace. The reason can be found in the half-hearted manner in which wars are usually waged. It cannot be taken in any scientific sense as the ultimate, universal objective of all defense.

Only one hypothesis remains: that the aim of the defense must embody the idea of waiting—which is after all its leading feature. The idea implies, moreover, that the situation can develop, that in itself it may improve, which is to say that if improvement cannot be effected from within—that is, by sheer resistance—it can only come from without; and an improvement from without implies a change in the political situation. Either additional allies come to the defender's help or allies begin to desert his enemy.

Such, then, is the defender's aim if his lack of strength prohibits any serious counterattack. But according to the concept of the defense that we have formulated, this does not always apply. We have argued that the defensive is the more effective form of war, and because of this effectiveness it can also be employed to execute a counteroffensive on whatever scale.

These two categories must be kept distinct from the very start, for each has its effect on the conduct of the defense.

The defender's purpose in the first category is to keep his territory inviolate, and to hold it for as long as possible. That will gain him time, and gaining time is the only way he can achieve his aim. The positive aim, the most he can achieve, the one that will get him what he wants from the peace negotiations, cannot yet be included in his plan of operations. He has to remain strategically passive, and the only success he can win consists in beating off attacks at given points. These small advantages can then be used to strengthen other points, for pressure may be severe at all of them. If he has no chance of doing so, his only profit is the fact that the enemy will not trouble him again for a while.

That sort of defense can include minor offensive operations without their altering its nature or purpose. They should not aim at permanent acquisitions but at the temporary seizure of assets that can be returned at a later date. They can take the form of raids or diversions, perhaps the capture of some fortress or other, but always on condition that sufficient forces can be spared from their defensive role.

The second category exists where the defense has already assumed a positive purpose. It then acquires an active character that comes to the fore in proportion as the scale of feasible counterattack expands. To put it in another way: the more the defensive was deliberately chosen in order to make certain of the first round, the more the defender can take risks in laying traps for the enemy. Of these, the boldest and, if it works, the deadliest, is to retire into the interior. Such an expedient, nonetheless, could hardly be more different from the first type of defensive.

One need only think of the difference between Frederick's situation in the Seven Years War and the situation of Russia in 1812. When war broke out, Frederick's readiness for it gave him some advantage. It meant he could conquer Saxony—such a natural extension of his theater of war that its occupation put no strain upon his forces, but augmented them. In the campaign of 1757 he sought to continue and develop his strategic offensive, which was not impossible so long as the Russians and the French had not arrived in Silesia, the Mark, and Saxony. But the offensive failed; he was thrown back on the defensive for the rest of the campaign, abandoning Bohemia and having to clear his own base of operations of the enemy. That required the use of the same army to deal first with the French and then[1] with the Austrians. What successes he achieved he owed to the defensive.

By 1758, when his enemies had drawn the noose more tightly round him and his forces were becoming seriously outnumbered, he still planned a limited offensive in Moravia; he aimed at seizing Olmütz before his adversaries were in the field. He did not hope to hold it, still less to make it a base for a further advance, but simply to use it as a sort of outwork, as a *contre-approche* against the Austrians, designed to make them spend the

[1] The first edition omits the phrase *die Franzosen, dann gegen* which appears in later editions and seems necessary to give point to Clausewitz's comment. Eds.

rest of the campaign, and possibly a second year's, in trying to retake it. That effort was a failure too, and Frederick now abandoned any thought of a serious offensive, realizing that it would still further reduce his relative strength. A compact position in the center of his territories, in Silesia and Saxony, exploitation of interior lines for quickly reinforcing any danger point, small raids as opportunities occurred, quietly waiting meanwhile on events so as to economize his strength for better times—such were the main elements in his plans. Gradually, his operations became more passive. Realizing that even victories cost too much, he tried to manage with less. His one concern was to gain time, and hold on to what he had. Less and less was he willing to give ground and he did not scruple to adopt a thorough-going cordon-system; both Prince Henry's positions in Saxony and those of the King in the mountains of Silesia deserve this description. His letters to the Marquis d'Argens[2] show how keenly he looked forward to winter quarters and how much he hoped he would be able to take them up without incurring serious losses in the meantime.

To censure Frederick for this, and see in his behavior evidence of low morale, would in our view be a very superficial judgment. Devices such as the entrenched camp at Bunzelwitz, the positions that Prince Henry chose in Saxony and the King in the Silesian mountains, may not seem to us today the sort of measure on which to place one's final hope—tactical cobwebs that a man like Bonaparte would soon have cleared away. But one must remember that times have changed, that war has undergone a total transformation and now draws life from wholly different sources. Positions that have lost all value today could be effective then; and the enemy's general character was a factor as well. Methods which Frederick himself discounted could be the highest degree of wisdom when used against the Austrian and Russian forces under men like Daun and Buturlin.

This view was justified by success. By quietly waiting on events Frederick achieved his goal and avoided difficulties that would have shattered his forces.

At the start of the 1812 campaign, the strength with which the Russians opposed the French was even less adequate than Frederick's at the outset of the Seven Years War. But the Russians could expect to grow much stronger in the course of the campaign. At heart, all Europe was opposed to Bonaparte; he had stretched his resources to the very limit; in Spain he was fighting a war of attrition; and the vast expanse of Russia meant that an invader's strength could be worn down to the bone in the course of five hundred miles' retreat. Tremendous things were possible; not only was a massive counterstroke a certainty if the French offensive failed (and how could it succeed if the Czar would not make peace nor his subjects rise against him?) but the counterstroke could bring the French to utter ruin. The highest wisdom could never have devised a better strategy than the one the Russians followed unintentionally.

[2] French author, and confidant of Frederick, who resided in Prussia during the Seven Years War. Eds.

No one thought so at the time, and such a view would have seemed far-fetched; but that is no reason for refusing to admit today that it was right. If we wish to learn from history, we must realize that what happened once can happen again; and anyone with judgment in these matters will agree that the chain of great events that followed the march on Moscow was no mere succession of accidents. To be sure, had the Russians been able to put up any kind of defense of their frontiers, the star of France would probably have waned, and luck would probably have deserted her; but certainly not on that colossal and decisive scale. It was a vast success; and it cost the Russians a price in blood and perils that for any other country would have been higher still, and which most could not have paid at all.

A major victory can only be obtained by positive measures aimed at a *decision*, never by simply waiting on events. In short, even in the defense, a major stake alone can bring a major gain.

CHAPTER NINE

The Plan of a War Designed to Lead to the Total Defeat of the Enemy

Having given a more detailed account of the various objects a war can serve, we shall now consider how the whole war should be planned with a view to the three distinguishable phases that can go with each particular aim. After everything we have so far said on the subject, we can identify two basic principles that underlie all strategic planning and serve to guide all other considerations.

The first principle is that the ultimate substance of enemy strength must be traced back to the fewest possible sources, and ideally to one alone. The attack on these sources must be compressed into the fewest possible actions—again, ideally, into one. Finally, all minor actions must be subordinated as much as possible. In short the first principle is: act with the utmost concentration.

The second principle is: act with the utmost speed. No halt or detour must be permitted without good cause.

The task of reducing the sources of enemy strength to a single center of gravity will depend on:

1. The distribution of the enemy's political power. If it lies in the armed forces of a single government, there will normally be no problem. If it is shared among allied armies, one of which is simply acting as an ally without a special interest of its own, the task is hardly any greater. But if it is shared among allies bound together by a common interest, the problem turns on the cordiality of the alliance. We have dealt with this earlier.

2. The situation in the theater of war where the various armies are operating. If all enemy forces are concentrated in a single army in one theater of war they in fact constitute a unity, and the question need not be pursued. But if the enemy in a single theater consists of separate allied armies, their unity is less than absolute; yet they will still be sufficiently integrated for a resolute attack on *one* to involve the rest. If the armies operate in neighboring theaters with no great natural barriers between them, one of them can still have a decisive influence on the others; but with theaters far apart, with neutral territory or mountain ranges in between, the influence in question will be doubtful—in fact, improbable—and if the theaters lie at opposite ends of the country under attack and

operations directed against them therefore have to take divergent lines, they will almost cease to be related.

Should Prussia be attacked by France and Russia simultaneously the effect on the conduct of operations would be as if there were two separate wars. Only at the peace negotiations might their essential unity become clear.

Conversely, in the Seven Years War the Austrian and Saxon forces were practically fused: they shared one another's fortunes, partly because from Frederick's point of view their theaters both lay in the same direction, and partly because of Saxony's total lack of political independence.

Numerous though the enemies were with whom Bonaparte had to contend in 1813, they all faced him from more or less the same direction. Their various operational zones were closely linked and interacted strongly on each other. Had he been able to concentrate his forces at one point and destroy his principal enemy, the fate of the rest would have been decided as well. Had he beaten the main allied army in Bohemia and pressed on via Prague to Vienna, Blücher could not with the best will in the world have remained in Saxony. He would have been summoned to help in Bohemia, and the Crown Prince of Sweden would certainly have lacked the will to remain in the Mark Brandenburg.

Should Austria, on the other hand, make war on France both in Italy and on the Rhine, she will always find it hard to produce a decision in both theaters by striking a successful blow in one of them. For one thing, the Alps are too great a barrier and besides, the roads from Austria to the Rhine and Italy diverge. France would have a somewhat easier task. In either case her lines of attack would converge upon Vienna and the core of the Austrian monarchy, and a decisive victory in one theater would be decisive for the other as well. We should add that if France were to strike a decisive blow in Italy it would have more effect on the Rhenish theater than the other way about. An offensive launched from Italy would threaten the center of Austrian power, while operations from the Rhine would threaten only one of its wings.

From this it follows that the concept of separate and connected enemy power runs through every level of operations, and thus the effect that events in a given theater will have elsewhere can only be judged in each particular case. Only then can it be seen how far the enemy's various centers of gravity can be reduced to one.

The principle of aiming everything at the enemy's center of gravity admits of only one exception—that is, when secondary operations look *exceptionally rewarding*. But we must repeat that only decisive superiority can justify diverting strength without risking too much in the principal theater.

When General Bülow marched into Holland in 1814 it was on the assumption that his 30,000 men would not only neutralize an equal number of the French, but would also enable the Dutch and English to put forces in the field that otherwise could not have been brought to bear.

The first task, then, in planning for a war is to identify the enemy's centers of gravity, and if possible trace them back to a single one.

The second task is to ensure that the forces to be used against that point are concentrated for a main offensive.

In this situation we may be faced with the following reasons for dividing our forces:

1. The original disposition of the forces—and therefore also the geographical location of the attacking states.

If concentration would entail detours and loss of time, and if the risks of advancing separately are not too great, one can justify such a course. If a laborious junction were effected needlessly and at great cost in time, and the first assault were, therefore, made with less than maximum *élan* and speed, it would violate our second general principle. This deserves particular consideration whenever there is a chance to surprise the enemy.

The argument has even greater weight if the attack is undertaken by allies who are placed not one behind the other but facing the enemy side by side. If Prussia and Austria were fighting France, to make both armies start from the same place would waste a great deal of time and strength. The natural route for Prussia to the heart of France is from the lower Rhine, and for Austria from the upper Rhine. It follows that no junction could be made without some sacrifice, and so in any given case the question is whether this sacrifice need be made.

2. An attack on separate lines may promise greater results.

As we are now discussing a divided advance against a *single* center, this implies a *concentric* attack. A divided attack on parallel or divergent lines would be classified as a *secondary operation,* which we have discussed already.

Both in strategy and in tactics a convergent attack always holds out promise of *increased* results, for if it succeeds the enemy is not just beaten; he is virtually cut off. The convergent attack, then, is always the more promising; but since forces are divided and the theater is enlarged, it also carries a greater risk. As with the attack and defense, the weaker form promises the greater success.

All depends, therefore, on whether the attacker feels strong enough to go after such a prize.

In 1757, when Frederick decided to invade Bohemia, he split his forces between Saxony and Silesia. He had two main reasons. First, that was where his forces had been deployed for the winter, and a concentration of forces would have deprived the attack of surprise. Second, his concentric advance threatened both Austrian theaters in flank and rear. The risk he ran was that one of his armies might be defeated by superior strength. If the Austrians *failed* to appreciate this, they would have either to accept battle in the center, or allow themselves to be maneuvered right off their line of communication by one flank or the other until they met disaster. That was the greatest success which this advance promised the King. In fact, the Austrians opted for battle in the center; but they took up positions at Prague, which was far too exposed to the enveloping attack, and their inactivity gave the attack the time it needed to produce its maximum effect. The result was that their defeat turned into a true catastrophe—as is proved by the fact that the commanding general and two thirds of the army was shut up in Prague.

This brilliant success at the start of the campaign was due to the willingness to risk a concentric attack. Who could criticize Frederick for trusting that the precision of his movements, his generals' vigor and his army's high morale, in contrast to the Austrians' obtuseness, were sufficient to guarantee success? It would be wrong to leave out these moral factors and imagine the geometric form of the attack was all that mattered. One only has to compare it with Bonaparte's no less brilliant campaign of 1796, when the Austrians were signally punished for their converging advance into Italy. Leaving aside the moral factor, the resources that the French general had at his command on that occasion were no greater than the Austrian had available in 1757. Indeed, they were less, for the Austrian commander, unlike Bonaparte, was not his enemy's inferior in strength. Hence, if there is reason to fear that a divided and convergent thrust will give the enemy a chance to equalize his strength by using his interior lines, it is better not employed. If the deployment of the forces makes it essential, it must be regarded as a necessary evil.

Seen from that point of view, one cannot possibly approve the way in which France was invaded in 1814. The Russian, Austrian, and Prussian armies were all assembled at Frankfurt, on the obvious and most direct route to France's center of gravity. They were then split up so that one army should invade from Mainz and the other should first pass through Switzerland. France's military strength at that time was so low that there was no question of her defending her frontiers: hence the only point of a convergent invasion was that if all went well one army would take Lorraine and Alsace while the Franche-Comté fell to the other. Was this meager advantage worth the trouble of marching through Switzerland? We know quite well that there were other equally bad grounds on which that march was ordered, but we mention the one most relevant to our discussion.

Bonaparte, on the other hand, had shown by his masterly campaign in 1796 that he knew exactly how to deal with a convergent threat, and though he might be seriously outnumbered, everyone was ready to admit from the start that morally he was far superior. Late in joining his army at Chalons and generally underrating his opponents, he almost managed to strike the two armies separately before they joined up. Yet how weak they were when he met them at Brienne! Out of 65,000 men, Blücher had only 27,000 with him, and of the 200,000 men of the main army only 100,000 were available.

They could not possibly have set the French an easier task. Moreover from the very moment the advance began the allied armies wanted nothing more than to link up again.

After all these considerations, we believe that while an attack on convergent lines is in itself a means toward success, nevertheless in general it should occur only as the result of the original deployment of forces, and it seldom justifies departing from the shortest and simplest line of advance.

3. The extent of a theater of war can constitute a ground for advancing with divided forces.

When an army starts an attack from a given point and succeeds in thrusting deeper into enemy territory, the area that it dominates is not strictly limited to the roads it uses but extends a certain distance on either side. But its breadth will very much depend (if we may use such a metaphor) on the solidity and cohesion of the opposing state. If the enemy country is rather loosely knit, if its people are soft and have forgotten what war is like, a triumphant invader will have no great trouble in leaving a wide swathe of country safely in his rear; but if he is faced with a brave and loyal populace the area of safety will resemble a narrow triangle.

To avoid that risk he must contrive to advance on a broader front, and if the enemy's strength is concentrated at one point the invader can only maintain this breadth until contact is made. On approaching the enemy position it has to be reduced. That is self-evident.

But if the enemy position itself extends over a certain breadth it would be sensible to extend one's own front to the same degree. We have in mind a single operational theater or several adjacent ones, and our remarks will therefore obviously apply no less to cases where the main offensive automatically settles all lesser issues.

But can that *always* be relied upon? Can one afford the risks that will arise if the influence of the main objective on the minor ones is not enough? Perhaps we should look more closely at this requirement, that a theater of operations ought to have a certain breadth.

As usual, it is quite impossible to cover every case that could conceivably arise; but we maintain that the decision on the main objective will, with few exceptions, carry the minor ones as well. That is the principle on which action should invariably be based unless there are obvious reasons to the contrary.

Bonaparte invaded Russia in the sound belief that if he destroyed the principal Russian army the success would sweep away the Russian forces on the upper Dwina too. Oudinot's corps was all he left initially to deal with these; but Wittgenstein attacked, and Bonaparte then had to send the Sixth Corps there as well.

Part of his forces, on the other hand, had initially been detached to deal with Bagration; but the withdrawal of the Russian center swept Bagration away and enabled Bonaparte to recall the force he had detached. Had Wittgenstein not been obliged to cover the second capital[1] he too would have followed the retreat of the main army under Barclay.

Bonaparte's victories at Ulm and Regensburg in 1805 and 1809 also settled the fate of Italy and the Tyrol, though Italy was a rather distant autonomous theater. Jena and Auerstädt in 1806 put an end to any threat that might have arisen in Westphalia, in Hesse or on the Frankfurt road.

Among the many factors that may influence resistance at subordinate points, two are particularly significant.

The first is that in a country so vast and relatively powerful as Russia, the decisive blow at the vital point may be long delayed and there is no need for a rapid concentration of all one's forces.

The second factor emerges when numerous fortresses confer unusual autonomy on a secondary area, as for example Silesia in 1806. Nevertheless Bonaparte imputed very little importance to it; and although he had to by-pass Silesia on his advance to Warsaw, he detached only his brother Jérome and 20,000 men to deal with it.

If it seems probable in any given case that the attack on the main objective will not shake the minor ones, or if it has already failed to do so; if the enemy has already committed forces at those points, it will then be necessary to despatch another and more adequate force to deal with them, since lines of communication cannot be left entirely unprotected.

One could be even more prudent. One might demand that the advance against the main objective should keep strictly in step with advances against the minor ones, so that whenever the enemy refuses to give way at other points, the main advance is halted.

This approach will certainly not directly contradict our principle of maximum concentration against the main objective; but the spirit underlying it is wholly contrary. It would impose such sluggishness on movement, such paralysis on the attack, create such opportunities for chance and waste so much time, as in fact to be wholly incompatible with an offensive aimed at defeating the enemy.

The difficulty becomes even greater if the enemy can withdraw his forces from these minor points along divergent lines. What would then become of the unity of our attack?

Consequently we must strictly oppose the principle that makes the main attack dependent on minor operations, and instead assert that an offensive intending the enemy's collapse will fail if it does not dare to drive like an arrow at the heart of the enemy state.

4. A fourth and final ground for advancing with divided forces may be to reduce the problems of supply.

No doubt it is a great deal more agreeable to take a small force through a prosperous area than a powerful army through a poor one; but the latter

[1] St. Petersburg. Eds.

is not impossible if proper arrangements are made and the army is accustomed to privation. The former option should therefore not have so much influence on plans as to justify taking greater risks.

We have now dealt with the grounds that justify dividing forces and splitting one operation into several. If such a division were made on one of those grounds with a clear idea of its purpose and after careful weighing of the pros and cons, we should not presume to criticize it.

But when the usual thing occurs, and a "trained" general staff makes such a plan as a matter of routine, and when all the various operational theaters have to be occupied, like squares on a chessboard, each with the appropriate units before the actual moves begin; when the moves themselves are made with self-styled expertise to reach their goal by devious routes and combinations; when modern armies have to separate in order to display "consummate art" by reuniting two weeks later at the utmost risk: then we can only say we abhor this departure from the straight, simple, easy approach in order to plunge deliberately into confusion. Such idiocy becomes the more likely, the less the war is run by the commander-in-chief himself in the manner indicated in our opening chapter: that is, as a single activity of an individual invested with huge powers; or, in other words, the more the plan as a whole is cooked up by an unrealistic general staff on the recipes of half-a-dozen amateurs.

The third part of our first principle has still to be considered: namely to keep each minor operation as subordinate as possible.

If one seeks to concentrate all military action on a single goal, and if so far as possible a *single* massive operation is envisaged as the means of gaining it, the other points at which the opponents are in contact must lose part of their independence and become subordinate operations. If absolutely everything could be concentrated into one action, those other contact points would be completely neutralized. But that is rarely possible; so the problem is to hold them strictly within bounds and make sure they do not draw off too much strength from the main operation.

We hold, moreover, that the plan of operations should have this tendency even when the enemy's whole resistance cannot be reduced to a single center of gravity and when, as we have once put it, two almost wholly separate wars have to be fought simultaneously. Even then one of them must be treated as the *main* operation, calling for the bulk of resources and of activities.

Seen in this light, it is advisable to operate *offensively* only in this main theater and to stay on the defensive elsewhere. There an attack will only be justified if exceptional conditions should invite it. Moreover the defensive at the minor points should be maintained with the minimum of strength; and every advantage that that form of resistance offers should be turned to account.

This view applies with even greater force to any theater of operations in which several enemy allied armies are engaged in such a way that they are all affected when the common center of gravity is struck.

Against the enemy who is the target of the main offensive there can therefore be no such thing as a defensive in subsidiary theaters of operations. That offensive consists of the main attack and such subsidiary attacks as circumstances make necessary. This removes all need to defend any point that the offensive does not itself directly cover. The main decision is what matters. It will compensate for any loss. If the forces are sufficient to make it reasonable to seek a major decision, then the *possibility of failure* can no longer be an excuse for trying to cover oneself everywhere else. For *this* would make defeat in the decisive battle that much more probable, and would thus introduce an element of contradiction into our actions.

But while the main operation must enjoy priority over minor actions, the same priority must also be applied to all its parts. Which forces from each theater shall advance toward the common center of gravity is usually decided on extraneous grounds; all we are saying, therefore, is that there must be an *effort* to make sure the main operation has *precedence*. The more that precedence is realized, the simpler everything will be and the less will it be left to chance.

The second principle is the rapid use of our forces.

Any unnecessary expenditure of time, every unnecessary detour, is a waste of strength and thus abhorrent to strategic thought. It is still more important to remember that almost the only advantage of the attack rests on its initial surprise. Speed and impetus are its strongest elements and are usually indispensable if we are to defeat the enemy.

Thus theory demands the shortest roads to the goal. Endless discussions about moving left or right, doing this or that, are otiose.

If we recall what was said in the chapter[2] on the aims of strategic attack, and the part in Chapter Four above about the influence of time, we believe no further elaboration is needed to show that this principle should be given the priority that we claim for it.

Bonaparte never forgot it. He always preferred the shortest road between one army and another, or between two capitals.

Now, what constitutes the main operation, which we have made central to all else, and for which we have demanded such rapid and straightforward execution?

In Chapter Four we explained what we mean by the defeat of the enemy, to the extent this can be done in general terms, and there is no need to repeat it. Whatever the final act may turn on in any given case, the beginning is invariably the same—annihilation of the enemy's armed forces, which implies a major victory and their actual destruction. The earlier this victory can be sought—that is, the nearer to our frontiers—the *easier* it will be. The later the main battle is fought—that is, the deeper in enemy territory—the more *decisive* its effect. Here, as everywhere, the ease of success and its magnitude are in balance.

In consequence, unless one is so much the stronger that victory is cer-

[2] Book Seven, Chapter Three. Eds.

tain, the enemy's main force must be sought out if possible. We say "if possible" because if it involved substantial detours, taking the wrong road and wasting time, this could easily prove a mistake. If the enemy's main force is not on our line of advance, and if other reasons make it impossible for us to seek it out, we are bound to find it later, since it cannot fail eventually to oppose us. Then, as we have just argued, the battle is fought under less favorable circumstances—a disadvantage we must accept. Nevertheless, if we win the battle, our victory will be the more decisive.

From this it follows that if in this hypothetical case, the enemy's main army lies across our line of advance, it would be wrong deliberately to by-pass him; at least, if our motive in so doing is to make our victory easier. On the other hand, the premises suggest that we can avoid the enemy provided we are massively superior, in order to make our ultimate victory more decisive.

We have been talking about a total victory—that is, not simply a battle won, but the complete defeat of the enemy. Such a victory demands an enveloping attack or a battle with reversed fronts, either of which will always make the result decisive. It is essential, then, that any plan of operations should provide for this, both as regards the forces it requires and the direction to be given them. We shall say more about this in our chapter on the planning of a campaign.[3]

It is not impossible, of course, for a battle to end in total victory even if fought with parallel fronts, and military history can show examples: but such cases are rare and are growing rarer as armies approximate to one another in training and in skill. Twenty-one battalions are not captured in a single village nowadays as they were at Blenheim.

Once a major victory is achieved there must be no talk of rest, of a breathing space, of reviewing the position or consolidating and so forth, but only of the pursuit, going for the enemy again if necessary, seizing his capital, attacking his reserves and anything else that might give his country aid and comfort.

Should the tide of victory sweep us past his fortresses, the question whether to besiege them or not will depend upon our strength. If our superiority is very great, we will lose less time by taking them as early as we can; but if we are not so sure that fresh successes lie ahead, we must invest them with the smallest possible forces that precludes all thought of regularly besieging them. From the moment when the siege of fortresses compels us to suspend the advance, the offensive has *as a rule* reached its culminating point. Therefore we demand that the main force should go on advancing rapidly and keep up the pressure. We have already disposed of the idea that an advance toward the main objective should be made to wait upon success at minor points. Hence the main force, as a rule, will leave no more than a narrow band of territory in its rear, which it can call its own and which forms its theater of operations. This can check momentum at the front, as we have

[3] The chapter was never written. Eds.

seen, and involve some risks for the attacker. It is a problem: might these tendencies not reach the stage at which further advance is brought to a halt? This is quite possible. But just as we have argued that it would be a mistake to try from the very start to avoid a narrow theater of operations and therefore rob the attack of its momentum, we continue to argue that so long as the general has not yet defeated the enemy, so long as he believes himself to be strong enough to gain his objectives, he must persevere. He may do so with increasing danger, but his success will be all the greater. Should he reach a point beyond which he dare not go, should he feel he must expand to right and left in order to protect his rear, so be it: very likely his attack has reached its culminating point. Its momentum is exhausted; and if the enemy is still unbroken, there is probably no future in it anyway.

Anything the general can do to develop his offensive by taking fortresses, passes and provinces, still means slow progress, but the progress is relative, no longer absolute. The enemy's precipitate retreat has stopped; he may be getting ready to renew his resistance, and it is now possible that even though the attacker is still improving his position the defender, by doing the same, is improving his chances every day. We repeat, in short, that once a pause has become necessary there can as a rule be no recurrence of the advance.

All that theory requires is that so long as the aim is the enemy's defeat, the attack must not be interrupted. If the general relinquishes this aim because he considers the attendant risk too great, he will be right to break off and extend his front. Theory would blame him only if he does so in order to facilitate the defeat of the enemy.

We are not so foolish as to suggest that history contains no example of a state being brought to the last extremity by degrees. Our suggested thesis is not an absolute truth admitting no exception, but is simply based on the normal and likely course of events. Further, we must establish whether the decline of a state was the gradual result of a historical process or was the outcome of a single campaign. We are here only dealing with the latter case, for only here are forces in such tension that they either overcome the load upon them or are in danger of succumbing to it. If the first year's fighting yields a slight advantage and the second year's increases it so that little by little one approaches the objective, the danger is nowhere very grave, but just for that reason it is all the more widespread. Every pause between one success and the next gives the enemy new opportunities. One success has little influence on the next, and often none at all. The influence may well be adverse, for the enemy either recovers and rouses himself to greater resistance or obtains help from somewhere else. But when a single impetus obtains from start to finish, yesterday's victory makes certain of today's, and one fire starts another. For every case of a state reduced to ruin by successive blows—which means that time, the defender's patron, has deserted to the other side—how many more are there in which time ruined the plans of the attacker! It is enough to cite the outcome of the Seven Years War, in which the Austrians sought their goal with such leisure, prudence and caution that they missed it completely.

In the light of this we cannot believe that concern for a secure and soundly administered theater of operations should go hand in hand with the offensive thrust and in a sense balance it. On the contrary, we regard the disadvantages that attach to the offensive as unavoidable evils that should not merit our attention until the advance promises no further hope.[4]

The case of Bonaparte in 1812, far from undermining our argument, merely confirms it.

His campaign failed, not because he advanced too quickly and too far as is usually believed, but because the only way to achieve success failed. Russia is not a country that can be formally conquered—that is to say occupied—certainly not with the present strength of the European States and not even with the half-a-million men Bonaparte mobilized for the purpose. Only internal weakness, only the workings of disunity can bring a country of that kind to ruin. To strike at these weaknesses in its political life it is necessary to thrust into the heart of the state. Only if he could reach Moscow in strength could Bonaparte hope to shake the government's nerve and the people's loyalty and steadfastness. In Moscow he hoped to find peace; that was the only rational war aim he could set himself.

He advanced his main force against that of the Russians'. They staggered back before him, past the Drissa camp, and never stopped till they got to Smolensk. He forced Bagration to withdraw as well, defeated both the Russian armies and occupied Moscow. He acted as he had always done. This is how he had come to dominate Europe, and this was the only way in which he could have done so. No one who admired Bonaparte as the greatest of commanders in his previous campaigns should feel superior to him with regard to this one.

It is legitimate to judge an event by its outcome, for this is its soundest criterion. But a judgment based on the result alone must not be passed off as evidence of human wisdom. To discover why a campaign failed is not the same thing as to criticize it; but if we go on and show that the causes could and should have been seen and acted on, we assume the role of critic, and set ourselves up above the general.

Anyone who asserts that the campaign of 1812 was an absurdity because of its enormous failure but who would have called it a superb idea if it had worked, shows complete lack of judgment.

Suppose that Bonaparte had waited in Lithuania, as most of his critics think he should have done, so as to make certain of its fortresses (of which, incidentally, Riga, lying to one side, is really the only one; Bobruisk is a wretched little place) it would have involved him in miserable defensive operations for the winter. The critics would then have been the first to exclaim, "That is no longer the old Bonaparte! He has not even forced his first great battle—the man who used to seal his conquests of enemy states by victories on their last ramparts, as at Austerlitz and Friedland. Moscow, the enemy's capital, is defenseless—ripe for surrender. Why has he failed

[4] Here we follow the text of the second edition since that of the first appears hopelessly corrupt. Eds.

to capture it and thus left it as a rallying point for fresh resistance? He had the incredible luck to surprise this remote giant as easily as a nearby city or as Frederick overwhelmed the small and neighboring Silesia—and he does not exploit his advantage. He breaks off his triumphant progress as if the devil was at his heels!" That is the sort of talk we should have heard, for that is the way most critics form their judgments.

We maintain that the 1812 campaign failed because the Russian government kept its nerve and the people remained loyal and steadfast. The campaign could not succeed. Bonaparte may have been wrong to engage in it at all; at least the outcome certainly shows that he miscalculated; but we argue that if he was to aim at that objective, there was, broadly speaking, no other way of gaining it.

Being anxious not to be committed to an interminable, costly defensive war in the East on top of the one he was already fighting in the West, Bonaparte tried the only means he had—a bold attack, which would compel his demoralized opponent to make peace. The risk of losing his army in the process had to be accepted; that was the stake in the game, the price of his vast hopes. It may have been his fault if his army was punished more severely than it need have been, but the fault did not lie in the depth to which he penetrated Russia; that was his object and was unavoidable. It lay in his being late in starting the campaign, in the lives he squandered by his tactics, his neglect of matters of supply and of his line of retreat. Lastly, he stayed too long in Moscow.

It is no great argument against us to point out that the Russians managed to bar the way at the Beresina in the hope of cutting off his retreat. The battle showed precisely how difficult it is to achieve such an object. The conditions were the worst conceivable, but the French contrived to fight their way through all the same. This whole episode deepened the catastrophe, but did not cause it. Second, it was only the unusual nature of the country that enabled the Russians to achieve as much as they did, for if the main road had not crossed the marshes of the Beresina with their wooded, inaccessible approaches, it would have been still less feasible to cut off the French army. Third, the only possible way of guarding against that risk would be by advancing on a certain breadth of front. To that we have already objected that once we are committed to an advance in the center while leaving armies behind as flank guards to left and right, any mishap to one of these would make us withdraw our center. Nothing much could then be expected of the offensive.

Nor can it be said that Bonaparte neglected his flanks. A superior force confronted Wittgenstein. At Riga there was an adequate investing-force—which, incidentally, was superfluous; and Schwarzenberg in the south had 50,000 men, outnumbering Tormasov's and almost equaling even Chichagov's force. In addition, there was Victor with 30,000 men as a central reserve. Even at the most critical period, in November, when Russian strength had grown while the French were already very much depleted, the Russians were not yet markedly superior in the rear of the Moscow army.

Wittgenstein, Chichagov, and Sacken had 110,000 men between them. Schwarzenberg, Reynier, Victor, Oudinot, and St. Cyr together still had 80,000 men. The most cautious general on the move would hardly have given his flanks more protection than that. Of the 600,000 men who crossed the Niemen in 1812 Bonaparte might have brought a quarter-million back instead of the 50,000 who recrossed it under Schwarzenberg, Reynier, and Macdonald, if he had not committed the mistakes we blame him for; but the campaign would have been a failure just the same. There would, however, have been nothing to criticize in theory, for the loss of more than half an army in such a case is not unusual. If it strikes us as such, the reason is simply the scale of the expedition.

So much for the main operation, the form it must assume and the risks inseparable from it. As for secondary operations, we would emphasize that all have a common aim, but this aim must be such as not to paralyze the activities of the separate parts. If anyone invaded France from the upper and the middle Rhine and from Holland, intending to join up in Paris, and if each army were ordered to take no risk and preserve itself so far as possible intact until it reached its rendezvous, we would call such a plan *calamitous*. A sort of balance between the three of them would be sure to come about and cause delay, timidity, and hesitation in each. It would be better to leave each army its own mission and only insist on united action at the point where their various activities naturally coincide.

The separation of forces in order to reunite again a few days later is a feature of almost every war, but basically it is senseless. If a force is detached it should know why, and the *purpose* must be met. This purpose cannot simply consist in a subsequent reunion as if one was dancing a quadrille.

So if armies do attack in different operational theaters, each should be given a distinct objective. What *matters* is that the armies everywhere expend their full energies, not that all of them should make proportionate gains.

If one army finds its task too difficult because the enemy's defensive scheme is not what it expected, or if it runs into bad luck, the actions of the others must not be modified or a general success will be unlikely from the start. Only if most of them are unfortunate or if the principal operations fail is it right and necessary that the others should be affected. Then it is the plan itself that has gone wrong.

That rule should also be applied to armies and detachments originally given a defensive role but set free by their success to take the offensive—unless one prefers to transfer their superfluous units to the main point of the offensive. The question will principally turn on the topography of the theater of operations.

But then what becomes of the geometric form and unity of the whole attack? What happens to the flanks and rear of columns adjoining operational units? It is exactly this kind of attitude which we are especially concerned to combat. The glueing together of a major offensive into a geometrical square is to get lost in a false intellectual system.

In the fifteenth chapter of Book Three we showed that the geometric element is less effective in strategy than in tactics, and at this point we need only restate the conclusion—that actual successes at particular points, especially in the offensive, deserve much more attention than the shape that may gradually emerge from the varying fortunes of the attack at one point or another.

In any case, seeing the large areas with which strategy is concerned, the commander-in-chief can properly be left to deal with the arguments and decisions that settle the geometric pattern of the parts, and so no subordinate commander has the right to ask what his neighbor is doing or failing to do. He can be told simply to carry out his orders. If serious dislocation should really result, the supreme command can still put it right. In this way the objection to separate operations is removed—that is, the obfuscation of realities by a cloud of fears and suppositions that seeps into the actual course of events, so that every mishap affects not just the part that suffers it but, contagiously, all the rest, and personal weakness and antipathies among subordinate commanders are given ample scope.

We do not think this point of view is likely to seem paradoxical to those who have spent much time and thought on the study of military history, learned to distinguish between essentials and inessentials, and fully realize the influence of human weaknesses.

As all experienced soldiers will admit, it is difficult even from the tactical point of view to make a success of an attack in several separate columns by smoothly coordinating every part. How much more difficult, or rather, how impossible the same must be in strategy, where intervals are so much greater! If then the smooth coordination of all parts is a precondition of success, a strategic attack of that kind ought to be avoided altogether. But, on the one hand, one is never wholly free to reject it since it may be imposed by circumstances that one cannot alter; while, on the other, the smooth coordination of every part of the action from start to finish is not even necessary in tactics, let alone strategy. From the strategic point of view, then, there is all the more reason to ignore it; and it is all the more important to insist that every part be given an independent task.

We must add an important comment concerning the proper division of labor.

In 1793 and 1794 the main Austrian army was in the Netherlands, with the Prussian army on the upper Rhine. Austrian troops then marched from Vienna to Condé and Valenciennes, crossing the Prussians' route to Landau from Berlin. Admittedly, the Austrians had their Belgian provinces to defend, and they would have welcomed any conquests made in French Flanders. But that concern was not adequate reason for these arrangements, and after Prince Kaunitz's death the Austrian Minister Thugut determined to relinquish the Netherlands altogether for the sake of a better concentration of his forces. Austria is indeed almost twice as far from Flanders as from Alsace, and at a time when troops were strictly limited and their supplies had to be paid for in cash, that was no small consideration. But

Thugut had yet another point in mind. He wanted to confront Holland, England, and Prussia, the powers that had the most interest in the defense of the Netherlands and the lower Rhine, with the urgency of the danger and the need for making greater efforts. He miscalculated, because at that time there was no way to make the Prussian government change its policy; but these events show the influence that political considerations had on the course of the war.

Prussia had nothing to defend or to conquer in Alsace. Her march in 1792 through Lorraine to Champagne had been made in a spirit of chivalry, but since as things turned out that operation promised little more, she pursued the war without enthusiasm. Had the Prussian troops been in the Netherlands, they would have been next door to Holland, which they almost looked on as their own, having occupied it in 1787; they would then have covered the lower Rhine and with it the part of Prussia that was nearest the theater of operations. Through her subsidies Prussia also had a closer alliance with England, and would thus less easily have become involved in the machinations of which at that time she became guilty.

It might have been far more effective, therefore, if the Austrians had placed their main force on the upper Rhine and the Prussians theirs into the Netherlands, where Austria would only have left a modest corps.

If General Barclay had commanded the Silesian army in 1814 instead of the enterprising Marshal Blücher and Blücher had stayed with the main army under Schwarzenberg, the campaign might well have broken down completely. Again, if the enterprising Laudon had not been given Silesia, the strongest part of Prussia, as a theater of operations, but had been with the army of the Holy Roman Empire, the whole of the Seven Years War might well have turned out differently.

For a closer look at the subject, let us examine the main characteristics of the following cases.

The first is when war is being jointly waged with other powers that are not only our allies but have independent interests of their own.

The second is when an allied army comes to our assistance.

The third is when all that matters is the personalities of the commanders.

In the first two cases, the question is whether the various allied troops are better mixed, so that armies have corps of different nationalities as was done in 1813–1814, or better kept as separate as possible so that each can play an independent role. Clearly the first is the better plan; but it assumes a rare degree of friendliness and common interest. With forces integrated in that way their governments will find it much more difficult to pursue their private interests; and as for their commanders' egoism, its harmful influence can, in the circumstances, only show among the subordinate commanders—that is, in the tactical realm, and even then less freely and with less impunity than if national contingents were completely separate. In the latter case it will extend to strategy, and crucial matters will be affected. But as we have said, a rare degree of self-effacement is required of governments. Sheer necessity drove everyone in that direction in 1813. Still one cannot speak too highly of the Czar of Russia. Although he commanded the largest army in the field and had had the greatest share in the reversal of our fortunes, he placed his forces under Prussian and Austrian generals and made no pretension to command an independent Russian force.

If forces cannot be integrated in that way, it is admittedly better to keep them completely rather than partially separate. The worst situation of all invariably results when two autonomous generals of different nationality share a theater, as was often the case with Russian, Austrian, and Imperial forces in the Seven Years War. If forces are wholly separate it is easier to divide the burdens; each army will then suffer only from its own. Circumstances will, therefore, stimulate each to greater efforts. But if they are in close contact with each other, or even in the same operational theater, that will not occur, and, what is more, if one of them does show bad faith the others will be paralyzed.

Total separation will do no harm in the first of the three cases I have sketched, for each state's natural interests will normally settle how its forces should be used. That may not be so in the second case, and in that event there is usually no choice but to place one's troops entirely at the disposal of the allied army, assuming the latter's size to be at all appropriate. The Austrians did this at the end of the campaign of 1815 as the Prussians had in 1807.

As for commanders' personal characteristics, everything depends on the individual, but one general comment must be made. Though it is often done, subordinate armies should not be put under the command of the soundest and most cautious men. The right men here are the most *enterprising* ones, for we must again insist that in separate strategic operations nothing is more important than that every part should do its best and develop its powers to the full. Any error made at one point can be set off against successes elsewhere. But maximum effort by everyone can only be ensured if all commanders are spirited, active, eager men, with a strong inner drive. Cool objective deliberation about the need for action is seldom enough.

Lastly, it remains to be said that wherever possible troops and commanders should be assigned to missions and areas appropriate to their special qualities. Regular armies, excellent troops, abundant cavalry, elderly, wise, and prudent generals should be used in open country; militia, national levies, hurriedly mobilized rabble, young and enterprising generals in wooded country, mountainous areas, and passes; and auxiliary forces in prosperous areas where they will enjoy themselves.

All we have said so far about the plans of operations in general and, in this chapter in particular, about plans intended to achieve the total defeat of the enemy, has been intended to emphasize their object and then to suggest principles to guide operational arrangements. We wished to gain a clear understanding of what we want and should do in such a war. We would emphasize the essential and general; leave scope for the individual and accidental; but remove everything *arbitrary, unsubstantiated, trivial, far-fetched*, or *supersubtle*. If we have accomplished that we regard our task as fulfilled.

Should anyone be shocked at finding nothing here about how to turn a river, command a mountain area from its heights, by-pass a strong position or find the key to a whole country, he has failed to grasp our purpose; we are afraid, moreover, that he has still not understood the essential elements of war.

In previous books we have dealt with these details in a general way, and reached the conclusion that they are apt to be a great deal less important than is usually thought. The part that they can or ought to play in a war intended to defeat the enemy is even slighter. It certainly cannot affect the general plan.

The structure of supreme command will occupy a special chapter at the end of the present book.[5] To conclude the present chapter we shall offer an example.

If Austria, Prussia, the German Confederation, the Netherlands, and England decide to make war on France, with Russia neutral—a case the last century-and-a-half has often seen—they would have capacity enough to wage an offensive war with the object of totally defeating the enemy. Large and powerful as France is, the greater part of her territory might well be overrun by hostile armies; Paris would be in enemy hands and France herself reduced to inadequate resources, with no other state but Russia able to give her really effective help. Spain is too far away and badly placed; the Italian states are still too weak and unstable. Not counting their possessions overseas the countries named have 75 million inhabitants to draw on while France has only 30 millions. At a conservative estimate the army that could take the field for a really serious attack on France could be composed as follows:

Austria	250,000 men
Prussia	200,000
The rest of Germany	150,000
The Netherlands	75,000
England	50,000
Total	725,000

If such a force were actually put into the field it would almost certainly be far superior to any that France could field against it. Under Bonaparte she never raised a force of comparable strength. Allowing for the troops required to man the fortresses and depots and to guard the coast, there can be little doubt that the allies would have a significant superiority in the principal theater; and this superiority would be the main consideration in their plan to bring about a French collapse.

The center of gravity of France lies in the armed forces and in Paris. The allied aim must, therefore, be to defeat the army in one or more major

[5] The chapter was never written. Eds.

battles, capture Paris, and drive the remnants of the enemy's troops across the Loire. The most vulnerable area of France is that between Paris and Brussels, where the frontier is only 150 miles from the capital. That is the natural concentration area for one group of allies—England, the Netherlands, Prussia, and the North German states—all of which have territories nearby, some being actually adjacent. Austria and southern Germany can conveniently operate only from the upper Rhine, and their natural direction of attack is toward Troyes and Paris, possibly also toward Orleans. Both invasion lines, the one from the Netherlands and the other from the upper Rhine, are perfectly natural, short, unforced, and effective; and the center of gravity of France's power is where the two lines meet. Between these two points, therefore, the whole offensive force should be divided.

Only two considerations qualify the simplicity of this plan.

The Austrians will not uncover their Italian provinces. They will always wish to control the situation there, and hence they will never let matters reach a point where Italy is only indirectly covered by forces engaged in attacking the heart of France. The state of Italian politics being what it is, this Austrian concern, though secondary, is real; but it would be a great mistake to let the old and oft-attempted scheme of attacking southern France from Italy be linked with it. Austrian strength in Italy would then be raised to a far higher level than security alone would require if the first campaign met with grave reverses. Only modest numbers should remain in Italy, and nothing more should be withheld from the main offensive if the precept of all precepts is to be observed—*unity of conception, concentration of strength.* One could as easily pick up a musket by the tip of the bayonet as conquer France by way of the Rhône. But even as a supplementary operation an attack on southern France should be condemned, for it could only stimulate fresh sources of resistance. Whenever an outlying province is attacked, one stirs up concerns and activities that would otherwise have remained quiescent. An attack on southern France from Italy would not be justified unless it were obvious that the forces left in Italy were more than its security required and were, therefore, destined to be idle.

Hence, we repeat that the force to be kept in Italy must be the very smallest that conditions permit. It is large enough if it will save the Austrians from losing the whole area in one campaign. For the purpose of our illustration we shall put it at 50,000 men.

The other consideration is the coast of France. England dominates the sea; France must, therefore, be extremely sensitive about her whole Atlantic coast and she must keep some forces to defend it. However weak their coastal defenses might be, they make her frontiers three times as long and hence she must withdraw substantial forces from the theater of war. If England has 20,000 to 30,000 landing troops available to threaten France, they might perhaps immobilize two or three times as many French; and this would involve not only troops but also money, guns, etc., for the fleet and the coastal batteries. Let us assume that for this purpose the English have 25,000 men.

The plan of operations therefore in the simplest terms would be as follows:

First, for assembly in the Netherlands, 200,000 Prussians, 75,000 Netherlanders, 25,000 English, 50,000 North-German Federal Troops—350,000 men.

Of these, some 50,000 would be used to garrison the frontier fortresses, which would leave 300,000 free to advance on Paris and fight a major battle against the French.

Second, 200,000 Austrians and 100,000 South-German troops would be assembled on the Rhineland. They and the Dutch would advance simultaneously toward the upper Seine and thence toward the Loire, and would also aim at a major battle. The two thrusts might perhaps be united on the Loire.

This outlines the main points. Our further remarks are chiefly intended to remove misunderstandings, and are as follows:

1. The main concern of the commanders-in-chief must be to seek the necessary major battle and fight it with such superiority of numbers, and under such conditions, as will promise decisive victory. Everything must be sacrificed to that objective and the fewest possible men should be diverted into sieges, investments, garrisons, and the like. If, like Schwarzenberg in 1814, they fan out as soon as they reach enemy soil, all will be lost. In 1814 it was only the impotence of France that saved the allies from complete disaster in the first two weeks. The attack should be like a well-hammered wedge, not a bubble that expands till it bursts.

2. Switzerland must be left to its own devices. If neutral, it forms a good *point d'appui* on the upper Rhine. If France attacks it, let it defend itself—which it can do very well in more respects than one. Nothing could be more foolish than to. think that Switzerland, as the highest ground in Europe, must dominate the geographical course of the war. That influence could only operate under certain very limited conditions, which do not exist in the present case.

While the heart of France is being attacked, the French cannot mount a powerful offensive based on Switzerland against either Italy or Swabia, and least of all can the altitude of Switzerland count as a decisive factor. Any advantage from this kind of strategic domination accrues in the first place primarily to the defense, and any importance it has for the attack can only operate in the first assault. If anyone does not understand this, he has not yet thought it through. If in a future council of war some learned general staff officer should solemnly serve up this kind of wisdom, we declare it in advance to be arrant nonsense, and we hope that some tough fighting soldier, full of commonsense, will be there to shut him up.

3. The space between the two offensives is hardly worth discussing. With 600,000 men assembled only 150 or 200 miles from Paris, poised to strike at the very heart of France, need one really think about covering the upper Rhine—which means covering Berlin, Dresden, Munich, and Vienna? There would be no point in doing so. Should the lateral communications be cov-

ered? This merits some attention; but one might then logically be led to give this cover the strength and significance of another offensive. Then instead of advancing on two lines, which the location of the allied states makes unavoidable, one would find them advancing along three, which is not necessary. The three could then turn into five, or even seven, and the whole sad business would begin again.

Each of the two attacks will have its own objective and there is no doubt that the forces detailed for them will be markedly superior to the enemy's. If each attack is pressed with determination it cannot fail to benefit the other. Should one of them run into trouble because the enemy's strength has not been equally divided, it should be possible to rely on success by the other automatically to repair the damage. This is the real connection between the two armies. Seeing how far apart they are, an interdependence covering day-to-day events would not be possible. Nor is it needed; so close or, rather, direct links between the two have little value.

The enemy, assailed at the very core of his being, can spare no strength worth speaking of to disrupt the cooperation of the two offensives. The worst that can happen is that the populace, supported by raiding parties, might try to do this, and save the French from diverting regular forces for this purpose. To counter them only a corps of 10,000 to 15,000 men, strong in cavalry, need be sent out from Treves in the general direction of Rheims. It will ride roughshod over any raiding party and can keep up with the main force. It should neither watch fortresses nor invest them, but by-pass them; it should not depend on any definite base, and should retire before superior force in any direction it pleases. No great harm can befall it, and even if it did, that would be no disaster for the whole. Under these conditions such a corps might usefully serve as a link between the two offensives.

4. The two subsidiary operations—the Austrian army in Italy and the English landing force—can pursue their purposes at their discretion. Provided they are not idle, their existence will be justified, and under no condition should either of the main offensives be in any way dependent on them.

We are quite convinced that in this manner France can be brought to her knees and taught a lesson any time she chooses to resume that insolent behavior with which she has burdened Europe for a hundred and fifty years. Only on the far side of Paris, only on the Loire, can she be made to accept the conditions which the peace of Europe calls for. Nothing else will demonstrate the natural relationship between thirty millions and seventy-five. But that will certainly not be done if France is ringed by armies from Dunkirk to Genoa, as she has been for a century and a half, while fifty different small objectives are pursued, not one of them important enough to overcome the inertia, the friction, and the outside interests that always emerge, especially in allied armies, and perpetually reappear.

The reader is unlikely to misunderstand how little such a scheme fits the provisional organization of the federal German armies. By this the federal part of Germany is to form the nucleus of German power; Prussia and Austria are thus weakened and lose the preponderance they should possess.

But a federal state is a poor sort of nucleus in war time, lacking unity and vigor, without any rational way of choosing its commander, bereft of authority or responsibility.

Two natural centers of power exist in the German *Reich*—Austria and Prussia. Theirs is the genuine striking-power, theirs is the strong blade. Each is a monarchy, experienced in war. Their interests are clearly defined; they are independent powers and are preeminent above all the rest. These natural lines, not the mistaken idea of "unity," define the lines that German military organization should follow. Unity is anyhow impossible under these conditions, and the man who sacrifices the possible in search of the impossible is a fool.

A COMMENTARY

Bernard Brodie

BERNARD BRODIE

A Guide to the Reading of *On War*

I observed in my introductory essay that many find Clausewitz difficult to read with comprehension, even though the ideas presented are not intrinsically difficult. The reasons are several, the most important being that whether because of the richness of example and qualification or the sometimes faulty organization, the main thread of the argument is often lost in the development. Also, Clausewitz is occasionally, though not often, metaphysical, and this worries readers more than it needs to. In addition, some long sections are of purely historical value, or, as some would hold, obsolete, while others are charged with the greatest significance for contemporary times, and it may help the reader to be alerted in advance to which is which. Finally, a reader not familiar with the field will have no means of knowing how distinctive are some of the points that the author makes, and he may profit from having this quality pointed out at the appropriate places. I have no desire to gild the lily, and if I am sometimes guilty of superfluous applause, I shall make amends by warning of sections which are relatively unrewarding.

In short, the main purpose of this guide is to enhance the reader's comprehension of the text at first reading. If he finds it failing in that purpose, or not sufficiently succeeding in it to be worth his time, he should avail himself of the nearest exit, which is as close as the next comma or period.

The reader will find the way that suits him best for using this guide. Some few may want to read it through before going to the text. Perhaps most will want to use it book by book, or even in smaller doses, in association with the text. It may help also as a quick review after reading the text, as well as for finding remembered passages of which one cannot recall the exact location.

Because this guide is not a synopsis we are free to dwell on the details of some chapters or passages and to touch others quite lightly or skip them entirely, in the latter instance without necessarily mentioning that we are doing so. It is not only that some portions are indeed more significant or arresting than others but also that some excellent chapters offer no special pitfalls.

The "Notice" of 10 July 1827 together with the undated but clearly later note which the author left behind with his manuscript tells of his unfulfilled plans for its revision. In the interval between writing these two papers he had entirely rewritten the first chapter of Book One, which is the only one he regarded as finished. His plans for further revision would seek mainly to make two ideas stand out in bolder relief throughout the work. These are (1) the importance of always distinguishing between the respective requirements of general and of limited war (applying modern terminology), and (2) the essential and all-pervading necessity of recognizing that "*war is simply the continuation of policy by other means*" (Clausewitz's italics). One senses that the latter idea, though approached early in the author's career, was still not crystalized in his mind, because it is developed only among the last portions that he worked on, that is, Chapter Six of Book Eight and the finally revised first chapter of Book One, and also because he talks of expecting the further working out of Book Eight to help clear his own mind on the subject. Possibly Book Eight as we now see it received some of that further working out after the 1827 "Notice" was written, but it was not yet in final form when the author left for Silesia in 1830, where he was to die a year later, and certainly he had not got round to suffusing the idea through the whole work.

Book One: On the Nature of War

We pass, without comment on the "Author's Preface," to Book One, Chapter One, the only chapter completed to the author's satisfaction. It is one of the most densely packed with ideas in the whole work, and their sweep and importance make this chapter a most comprehensive introduction. It may also be for some readers somewhat confusing without the appropriate key, which I shall here attempt to provide—in more detail than will apply to other chapters.

The *object* of war, Clausewitz says in a three-tiered statement, is (a) to impose our will on the enemy, to do which (b) we use the *means* of maximum available force, with (c) the *aim* of rendering him powerless. We thus note at the outset the distinction between military aim and political object.

Distress at the brutality of war must not be allowed to inhibit the use of means for "war is such a dangerous business that the mistakes that come from kindness are the very worst." Civilized nations may practice inhibitions, but this is due to social forces which "are not part of war." And then: "To introduce the principle of moderation into *the theory of war itself* would always lead to logical absurdity" (italics added).

The meaning will be a little clearer if one observes that in these opening pages Clausewitz is using word-images made fashionable by the great German philosophers of his day, especially Kant and Hegel. They had powerfully revived the ancient metaphysical school which bears the somewhat deceptive name of "idealism." Thus, in the language of this school, war is simply another form of being that, like any other form of being, derives from an essential pattern or "idea" in which it has its true reality. To understand war properly, one must first see it in its "absolute" or "ideal" form, which Clausewitz calls the "pure concept of war." It is in this sense that he speaks of "the theory of war [in] itself," which recalls Kant's famous phrase, *das Ding an sich* ("the thing in itself"). Fortunately, Clausewitz was of much too pragmatic a fiber to lose himself either deeply or for long in this brand of idealism, but here as the curtain rises on his masterwork, this man who never graduated from a university enters wearing the gown of academe.

Still, he is at the same time being realistic in the normal sense of the word. Certainly he believes profoundly in what he is saying. He has witnessed some of the worst horrors in the long grim history of war, including the disastrous crossing of the Berezina River by the French in their retreat from Moscow, which he saw with his own eyes from the Russian side and which he described with the most deep-felt, shuddering anguish in a letter to his wife. Yet he was later also to write, in criticism of the Russian general nearest the scene, Wittgenstein, that if he had attacked more forthrightly "he might have made the French loss much greater." A sensitive man unquestionably, Clausewitz had long since come to terms, at least intellectually, with the harsh demands of his profession. He was quite conscious that he had to steel himself emotionally to meet these demands, not once and for all but repeatedly, and he loses no time in imposing upon the reader a comparable obligation. It is noteworthy that he feels the need to mention these unpleasant and disturbing things, which other writers on strategy have not.

In these opening pages he also touches insightfully on the part played in war by passion, which inevitably distorts the clear conception of the object. "If war is an act of force, the emotions cannot fail to be involved." Since the thrust of the main argument of the work will be in the opposite direction, it is necessary to get these matters straight at the outset.

Having made his bow to the "extremes" of war, in which he has not failed to set down some important practical ideas, he makes an epochal shift to the ground he will occupy for virtually all the rest of the work. He gives to Section Six the title "Modifications in Practice." When we move from the abstract to the real world, "the whole thing looks quite different." War in its abstract or "perfect" form could exist in the real world only if certain conditions were met, the most important being that war should be "a wholly isolated act, occurring suddenly and not produced by previous events in the political world." But, he says, "war is never an isolated act." He notices also that "even the ultimate outcome of a war is not to be regarded always as the final one" (his own Prussia had been virtually annihilated as a military power in the Jena campaign of 1806, only to come back strong in the campaigns of 1813 and 1814 and at Waterloo in 1815, and Clausewitz had been deeply involved in all these events).

As he goes on, the political aim comes to the fore. It is the original motive for the war, now perhaps modified by operating in a war context. In some cases the political and the military objectives are the same, as in the conquest of a province intended to be kept; in other cases "another military aim will be adopted that will serve the political purpose and symbolize it in the peace negotiations." Here Clausewitz has in mind, for example, seizing and holding a territory only for bargaining purposes. But in these decisions one must tread carefully; "attention must be paid to the character of each state involved." What will induce one nation to yield will cause another to resist more fiercely.

There follows a discussion on the factors that make for interruption of activity in war, where he develops a concept which he calls the "principle of polarity" but which turns out to be exactly the idea for which we should now use the term "zero-sum game" (that is, a contest in which the gain of one is directly proportional to the loss of the other). Many war situations, he points out, do not fall into this pattern; the loss of one is not always the gain of the other. He introduces the idea, to be developed later, that "defense is a stronger form of fighting than attack." Clausewitz still makes occasional references to "the Absolute," that is, to the perfect or pure form of war, but increasingly he emphasizes and dwells upon those qualities which remove war from that realm.

He now moves toward a more precise definition of war. "Were it a complete, untrammeled, absolute manifestation of violence (as the pure concept would require), war would of its own independent will usurp the place of policy the moment policy had brought it into being." Actually, that is exactly what many if not most modern generals have felt should happen. There is more than a smell of that idea in General MacArthur's phrase: "There is no substitute for victory." And this is also the view that Clausewitz commonly experienced in his own time, but he quite unflinchingly calls it "thoroughly mistaken." He now leads up to his great and famous formula in the following words: "If we keep in mind that war springs from some political purpose, it is natural that the prime cause of its existence will remain the supreme consideration in conducting it. . . . It must adapt itself to its chosen means, a process which can radically change it; yet the political aim remains the first consideration. Policy, then, will permeate all military operations and, in so far as their violent nature will admit, it will have a continuous influence on them." This naturally brings him to the pronouncement which is the title of the next section: "War Is Merely the Continuation of Policy by Other Means."

Thus, before his first use of that great dictum Clausewitz has already told us exactly what he means by it. It is a specific meaning not likely to be sensed correctly by the person who has simply heard the phrase quoted to him. The conception is age-old and yet forever neglected or rejected. It suffers this fate for a number of reasons, one being that war does arouse passions, usually very strong ones, and another being that generals like to win decisively whatever contests they are engaged in, and do not like to be trammeled by a political authority imposing considerations that might modify that aim.

Clausewitz concedes that the general has a right to demand that the trends of policy shall not be inconsistent with the means he is being called upon to use. That is no small demand, he adds, but it will never do more than modify the political aims. Then he says again: "The political object is the goal, war is the means of reaching it, and means can never be considered in isolation from their purposes."

The strain between political purpose and military aim, Clausewitz then points out, is minimal in what we would now call general war. "The more powerful and inspiring the motives for war . . . the closer war will approach its abstract concept . . . the more closely will the military aims and the political objects of war coincide, and the more military and less political will war appear to be." On the other hand, where motives are less intense, as in limited war, "the political object will be more and more at variance with the aim of ideal war, and the conflict will seem increasingly *political* in character." However: "While policy is apparently effaced in the one kind of war and yet is strongly evident in the other, both kinds are equally political." The reason is that "among the contingencies for which the state must be prepared is a war in which every element calls for policy to be eclipsed by violence." He is describing total war, which today would be nuclear, but the acceptance of such a state of affairs is still policy.

Clausewitz seems intent not only on informing us what war is all about but also on obliging us to keep it everlastingly in our minds. He insists and reiterates that war is always *an instrument of policy* because he knew, and we know today, that the usual practice is rather to let war take over national policy.

Book One, Chapter Two: In this chapter Clausewitz considers why some wars are and ought to be kept limited, in duration and object as well as in intensity and means. In reading it (and later chapters) one remembers wryly that Clausewitz has been often called the "apostle of total war." This opinion is likely to be confined to those who have never touched the book. However, even the late Colonel Joseph I. Greene, in writing a foreword to a previous translation of *On War*, attributes to Clausewitz the central thesis that the aim of war is the overthrow of the enemy's powers of resistance, and then adds: "He had witnessed the whole course of Napoleon's campaigns, and had seen that Napoleon fought his wars to win them. There might have been men or nations who went into wars with some lesser intention in mind. But war as a whole could not be seriously discussed on any such half-way basis."[1] Either Colonel Greene had never got to the second chapter of the book for which he was writing a foreword (actually Clausewitz had already signaled his views on the subject in the first chapter), or, more likely, a casual reading (though he warned against such casualness) did not touch his prepossessions.

After opening the chapter with a brief discussion of wars that are *not* limited in aim, in which the author reminds us again that even the successful accomplishment of such aims is not necessarily final (Napoleon's victories are relevant here, as are Hitler's), he goes on to consider examples where one or both sides aim at something less than the complete disarming of the enemy.

Though this kind of conflict does not conform to the pure theory of war, so much the worse for pure theory. "On no account," he says, "should theory raise [the aim to disarm the enemy] to the level of law."

Anyway, the aim of totally disarming the enemy is obviously unrealistic in all those cases where he is substantially stronger. Even where it is possible to carry on the struggle, grounds for making peace exist where, first, victory is improbable, and second, where it is unacceptably costly. Whether it is either depends not only on strength but on motivation: "Since war is not an act of senseless passion but is controlled by its political purpose, the value of this purpose must determine the sacrifices to be made for it in *magnitude* and also in *duration*. Once the expenditure

[1] Col. Joseph I. Greene, "Foreword" to Karl von Clausewitz, *On War*, trans. O. J. Matthijs Jolles (New York: Modern Library, 1943), p. xiii.

of effort exceeds the value of the political purpose, the purpose must be renounced and peace must follow."

Clausewitz is of course aware that "the original political aims can greatly alter in the course of the war, and may finally change entirely, since they are influenced by events and their probable consequences." He seems, however, to be clearly rejecting any sense of inevitability about such basic changes, especially as they might arise out of the war itself. For to admit even a high probability of such a feedback effect would be to destroy his basic contention that war is the instrument of policy and not the reverse. Clearly he is implying the necessity of control, both on passions and on notions of prestige.

"In war," says our author, "many roads lead to success, and . . . they do not all involve the opponent's outright defeat." He reviews what these many roads are, including attrition and "merely passive waiting for enemy attacks," which may wear down the will of an enemy who has no great will to begin with, and he notices that it is important always to bear in mind the *ad hominem* circumstances. "The personalities of statesmen and soldiers are such important factors that in war above all it is vital not to underrate them."

Although the political object of war may vary infinitely, from the will to annihilation to the reluctant fulfillment of an alliance obligation, the *means* of war is always the same—combat. However the forms of combat are also widely variable, from that which seeks destruction of the enemy forces to mere passive resistance. But the commander who prefers any strategy other than destruction of the enemy's armed forces "must first be sure that his opponent either will not appeal to that supreme tribunal . . . or that he will lose the verdict if he does."

Book One, Chapter Three: Inasmuch as the author will subsequently be attaching great importance to the talent or genius of the commander, he devotes this early chapter to analyzing the character of military genius. It is quite distinctive from what we call genius in other callings. For one thing, sheer physical courage is "the soldier's first requirement." No doubt it mattered more in his own day than it does in ours—one thinks of Wellington constantly in danger at Waterloo, or old Blücher injured two days earlier at Ligny when his horse was killed under him—but even in our own time the top commander rises to his post by showing his mettle in combat. Similarly, the commander must make himself indifferent to the physical exertion and suffering of others.

But the difficulty is in describing the qualities of intellect required. Clausewitz considers the matter with great care, and he concludes that inasmuch as quick intuitive judgment and determination are much more necessary than great powers of meditation, what is required is "a strong mind rather than a brilliant one." The response to the unexpected need not be exceptional, "so long as it meets the situation." But qualities of temperament, such as those which make for strength of will and energy in action, are equally integral to the commander's genius.

There may also be something self-revelatory in his insistence that "of all the passions that inspire man in battle"—which give him above all the requisite energy in action—"none . . . is so powerful and so constant as the longing for honor and renown." Other emotions, he says, may be more common and more venerated, "but they are no substitute for a thirst for fame and honor." In this respect introspection might have got the better of his objective judgment, for there is no doubt that Clausewitz is describing a thirst which affected him deeply. Still, we should also observe that in dealing with this provocative and elusive subject, he never let himself slip into romanticism, as so many others have done.

He is curiously negative about the quality we call "imagination." After stressing the importance of what he calls a sense of terrain and of locality, he says: "We attribute this ability to the imagination; but this is about the only service that war can demand from this frivolous goddess, who in most military affairs is liable to do more harm than good." He leaves the comprehension of this somewhat cryptic statement to the reader's imagination, but in summing up the qualities of true genius in the supreme commander (just after mentioning the name of Bonaparte), he says: "What this task requires in the way of higher intellectual gifts is a sense of unity and a power of judgment, raised to a marvelous pitch of vision, which easily grasps and dismisses a thousand remote possibilities an ordinary mind would labor to identify, and wear itself out in doing so. Yet even that superb display of divination, the sovereign eye of genius itself, would still fall short of historic significance without the qualities of character and temperament we have described." Nor has he failed to call attention to the importance of the commander's having a thorough grasp of national policy. "On that level strategy and policy coalesce: the commander-in-chief is simultaneously a statesman." And then, at the end, he identifies military genius with "the inquiring rather than the creative mind, the comprehensive rather than the specialized approach, the calm rather than the excitable head."

Book One, Chapters Four–Eight: These five brief but brilliant chapters form a unit in that their central concern is with that characteristic of war which Clausewitz calls "friction" and which he mainly elucidates in Chapter Seven. Friction is what so frequently makes difficult and hence precarious of execution what looks so simple and easy on paper; it is what makes so much depend on the will as well as the intellect of the commander.

Chapter Four, "On Danger in War," suggests in a few paragraphs the impediments to thinking that result from the sound of bullets hissing round one's head and from the sight of men being killed and mutilated. In such an atmosphere "the light of reason is refracted in a manner quite different from that which occurs in academic speculation."

Chapter Five tells of the friction that derives from the often unbelievable physical effort in war. It concerns both the efforts that a general can demand of his troops, who on forced marches or in battle are usually beset with many forms of misery, and the effects of weariness upon himself and upon his sensibilities.

Chapter Six introduces the element that others have called "the fog of war," the perennial inadequacies and inaccuracies of intelligence. Clausewitz tells us that "as a rule most men would rather believe bad news than good," and that if a commander does not have a buoyant disposition, "he had better make it a rule to suppress his personal convictions, and give his hopes and not his fears the benefit of the doubt." One wonders again whether Clausewitz is here telling us something about himself. He was not of a "buoyant disposition," and perhaps for that reason he was as a senior officer invariably in a staff rather than a command position.

Chapter Seven deals directly with the subject which the three preceding chapters were leading up to, "Friction in War." This is one of the most taut and inspired of his chapters. In war, he says, everything is simple, "the strategic options are so obvious, that by comparison the simplest problem of higher mathematics has an impressive scientific dignity." Yet what is simple is also difficult. "Action in war is like movement in a resistant element. Just as the simplest and most natural of movements, walking, cannot easily be performed in water, so in war it is difficult for normal efforts to achieve even moderate results." Also, "every war is rich in unique episodes; each is an uncharted sea beset with reefs. The commander may suspect the reefs' existence, without ever having seen them; now he must steer past them in the dark." Though his progress may be only by virtue of overcoming great difficulties, "from a distance everything may seem to be proceeding automatically." Resistance is "a force that theory can never quite define," and that makes every precise theory irrelevant. The author promises to return often to this subject.

Chapter Eight, which concludes Book One, deals with the problems of gaining during peacetime relevant experience for all ranks but especially for commanding officers. "It is immensely important that no soldier . . . should wait for war to expose him to those aspects of active service that amaze and confuse him when he first comes across them."

It is difficult to guess whether Clausewitz in his planned revision would have expanded appreciably the last five chapters of Book One. Each is only a few paragraphs long, but each, except perhaps for the concluding Chapter Eight, seems to say all that needs to be said on its subject. It remains to be added that few other writers on strategy have even mentioned the "frictions" caused by danger and personal hardship, though many have been obsessed with the difficulties caused by faulty intelligence. The chapter devoted directly to the subject of frictions is a real gem, and explains, among other things, why Clausewitz was so averse to the compounding of maxims and axioms and to the people who engaged in it.

Book Two: On the Theory of War

When Clausewitz, in the "Notice" of 10 July 1827 found with his manuscript after his death, described the first six books as they then existed as being "a rather formless mass that must be thoroughly revised again," he was not indulging in vain self-depreciation. The standard by which he was measuring himself was undoubtedly high, but not unrealistic, for we see it realized in most of Book One and especially the only chapter he regarded as finished. In that chapter, and to a moderately less extent in the rest of Book One, the structure is taut, the ideas clear but densely packed in with no excess of verbiage, and they flow one from the other with the utmost logical coherence. There is not in those portions the "good deal of superfluous material" that he speaks of in the "Notice." In Book Two, however, as in various of the succeeding books, we see a different kind of writing. We are most decidedly still dealing with a Clausewitz, who has set down "the fruit of years of reflection on war and diligent study of it," but we are also dealing with an unfinished manuscript. It is not the "formless mass" he called it, but it is less well organized than it could be, and we do occasionally find it repetitious. Unquestionably these characteristics have contributed to the difficulty that some have found in reading or understanding Clausewitz, and this difficulty is best overcome by facing frankly the reason for it.

Book Two, Chapter One: This chapter is, for example, given over to some definitions and distinctions which do not much advance the argument. There are a few well-etched and highly quotable sentences, such as the one in the first paragraph which reads: "Fighting . . . is a trial of moral and physical forces through the medium of the latter." Also, he sets down twice, in related though different language, the distinction between tactics and strategy. In his first try he calls tactics *"the study of the employment of fighting forces in battle,"* and strategy he describes as *"the study of the employment of battles for the object of the war"* (his own italics).

Book Two, Chapter Two: This chapter, however, becomes immediately more interesting. It describes in broad and general terms the unsatisfactory nature of strategic studies prior to his own. He does not mention names, but he seems at different times to be speaking of Louis XIV's great master of siege warfare, the Marquis de Vauban; of Louis XV's Marshal Maurice de Saxe; of Frederick the Great, of Clausewitz's immediate predecessor, Count Dietrich von Bülow; and of his contemporary, Jomini. Among them and others he sees differences, and even advances, but in general he finds their ideas confining, that is, not taking account of the "endless complexities" involved in war, and also "pertaining only to physical matters and unilateral activity." It is particularly important, he says, to take account of the fact that in war "everything is uncertain and calculations have to be made with variable quantities," that "all military action is intertwined with psychological [or "moral"] forces and effects," and that war consists of "a continuous interaction of opposites"—of opposites, moreover, representing very different degrees of talent.

One must, he insists, eschew doctrines which presume to provide a manual for action. He indeed believes in theory, but only in the kind that develops out of long study of the history of war. Theory exists, he says, to "distinguish precisely what at first sight seems fused." With good theory one absorbs new material and experience without having to start afresh each time. It is, in other words, a way of promoting insights by organizing experience in one's mind. This kind of theory serves the future commander by guiding him in his self-education but it is not suitable for accompanying him to the battlefield. In short, it is a matter of training the mind rather than filling it with doctrine.

If all this seems nebulous and impractical, it should suffice to point out that Clausewitz is really describing an experience which many undergo but few perceive—the subtler effects of education on the mind. Education, almost regardless of the field of specialization, undoubtedly enhances our intellectual sensibilities, and one of the ways in which it does so is by expanding our awareness of connections between events or insights remote in time and in circumstance. In any field, theory is valuable to the degree that it promotes such expanded awareness in a specialized form and avoids falling into pedanticism or dogma. Clausewitz says almost the same thing: a sensible teacher is one "who guides and stimulates a young man's intellectual development, but is careful not to lead him by the hand for the rest of his life." He is not averse to principles and rules *if* "truth spontaneously crystalizes into these forms." It is the imposed kind of crystalization that alarms him.

He proposes to begin a theory of war by studying the relationship of means and ends, first in tactics and then in strategy. The purpose of doing so is mostly to exclude irrelevancies. The conduct of war, for example, has to do with the use of guns, but not with how to make them. Similarly, "strategy uses maps without worrying about trigonometric surveys." This simplification of relevant areas incidentally explains, in Clausewitz's view, "why in war men have so often successfully emerged in the higher ranks, and even as supreme commanders, whose former field of endeavor was entirely different." Clausewitz does not offer us names, but he could have been thinking of Cromwell, of Cromwell's sea commander Blake, and of any number of warrior kings from Alexander the Great to the great monarch of his own Prussia, Frederick II. The last named until his accession to the throne had been far more interested in music and in French literature than in things military. This bemusement with men who became great commanders with the barest minimum of preparation brings Clausewitz perilously close to denying, or at least appearing to deny, the value of any kind of intellectual preparation. As one writer has said of his work, "the expressed aim of these hundreds of pages of vast erudition is to impress on the mind of the reader the futility of all book-learning."[2] Well, it is not quite so. Before this chapter ends Clausewitz has insisted, first, that "no great commander was ever a man of limited intellect," and second, that the natural talent, where it exists, must develop the capability for making appropriate decisions, which means that it must be "trained and educated by reflection and study."

Book Two, Chapter Three: Clausewitz takes up the now familiar question whether the conduct of war is an art or a science, and he concludes that while it is more the former than the latter it is really neither. It is, he says, a clash of interests, and then he curiously compares it to commerce, which he also calls "a clash of human interests and activities"! He is, of course, expressing pure mercantilist sentiment. He had not got round to reading Adam Smith's great work, published four years before his birth. Nevertheless, his essential point is valid: war is not an exercise of will directed at inanimate matter (which presumably is his definition of art) but rather will directed "at an animate object which

[2] Jolles, "Introduction," *On War*, p. xxiii.

reacts." The same basic tenet is what distinguishes modern game theory from simple probability theory.

Book Two, Chapter Four: Following some brief definitions of terms, Clausewitz points out that in the theory of war, "there can be no formulation general enough to deserve the name of law." However, principles and rules do apply, though more to tactics than to strategy. They are useful to bear in mind as supports, but they are not to be dogmatically applied to every situation. Routine, too, can be useful, especially with respect to junior officers, for routine methods "will steady their judgment, and also guard them against eccentric and mistaken schemes which are the greatest menace in a field where experience is so dearly bought." In the absence of good theory, routine will affect even the major decisions of the highest ranks. Although a great commander is entitled to his personal style, when other commanders ape that style they may be adopting a meaningless routine. As an example Clausewitz offers the defeat which had been a bitter personal experience for him. He was present at Jena in 1806, when the Prussian commanders, following the tactical routines of Frederick the Great, presented their army for destruction by Napoleon.

Book Two, Chapter Five: This important chapter presents at the beginning some slight extra dimensions of difficulty in the reading, because Clausewitz is philosophically sophisticated enough to know that questions of cause and effect are more complex than laymen usually assume them to be. The subject is criticism of past campaigns or battles, which the author finds a good way of studying military theory, partly because the theoretical ideas upon which any criticism has to be based become more familiar through repeated application. However, because criticism is fundamentally concerned with the clear tracing out of causes and effects, it is worthless unless based on an accurate and adequately comprehensive narrative of the events under study. The critic must use his theoretical concepts as the soldier should use them—as aids to judgment and not as laws. Clausewitz does not fail to ring in once again the idea that inasmuch as all military acts are designed to produce certain ends, in "those involving great and decisive actions, the analysis must extend to the *ultimate objective*, which is to bring about peace."

In any good criticism, "things that did not actually happen but seemed possible . . . cannot be left out of account." The author takes as an example Bonaparte's campaign of 1797 through Italy into Austria, and in passing makes the fundamentally Clausewitzian observation that in 1797, as contrasted with the situation Bonaparte encountered in Moscow in 1812, "the secret of the effectiveness of resisting to the last had not yet been discovered." He uses another example, Bonaparte's lifting the siege of Mantua in 1796 in order to fall on some relieving columns, to show how the tyranny of fashion can prevent one's mind from considering valid alternative possibilities. The tyranny of bias can have the same effect, as proved with another example from Bonaparte's campaign, this one in 1814.

The chief purpose of the criticism just described is not to ascribe praise or blame to the commander, which is difficult or impossible to do with justice because the critic cannot put himself mentally into the situation of the participant. One conspicuous difference is that he knows the outcome. "An analyst should . . . not check a great commander's solution to a problem as if it were a sum in arithmetic." It may be useful, even unavoidable, to discuss particular actions or decisions in terms which suggest praise or blame, but, particularly with respect to an outstanding commander, the critic must avoid implying that he would have done better.

The difference between Bonaparte's disastrous failure in the Moscow campaign of 1812 and his successes in earlier campaigns such as Austerlitz (1805), Friedland (1807), and Wagram (1809) was less a matter of varying strategies than of the fact that in the earlier campaigns he had gauged his enemy correctly while in that of 1812 he had not. This is proved simply by the outcomes (Clausewitz will return to this theme in Chapter Nine of Book Eight). Still, in war all action is aimed at probable rather than at certain success, and there are times when the utmost daring is the height of wisdom.

Clausewitz closes this chapter by reviewing three common errors of critics. The first he states only after once again reviewing the fallacies of using principles or rules as rigid external guides to conduct rather than as aids to training a commander's (or critic's) mind. "One should never elaborate scientific guidelines as though they were a kind of truth machine." The second is the overuse of jargon, technicalities, and metaphors, where the critic "no longer knows just what he is thinking and soothes himself with obscure ideas which would not satisfy him if expressed in plain speech." The third is the misuse and abuse of historical examples, where the critic drags in three or four alleged examples from remote times and places just to show off his erudition. Obscure and usually at least partially false, these examples make only a negative impression on the practical military man whom they are presumably intended to benefit.

Theory must stay with simple terms and straightforward observation of the conduct of war; it must avoid spurious claims and unseemly displays of scientific formulae and historical compendia; and it must stick to the point and never part company with "those who have to manage things in battle by the light of their native wit." We observe that Clausewitz, for all the charges made against him of being over-philosophic, was determined to write a book that was practical.

Book Two, Chapter Six: In this concluding chapter of Book Two, Clausewitz develops a point made negatively in the previous chapter, where he says one of the three cardinal errors common to critics is their misuse of historical examples. Here he proposes to demonstrate their correct use.

"The knowledge which is basic to the art of war," he says, "is empirical." Thus, the nature of war is revealed to us only by historical experience. We can study means, like the physical effects of shot fired from guns, but this will not inform us, as historical example will, of the differences between seasoned troops and raw troops in the steadiness with which they stand up to artillery fire. To be sure, we do not always have a sufficient range of experience to inform us sufficiently about all the things we should like to know. It is incidentally interesting that Clausewitz, who did not know rapid technological change in the tools of war, rather inverts the point we should be inclined to make today concerning the utility of historical example to justify a change from accustomed usage. Where methods have proved highly effective and thus become fashionable, he says, we should want to have good reasons, meaning such as can be drawn from history, for changing them. The argument has an odd ring in an age of nuclear weapons—though it may well be that for changes less drastic his point still would hold. As we have seen, in an age of steam-driven, steel warships Mahan compounded a cogent and valid body of naval theory derived almost exclusively from naval warfare in days of sail.

There are, says Clausewitz, four different uses for historical examples. The first is simply to *explain* an idea which is otherwise too abstractly stated, and the second is to show the *application* of an idea so that we can see the play of minor circumstances such as would not be included in a general formulation. These two have in common the fact that the historical example need not be altogether accurate or authentic, because there is no desire to *prove* anything. In these cases the example is simply an expository aid.

In the third case one wants to prove the *possibility* of some phenomenon or effect, and in the fourth one wants to deduce a rule or doctrine, where the proof can only lie in the historical evidence itself. Here one needs accurate history, but in the third case a single example may suffice while in the fourth the demands are much more stringent. In the fourth case one needs a fair number of supporting cases and an absence or paucity of opposing ones, and moreover, in each case "care must be taken that every aspect bearing on the truth at issue is fully and circumstantially developed."

The main trouble with the common practice of touching lightly upon many different historical events is less that readers do not know enough about them to evaluate the author's judgment than that the author himself has never mastered the events he cites. "Such superficial, irresponsible handling of history leads to hundreds of wrong ideas and bogus theorizing." Better a single thoroughly detailed event than ten that are only touched upon. For these and related reasons, it is also better to use events from recent rather than remote military history. Not only is recent history better known, but the conditions, including those respecting armaments, are also more like the present. For Clausewitz this means examples beginning with the War of the Austrian Succession (1740–1748), which happened also to be the first of the wars of Frederick the Great. This encompasses only about seventy-five years of military history (which perforce ends with Waterloo), but he admits that more remote times, including antiquity, are not absolutely excluded, depending on what it is we want to derive from that history.

One notices Clausewitz's concern with the honesty of the process he is considering. Those who refer lightly to events remote in history rarely demonstrate "any honesty of purpose, any earnest attempt to instruct or convince." They are usually just showing off. But what would Clausewitz have made of most of the post-Second World War writers on strategy, who, far from making excessive pretensions to knowledge of military history, seem not to feel naked in the absence of any knowledge at all of such history?

Book Three: On Strategy in General

Book Three, Chapter One: The author now penetrates further into the deeper meaning of strategy as well as of military genius. The important thing is to keep one's eye on the goal, both as a participant and as a critic, and to see by what course of behavior and by what gifts of insight and decision the talented commander achieves his goal.

"Everything in strategy," Clausewitz says again, "is very simple; but that does not mean that everything is very easy. Once it has been determined, from the political conditions, what a war is meant to achieve and what it can achieve, it is easy to chart the course. But great strength of character, as well as great lucidity and firmness of mind, is required in order to follow through steadily, to carry out the plan, and not to be thrown off course by thousands of diversions."

In Frederick the Great's campaign of 1760 (mid-course in the Seven Years War) it is his rapid marches and maneuvers that are always praised, when the real proof of his wisdom is the fact that in "pursuing a major objective with limited resources, he did not try to undertake anything beyond his strength." The maneuvers are indeed worthy of praise, though not for the concepts they exposed but rather for the boldness, resolution, and strength of will that enabled this king to carry them out. Such miracles of execution, he says, can be spontaneously appreciated only by those with actual experience in war.

The latter thought, we may add, is the key to Clausewitz's constant emphasis on the importance of practical experience, an emphasis that has made some of his interpreters wonder whether he thought that anything of value for the theory of war could be got out of books! He expected most of his readers to have had the experience, and the others to understand what is lacking from the absence of it. "Those who know war only from books or the parade-ground cannot recognize these impediments to action; so we must ask them to accept on faith what they lack in experience."

The taking of a town or province may have little military value, but it adds to the constraints upon the enemy. On the other hand, if we lose sight of the ultimate goal and succumb to the idea that the captured provinces are of value in themselves, we may ignore the possibility that their possession may lead later to definite disadvantages. What is important is not the isolated advantage but the final balance. Clausewitz undoubtedly had many examples in mind from his own observation—like Napoleon's success in taking Moscow—but we have also seen in our own day the truth of his observations. The islands and other territories seized by Japan following her Pearl Harbor attack in December, 1941 turned out to be liabilities in the latter stages of the Second World War, when her garrisons were isolated and immobilized. More recently, Israel's conquest of the Sinai in the 1967 war resulted in a grave disadvantage in the 1973 war, for it meant, among other things, taking over extra logistics problems that had previously been solely the enemy's burden. Besides, the 1967 war had suggested that during battle it is better to have a terrible desert at the enemy's back than at one's own. By sending his army across that desert, President Nasser had too clearly revealed his hostile intentions, and when his forces broke under attack it was the desert that annihilated them.

Book Three, Chapters Two–Seven: The very brief Chapter Two must be interpreted simply as a disdainful rejection of some of Clausewitz's predecessors and contemporaries, especially von Bülow. Chapter Three, "Moral Factors," restates a theme which Clausewitz reemphasizes throughout his work and on which he knew he did not stand alone. However, he complains that other theorists neglect it, and it is true that his contemporaries, including Jomini, threw the emphasis on "principles," which is to say on procedures. Nevertheless, the glory always reaped by successful leaders in war has been a spontaneous recognition of the qualities that Clausewitz and other military writers have called "moral"—which, of course, in this sense has nothing to do with ethics.

Certainly, too, the great majority of writers on strategy since Clausewitz have not failed to recognize the importance of this quality even when discussing other things—sometimes, as in the case of Foch, making rather too much of it as compared with other things. Still, it was not vainglory that made Wellington exclaim to his friend Thomas Creevey on the morning after Waterloo that the battle would not have been won "if I had not been there." Nor did he doubt that with a partner less determined than Blücher to come to his assistance it would also not have been won. Clausewitz had witnessed that determination on that occasion—he was then chief of staff to Thielmann, one of Blücher's corps commanders—as he had witnessed it on many other occasions in the preceding two years.[3]

The author's discussion of the military spirit of the army amazes us for its modernity. In matters psychological Clausewitz is always the keenest and also the most measured of observers. "An army's efficiency," he says, "gains life and spirit from enthusiasm for the cause for which it fights, but such enthusiasm is not indispensable." Then, after delineating what it is that creates the "true military spirit" and describing its importance, he goes on to say, as only Clausewitz would among all the classic writers on strategy, "It cannot be maintained that it is impossible to fight a successful war without these qualities. We stress this in order to clarify the concept, so as not to lose sight of the ideas in a fog of generalities and give the impression that military spirit is all that counts in the end. . . . The spirit of an army . . . is a tool *whose power is measurable*" (italics

[3] During the great battle Clausewitz was nearly sacrificed to Blücher's spirit and to the strategic insight of the latter's chief of staff, General von Gneisenau, who was Clausewitz's friend and sponsor. Responding wholeheartedly to Wellington's appeal for assistance, Blücher left behind at Wavre, some fourteen miles east of Waterloo, only General von Thielmann's corps (with Clausewitz) to confront Marshal Grouchy. When the latter proved of greatly superior strength, Thielmann sent urgently after his chief for reinforcements; but Gneisenau, then approaching Waterloo, turned him down with the words, "It doesn't matter if he's crushed so long as we gain the victory here."

added). By "measurable" the author naturally does not mean that he can quantify it with numbers, but simply that the significance of even the most important things can be exaggerated. The same idea can be put truistically: where several qualities (in an army and its leadership) are important, no one quality can be all-important. However, other writers do not keep straight these simple truisms.

There are only two sources for military spirit: one is a series of victorious campaigns, and the second, "frequent exertions of an army to the utmost limits of its strength." The latter seems startling, but "a soldier is just as proud of the hardships he has overcome as of the dangers he has faced." We are hearing again the voice of long experience. Clausewitz knew very well what it meant to be in armies driven to the limits of their endurance, which by virtue both of their exertions and their victories could "survive the wildest storms of misfortune and defeat." We should take care, he says finally, "never to confuse the real spirit of an army with its mood."

In Chapter Six, "Boldness," however, he seems to have trouble making up his mind. He does not doubt that boldness is desirable in the lower ranks where it is also most readily found. There it can be held in check, but meanwhile it "acts like a coiled spring, ready at any time to be released." With the top ranks the situation is different, and here the author displays a quite unwonted ambivalence. Obviously, we have a Clausewitz who is struggling with his own temperament.

He knew the value of boldness in a commander, but he could not help wanting to see that boldness "disciplined by reflection." The result is a series of fine distinctions which are hardly practical even if meaningful—for example, the threefold distinction between "boldness," "deliberate caution," and "timidity"—and another series of statements which are plainly contradictory. In the net he wants to be recorded in favor of boldness, and he makes some eloquent statements in favor of that disposition. But then he says: "Boldness governed by superior intellect is the mark of a hero. This kind of boldness does not consist in defying the natural order of things and in crudely offending the laws of probability." What it does consist of is then described in somewhat mystical terms and generalities, a quality normally much more characteristic of other nineteenth and twentieth century writers on strategy than of Clausewitz. In the end we know that he believes "a distinguished commander without boldness is unthinkable," and he also calls this quality "the first prerequisite of the great military leader." It is at any rate remarkable that he should have been able to pay such tribute, albeit modestly qualified, to a quality he probably knew he did not possess.

The very brief chapter, "Perseverance," however, is free of ambiguities. The man who had taken part in the Russian campaign in 1812 and in the campaigns in western Europe in 1813, 1814, and 1815 knew what he was speaking of when he said "there is hardly a worthwhile enterprise in war whose execution does not call for infinite effort, trouble, and privation."

Book Three, Chapter Eight: In this chapter, "Superiority of Numbers," we are clearly back to the Clausewitz of the sure touch. He first tells us what we already know, that a brilliant commander has often won an engagement over an inferior opponent even though greatly outnumbered. But, he points out—what we may too easily forget—there is a limit to which brilliance of leadership can compensate for inferiority in numbers, at least among European armies which tend to be comparable in other respects. Thus, examples show that "even the most talented of commanders will find it difficult to defeat an opponent twice his strength." We must, therefore, acknowledge that in ordinary cases "a significant superiority of numbers . . . will suffice to assure victory, however adverse the other circumstances." This means that the first rule of strategy should be to put the largest possible army into the field.

But why are we treated to such a platitude? Because, Clausewitz assures us, in his time it was not really a platitude. He gives us the intriguing information that to most military historians down to the end of the eighteenth century, the strength of armies seemed not to be of major significance. They rarely mentioned it, even when they offered all sorts of other details. Some authors seemed even to believe "that there is a certain optimum size for an army, an ideal norm, and that any troops in excess of it are more trouble than they are worth." Besides, we know that there were battles in which not all forces available were used "because numerical superiority was not given its due importance."

Thus we have the dual principle that one should operate with the largest forces possible, and that the forces available must be employed with such skill "that even in the absence of absolute superiority, relative superiority is attained at the decisive point." This object obviously requires appropriate "calculations of space and time," but obviously too, such phrases had already become clichés in Clausewitz's time. "Let us not," he says "confuse ourselves unnecessarily by conventional jargon."

Then he finishes with the characteristic measured appraisal. Superiority of numbers is of fundamental importance, but even that degree of importance is still only relative. "The principle is served if we use the greatest possible force; the question whether to avoid a fight for lack of strength can be decided only in the light of all other circumstances."

Book Three, Chapter Nine: Clausewitz is in agreement with all other writers on strategy concerning the universal desirability of achieving surprise, but he nevertheless thinks that the common emphasis on it is somewhat overdone. The reason is that it is much more difficult to achieve than is generally assumed. "The principle is highly attractive in theory, but in practice it is often held up by the friction of the whole machine." Its success is often due to favorable circumstances beyond the control of the commander, and it is frequently at the mercy of chance.

Surprise is also primarily a tactical device. "In strategy surprise becomes more feasible the closer it occurs in the tactical realm, and more difficult the more it approaches the higher levels of policy." Thus, while "surprise lies at the root of all operations without exception," it does so "in widely varying degrees depending on the nature and circumstances of the operation."

The author presents some examples from his reading of military history—which was often significantly different from that of the military historians of his time, who tended always to inflate the instances of surprise—but we can use some examples from our own time which bear out convincingly what Clausewitz had to say on the subject, especially with respect to the greater ease of achieving surprise as we go from strategy to tactics, and also the bearing of favorable circumstances.

In the spring of 1944 nothing was more obvious than that the British-American Allies would soon land on the northern coast of France. The Allied High Command had gone to great lengths to conceal their intentions, but it was impossible for many months before the event to conceal the fact that the invasion was going to occur, and the Germans had good grounds for guessing the approximate time and the approximate area (that is, the Channel coast rather than the Atlantic coast of France). Thus, in the higher strategic realm there was no surprise. The Allies did achieve some surprise in the exact timing of their landings (otherwise Field Marshal Rommel would not have been away from his post) and also the exact locations; but the most important surprise achieved was in the fact that the original simultaneous landings on the Normandy coast were not the feint that the Germans thought they might be but the whole affair. Thus, the Germans held in reserve too long some divisions that would have been much more effective if thrown earlier into the fight.

The Japanese attack at Pearl Harbor was indeed a great surprise, strategic as well as tactical—more so than it should have been. The United States Government knew, and its senior commanders in the field were warned, that war was imminent. It was taken as likely that the Japanese would initiate the war with a sudden attack somewhere, as they had initiated the Russo-Japanese War in 1904 with a surprise attack upon the Russian Far Eastern Fleet at Port Arthur. Why, then, should not reasonable precautions have been taken against exactly the kind of attack that did occur? The tactical method of attack was a surprise, though it should not have been. Such an attack had even figured in a U. S. fleet exercise some years earlier but had effectively been put out of mind.

The best safeguard would have been to keep the major units of the fleet as much as possible at sea, as the American carriers happened to be on the day of the attack (in this respect chance operated against the Japanese). But the American fleet commander and his aides apparently could not think in terms of Japanese aircraft over a naval base so far from Japan. Thus, even disregarding the remarkable chance circumstance that a radar sighting of incoming aircraft was ignored, the Japanese reliance on surprise was a fairly desperate gamble which they happened to win.

As it turned out, the Japanese could hardly have more effectively inspired a warlike determination and unity in the theretofore divided United States. That was an excessive price to pay for destroying or disabling some old battleships that were relatively soon repaired or replaced —but now we are in the realm of the connection between strategy and politics, which Clausewitz treats in other chapters.

Book Three, Chapter Ten: In his brief chapter, "Cunning," or the use of stratagems, Clausewitz is again strikingly negative, for reasons which have to do with the economy of war. He is speaking of strategic rather than tactical stratagems, and he feels that the only kind that is likely to have much effect by way of deception is one that deploys not words but a substantial force. But, he says, "it is dangerous . . . to use substantial forces over any length of time merely to create an illusion; there is always the risk that nothing will be gained, and that the troops deployed will not be available when they are really needed."

He seems to make some curious connection between capacity to deceive in war and the general's character. We notice it, for example, in his conclusion "that an accurate and penetrating understanding is a more useful and essential asset for the commander than any gift for cunning— though the latter will do no harm so long as it is not employed, as it all too often is, at the expense of more essential qualities of character." He does, to be sure, seem to commend cunning when the commander is so weak as to be otherwise without hope.

Although his argument is unquestionably sound so long as he is talking about using substantial forces separately for diversion, not all stratagems have been of this nature. Clausewitz was certainly familiar with instances where Marlborough and Frederick the Great successfully used deceptive movements of whole armies, and other ruses as well.

An outstanding modern example of a clever and successful stratagem, conforming with though probably not inspired by Clausewitz's ideas, occurred at the naval Battle of Leyte Gulf in October 1944. Admiral Ozawa presented a decoy force to the north of the Philippine Archipelago which lured Admiral Halsey and his entire Third Fleet away from the strait 300 miles to the south through which Admiral Kurita, who had already been sighted and attacked, would sortie on the following morning. Ozawa's mission was apparently based on the hope that Halsey would be ruled by two clichés then dominant in the American fleet—and he was. The first of these, valid in the earlier stages of the war but now obsolete, was the maxim: "the enemy's main force is where his carriers are." The carriers Ozawa had were the only four left in the Japanese navy, and of these three were very small and the fourth not large; the aircraft they carried were few and manned by poorly trained pilots. Moreover, Halsey's staff had the information to deduce these facts. Ozawa's force was thus too weak to serve as anything but a decoy, and Ozawa had no confidence it could do even that. Kurita's seven powerful battleships and eleven heavy cruisers as well as other ships undoubtedly constituted the main Japanese force.

The other cliché to which Halsey fell victim was that against dividing the fleet in the presence of the enemy—the old principle of "concentration of force." With his overwhelming strength he could easily have divided his fleet and still presented a vastly superior force to each of the forces led by Ozawa and Kurita, as two of his subordinates tried to hint he should do. Kurita did sortie through San Bernardino Strait and was in a position to wreak havoc with the American shipping and moderate naval forces covering the landing on Leyte, but he lost his resolve and at the last moment turned back. Thus, despite Ozawa's success in his ruse, Halsey escaped the full penalty for his error; but his cost was that of failing to destroy either Ozawa's force (he had been obliged to turn around before reaching him) or Kurita's.

Book Three, Chapters Eleven–Thirteen: The extraordinary brevity of Chapter Eleven, the "Concentration of Forces in Space," suggests that this was an initial statement which Clausewitz intended to fill in with his final revision; but it is possible also that he considered this subject so well known and agreed upon that he did not wish to waste undue space on it even in a finished version. Here we have Clausewitz, normally so disdainful of "principles of war," accepting one of them without hesitation or much qualification as basic. The qualification which does appear is, however, extremely important. Although "there is no higher and simpler law of strategy than that of *keeping one's forces concentrated*," that does not mean keeping them concentrated when there is a definite and urgent need to do otherwise—a subject he promises to amplify in a later chapter (he does, modestly, in several chapters, but especially Chapter Nine of Book Eight).

The next chapter, "Unification of Forces in Time," takes up on the tactical level the importance in battle of keeping some forces in reserve. Thus, one has one's forces concentrated on the field of battle, but one does not use them all at once. Later, in the discussion of "economy of force," we shall learn that it is equally important that one should see that all are used. Clausewitz's discussion of concentration or unification of forces on the strategic level lacks clarity, which may be due to its being unfinished. He seems to be saying that inasmuch as a comparable need for keeping fresh troops in reserve does not exist on the strategic level, all forces available should be put in motion toward the strategic goal from the onset of the war. His meaning gains some clarification from his example of Napoleon's march into Russia in 1812, where the author is apparently answering criticisms that among Napoleon's errors was that of using too large an army. Clausewitz argues that it was the method of advance, along a narrow rather than a broad front, that should be criticized rather than the size of the army. Napoleon could not know the minimum level of forces that would be clearly sufficient, so he was right to send in as large an army as he could gather. Anyway, as the author points out in another chapter, the force that finally arrived at Moscow was much diminished from that which crossed the Niemen.

There is a certain superficial parallel between the argument Clausewitz is making here and the criticisms voiced by some military persons concerning the slow buildup of United States military forces in Vietnam following the decision in March 1965 to send in combat forces. There critics say that the reserves should have been called out at once and that the maximum level of over half a million men in Vietnam reached only in 1968 should have been reached much sooner. This argument lacks merit, partly because it implicitly denies the legitimacy of domestic political constraints affecting United States participation in Vietnam and even more because it makes no effort to explain what that number of men could have accomplished in 1965 that it failed to accomplish in 1968. Nevertheless, the Clausewitzian idea described above deserves notice.

The following chapter, "The Strategic Reserve," seems in the beginning to be contradictory to the previous one, but this impression is quickly dispelled. There can, the author says, be some use for a strategic reserve, but only when emergencies are conceivable. When, one may ask, are they not conceivable in wartime? The author answers that "uncertainty decreases the greater the distance between strategy and tactics; and it practically disappears in that area that borders on the political." Again his meaning is clarified by his example—and we sense again what a searing experience for him was the Jena defeat of 1806. During that campaign, in which the identity of the enemy and his purpose were clear and not subject to change, the Prussians kept two substantial forces in strategic reserve which never saw action.

Book Three, Chapter Fourteen: The example just given from the Jena campaign could serve also to illustrate the argument of the following and very brief chapter, "Economy of Force." We see again that Clausewitz's rejection of the idea that a theory of war can rest on a body of "principles" of near-universal application does not prevent him from including those principles in his own theory.

The "principle of economy of force," treated at some length by Jomini, has continued to appear in the standard lists of principles down to the present day. Significantly, however, recent writers usually show their lack of understanding of the matter by twisting the meaning of the word "economy" almost to the opposite of that originally intended, which was not that of economizing or of niggardly use but rather of effective use. In war, Clausewitz says, one must "make sure that all forces are involved . . . that no part of the whole force is idle." Forces which are not busy with the enemy "are being wasted, which is even worse than using them inappropriately," for forces which are used inappropriately are at least occupying some of the enemy forces and reducing his overall strength. Most writers in our time have mistakenly interpreted the old phrase "economy of force" to mean applying only the minimum necessary strength for the task—a concept that makes sense only for diversionary or holding operations, where the object is to maximize concentrations elsewhere.

It seems that every major war presents its classic examples of violation of the principle of economy of force. In the battles of Ligny and Quatre Bras two days before Waterloo, Count D'Erlon with 20,000 men marched back and forth between the two battlefields because of a confusion of orders and took part in neither action. Napoleon desperately wanted him at Ligny. During the Waterloo battle Wellington never recalled Prince Frederick of the Netherlands with some 17,000 men whom he had sent to a position some ten miles to his right because he had originally expected Napoleon to come from that direction. However, Napoleon outdid him in wasting troops by again dictating a confusing order which detached 33,000 men, later additionally reinforced, to march under Grouchy against Wavre, from which Blücher and most of his forces had already departed to march on Waterloo. The famous example from the First World War was von Moltke's detaching two army corps from his right wing advancing into France in order to send them to the forces resisting the Russian incursion into East Prussia, but these large forces were still on their trains moving across Germany when the decisive battles were fought both in the East, where without them the Germans annihilated two Russian armies at Tannenburg and the Masurian Lakes, and in the West, where the Germans felt their absence in suffering the defeat of the Marne which spelled the ruination of the Schlieffen Plan. We have already given a naval example from the Second World War, when Admiral Halsey took his great Third Fleet on a high-speed dash three hundred miles to the north against Ozawa, but stopped before reaching him and turned to pursue Kurita, whom he also never reached. The Third Fleet on that occasion used up plenty of fuel but no ammunition, and as an organized force it never got another chance. Some of these examples border on the tactical rather than the strategic, but the demarcation is not hard and fast and anyway the principle is the same. There are all kinds of ways of failing to make one's available forces fully effective, and resorting to movements in the wrong direction is only one of them.

Book Three, Chapters Fifteen–Eighteen: The brief chapter, "The Geometrical Factor," is little more than a rejection of the notions of von Bülow, who wanted to make strategy "more scientific" by discussing it in terms of geometry. Clausewitz allows that it may have some relevance to tactics but virtually none to strategy. "We believe," he says, "that it is one of the chief functions of a comprehensive theory of war to expose such vagaries."

In the chapter that follows, "The Suspension of Action in War," Clausewitz is developing further a theme he has already expressed in the great opening chapter of Book One. How does it happen that war, which inherently calls for relentless progress toward the goal which is its object (at least for the side initiating it), is so often marked by inactivity? Part of the reason has already been expressed in that earlier chapter—that is, the decision of the stronger side to postpone its offensive does not necessarily make it appropriate for the weaker to shift to the offensive—but there are also more fundamental reasons. In the past the absence of movement has been mostly due to the absence of energy, that is, to the "chains of human frailty." With Napoleon we saw how much energy it was possible to achieve, "and if it is possible, it is necessary."

The main reasons for the failure of the requisite energy have to do, on the one hand, with the fear and indecision that are native to the human mind when confronted with risk, and on the other hand to the greater strength of the defensive. However, there have also been wars in which both sides lacked spirit because there was no great impetus of interest. But it is a great mistake to regard these half-hearted kinds of endeavor as representing "the real, authentic art of war." To some of Clausewitz's contemporaries the Napoleonic wars were already being regarded "as crude brawls which can teach nothing, and which are to be considered as relapses into barbarism." These people did not know what war truly is. "Woe to the government, which, relying on half-hearted politics and a shackled military policy, meets a foe who, like the untamed elements, knows no law other than his own power!"

The next chapter carries this argument further. Napoleon showed the world the real nature of war, and he prompted his opponents to rise to the same heights of decision. Before his time the diplomats used to rush to conclude a peace, however bad, whenever their side had lost a few battles, but Russia in 1812 and Prussia and other nations in 1813 showed "what an enormous contribution the heart and temper of a nation can make to the sum total of its politics, war potential, and fighting strength. Now that governments have become conscious of these resources, we cannot expect them to remain unused in the future." This is a theme to which Clausewitz will return several times.

And in the final chapter of this book Clausewitz analyzes what can be achieved through understanding the dynamic nature of the changes between tension and rest in war. "Any move made in a state of tension will be more important, and will have more results, than it would have if made in a state of equilibrium." Again the author uses the Jena campaign of 1806 as an example. "During that period of enormous tension, events were pressing towards a major decision which, with all its consequences, ought to have absorbed the full attention of the [Prussian] commander"; yet at that very time the Prussian leaders were dissipating their energies on confusing schemes.

"The state of crisis," Clausewitz says at the end, "is the real war; the equilibrium is nothing but the reflex."

Book Four: The Engagement

Book Four, Chapters One–Two: Where Book Three was concerned with the "operative elements" in war, Book Four is concerned with its "essential military activity," which is the engagement. One could say that the shift is from strategy to tactics, but it is obvious that Clausewitz does not sharply separate the two, and anyway he is interested in the analysis of war and not just of strategy. It is thus easier for him to conclude that "a change in the nature of tactics will automatically react on strategy," a view that separates him considerably from Jomini and the latter's followers, who have been so fond of repeating the axiom that "methods change but principles are unchanging." But we have already seen that the principles that Clausewitz would grant to be little affected by changes in methods are only those of the most fundamental kind, like the principles of economy of force and of concentration, each of which he was able to dispatch with a few paragraphs. Clearly in his mind such concepts are only the linchpins of strategy and not the substance.

When one considers the changes in strategy later effected by the breech-loading rifle, which because it permitted firing from a prone position increased the value of cover and hence the ability of small forces to delay or hold back larger ones; of the machine-gun, which carried the process much further and completely transformed the First World War; of submarine warfare, which had a comparable effect on the seas in the same war, an effect which Mahan among many others completely failed to foresee; and of aircraft, whether carrier-based or land-based, which had a similar effect on the Second World War (mention of nuclear weapons in this context is really superfluous), one has to agree that it was Clausewitz who was right on this matter and one marvels that he was able to see it so clearly even from the relatively minor changes in tactics that occurred during his lifetime.

Book Four, Chapters Three–Four: The first problem that Clausewitz takes up concerning the engagement is the question: What do we mean by the defeat of the enemy? The answer for the engagement is the same as for the war of which it forms a part—the destruction of the enemy forces. "Admittedly, an engagement at one point may be worth more than at another. . . . That is what strategy is all about." Still, he wants to establish at the outset the "dominance of the destructive principle"—which he feels needs being established because there have been some long-held contrary notions.

The destruction of the enemy's forces must of course be disproportionate in some meaningful way to the destruction of one's own. Clausewitz finds that during the course of an engagement "the winner's casualties . . . seldom show much difference from the loser's"—one of the characteristics which may have distinguished his time from our own, when there is greater likelihood of significant differences in equipment and tactics. In his time, "the really crippling losses which the vanquished does not share with the victor only start with his retreat." It is the pursuit and the fruits which fall to it that make the difference.

During battle the losses are mostly in dead and wounded, which winner and loser may share fairly evenly, but after battle they are mostly in guns and prisoners, which are far greater on the side of the loser. "That is why guns and prisoners have always counted as the real trophies of victory; they are also its measure, for they are tangible evidence of its scale."

The question then is what causes one side or the other to turn from battle to retreat, thus exposing itself to the heavy penalties which enemy pursuit will bring? The loss of ground and the failure to produce fresh reserves may bring the conviction of defeat, which means finally that "the loss of morale has proved the major decisive factor." The loss of morale is made worse by retreat or flight and the enemy's pursuit. But here much depends on circumstances.

At one point in his brilliant and detailed discussion of these matters, which rests mainly upon his great personal experience, Clausewitz observes: "Casualty reports on either side are never accurate, seldom truthful, and in most cases deliberately falsified." That is one respect in which his times differed not at all from ours.

Clausewitz finally distinguishes between the kind of victory which represents for the defeated side a measured and redeemable loss, and that rarer kind of victory on a grand scale which results in the utter debacle of major enemy forces. As examples of the latter, he names his unforgettable Jena and also Waterloo (which the Prussians named after the inn called "Belle-Alliance" near the battlefield), and as an example of a large-scale engagement which ended with only nominal defeat for one side he mentions Borodino, the battle on the road to Moscow where Marshal Kutuzov shifted rather than yielded his position (and which Tolstoy in his *War and Peace* glorified as a victory for Kutuzov).

Later in the nineteenth century the French officer Ardant du Picq, who was to lose his life in 1870 at the head of his regiment, carried much further the ideas that Clausewitz expressed concerning the risk of high penalty in yielding to an enemy in battle and turning to retreat or, worse, to flee. Du Picq arrived at his views independently from studying famous battles of antiquity, where the loser seemed always to suffer enormously greater losses than the victor. He decided that it was because in turning to flee the losers presented their defenseless backs to their opponents, who pursued and slaughtered. Quotations from du Picq's conclusions were later encapsulated into mottoes, like: "He will win who has the courage to advance." At the turn of the century the French school of Foch and his followers fed these mottoes into the doctrine of *l'offensive à outrance* with which they fought the First World War—under conditions of engagement vastly different from those of antiquity.

Book Four, Chapters Five–Ten: These chapters require little comment, being clear enough and also not particularly memorable. Clausewitz continues to describe the battle that was characteristic of his time, where a single day was enough to use up reserves and where nightfall provided cover for the retreat of the loser. The battle wears down both sides, but one side has in addition the agony of defeat, which is no doubt why Clausewitz finds that "the outcome of a major battle has a greater psychological effect on the loser than on the winner."

Book Four, Chapter Eleven: This chapter, "The Use of the Battle," must be considered in the light of some of the military philosophies of the pre-Napoleonic era, which to Clausewitz looked fairly recent. The great Marshal de Saxe, who died in 1750, had written in his posthumously published *Mes Rêveries*: "I do not favor pitched battles, especially at the beginning of a war, and I am convinced that a skillful general could make war all his life without being forced into one." Frederick the Great, too, became in his later years less given to offensive ardor and to the major battle, which he found left too much to chance. Napoleon was the apostle of the decisive battle, but the one he fought in 1815 proved his utter ruin. Obviously, what Clausewitz was reading and hearing from some contemporaries made him fear the renewed ascendancy of the notion that waging campaigns without battles was "evidence of higher skill." This line of thought, he says, "has brought us almost to the point of regarding . . . a battle as a kind of evil brought about by mistake."

He recognized that "the character of the battle . . . is slaughter, and its price is blood," and that for that reason the commander as a human being will recoil from it. Still, inasmuch as the military aim in war is to destroy the enemy's forces, the battle is the only way to accomplish it. It is similarly wrong to try to restrict military action to a series of minor engagements, for these tend to equalize losses and draw out the affair. Clausewitz in this chapter is justifying not simply continued fighting but the *great* battle, the kind that tends to decide the outcome of the war. No other factor in war rivals the battle in importance, "and the greatest strategic skill will be displayed in creating the right conditions for it, choosing the right place, time, and line of advance, and making the fullest use of its results."

Book Four, Chapter Twelve: Here Clausewitz goes deeper into an issue he has already treated in Chapter Four of this book, the imperative need for pursuit after victory, and the reasons why generals so often fail to accomplish it.

One of the major reasons is the fatigue and disorder which the victors share almost equally with the losers. Clausewitz, one notices again, is more than almost any other writer on strategy aware of the importance of exhaustion, hunger, and general misery in the soldier—one of the factors on which he bases his elemental conception of "friction" in Book One. The general's own energies have been sapped by physical and mental exertion, which inclines him to yield to entreaties for rest and recuperation. What does get accomplished, he says, "is due to the supreme commander's *ambition, energy*, and quite possibly his *callousness*."

Also, in earlier wars, "smaller in scope and more narrowly circumscribed," conventions had developed which restricted all kinds of operations and especially pursuit. "*The very idea, the honor* of victory, appeared to be the whole point. . . . Once a decision had been reached, one stopped fighting as a matter of course, further bloodshed was considered unnecessarily brutal." But such a view could prevail "only where the fighting forces were not considered to be the vital factor," for nothing is clearer than that during pursuit the fleeing forces suffer disproportionately.

In describing examples from history, Clausewitz makes a characteristic detour to explain why Napoleon's action in *not* pursuing Kutuzov's army after Borodino was justified. At that point his overriding concern had to be in reaching Moscow without more losses to his already considerably depleted army. But this was a thoroughly exceptional situation.

Also by way of indicating a justified exception to a rule, Clausewitz in his long discussion of the various forms of pursuit takes occasion to assert that at such a time "the victor must not be afraid to divide his forces in order to envelop everything within reach of his army. . . . He may do whatever he wants until the situation changes; the more liberties he takes, the later that moment will come."

Again his historical examples include Jena and Waterloo (Belle-Alliance), the latter being one of the classic pursuits in history, where the tired Prussians under Blücher took over from the utterly spent British-Dutch army under Wellington, and in the night following the battle Napoleon's army disintegrated. An example of a great opportunity missed came some thirty-two years after Clausewitz's death in the Battle of Gettysburg, when after three days of fighting Meade permitted Lee to withdraw without pressing him, though the swollen condition of the Potomac held up the latter from crossing for seven days. Meade followed but though reinforced and greatly superior did not attack, and Lincoln, in agony at Lee's escape, began again his search for an aggressive commander, looking now at the general who had taken Vicksburg.

Book Four, Chapters Thirteen–Fourteen: These concluding chapters of Book Four are rather less consequential. Chapter Thirteen considers the strategy of the side forced into a retreat after a lost battle. First of all, one takes measures to disengage before one's fighting capital is all used up, so that the withdrawal can be orderly and full of menace for the pursuer. What is then most imperative is not putting maximum distance between oneself and the superior enemy but rather preventing the retreat from becoming a rout. Chapter Fourteen is in general a warning against attempting large scale attacks at night. They are too risky and too difficult to execute.

In these chapters, as indeed throughout the whole of Book Four, the reader is called upon rather more than in the three previous books to distinguish between Clausewitz's own times and our own, and this element will continue to characterize the work until we arrive at Book Eight.

Book Five: Military Forces

Book Five, Chapters One–Five: Like the preceding book only more so, Book Five deals with some of the narrower technical issues of war and hence will appear somewhat dated in its subject matter. It will thus appeal more to the military historian than to the modern student of war. Still, as we shall see, it is not without passages of extraordinary interest to the latter.

In Chapter Three Clausewitz returns again to the idea that even the best of generals rarely produces a victory when he brings inferior numbers to the battle. Nevertheless, he says, "war is not always the result of a voluntary policy decision," and if one has to fight with inferior numbers, "it would be a peculiar theory of war if it broke off just where the need for it was greatest." But all he has to come up with at this point is the admonition that the more restricted the strength, the more restricted the goals and the duration must be. "If an increase in vigor is combined with wise limitation in objective, the result is that combination of brilliant strokes and cautious restraint which we admire in the campaigns of Frederick the Great."

Chapter Four, "Relationship Between the Arms of the Services," is fascinating because of the author's groping toward that which is one of the most modern developments in military studies—what we call "systems" or "cost-effectiveness" analysis, which is somewhat related to what the economist calls "marginal-utility" analysis. Granted that infantry is the most versatile and indispensable of the arms, one nevertheless also has need for artillery and cavalry. The question then arises, what constitutes the optimum relative proportions? Then follows a dramatically modern insight: "If one could compare the cost of raising and maintaining the various arms with the service each performs in time of war, one would end up with a definite figure which would express the optimum equation in abstract terms." But, he adds, "this is hardly more than a guessing game." And we may add that despite some interesting and useful modern refinements concerning more limited areas of decision, the kind of problem that Clausewitz poses is still today a guessing game. Clausewitz, however, goes on with the notion that money is the common unit of account in seeking the "optimum equation" referred to above, thus coming ever closer to the modern conception of cost-effectiveness analysis:

"But since . . . we cannot quite dispense with all standards of comparison . . . we shall simply make use of the only ascertainable factor: the money cost. For our purpose it will suffice to state that, according to common experience, a squadron of 150 horses, a battalion of 800 men, and a battery of eight six-pounders cost approximately the same as to equipment and maintenance." Unfortunately (as is true today), the effectiveness side of the equation is far more difficult to work out. "It might conceivably be possible if destructiveness were all that had to be measured; but each branch has its own particular use and thus a different sphere of effective action."

Then, to add to the delight of the modern theorist, who is often a youthful civilian attempting to persuade a senior military officer that he has a superior solution to some of the problems that the latter has tended to solve by "mature military judgment": "People often talk of the lessons of experience in this context, in the belief that the history of war provides sufficient grounds for a definite answer. But that is obviously a matter of empty phrases which, since they cannot be traced back to any fundamental and compelling basis, are not worth considering in a critical investigation."

Clausewitz, however, is tenacious, and the rest of Chapter Four is given over to a carefully reasoned approach to that theoretical problem which he has just admitted defies any kind of comprehensive solution. His method is as interesting as his conclusions, which are summarized at the end of the chapter in four numbered propositions. Most interesting is his insight on the declining value of cavalry relative to infantry.

Chapter Five, "The Army's Order of Battle," is also dated in particulars but timeless in the nature of the problems that the author seeks to solve. Here he concerns himself with the nexus between tactics and strategy respecting primarily two problems: the need to increase the flexibility of an army, and the desirability of shortening the chain of command. In his own lifetime he had seen enormous changes in the organization of armies, the trend being from huge, unwieldy masses into smaller units of which each had its own component of the three arms—that is, infantry, artillery, and cavalry. An army organized into corps and divisions is more easily maneuvered, and portions thereof lend themselves more readily for detachment on separate assignments. The questions then are: how many corps and divisions should there be, and what should be their respective sizes? To Clausewitz the answers are found in the characteristics of the command problem, rather than in any fancied optimum size for the units. The supreme commander would prefer to deal directly with division commanders, but in a large army he may have to insert corps commanders between himself and the division commanders, lest the division and its component brigades become too unwieldy. Thus, the key sentence of the chapter holds "that the number of subdivisions with equal status should be as large as possible, and the chain of command as short as possible; the only qualification being that command is difficult to exercise over more than eight to ten subdivisions in an army. . . ."

Book Five, Chapters Six–Thirteen: These chapters have little for the modern student of war, and Clausewitz himself apologizes at the end of Chapter Thirteen for having put forward considerations which "are obviously more of a tactical than a strategic nature"; but, he says, he "thought it better to stray into the tactical field than to take the risk of not making ourselves clear." Still, one remembers that the army dispositions and encampments he talks about in Chapters Six–Nine, and even more so the marches treated in Chapters Ten–Twelve, give us insights about some of the problems of disposition and mobility confronting armies in the field up to and including the opening battles of the First World War. Thus, any student of military history who is not already familiar with these matters will gain from reading them. We learn enough here about forced marches and about the assembly of troops at the opening of great battles to comprehend what these phrases mean when they are tossed off by historians, who often themselves do not fully understand them.

We also learn, as usual, the relevant changes that Clausewitz was witnessing in his own times. It was only a generation before his time that "the artillery went its own way in order to travel on safer and better roads, while the cavalry generally alternated on the flanks in order to give every unit in turn the honor of riding on the army's right." We also notice the author's sensitivity to the consequences of such changes as giving up that part of the baggage train which had previously carried tents. It meant more mobility for the army, and the horses previously used to draw the wagons could draw more guns or carry more cavalry men. However, "the protection afforded by a roof of cheap canvas may not be much, but over a period of time it is a relief that will be missed by the troops when it is not there." The difference is slight for a single day, but over many days it begins to matter, and "increased losses due to sickness will naturally result."

The same kind of disciplined sensitivity is evident also in his discussion of the heavy losses in the army's strength that inevitably follow from forced marches extending over more than a very few days, and sometimes even within a quite short period. After poignantly describing the lot of the soldier who falls ill on the road or who suffers parching thirst and exhaustion from a march in the heat of summer, he adds: "None of this is meant to say that there should be any less activity in warfare; tools are there to be used, and use will naturally wear them out. Our only aim is clarity and order; we are opposed to bombastic theories which hold that the most overwhelming surprise, the fastest movement, or the most restless activity costs nothing." One thinks about the execution of the Schlieffen Plan in August of 1914 and wonders about the state of the German reservists by the time they had walked, each with a 65-pound pack on his back and a rifle, through Belgium and northeast France to the Marne. One wonders also whether their condition, and the losses in stragglers of the German armies by the time they reached the latter position, had been adequately taken account of in the Plan. How many historians have raised these questions? Fortunately for the Germans when they met the French and British along that river, the latter had been on some excursions too.

The latter part of Chapter Twelve has some remarkable figures on the losses in numbers of troops from forced marches during the campaigns of 1812 and 1813, in which Clausewitz took part; and gathering added interest also because of his participation (under Thielmann) is his description in the latter part of Chapter Thirteen of the assembly of the widely spread Prussian army on the eve of the Battle of Ligny, which preceded that of Waterloo by two days.

Book Five, Chapter Fourteen: Though still strictly of historical value, this long chapter on the provisioning of armies in the field has a special interest. We have heard so much of great armies living off the land, of their "moving on their bellies," and of the great changes particularly in this respect that distinguished the campaigns that Clausewitz saw from those of earlier times, that we welcome the opportunity to learn something of what was involved.

Since the time of Louis XIV (who died in 1715, just a century before Waterloo), armies had become very large. More important, the campaigns within any war had only recently become much more closely interconnected, no longer separated by long periods of inactivity. Thus, the old system of relying on depots while at rest and on huge wagon trains during the campaigns simply could not work. The new emphasis on rapid movement required systems of requisitioning that themselves depended on movement. Clausewitz barely hints at one of the other things it depended on—a certain ruthlessness toward the foreign population through which one's armies passed. He speaks, however, of another kind of ruthlessness, that of the commander toward his own troops. "What can be more moving," he asks, "than the thought of thousands of soldiers, poorly clad, their shoulders bent under thirty or forty pounds of pack, plodding along for days on end in every kind of weather and on every kind of road, continuously endangering their health and their lives, without even a crust of bread to nourish them? When one knows how often this happens in war, one must marvel at the fact that heart and strength do not give out more often." Later he observes that "a horse will perish from want much sooner than a man."

In the same vein he adds: "If war is to be waged in accordance with its essential spirit—with the unbridled violence that is at its core, the craving and need for battle and decision—then feeding the troops, though important, is a secondary matter." He even quotes with approval Napoleon's impatient insistence *qu'on ne me parle pas des vivres!* ("let no one speak to me of provisions!").

Yet in the next sentence he admits that the latter's Russian campaign "proved that such neglect can go too far." For it is undeniable that "the lack of care over supplies must be held responsible for the unprecedented wastage of his army on the advance, and for its wholly calamitous retreat." Napoleon forgot "how vast a difference there is between a supply line stretching from Vilna to Moscow . . . and a line from Cologne to Paris."

Book Five, Chapters Fifteen–Sixteen: These chapters bear an integral relationship to the important one on provisioning which immediately precedes. They are also less dated, because in them we move away from a conception of living off the land by requisitioning, which today is done only by guerillas or partisans. In these chapters as in the previous one we also observe something of the experience of the author as a staff officer.

Chapter Fifteen, "Base of Operations," brings us back to the realization that even where provisions can be taken or requisitioned, other military supplies and replacements cannot be. These can only be brought forward from a base, which thus becomes an integral part of the army, the relationship being, in Clausewitz's view, comparable to that of the tree with its roots. On the other hand, the army must be mobile (which the tree is not), and the base must not be too far behind it. Neither must it be too vulnerable to enemy attack, which in deep penetrations into hostile territory obviously makes for some problems.

Chapter Sixteen, "Lines of Communications," deals with the linkage of the army to its base. Such lines provide the means of advance and also the lines of retreat. The relevant problems deal naturally with roads—that is, with their length, direction, and quality, but also with the terrain through which these roads run, the condition and temper of the local inhabitants, and finally the amount of protection to communications that can be afforded by fortresses, where available, and by garrisons. The implication throughout is that the enemy's army is always somewhere out in front of one's own, so that unless he carries out successful flanking operations—"which have always been more popular in books than in the field"—the protection accorded to depots set up along the lines of communication is sufficient if it can cope with modest detachments from the enemy's main force or with partisans.

Clausewitz would no doubt have modified his tone if he could have foreseen Stonewall Jackson's march around Pope's army to destroy the vast federal depot of Manassas in the events leading up to the second battle of that name.

Book Five, Chapters Seventeen–Eighteen: These two chapters, "Geography and Ground" and "The Command of Heights," reveal an author apparently less interested in the subjects he is treating than he had been in the preceding chapters. An occasional memorable statement will be a digression, as for example: "In war the sum-total of individual successes is more decisive than the pattern that connects them." In Chapter Seventeen we learn little besides the fact that rough country makes for fragmentation of forces and hence favors the side that can depend more on individual initiative. It is therefore to be avoided by forces which are at an advantage when fighting as a concentrated mass. Also, in difficult terrain infantry is undoubtedly the ascendant arm.

Chapter Eighteen finds the author again debunking old shibboleths. High ground is definitely an advantage, but the advantage is too often overrated. "Its reality is undeniable. But when all is said and done, such expressions as 'a dominating area,' 'a covering position,' and 'key to the country' are, insofar as they refer to the nature of higher or lower ground, for the most part hollow shells lacking any sound core." Following a renewed stress on the relative qualities of the opposing armies and their commanders, his final sentence in this final chapter of Book Five is: "The part played by terrain can only be a minor one."

In this chapter Clausewitz makes one curious error. Among the advantages, he says, of being on high ground is that "shooting downward, considering all the geometrical relations involved, is perceptibly more accurate than shooting upward." It is quite impossible to see how it should be so on the grounds he invokes, either in his time or in ours.

Book Six: Defense

Book Six, Chapters One–Four: Clausewitz's conception of the defense being the *stronger form* of war is likely to be treated with some reserve if not suspicion by most members of the military profession today, as it apparently was also in his own time. We see him complaining in Chapter One of his view being "at odds with prevalent opinion," and in Chapter Two of the persistence of an "older view" by which "a battle accepted [that is, initiated by the opponent] is regarded as already half lost." In a later chapter (Chapter Eighteen) we will see him speaking scornfully of "the clamor of those whose vague emotions and still vaguer minds impel them to expect everything from attack and movement, and whose idea of war is summed up by a galloping hussar waving his sword."

The modern soldier may know as a matter of practical experience that it is sometimes necessary to assume a defensive posture (and he may even be accustomed to seeing a "principle of security" included among the various listings of "principles of war") but that occasional necessity he is reluctant to sanctify with any philosophy which imputes a special *virtue* to the defensive. His indoctrination impels him to see a monopoly of virtue on the other side. The purpose of such indoctrination is no doubt to stimulate the strategic or tactical aggressiveness of commanders, for long experience does suggest that few would otherwise tend on their own initiative to provoke heavy risks, or dangers, or even exceptional effort.

The extraordinary apotheosis of the offensive that marked the thinking of the general staffs of the First World War among the Western Allies was only a more aberrant form of a much longer-lived conviction. The ascendancy of this school, expressed in the prewar writings of Ferdinand Foch among others,[4] happened most tragically to occur during that one

[4] The most accessible book in English is the Morinni translation of Ferdinand Foch's 1903 *Principles of War* (New York: H. K. Fly, 1918). See also my *Strategy in the Missile Age* (Princeton: Princeton University Press, 1959), ch. 2.

great war in which tactical conditions made it as absurd and grotesque as it could ever have been. In Clausewitz's time the tactical conditions were vastly different, and we find, indeed, that most of the advantages he attributes to the defense tend to be strategic rather than tactical.

One surmises that the argument in the first four chapters of this book rises from a deep felt conviction based on the author's personal experience, and he does indeed mention two campaigns in which he participated. The 1812 campaign in Russia, to which he refers again in Chapter Three, was one of the greatest triumphs of defensive strategy in all military history. In the Waterloo campaign of 1815, too, the British-Dutch-(Belgian)-Prussian allies successfully relied on a defensive strategy. Knowing that Napoleon was bound to come to them, they organized and augmented their power while awaiting him, in positions near their bases which they had thoroughly reconnoitered and which suited their needs in other respects.

The awaiting of the enemy blow, he tells us at the outset of this book, is what characterizes the defensive. The defender has many opportunities for tactical surprise, but it is the opponent who has moved against him rather than the reverse. The object of the defensive is to preserve, which is a negative object, and it therefore follows "that it should be used only so long as weakness compels, and be abandoned as soon as we are strong enough to pursue a positive object." This assurance Clausewitz is already giving us in Chapter One, but he insists on establishing that the weaker commander adopts the defensive because its inherent strength tends to offset his weakness. He gives various strategic reasons for this, mostly having to do with the fact that the defender usually enjoys optimum lines of communication and retreat while the attacker is extending his and usually suffering a wastage of strength as he moves forward, and also that the defender chooses to his own advantage the place of contact or engagement. His clinching rhetorical argument is that if it were not the stronger form there would never be any reason for resorting to it.

At the end of Chapter Three he grants that there is a "sense of superiority in an army that springs from the awareness that one is taking the initiative"—something that Foch and others were to emphasize to the extreme—but he quickly adds that this feeling "is soon overlaid by the stronger and more general spirit that an army draws from its victories or defeats, and by the talent or incompetence of its commander." He will put the matter even stronger in Chapter Fifteen of Book Seven.

In Chapter Five Clausewitz returns to give more emphasis and eloquence to the point he has already made in the first chapter, that defense must be looked upon as a temporary expedient while preparing an improved basis for going over to the attack. "A sudden powerful transition to the offensive—the flashing sword of vengeance—is the greatest moment for the defense." Why then make so much of his special position on this question? The answer comes from the contrast between his position and that of other authors, especially the afore-mentioned Foch, who with his colleagues built up a veritable mystique of the offensive, to which they attributed every conceivable advantage *including lower casualties*. In their successive, deadly, but futile "pushes" throughout the First World War, the allied commanders could never get over the wholly erroneous idea that they were inflicting more casualties on the enemy than they were suffering themselves. They and their followers continued to insist on this point even after the war, some of them even resorting to doctoring the figures to shield their colleagues from embarrassing exposure.[5]

However, apart from these aberrant schools, what distinguishes Clausewitz from most of his colleagues in discussing these issues is mostly a matter of tone. Although he accepts whole-heartedly the need for going over to the offensive if and when it becomes feasible, he does not want the defensive posture to be despised. That it was indeed being despised, in his time as in ours, is apparent again from his last sentence in Chapter Five: "Thus constituted, defense will no longer cut so sorry a figure when compared to attack, and the latter will no longer look so easy and infallible as it does in the gloomy imagination of those to whom attack is simply courage, determination, and movement, and defense no more than impotence and paralysis."

Book Six, Chapters Six–Seven: Chapter Six expands upon the reasons mentioned in Chapters Two and Three for the superior strength of the defensive. For one thing, certain kinds of forces can be drawn upon, like the militia, which are not generally available to the professional army. But this is only one example of the fact that the support of the people is more immediately available during operations in one's own country, which are automatically conceived as defensive. Clausewitz mentions as one outstanding example the war in Iberia (1808–1814), the so-called Peninsular War, in which virtually the whole population became involved in the struggle, furnishing innumerable guerilla bands, both in Spain and in Portugal.

Most interesting is his commentary on the greater degree of support from allies that one may expect on the defensive. There will be some states that will be deeply interested in maintaining another country's integrity simply because their rulers feel more secure with the maintenance of the *status quo*. To Clausewitz this attachment to the *status quo* ex-

[5] See Sir B. H. Liddell Hart, "The Basic Truths of Passchendaele," *Journal of the Royal United Service Institution* (London), 104, 616 (November, 1959): 1–7.

plains the often spontaneous development of balance-of-power involvements. "If it were not," he says, "for that common effort toward maintenance of the *status quo*, it would never have been possible for a number of civilized states to coexist peacefully over a period of time. . . . The fact that Europe as we know it has existed for over a thousand years [instead of becoming one unified state] can only be explained by the operation of these general interests." There have of course been great territorial changes, and Poland is a special example of a considerable nation which had been (only recently in Clausewitz's time) eliminated as a political entity, but there were special and instructive reasons for that. In any case, the defender can usually count more on foreign aid than the assailant, the more so as his political and military condition is sound.

In Chapter Seven the author expands a point he has already made—that the war takes its form and character from the defender's denying the assailant possession of something that the latter would normally be content to have without war. The defender is thus "the first to commit an act which really fits the concept of war." Clausewitz emphasizes at this point that he is speaking only from the point of view of theory; he naturally understands full well that the assailant usually attacks with the assumption that his aggressive act will provoke a military response.

Book Six, Chapter Eight: This long and somewhat involved chapter, "Types of Resistance," in good part carries forward ideas already expressed in the preceding chapters of this book. Defense, the author says, is composed of two distinct parts, waiting and action. But, especially in a large-scale defensive action such as covers a whole campaign or a war, the waiting and the action will not be separable into two distinct phases; emphasis will alternate between the two. Clausewitz is bent on making two fundamental points: (a) the waiting, with all the dynamics which the term is supposed to encompass such as obliging the enemy to spend part of his power in his advance, deserves status as an independent concept or "principle"—it is "such a fundamental feature of all warfare that war is hardly conceivable without it"—and (b) the benefits of the waiting rarely if ever accrue without the action, actual or threatened.

The author then indicates four ways in which the defender can elect to carry on his defense. The first three, however, have in common that they occur at or near the frontiers of the country, while the fourth has the defender withdrawing into the interior of the country and resisting there. It is the fourth that interests him the most and it is apparent that he has much in mind two historic episodes, both of which he mentions and in one of which he took part. The first is the campaign of Torres Vedras in 1810–1811, where Wellington withdrew before Marshal Massena to the fortified lines he had prepared in the mountains above Lisbon. There

Massena besieged him at the end of a line of communications extending the full length of a violently hostile Spain teeming with partisan bands, only to withdraw finally when starvation overtook his forces. The other is the epic case that occurred in the following year, when Napoleon's *grande armée* went to Moscow and perished from starvation and cold in the retreat.

In both cases the defending forces, including the partisans, played as great a role as distance and the hostile physical environment in causing the destruction of the invader—both in actual attacks and in threats of attack. Otherwise, the invading armies would have been quite able to provision themselves and would thus not be forced to retreat. Also, especially in Russia, the retreat to which the invading army was finally compelled was made enormously more costly by the haste and the disorganization that resulted from the incessant Russian attacks.

But Clausewitz is also bemused by the cases "where there is no actual fighting but the outcome is affected by the fact that there could be." He considers the situation where the invader, losing strength as he presses forward, begins to fear that the defending opponent has become tactically superior, and his resolution departs. He mentions several such situations; but one thinks also of the German liquidation of the Schlieffen Plan in 1914 and the retreat to the Aisne after a battle at the Marne which was neither of climactic ferocity nor a clear tactical victory for the French. Because of its consequences, however, it became the "miracle of the Marne."

Book Six, Chapter Nine: After describing an imaginary battle, based on contemporary tactics, to show the defender's advantages which he can turn to good account if he is not too grossly inferior—and thus to show (against a basic assumption that he finds prevalent in his time) that the side on the tactical defensive is not inherently less likely to win a decisive victory than the side that initiates the attack, Clausewitz goes on to explain why such defensive victories have rarely happened in history. In most cases the defender is either markedly the weaker of the two or thinks he is. Curiously he indicates what *might* have happened if Napoleon on the defensive at Leipzig had been victorious, but he makes no mention of what *did* happen in the great defensive victory at Waterloo. We have seen that Clausewitz was himself at nearby Wavre on the day of that battle and taking part in a heated defensive action against Marshal Grouchy. One can only wonder why he does not mention the greater one of those two actions, in which the side that was on the defensive until the very end of that long day won one of the most decisive victories of all time.

Book Six, Chapters Ten–Fourteen: The two chapters on fortresses are less dated than one would suspect from their titles. After describing the difference in functions between the medieval fortress and that existing in his own day, Clausewitz goes on to describe in Chapter Ten the several purposes which a modern fortress can serve and in Chapter Eleven the considerations which should govern the selection of sites for fortresses. Obviously, these purposes and considerations have varied with time, but less so than the reader might imagine. The Schlieffen Plan executed in 1914 was devised to outflank the great French chain of fortress cities on France's eastern frontier. In 1940 the German attack through Belgium and the Ardennes was designed to outflank the Maginot Line, which was a fortified line rather than a chain of fortresses. In each case the system channeled the enemy attack, and if the French failed to get more profit than they did from that fact, the fault was not altogether in the fortress idea.

No doubt fortifications have since the Second World War become less elaborate and they have ceased to fulfill all the purposes Clausewitz describes, but only gradually. We are frequently made aware of the failure of fortress systems, including the examples mentioned above, but as Clausewitz points out, whether in their direct or indirect effects fortresses "do not make the enemy's advance impossible; they only make it more hazardous—in other words less likely and less dangerous to the defender." That too much has been expected of particular fortresses or fortifications is a matter of historical record; but the record has to be read with care before one can judge the utility of those systems that failed. Besides, not all systems have failed under attack. The Turkish forts along the Dardanelles that helped prevent the penetration of a British fleet in 1915 had weighty consequences on the subsequent course of the First World War.

Clausewitz's sage advice about the use of fortresses in considerable depth rather than solely along the frontier has rarely in recent times been followed except by small nations like Belgium—which had little enough depth to dispose of—because the larger ones were not prepared to admit to themselves or others that an enemy might not be stopped at their frontiers.

Chapters Twelve to Fourteen on defensive and flanking positions tend to be more dated because they are somewhat more particularized about tactics than the chapters on fortresses.

Book Six, Chapters Fifteen–Seventeen: In these three chapters, "Defensive Mountain Warfare," Clausewitz reveals again his pleasure in knocking down that obvious conclusion which turns out to be full of exceptions or plainly wrong. In this case it is the conception reflected in the expression used by those "who speak of a gorge so narrow that a handful of men could hold off an army." This impression that a mountain range offers favorable terrain for one's major defensive effort Clausewitz denies, exploring the matter on a tactical and strategic level with an attention to detail which indicates on his part a close study of the matter both in historical literature and in personal reconnoitering. Moreover, what he says about mountain defense seems to be little altered by tactical changes since his time.

The central question to be decided, he says, is "whether resistance in defensive mountain warfare is intended to be *relative* or *absolute*." Small bodies of men can certainly slow down the progress of large forces of the assailant, which is what he has in mind by the term "relative," but by "absolute" he means totally stopping the enemy or of winning a decisive victory over him, and for this purpose mountains are generally not suited at all.

The problems for the defense are several, which Clausewitz describes in detail. They are mostly centered round the extreme passivity of the defense posts. The assailant adjusts to them, and not they to the assailant. Besides, the position of these posts means usually that they can be held only by infantry, which are limited to the short range of small arms fire (this factor would be altered by modern weaponry, and yet the general proposition would still hold). Defense forces do not occupy the ridges, which in lofty mountains are too inaccessible, but only the valleys. Their positions there are isolated and usually susceptible to being turned.

However, a mountain barrier can be of great aid to the defender strategically provided he does not deploy the major part of his army within its passes, where they are inevitably fragmented, immobile, and passive. He should indeed make the most of what small detached forces can do to impede the enemy within those passes but should mass his major forces in the open country behind the barrier. "We do not maintain," he says, "that Spain would be stronger without the Pyrenees, but we do hold that a Spanish army which felt it had the strength to risk a decisive battle would be wiser to make a concentrated stand behind the Ebro than to split up among the fifteen passes of the Pyrenees. This is not going to obviate the effect of the Pyrenees on the war. We believe the same to hold true of an Italian army. . . . No one will be ready to believe that an attacker likes to march across a mountain massif like the Alps and to leave it in his rear." In his rear it makes for difficult, constricted, and precarious lines of communications and of retreat.

Book Six, Chapters Eighteen–Nineteen: In these two long chapters, on "Defense of Rivers and Streams," Clausewitz is writing from more direct personal experience than he could possibly have had with mountains, and his treatment does indeed become more detailed. However, his general strategic conclusions are quite similar. Again he asks us to distinguish between relative and absolute defense. Like mountains, rivers can reinforce a limited defense; "but their peculiar characteristic is that they act like a tool made of a hard and brittle substance: they either stand the heaviest blow undented, or their defensive capacity falls to pieces and then ceases completely." Historical examples of the successful defense of rivers, he adds, are fairly rare.

He mentions three governing factors: (a) the width of the river; (b) the means available for crossing it; and (c) the strength of the defending force. The attacker's overall strength, he says, "is not relevant at this stage," because obviously he can get only portions of his force across initially and the issue hangs on whether the defender is strong enough to annihilate these portions before they are built up to a local superiority. The most significant factor here is that there is normally little use in massing a large force to defend a relatively short stretch of river; "any direct defense of a river must always be extended till it amounts to a kind of cordon system." Such an effort can absorb large forces. After considering the problem in some depth, he concludes that "the direct defense of a river is suitable as a rule only for the very largest European rivers, and only on the lower half of their courses." On the other hand, as in the case of mountains, an advancing army is disadvantaged from having a wide river in its rear, for its lines of communications and of retreat may be limited to one or a few crossing points.

In the Second World War there is no record of a river forming a significant barrier to an invading force, for exactly the reasons Clausewitz describes. However, the English Channel served the British very well for four long years, the Germans lacking the power and the means to cross. On the other hand, the British and their American allies ultimately crossed it, at which time the German defense of the Channel shore and the failure of that defense can be described strategically in exactly the terms Clausewitz uses for the crossing of rivers. The Germans had to spread their defense along the whole length of the Channel, which meant that they were not strong enough locally to destroy the forces that the allies began to put down on selected portions of the Normandy coast on June 6, 1944.

To consider a case at the other extreme, one wonders how the Israelis permitted themselves to think that the Suez Canal, only 200 feet wide, could be a meaningful barrier to the crossing of Egyptian troops, which it conspicuously failed to be in October 1973. It argued a grotesquely low opinion of Egyptian capabilities, which events quickly proved to need some correction.

Book Six, Chapters Twenty–Twenty-Two: Here the author considers the defense of swamps and of forests, and the defense of a territory by a cordon. They have in common with the defense of mountains and of rivers the factor of passiveness and also that defense maneuvers are highly localized, and as the author says (Chapter Twenty), "there is always something very insidious and dangerous about local defense." Nevertheless, there are also important differences. Swamps are usually much wider than rivers, and also much more difficult to cross, especially with heavy equipment. On the other hand, if a means of crossing is established, it is much more difficult for the opponent to demolish than is a bridge across a river. Clausewitz finally admits, apparently somewhat reluctantly, that wide swamps and marshes "are among the strongest lines of defense possible."

After giving special and rather considerable consideration to the special case of the Netherlands and their means of defense by inundation (the Netherlands, after all, border on Prussia), the author goes on to consider the defense of forests. The direct defense of a forest is risky, because the defender needs above all to be able to see. To have a forest in front of him is the worst of all situations, because the attacker can see without being seen, but to have a forest in the defender's rear may be an asset in retreat.

The cordon-defense may make sense when the object is to withstand a slight attack—"slight either because the attacker is easily discouraged or because the attacking force is small." On the other hand, it would be really absurd "if the main force, which is meant to defend the country, were to be strung out in a long series of defensive posts against the enemy's main force—in fact, a cordon. It would be so absurd that one would have to investigate the immediate circumstances which would accompany and explain such an occurrence."

What Clausewitz could not foresee, no more than the generals of Europe on the very eve of its occurrence, was the kind of cordon-defense that characterized virtually all of the First World War, especially on the western front. What caused it to come into existence was (a) the enormously greater size of the armies on both sides than had ever been mobilized before, and (b) the equally striking increase in defensive firepower, especially because of the machine gun. In the Second World War the tactical situation had again changed, because of the heavy use of tanks and of tactical aircraft. The enormous size of armies again made some degree of cordon defense almost inevitable, but only during pauses between the main fighting efforts were the lines relatively stationary.

Book Six, Chapters Twenty-Three–Twenty-Four: The brief Chapter Twenty-three is almost a digression and notable only for the heavy scorn with which Clausewitz exposes cant among his fellow theorists. In this case the offending phrase is "key to the country," which he calls an "incantation" and its meaning "obviously something of a mystery, beyond the bounds of normal understanding, and requiring the magic of the occult sciences." Moreover, he points to examples where the concept was responsible for some senseless moves. "The real key to the enemy's country," he says, "is usually his army." Though he allows some justification for using the term to denote "an area which one must hold before one can risk an advance into enemy territory," he prefers to reject the concept altogether.

The long chapter which follows considers the problem of strategic defense by operations against the flank of an invading army. The issue is presented most forcefully in considering the campaign to which Clausewitz almost immediately refers, Napoleon's march to Moscow in 1812; for apart from the author's personal experience with it, this campaign raises the relevant questions in their most extreme form. How could it happen that Napoleon could take so huge an army 600 miles deep into a hostile country, while moving along a narrow front, confident that his line of communications would not be seriously threatened? And indeed it was not, at least during his advance. The number of troops that he had to assign to guarding his communications as he proceeded was not large—as we have learned in a previous chapter—being far smaller than the numbers lost simply as stragglers (see Chapter Twelve of Book Five).

Clausewitz seems to have some difficulty himself in explaining the matter on grounds other than experience. Even upon positing a case where a *superior* army is attempting flanking operations against its opponent, he says: "It would seem that the latter would be hard-pressed to protect its rear. True enough, if only war were as predictable in practice as it is on paper!" Well, obviously it is not, but, in this particular matter, why not?

First of all, he tells us, "forces sent to operate against the enemy's rear and flank are not available for use against his front." In the usual case where the defender does not enjoy superiority, as Kutuzov did not in falling back before Napoleon, this is a vital consideration. Also, falling upon the enemy's communications is not much of an accomplishment in isolation. What is there to strike, and for what purpose? Besides, good intelligence is lacking, and "a party sent past the enemy's wing to raid his rear is like a man in a dark room with a gang of enemies."

This concerns the enemy advancing with great power. However, when he lacks either purpose or capability for advancing further—as was Napoleon's situation when he reached Moscow—the situation changes.

"If the enemy is blocked from making further progress by something other than our own defense—no matter what it may be—we need no longer be afraid of weakening our forces by sending out strong detachments. Even if the enemy hopes to make us pay by launching an attack, we can simply yield some ground and decline the battle," as did the main Russian army before Moscow.

Clausewitz distinguishes sharply between interfering with the enemy's communications, meaning mostly an occasional convoy or courier, and cutting his line of retreat. The former may not mean much when the enemy is advancing in great power. The latter may not mean much when he has no intention of retreating. But when he has to think seriously of retreating or is actually embarked upon a retreat, the situation changes drastically. Then fear of being cut off, augmented by attacks and denial of provisions, may induce desperation or panic.

Thus, flanking operations in defense are most useful towards the end of a campaign, "when the attack has spent its force," and in conjunction with an uprising of partisans (what Clausewitz calls "insurrection"). The latter do not have to be spared from an army to which they do not belong.

One might ask how the situation would be changed today. Attacks on the enemy's rear are mostly by aircraft, and in some instances partisans, and it is still true that unless it is done in conjunction with major activity on the fighting front, "interdiction" attacks accomplish little. Ground forces advancing today usually do so on much broader fronts than in Clausewitz's time, and attacks on their flanks or rear will, therefore, have to be from different forces than those they are pushing before them, as in the case of Gallieni's Paris garrison in 1914, and MacArthur's Inchon landing in 1950. However, such instances are rare.

When the Germans under Hitler invaded the Soviet Union in June 1941, thereby apparently inviting a replay of Napoleon's disaster of 1812, they attacked on a broad front with great power, and the totally surprised Red army, by offering battle near the frontiers, suffered tremendous losses in prisoners as well as casualties. The Germans also visited such savage treatment on the peoples of the conquered territories as to minimize through terror partisan activities against their communications. Though in the end the task proved too much for the Germans and the result was a disaster comparable to Napoleon's, it took almost three years instead of just a few months for that result to be accomplished. And except for local enveloping operations, the German rear was never significantly threatened.

A breakthrough on a narrow front such as that accomplished by the German army against the French facing the Ardennes forest in May

1940 should in principle have proved vulnerable to attack on its flanks, but the French army lacked the mobility, the reserve power, and, above all, the spirit to counteract effectively. On the other hand, when the Germans did the same thing in almost the same sector against the American army in December 1944—the so-called Battle of the Bulge—the Americans and British did have the power to strike decisively against the flanks of the deep German salient, a power augmented, when the weather finally cleared, by very heavy air attacks. The quick collapse of the German offensive proved only that their objective was well beyond their means.

Book Six, Chapter Twenty-Five: In this substantial chapter on "Retreat to the Interior of the Country" Clausewitz analyzes in a somewhat generalized fashion the Russian defensive campaign of 1812, from which he derives most of his illustrations. The Russian defense, not preplanned as such, took the form "that destroys the enemy not so much by the sword as by his own exertions." It profited from the fact that the attacker's strength always diminishes as he advances, provided the retreat of the defender which both invites and obliges that advance can be sufficiently deep. It is important that the defender's retreat not follow a severe defeat in battle, for then it is likely to be more costly to him than to the invader, and this means not accepting battle prematurely. The Russians were spared making that mistake in 1812 not out of prescience but simply because the French were so overwhelmingly superior in the early part of their advance.

The forces in retreat preempt local supplies as they move, using bridges but destroying them in their wake, and carrying out generally what has since come to be called "scorched earth" policy. Depending on the original differences in strength and on various other factors the defender can ultimately offer battle with much improved prospects of success. "Indeed, the condition of the invader at the end of his course is often such that even a *victory* can force him to retreat," if he has used up reserves which he cannot replenish.

There are two main drawbacks to this kind of defense. The first consists of the loss of country to the invader, the importance of which depends on how rich and how populated is the portion temporarily lost. The basic point, however, is that "it cannot be the object of defense to protect the country from losses; the object must be a favorable peace." The second and usually more important drawback is psychological. Normally the army and population cannot be expected to "tell the difference between a planned retreat and a backward stumble; still less can they be certain if a plan is a wise one, based on anticipation of positive advantages, or whether it has simply been dictated by fear of the enemy." Then there are inevitable considerations of national pride and of honor, requiring "that an enemy who violates a frontier will be made to pay a penalty in blood." Such considerations compelled the Russians to accept battle at Borodino rather than surrender Moscow without a fight, though by that time their inferiority was much ameliorated. Thus, the commander's major problem in a prolonged retreat is to maintain not only the morale of the retreating force but also the support of the people and perhaps also of the government.

An invasion of a large country requires not only numerical superiority over the defender but a force large in absolute numbers—"One may be able to march on Moscow with 500,000 men, but never with 50,000"—and that increases the debilitating effect which a long march will impose on the invader. The more ponderous mass moves more slowly and thus forfeits the chance of overtaking and destroying the retreating army; the problems of supply and billeting are increased; and the losses are proportionately greater for each measure of distance covered.

Clausewitz also considers at length the degree to which the retreating army can determine the direction of movement of the invader, being by itself both a bait and a threat, and he decides that much depends on specific circumstances.

How different is the situation in our own times as compared with that which Clausewitz observed in 1812? Motor-driven tanks, guns, and vehicles of all kinds have of course made an enormous difference, especially in the speed of penetration of an invading force, so have aircraft and radio communications. But there are compensating costs in the huge requirements of supply, especially of the liquid fuel by which all things move. Contrasting markedly with the ponderous movement of the German army in August 1914 in its abortive effort to execute the Schlieffen Plan was the impetuous and wholly successful breakthrough against the French in the spring of 1940, but the larger-scaled attack against the Russians a year later ultimately failed, as we have seen, despite spectacular German successes at the outset which brought about horrendous Russian losses. By November 1941 the German army was deep in Russia over a vast front but already panting desperately at the end of its long lines of supply, while the Russians were bringing fresh troops from Siberia and beginning to achieve local successes. In the following summer the Germans launched a strong new offensive, especially in the south, but despite some anxious months for the Russians and their western allies the Germans failed to achieve their major objectives. Before the year 1942 was out their Sixth Army was already encircled at Stalingrad and disaster was impending. Clausewitz was vindicated by the events of

1941–1944, on a scale that he could hardly have dreamed of, though the Russians could have done far better for themselves in following his implied admonitions against a rigid defense at the frontier. Similar lessons were underlined in other areas of that farthest-spread of all wars, and on a much lesser scale in the Korean War of 1950–1953, and again in the Vietnam War.

One should notice too that most of what is supposed to be distinctive and original about Mao Tse-tung's strategic theories, which make much of the concept of "protracted war," is in fact an exploration of modes of waging a strategic defensive akin to Clausewitz's. Mao may indeed have got his ideas, which enabled him to defeat the more numerous and vastly better equipped forces of Chiang Kai-shek, from his reading of the ancient Chinese strategist Sun Tzu (from whom there is indeed much to learn) rather than from Clausewitz. Or it is possible that he developed his ideas simply on the basis of his sturdy common sense. For it does not require profound strategic insight to understand the uses and strength of the strategic defensive under appropriate circumstances. It requires, on the contrary, a perverse kind of teaching to cause one to scoff in principle at defensive theories.

Book Six, Chapter Twenty-Six: In this chapter, "The People in Arms," Clausewitz deals with a phenomenon new to Europe in his lifetime—a result not so much of the French Revolution as of the Napoleonic wars. Irrelevant to his purposes are the fears others are expressing about the uprising of partisans being "legalized anarchy . . . as much a threat to the social order at home as it is to the enemy," but he *is* interested in its military value.

By its very nature, the scattered resistance of guerilla bands "will not lend itself to major actions, closely compressed in time and space." However, given sufficient time and broad enemy exposure to its effects, "it consumes the basic foundations of the enemy forces." By itself the general insurrection will not do much, but within the framework of a war conducted by the regular army, its effects can be critical.

Clausewitz then lays down five conditions under which a general uprising can be effective, all of them together describing the conditions of the Peninsular War, which for some reason he does not mention. Despite the newness of this kind of war in his time and thus the general lack of experience, he clearly delineates its character and sets down its requirements. "It does not matter much if a body of insurgents is defeated and dispersed—that is what it is for—but it should not be allowed to go to pieces through too many being killed, wounded, and taken prisoner: such defeats would soon dampen its ardor." However, he acknowledges himself to be simply groping for the truth. This kind of war is new, and "those who have been able to observe it for some time have not reported enough about it." The latter statement indicates that the risings of the people played little if any part in the campaigns he personally observed, including that in Russia in 1812, which was among other things too brief.

Toward the close of the chapter the author approaches a new subject, though he hinges his discussion of it upon the calling out of the home guard after loss of a decisive battle. "Even after a defeat," he says, "there is always the possibility that one's fortune can be turned by developing new sources of internal strength, or through the natural decimation suffered by all offensives in the long run, or by means of help from abroad. There will always be time enough to die." Then later: "No matter how small and insignificant a state may be in comparison with its enemy, it must not forsake these last efforts, or else one will conclude that its soul is dead." In this respect one thinks of the Netherlands, Belgium, and Norway in the Second World War, and of the contrasting example of a France that perhaps gave up too easily. But one thinks also of the France of 1870–1871, which after the decisive defeat at Sedan that occurred only six weeks after the outbreak of war continued to fight on, mostly through the raising of *levées en masse*, until the capitulation of Paris some five months later. Determining when a desperate resistance is the ultimate wisdom or on the contrary is a futile waste of life is a question which also is not covered by an airy formula—not even Clausewitz's.

Book Six, Chapters Twenty-Seven–Thirty: In this long section, where four chapters deal with the single topic, defense of a theater of operations, Clausewitz is not at his inspired best. Partly, it is a matter of his writing, which verges occasionally on the obscure and sometimes on the tedious, but it is more a matter of his having made his major points in previous chapters, leaving to this place an accounting of some points which deserve also to be covered.

He restates at the outset what he has already established—that it is always more important to preserve one's armed forces and to destroy those of the enemy than to hold on to territory. Though loss of territory will in the long run weaken one's capabilities, it will not usually do so "within the decisive phase of the war." However, the enemy forces will not always present themselves in a way that will make a single decisive blow against them possible. For one thing, the invaders may represent a coalition of enemies invading from widely different directions. It is all very well to argue for concentrating one's own forces and attacking opposing forces in turn, but that is not always certain to be accomplished.

One may be compelled to divide one's forces. There are other problems, even where the opponent is not multiple but singular. He may, for example, decline to advance upon one's ideally selected and well-prepared position, choosing rather to bypass one's own forces for a different objective. Or he may in other ways compel the defender to assume the tactical offensive despite the latter's preference for fighting a defensive battle.

Clausewitz rather uncharacteristically goes through a long checklist of possibilities, offering for each some suggestions on how they may be dealt with.

He is here demonstrating again his refusal to let his judgment be bound by some formulae, however meritorious the latter may be as general propositions. Unfortunately his historical examples, taken so often from the campaigns of Frederick the Great, will not help bring alive to the modern reader some of the points he is making. In recent times we have seen several instances when the commander-in-chief felt compelled to reject the simplistic ideal in order to protect territories. The younger von Moltke, for example, has been almost universally condemned for "vitiating" the Schlieffen Plan by strengthening the southern part of the line facing France and also by taking steps to guard against too deep a penetration of East Prussia by the Russians. But his reasons were not frivolous, and one would have to examine carefully and in detail his decisions against the circumstances existing at the time to determine that he was mistaken. That, at least, would be Clausewitz's approach. Similarly, in the Second World War Roosevelt and Churchill correctly decided shortly after Pearl Harbor that they must jointly concentrate on defeating Germany first, but that did not mean that they could let Japanese advances in the Pacific and in East and Southeast Asia go unopposed. The formula of "concentration of force" may represent the ideal, but not necessarily what is feasible or invariably correct.

Book Seven: The Attack

Book Seven, Chapters One–Seven: Most of these chapters have a tentative, preliminary tone. It is not alone their brevity, for some of the brief chapters in other books appear highly finished, and indeed the first chapter of the present book, which is also the briefest, says in its two paragraphs all that needs saying by way of transition between the preceding book and the present one. But, apart from the unsatisfactory presentation of some of the ideas, one notices the shorthand way in which illustrative campaigns are merely noted at the end of Chapter Four and again in Chapter Seven, and of course the footnote by Marie von Clausewitz at the end of Chapter Five makes evident what is otherwise merely sensed. And we recall that in his "Notice" of 1827 the author refers to the chapters of this book as existing in "rough draft."

The author observes that much of what would otherwise have to be said about attack has already been stated or implied in the preceding book on defense. However, there are some special features of attack which do not directly arise from the defense. For one thing, where counterattack is an *inherent* part of the defense, the offensive thrust is complete in itself. Although need for defense does intrude into the offensive, it does so only as a "necessary evil," first, because the offensive thrust has to be interrupted for periods of rest, during which defense takes over automatically; and second, because the area left in the rear of the advancing forces is not necessarily covered by the attack and may need special protection. In short, defense is an impeding burden to the attack, "its original sin, its mortal disease."

Also, although the defense can be implemented by elements other than the fighting forces—like fortresses, advantages of terrain, and even a greater likelihood of gaining allies—the offense usually has to depend on its fighting forces alone.

Chapter Four, "The Diminishing Force of the Attack," has already been covered in the preceding book, though the author here contributes an itemization of seven ways in which the overall strength of the invading force may be depleted in the advance. Chapter Five, "The Culminating Point of the Attack," merely states a subject which will be discussed in greater depth later on. Chapter Six, however, "Destruction of the Enemy's Forces," though also brief and tentative, presents some ideas that are both new and significant. Repeating his own formula that destruction of the enemy's forces is the military objective, Clausewitz raises the question: What do we mean by "destruction of enemy forces?" And, at what price? These do not seem like striking questions until one recalls that few other writers on strategy have thought to raise them, let alone attempt to answer them. Clausewitz mentions four quite dissimilar ways in which one can view the issue of "destruction of enemy forces," and then mentions some indirect ways of accomplishing this goal, like seizing a strip of territory or an enemy fortress. He admits that such indirect means are generally overrated, that they are tempting only because they cost so little, but then adds significantly that they are "obviously preferable to pointless battles." By "pointless" he clearly means indecisive but costly. What is more obvious than the need to consider price in the pursuit of some gain, whether strategic or other? Yet his constant readiness to do so is not the least of the qualities that distinguish Clausewitz from virtually all other writers in the field.

Chapter Seven, "The Offensive Battle," presents really but one new and useful point—that "a peculiarity of most offensive battles is doubt about the enemy's position," which makes it all the more necessary to concentrate one's forces. It is for this reason that he urges attempting to outflank rather than to envelop the enemy. This leaves a good deal unsaid about the feasibility of outflanking an enemy whose position is not established, and who presumably has taken precautions against being outflanked, but it does at least signal Clausewitz's distaste for frontal attacks.

Book Seven, Chapters Eight–Twenty: In these chapters the tone has become somewhat less tentative but still cursory, still burdened by the author's feeling that inasmuch as this book is but the obverse of the preceding book, most of what would need saying under the various chapter headings on attack has already been said in the related chapters on defense. However, though he does not wish to repeat himself unduly, he seems less concerned about appearing occasionally inconsistent.

Thus, in Chapter Eight, "River Crossings," one might think he has forgotten his previous low valuation of rivers for strategic defense. His very first sentence reads: "A major river which cuts across the line of attack is a great inconvenience to the attacker." The inconsistency, however, is superficial. He is at this point speaking more at the level of tactics than of strategy, and one remembers that Clausewitz is not always careful to make clear to which of those levels he is at the moment addressing himself. Later he says that the defended river will actually favor the attacker "if the defender makes the mistake of staking his whole future on this defense." So there is not much inconsistency after all. Also, Clausewitz is again speaking, albeit somewhat obliquely, in terms of reasonable measure or proportion. Thus, although "river crossings as such seldom present major difficulties," the attacker is likely to have "misgivings" about them "unless a great decision is involved." He had in the previous book spoken of the width of the river and of available means of crossing as important factors to consider; now he is also speaking of the issue at stake, which is again a variable. One does not take large risks for modest ends.

In Chapters Nine and Ten the author is again expressing his deep aversion to attacking an able enemy who is in a strong defensive position or entrenched camp. Not only reason, he says, but "hundreds and thousands of examples show that a well-prepared, well-manned, and well-defended entrenchment *must generally be considered as an impregnable point*." The "thousands" must certainly be an exaggeration—unusual for

Clausewitz—but it shows how strong his feelings are on this matter. And that in view of the armaments of his time, with muzzle-loading muskets so slow to reload! The generals of the First World War could well have used some of this caution.

Like Chapter Eight, Chapter Eleven, "Attack on a Mountainous Area," is a supplement to the related chapter of Book Six, there being even less reason here to speak of inconsistency. Chapter Twelve, "Attack on Cordons" is slight as well as brief, but the following chapter, "Maneuver" is more interesting. The modern student reads and hears so much about eighteenth-century armies (and fleets) maneuvering in the presence of each other—the maneuvering being an apparent substitution for or at least postponement of the fighting—that he cannot help but wonder what it was all about. It was in large measure a by-product of the very short ranges of the weaponry used, so that hostile forces could be in full view of each other and still be quite uncommitted to battle. Engagement would require on one side or the other the will to attack, which might await the development of some clear advantage. Thus, in 1812 Wellington and Marshal Marmont before Salamanca maneuvered for three weeks in sight of each other before Wellington saw the opportunity prompting him to attack. Modern forces will also maneuver for advantage, but an opponent in view will be within range and thus firing and seeking cover. Clausewitz gives five possible objectives for maneuver, but concludes that there can be no rules of any kind that govern the value of the action other than possession of the same superior military qualities that generally determine the outcome of a battle.

Chapter Fourteen, "Attacks on Swamps, Flooded Areas, and Forests," adds virtually nothing to what has been discussed under the same headings in the previous book. The same could be said about the two chapters that follow concerning attack on a theater of war, except that we find a comment here and there that deserves special notice. For example, in Chapter Fifteen the author refers again to the advantage that an army derives from the knowledge that it is on the attacking side, but this time he is even more explicit in calling the advantage "modest" and usually "much overrated." "It is short-lived and will not stand the test of serious trouble." We have already noticed (in Chapter Three of Book Six) how this view contrasts with that of romanticists like Foch. There is also a somewhat confusing paragraph about whether the attacker should ever divide his forces when he is bent on a major decision; what seems at the beginning like a stern injunction against it becomes within a few lines a license to do so under certain circumstances, especially when enemy forces are also divided. It is a pity that Clausewitz did not devote more space to this matter, on which there is so much rigidity. We learn also that in his day a large army always advanced on a front at least a day's march in width, so that if its lines of communication and retreat did not deviate too much from the perpendicular the front itself would usually provide all the cover necessary to those lines—a point which both supplements and clarifies the related discussion in Chapter Twenty-Four of Book Six. More significant for modern times are the statements in the last paragraph, again concerning the protection of the attacker's rear: "If everything is subordinated to the pressure of an imminent major decision, the defender will have little scope for auxiliary operations, and the attacker will therefore ordinarily not be in great danger. But once the advance is over and the attacker gradually goes over to a state of defense, the protection of the rear assumes increasing urgency and importance." Whatever campaigns Clausewitz was thinking of as he penned these two sentences, they apply equally to campaigns of the Second World War.

Chapter Sixteen is interesting for exposing the objectives—or lack of them—that in his own day might lie behind the temporary seizure of a strip of territory or a fortress. It might be for a bargaining token in a peace conference, but it might equally be for the sake of glory or honor, for trophies, "and at times merely to satisfy a general's ambition." Sometimes it is simply a kind of shadow-boxing; "as the attacker, one is after all bound to do something"! Again, the difference between his own times and ours in these respects would be mostly a matter of degree.

Chapter Eighteen, "Attack on Convoys," also supplements and helps clarify the mystery referred to in the discussion of Chapter Twenty-Four in Book Six. Convoys carrying supplies to an army penetrating deep into hostile territory seem so vulnerable, especially inasmuch as their escorts appear to offer such meager protection. But, Clausewitz points out, it is usually the strategic situation, not the tactical, which protects them.

Chapter Nineteen, "Attack on an Enemy Army in Billets," is utterly dated insofar as it might apply to major forces, and was apparently already so in Clausewitz's own time, inasmuch as the only successful examples he offers are from an earlier generation. It is curious that he spends so many pages on the subject and so relatively few on the subject of the following chapter, "Diversions," which is timeless in its application. A successful diversion, he says, must provoke the enemy to use more force in coping with it than one uses oneself, and must thereby be of net benefit to the main objective. However, he permits the term to be stretched to cover attacks where no larger effort elsewhere is for the time being contemplated, in which case it is not simply a feint but rather a limited and harassing operation on the periphery of the enemy's power in lieu of a greater one aimed at a decisive result. For both meanings of the term we can find examples from all major wars, certainly including the Second World War.

Book Seven, Chapter Twenty-One: This chapter deserves consideration by itself. After a long chain of dated, abbreviated, or otherwise unsatisfactory chapters, we come at last to one of basic importance, which is also the concluding chapter of this book. It is one which Clausewitz has earlier mentioned in anticipation, for the central issue with which it deals is what he calls "the culminating point of victory." There are times when an invading army grows stronger as it advances, but usually the reverse is true, for reasons which Clausewitz again outlines though he has in fact dwelt upon them in some earlier chapters, especially of Book Five. Actually, one notices some significant inconsistencies between the present and the relevant earlier chapters, especially in what he says here about the heavy requirements upon the invading army for protecting its rear and flanks and in what he neglects to say here but has said earlier about losses due to fatigue and sickness (see especially Chapter Twelve of Book Five).

Anyway, unless an offensive results in the defender's collapse, there will be a "culminating point" at which the attacker is about to lose effective superiority. To push beyond this point without a good chance of an imminent favorable decision is dangerous. Thus, "every attack which does not lead to peace must necessarily end up as a defense." Also, it is likely to be an awkward defensive position. And it is usually the generals of high courage and enterprising spirit, normally deserving of so much credit, who will overshoot the mark.

Clausewitz calls Napoleon the one who revolutionized war by pushing his offensives to complete victory, but he refrains from mentioning the 1812 march to Moscow, which might be in the reader's mind as a prime example of "enterprising spirit" pushing too far. Apparently Clausewitz does not so regard it. As he will point out in the last chapter of Book Eight, Napoleon's invasion of Russia was inevitably a commitment to going to Moscow—unless the Czar did him the favor of accepting battle near the frontier, which he was most unlikely to do in view of Napoleon's overwhelming superiority at that stage. As we shall see in that later chapter, the author attributes Napoleon's disaster not to going beyond the "culminating point" but to some basic miscalculations.

It is not clear what modern campaigns Clausewitz would accept as illustrating the doctrine he is expounding here. Would they include the Japanese attacks in the western Pacific and in southeast Asia beginning in December 1941? Japan was bound to reach the limits of her capabilities before she could register a decision against her chief adversary, and her strategy was therefore wrong from its basic conception—as Admiral Isoroku Yamamoto had tried to warn his colleagues before they plunged. Similarly, the German attempt to conquer the Soviet Union beginning in 1941 proved inherently beyond German capabilities, and thus the line of maximum advance was hardly a "culminating point of victory," the Soviet government being as clearly determined as Alexander I not to negotiate a peace.

Though the principle Clausewitz is describing in this chapter undoubtedly holds good today, its application is bound to be affected by some important differences in the way armies operate. In his day infantry moved exclusively on foot, and in long forced marches the major losses were from stragglers dropping out from exhaustion and sickness. The moving army lived mostly off the land, and the wagon train attending it needed to carry only a few days supplementary provisions plus munitions enough so that it would not be disarmed by fighting a major battle. Thus the supply problem, though certainly not insignificant, was also not the chief cause of the steady drop in combat effectiveness as the army moved forward. Today, the fatigue problem would be greatly lessened because of the abundance of motor vehicles, but the supply problem would be correspondingly increased. A swiftly moving army might temporarily outrun its supplies, especially of liquid fuel, and its capacity to recover its momentum would depend heavily on its total logistics capability and the vulnerability of that capability to enemy attack from the air and otherwise.

Eisenhower has often been criticized for halting his headlong rush to the east in the late summer of 1944 largely for fear that his tanks would run out of fuel, and the pause did permit the retreating and disorganized Germans to regroup and restore their battle line. But when the Germans counterattacked in the Ardennes in December of that year their tanks *did* run out of fuel and became sitting ducks to Allied air power. Eisenhower played it safe and paid a price for doing so, but the price was an insurance against catastrophe. He refused to outrun his "culminating point of victory," and stopped to prepare the base for a new offensive which did in fact end the war.

Book Eight: War Plans

Book Eight, Chapters One–Two: With Book Eight we are back in the realm of pure gold. Book Seven was not exactly wandering in the wilderness, but it dealt mostly with the specific and thus inevitably the dated, and Clausewitz himself seemed to be eager to hurry through it. "We now revert," he says in opening Book Eight, "to warfare as a whole . . . which means returning to the ideas put forward in Book One." The crucial area we are now entering, he tells us, is the central one "on which all other threads converge"—and he admits to entering it with some diffidence. The diffidence is warranted by the boldness of the concept, but the author will again prove himself totally equal to the challenge. The intervening books have also reflected the greatness of his mind and have carried us far forward in our understanding both of the nature of war and of what distinguishes Clausewitz from other writers on that subject, but it is primarily Books One and Eight which give him the unique place he continues to hold and which account for his claims on our attention today.

In his brief introduction he returns to a subject he has dwelt on in Book One but which he feels worth some further reflection. If there is no substitute for talent in the commander and if one of the marks of his talent is his ability to see things simply, why indeed should we—and he—study the theory of strategy and of war, which threatens always to lapse into "dreary pedantry"? He tries again to help us with an answer: "Theory should cast a steady light on all phenomena so that we can more easily recognize and eliminate the weeds that always spring from ignorance, show how one thing is related to another and keep the important and the unimportant separate." And so far as its effects on the commander are concerned: "Theory cannot equip the mind with formulas for solving problems, nor can it mark the narrow path on which the sole solution is supposed to lie by planting a hedge of principles on either side. It can, however, give the mind an insight into the great mass of phenomena and of their relationships, then leave it free to rise into the higher realms of action. There the mind can use its innate talents to capacity, combining them all to seize on what is right and true—as though it were a response to the immediate challenge rather than a product of thought."

Chapter Two returns again to the distinction between "absolute war and real war," and presents in its first paragraph a priceless axiom: "No one starts a war—or rather, no one in his senses ought to do so—without first being clear in his mind what he intends to achieve by that war and how he intends to conduct it." What could be simpler and more obvious —and yet so often disregarded!

He then goes on to ask again why there is so great a gap between the pure concept of war and the concrete shape which war generally assumes. His previous answers to this question, he acknowledges, were only partial ones. There is in fact a vast array of factors, forces, and conditions in national affairs which are affected by war. "Logic comes to a stop in this labyrinth; and those men whose habit in great things and small is anyway to act on particular dominating impressions or feelings rather than according to strict logic are hardly aware of the confused, inconsistent, and ambiguous situation in which they find themselves." Some, including perhaps the man in overall command, may see the objective clearly and the requirements for achieving it, but the many will not. Energy to overcome the opposition of the latter will often be lacking. Thus, war (until Clausewitz's time) has usually been "something incoherent and incomplete." His contemporaries would have wondered if there were ever any basis for the concept of absolute war had Napoleon not demonstrated that it could be fairly closely approximated.

The question that Clausewitz now faces is: Will the new pattern be that of the future, or must he as a theorist reckon also upon some return of the "incoherence" which after all characterized most of the wars between Alexander the Great and Napoleon? As he will say in the next chapter: "When barriers which in fact consisted only in ignorance of what was possible are broken down, it is not easy to build them up again." On the other hand, it would be presumptuous to assume that all wars will henceforward be of the unfettered kind. Of war in the future: "We must allow for natural inertia, for all the friction of its parts, for all the inconsistency, imprecision, and timidity of man; and finally we must face the fact that war in all its forms results from ideas, emotions, and conditions prevailing at the time—and to be quite honest we must admit that this was the case even when war assumed its absolute state under Bonaparte."

Clausewitz proved on the whole a good prophet. That future that his mind was trying to penetrate was to contain wars reaching a degree of "absoluteness" which the Napoleonic pattern only dimly foreshadowed, but not all of them would be of that nature. As of this writing the United States has but recently extricated itself from a war in Vietnam that had every possible kind and degree of incoherence both of objective and of method.

Book Eight, Chapter Three: This long chapter has no chapter title, but its two parts have their individual headings. Part A, "Interdependence of the Parts in War," approaches yet again from another angle the difference between absolute war and those considerable modifications of the absolute which have been observed in history. War consists of a host of interactions and usually of a whole series of engagements. What, then, is the relation between these separate parts? The more war approaches the absolute, the more clear it is that the only result that counts is final victory. In such a war a general must have with his first move "a clear idea of the goal on which all lines are to converge."

In the kind of wars that were so common in the eighteenth century and earlier, however, the separate campaigns and actions were seldom directed towards such an ultimate goal. In those wars it was "legitimate to pursue minor advantages for their own sake and leave the future to itself." Again, Napoleon changed all that, but again we have to consider whether the change must apply to all cases in the future. Clausewitz concludes that at the outset of a war, "its character and scope should be determined on the basis of the political probabilities," and if those probabilities drive war towards the absolute, it becomes imperative "not to take the first step without considering the last."

Part B, "Scale of the Military Objective and the Effort to be Made," follows directly from the question posed in Part A. Absolute war would demand not only a unified purpose but also a total effort. However, in actuality we see all kinds of factors which modify the effort. Again we have to consider the scale of political demands on either side, and also the fact that the contenders have quite different characteristics as well as different degrees of power. Thus, for a variety of reasons, both sides are not likely to exert themselves to the same degree. A dilemma arises, however, from the fact that "in war too small an effort can result not just in failure but in positive harm." This factor tends to push one toward maximum effort, but in such a case "all proportion between action and political demands would be lost," which would obviously be irrational and hence unacceptable. What is required, then, is "the faculty of using judgment to detect the most important and decisive elements in the vast array of facts and situations."

The prescription which follows reminds one again of the problem the United States faced in Vietnam: "To discover how much of our resources must be mobilized for war, we must first examine our own political aim and that of the enemy. We must gauge the strength and situation of the opposing state. We must gauge the character and abilities of its government and people and do the same in regard to our own. Finally, we must evaluate the political sympathies of other states and the effect the war may have on them." Clausewitz admits that rapid and correct appraisal of all these factors and their ramifications "clearly calls for the intuition of a genius." But again he warns that "the vast, unique importance of war, while not increasing the complexity and difficulty of the problem, does increase the value of the correct solution."

Then follows for the rest of this chapter a masterful résumé of the relevant changes in war from the times of Rome and of Alexander to his own time. Clausewitz here reveals his wide and thorough grasp of political as well as military history. It is an austerely condensed intellectual tour-de-force. Perhaps the most telling part is that where the author describes the changes that have occurred within his own lifetime, in which he raises again the question: what does this mean about the future? The chapter ends with the following summation: "The aims a belligerent adopts, and the resources he employs, must be governed by the particular characteristics of his own position; but they will also conform to the spirit of the age and to its general character. Finally, they must always be governed by the general conclusions to be drawn from the nature of war itself."

Book Eight, Chapters Four–Five: These two chapters share a common title, "Closer Definition of the Military Objective," Chapter Four being concerned with what it really means "to defeat the enemy," and Chapter Five taking up the question of what to do when defeat of the enemy appears beyond one's means.

Defeating the enemy, Clausewitz says, normally means smashing his army, and if feasible it is always "the best way to begin." However, there may be circumstances which modify this simple axiom. "One must keep the dominant characteristics of both belligerents in mind. Out of these characteristics a certain center of gravity is formed, the hub of all power and movement, on which all depends. That is the point against which all our energies should be directed." Thus, seizing the enemy's capital may sometimes be more significant than destroying his army, and if the enemy has a stronger ally an effective blow against that ally may carry one's purpose further than dealing with the weaker party. "If we can vanquish all our enemies by defeating one of them, that defeat must be the main objective of the war."

But we must consider in advance whether we can accomplish the defeat of the enemy. To do so requires forces adequate to scoring a decisive victory over the enemy's and to pursuing our victory "to the point where the balance is beyond all possible redress." But Clausewitz also adds a political requirement: "We must be certain our political position is so sound that this success will not bring in further enemies against us who could force us immediately to abandon our efforts against our first opponent."

The author stresses the importance of momentum. Time tends usually to favor the side that has suffered the first defeats, partly because its plight may alarm other states and prompt them to come in on its side. Also, there will, at best, be insufficient time to derive military benefit from the territories that have been conquered. Thus, the conquest must be carried through as quickly as possible, and not in stages but by one continuous advance.

Clausewitz's insistence on the latter point is clearly a reaction against the kind of highly patterned procedures generally followed in pre-Napoleonic times, and he argues that pauses are likely to be at least as favorable for the partially defeated opponent as for oneself. "Our belief, then, is that any kind of interruption, pause, or suspension of activity is inconsistent with the nature of offensive war. When they are unavoidable they must be regarded as necessary evils, which make success not more but less certain."

One should again remember that he is talking about the armies of his time. In one place he says: "Given the way in which armies today are supplied, they need depots more when halted than when on the move." His argument would have to be much modified for recent wars, when advancing armies have been heavily dependent upon a constant stream of supplies, especially of liquid fuel. As we have seen from some examples in the Second World War, even though a pause in the advance will be exploited by the retreating side, the side on the offensive is sometimes compelled to it by the supply situation, or sometimes by the need to broaden its front or otherwise consolidate its position. Even the defeat of France in 1940 was not accomplished by a single German thrust but was actually done in two main stages. Thus, the uninterrupted offensive should be regarded as the ideal which today must sometimes be compromised in practice.

Chapter Five, subtitled "Limited Aims," takes up the question of what is to be done if circumstances rule out defeat of the enemy. The discussion seems based on the premise that the option of avoiding the war altogether is ruled out. The choice, he says, is between seizing a portion of the enemy's territory and holding on to it, or attempting to hold on to one's own "until things take a better turn." But the latter phrase implies that there are grounds for expecting this to happen. The second possibility is that the future affords better prospects to the enemy than it does to us. In that case, Clausewitz argues, one must take the offensive, which means "exploiting the advantages of the moment." The third possibility, and the most usual, "arises when the future seems to promise nothing definite to either side and hence affords no ground for a decision." In that case the offensive should be taken by the side which possesses the political initiative, which is to say by the side which had an active purpose or aim for which it went to war.

The second of the three possibilities just mentioned fits the case of Japan choosing to go to war against the United States in 1941, not because its leaders thought they had the means to win but because they felt they could not afford to wait. Time was against them, mostly due to the American oil embargo, imposed by President Roosevelt in 1940, which directly affected the fuel reserves of the Japanese fleet. The results for Japan of accepting a "now or never" option suggest something of the perils of this course of action.

We might notice that Clausewitz is here brushing away a weighty issue. The fact that time appears to be on the side of the opponent does not itself indicate that one can defeat him now. It is interesting that Lenin, who like Engels studied Clausewitz, made much of this point; in his precepts to his followers he repeatedly rejected the "now or never" philosophy wherever the "now" was still an unfavorable situation for offensive action.[6] Clausewitz, on the other hand, adhering to the "now or never" philosophy, was in despair in 1809 because his sovereign, Frederick William III, refused in the absence of Russian support to go to Austria's aid against Napoleon in the campaign that ended with Wagram; and when the king two years later chose again not to fight Napoleon but to accept a limited alliance with him in anticipation of the latter's invasion of Russia, Clausewitz fulminated in a letter to a friend: "One is obliged to undertake an operation against chance of success when it is impossible to do anything else." By "impossible" he probably meant "dishonorable," but possibly also "missing an opportunity." Clausewitz then left the Prussian service to enter that of Russia, soon to be the nominal enemy of his own king. However correct his personal action in view of his abhorrence of Napoleon, his attitude about what the Prussian state ought to do was colored by a certain romanticism. Prussia, with its army reduced by the terms of the peace of 1806, had not the slightest chance of withstanding the armies of Napoleon, and the king's decision turned out in the end to be the wisest for the interests of the state.

Book Eight, Chapter Six: This important chapter is pitched on the heroic level—at least after one gets through Part A, with the subtitle "The Effect of the Political Aim on the Military Objective," which provides in advance some illustrations of what the author means to say in Part B but which carries little meaning in itself. We therefore go immediately to Part B, with its pregnant subtitle, "War As An Instrument of Policy." Here we have a return to the argument of the great opening chapter of Book One, where we first encountered the doctrine that "war is a continuation of policy by other means." Now we have an additional illumination of the meaning of that now famous phrase, with some elucidation of its ramifications. We must, however, also recall that Clausewitz planned considerably further development of the idea in his proposed revision.

Absolutely inadmissible, he says, is the common notion that war suspends political intercourse between the contestants and replaces it with

[6] On this point see Nathan Leites, *A Study of Bolshevism* (Glencoe, Ill.: The Free Press, 1953), pp. 512–524.

a wholly different condition, ruled by no law but its own. War may have its own grammar, but not its own logic. The logic is determined by the political aim, and acts of war merely replace the usual exchange of diplomatic notes. If this were not so, war would be "something pointless and devoid of sense."

The reason that war does not advance relentlessly toward the absolute is precisely that "it cannot follow its own laws but has to be treated as a part of some other whole, the name of which is policy." However, if war is part of policy, then policy must determine its character. That does not mean it will influence operational details—"political considerations do not determine the posting of guards or the employment of patrols"—but it will influence "the planning of the war, of the campaign, and often even of the battle." And somewhat later: "Under no circumstances can the art of war be regarded as the preceptor of policy."

The point is too important to worry about repetition, and Clausewitz keeps hammering at it: "That the political view should wholly cease to count on the outbreak of war is hardly conceivable unless pure hatred made every war a struggle for life or death." And again, "Policy is the guiding intelligence and war only the instrument, and not *vice versa*." The military point of view, then, must *always* be subordinated to the political.

Of course, statesmen must understand the language of war, so that they do not use it incorrectly. However, if that happens it is the policy that is wrong, not the fact that policy is influencing the war. But if the political leader is a man of distinguished intelligence and strength of character, he can always get the military information he needs "somehow or other." Apparently Clausewitz thought the reverse to be unlikely, or at any rate not appropriate.

The chapter finishes with the reminder that the vast changes wrought in the wars of the Napoleonic era arose from "the new political life which the revolution created for Europe as well as for France. Other means and forces were thus called forth, which conferred on warfare a degree of energy inconceivable without them." And then: "It follows that the transformation of the art of war resulted from the transformation of politics."

This is a breathtaking chapter. Clausewitz addresses his argument with a fervor which stems from his conviction that *nothing* he has been saying throughout his whole work is more important or more basic—or more likely to be ignored. The First World War was conspicuously not fought according to these precepts. But Clausewitz did not deny that war could become "something pointless and devoid of sense." He only argued that it should not.

Book Eight, Chapters Seven–Eight: These two short chapters, bearing the common title "The Limited Aim," are much below the high level established in Chapter Six. For one thing, they are dated in a way that the previous chapter was not. It appears also that Clausewitz is not happy with the premises he establishes for these chapters, which are basically twofold: the expressed premise that one lacks the means to impose a decisive defeat on the enemy, and the implied premise that one is nevertheless not too grossly inferior to him. Under those conditions one may still (Chapter Seven) offensively seize a portion of the enemy's territory, thereby reducing his national resources and holding in one's own hand an asset for the peace negotiations. The important question in this case is whether there is a good chance of holding on to the territory for the remainder of the war, and if not, whether temporary occupation would be worth what the occupation costs. "In general," Clausewitz says, "one tends to lose more from occupation by the enemy than one gains from occupying his territory, even if the value of both areas should be identical." More important, in assuming the defensive over a larger area, one surrenders the initiative without having ever attempted a critical blow at the heart of the enemy's power. Clausewitz seems at this point to be drifting away from his primary premise, and he finishes this chapter with a reaffirmation of how noble it is to be willing to accept large risks.

In Chapter Eight the author takes up a more strategically passive form of defense, where the main object is to gain time while keeping as much as possible of one's territory out of enemy hands. One can resort to beating off attacks wherever they occur, and can even undertake minor offensive operations in the form of raids or diversions, but only if the basically defensive purpose is not compromised. The purpose of bidding for time is to await political changes, such as the enemy's wearying of his effort, or of allies adhering to one's own side or departing that of the enemy.

If, however, the enemy's offensive has the effect of so reducing his strength that one becomes superior to him a counter-offensive is naturally in order. At this point Clausewitz shifts away from examples describing Frederick the Great's strategy to still another consideration of the 1812 campaign, and he offers us a revealing observation: "The highest wisdom could never have devised a better strategy than the one which the Russians *happened to follow*" (italics added). Clausewitz was throughout that campaign in a good position to know.

We presume to call these chapters dated partly because Clausewitz himself considered them already to be so in his time. After describing Frederick's successful defensive strategies, he observes: "But one must remember that times have changed, that war has undergone a total transformation and now draws its life from wholly different sources. Positions which have lost all value in our time could be effective then; and the enemy's general character was a factor as well." The latter remark refers to the lack of aggressiveness of Frederick's opponents. The Russian example, which Clausewitz did not consider dated, refers, of course, to a very large country. As we have seen, for the same reason it would not be dated even today.

Book Eight, Chapter Nine: The last and longest chapter in the work, "The Plan of a War Designed to Lead to the Total Defeat of the Enemy," returns to the monumental scope of Chapter Six. But where Chapter Six concentrates on the level of statesmanship, Chapter Nine focuses on the level of grand military strategy. The object laid down in the chapter title is clearly more congenial to the author's spirit than that of the two chapters immediately preceding, and we encounter an epitome of the military ideas considered typically Clausewitzian.

Having with little wavering disavowed thus far any allegiance to "principles" as firm guides to action, the author starts this chapter by indulging in the enunciation of two. The first principle is to act with the utmost concentration of aim and of force, and the second is to act with the utmost speed. This does not mean that forces must never be divided both in space and in mission, and Clausewitz proceeds to investigate in depth four basic reasons for dividing them. The reasons in each case must be strong ones, to be weighed circumspectly against the primary demand for concentration. Neverthless, it is characteristic of Clausewitz to be exploring exceptions even to his most favored rules.

The decision against the major objective will carry the minor ones as well. Again there are likely to be exceptions, but in any case each minor operation embarked upon must be kept as subordinate as possible.

As he develops these points one suddenly notices that Clausewitz is laying down the intellectual framework of what nearly a century later will be the famous Schlieffen Plan. There are times he says, when "two almost wholly separate wars have to be fought simultaneously. Even then one must be treated as the *main* operation, calling for the bulk of resources and of activities. Seen in this light, it is advisable to operate *offensively* only in this main theater and to stay on the defensive elsewhere. There an attack will be justified only if exceptional conditions should invite it. The defensive at the minor points, moreover, should be maintained with the minimum of strength; and every advantage which that form of resistance offers should be turned to account." This concept describes the German division of forces between the eastern and western fronts in August 1914, and also the division of forces along the western front itself. But he goes on, concerning the enemy who is the target of the main offensive: "The main decision is what matters. It will compensate for any loss. If there are sufficient forces to make it reasonable to seek a major decision, then the possibility of failure can no longer be an excuse for trying to cover oneself everywhere else. For this would make defeat in the decisive battle that much more probable."

In the same vein he returns to his second principle which is that of speed. "Almost the only advantage of attack," he says, "rests on its initial surprise. Speed and impetus are its strongest elements and are usually indispensable if we are to defeat the enemy." Another advantage of speed is that it will bring one quickly into the heart of the enemy country, for though a victory is easier to achieve if the battle is fought near one's own frontiers, it is more likely to be decisive in its effect if it is fought deep in enemy territory. And if we are talking about the complete defeat of the enemy, "such a victory demands an enveloping attack or a battle with reversed fronts." And naturally such a victory must be followed by unrelenting pursuit.

Also, should the tide of victory sweep us past his fortresses (as in Belgium in 1914), we must invest them with the smallest possible forces. "From the moment when the siege of fortresses compels us to suspend the advance, the offensive has as a rule reached its culminating point. Therefore we demand that the main force should go on advancing rapidly and keep up the pressure." And one must not worry about the front of advance being too narrow so long as the momentum is continued: "So long as the general has not yet defeated the enemy, so long as he believes himself to be strong enough to gain his objectives, he must persevere. He may do so with increasing danger, but his success will be all the greater. Should he reach a point beyond which he dare not go, should he feel he must expand to right and left in order to protect his rear—so be it: very likely his attack has reached its culminating point. Its momentum is exhausted; and if the enemy is still unbroken there is probably no future in it anyway."

This confluence of the Clausewitzian vision with that of Schlieffen, which we have only outlined here, is too detailed and too sharply distinguished from the ideas of other writers to be an accident, especially inasmuch as we know that Schlieffen was a keen student of Clausewitz.

Then follows an extraordinary analysis of Napoleon's 1812 campaign in Russia, the failure of which, the author is constrained to show, was *not* due to the advance being too fast and too far. Nor was his going to Moscow a mistake. "Only if he could reach Moscow in strength could Bonaparte hope to shake the government's nerve and the people's loyalty and steadfastness." The campaign failed "because the Russian government kept its nerve and the people remained loyal and steadfast." The outcome shows that Napoleon miscalculated. It may have been wrong for him to risk such a miscalculation, but his estimate of how the government and people would react to his taking Moscow was not so predictably wrong as to make the campaign an absurdity. If Napoleon's army was punished more severely than it need have been, the fault which can be laid to him was not in the depth to which he penetrated Russia. "It lay in his being late in starting the campaign, in the lives he squandered by his tactics, his neglect of matters of supply and of retreat," and in his staying too long in Moscow.

Most significant are the criteria by which Clausewitz judges this as well as other events: "It is legitimate to judge an event by its outcome, for this is its soundest criterion. But a judgment based on the result alone must not be passed off as evidence of human wisdom. . . . Anyone who asserts that the campaign of 1812 was an absurdity because of its enormous failure, but who would have called it a superb idea if it had worked, shows complete lack of judgment." In other words, there is never only one reason for success or for failure, and good critical judgment must take account of the several important reasons for each.

This great chapter continues with a discussion of attack on a common objective being made by widely separated armies as distinct from a single concentrated attack. Clausewitz prefers the latter, but the former "may be imposed by circumstances which one cannot alter." The author is of course talking about an era which knew nothing of modern electronic communications, but the wisdom he brings to *his* problem is none the less piquant. If forces are operating in different theaters, perhaps widely separated, against a common enemy, they should not worry too much about smoothly coordinating their efforts. "The smooth coordination of every part of the action from start to finish is not even necessary in tactics, let alone strategy." Each of the separated forces should therefore be given an independent task. Actually, even in modern times this advice still holds.

One notices ruefully in passing that in the Prussian campaign in France in 1793—in which Clausewitz participated as a boy ensign of thirteen—"Prussia had nothing either to defend or conquer in Alsace." The discovery that this province was a long-lost part of Germany was to come much later. We also notice, perhaps with surprise, that in the campaigns of 1813–1814, the Russian Czar did not insist on his army operating independently but put his forces, at the level of individual corps, under Prussian and Austrian commanders.

The real peroration of the book comes not at the end but rather in the few paragraphs of summary about plans aimed at the total defeat of the enemy, where he says in part: "We would emphasize the essential and general; leave scope for the individual and accidental; but remove everything *arbitary, unsubstantiated, trivial, far-fetched,* or *supersubtle.* . . . Should anyone be shocked at finding nothing here about how to turn a river, command a mountain area from its heights, by-pass a strong position or find the key to a whole country, he has failed to grasp our purpose; we are afraid, moreover, that he has still not understood the essential elements of war. In previous books we have dealt with these details in a general way, and reached the conclusion that they are apt to be a great deal less important than is usually thought."

The plan that follows for some future invasion of France—in the event she should "resume that insolent behavior that has been the curse of Europe for a hundred and fifty years"—is interesting but of no real significance except for what it says about armies invading separately from different theaters, and the paragraph which again calls to mind the Schlieffen Plan: "The center of gravity of France lies in the armed forces and in Paris. . . . The most vulnerable area of France is that between Paris and Brussels, where the frontier is only 150 miles from the capital."

Rosalie West

CHRONOLOGICAL INDEX OF WARS, CAMPAIGNS AND BATTLES

Wars

Campaigns and Battles

Books Written Under the Auspices of the Center of International Studies Princeton University

Gabriel A. Almond, *The Appeals of Communism* (Princeton University Press 1954)
William W. Kaufmann, ed., *Military Policy and National Security* (Princeton University Press 1956)
Klaus Knorr, *The War Potential of Nations* (Princeton University Press 1956)
Lucian W. Pye, *Guerrilla Communism in Malaya* (Princeton University Press 1956)
Charles De Visscher, *Theory and Reality in Public International Law*, trans. by P. E. Corbett (Princeton University Press 1957; rev. ed. 1968)
Bernard C. Cohen, *The Political Process and Foreign Policy: The Making of the Japanese Peace Settlement* (Princeton University Press 1957)
Myron Weiner, *Party Politics in India: The Development of a Multi-Party System* (Princeton University Press 1957)
Percy E. Corbett, *Law in Diplomacy* (Princeton University Press 1959)
Rolf Sannwald and Jacques Stohler, *Economic Integration: Theoretical Assumptions and Consequences of European Unification*, trans. by Herman Karreman (Princeton University Press 1959)
Klaus Knorr, ed., *NATO and American Security* (Princeton University Press 1959)
Gabriel A. Almond and James S. Coleman, eds., *The Politics of the Developing Areas* (Princeton University Press 1960)
Herman Kahn, *On Thermonuclear War* (Princeton University Press 1960)
Sidney Verba, *Small Groups and Political Behavior: A Study of Leadership* (Princeton University Press 1961)
Robert J. C. Butow, *Tojo and the Coming of the War* (Princeton University Press 1961)
Glenn H. Snyder, *Deterrence and Defense: Toward a Theory of National Security* (Princeton University Press 1961)
Klaus Knorr and Sidney Verba, eds., *The International System: Theoretical Essays* (Princeton University Press 1961)
Peter Paret and John W. Shy, *Guerrillas in the 1960's* (Praeger 1962)
George Modelski, *A Theory of Foreign Policy* (Praeger 1962)
Klaus Knorr and Thornton Read, eds., *Limited Strategic War* (Praeger 1963)

Frederick S. Dunn, *Peace-Making and the Settlement with Japan* (Princeton University Press 1963)

Arthur L. Burns and Nina Heathcote, *Peace-Keeping by United Nations Forces* (Praeger 1963)

Richard A. Falk, *Law, Morality, and War in the Contemporary World* (Praeger 1963)

James N. Rosenau, *National Leadership and Foreign Policy: A Case Study in the Mobilization of Public Support* (Princeton University Press 1963)

Gabriel A. Almond and Sidney Verba, *The Civic Culture: Political Attitudes and Democracy in Five Nations* (Princeton University Press 1963)

Bernard C. Cohen, *The Press and Foreign Policy* (Princeton University Press 1963)

Richard L. Sklar, *Nigerian Political Parties: Power in an Emergent African Nation* (Princeton University Press 1963)

Peter Paret, *French Revolutionary Warfare from Indochina to Algeria: The Analysis of a Political and Military Doctrine* (Praeger 1964)

Harry Eckstein, ed., *Internal War: Problems and Approaches* (Free Press 1964)

Cyril E. Black and Thomas P. Thornton, eds., *Communism and Revolution: The Strategic Uses of Political Violence* (Princeton University Press 1964)

Miriam Camps, *Britain and the European Community 1955–1963* (Princeton University Press 1964)

Thomas P. Thornton, ed., *The Third World in Soviet Perspective: Studies by Soviet Writers on the Developing Areas* (Princeton University Press 1964)

James N. Rosenau, ed., *International Aspects of Civil Strife* (Princeton University Press 1964)

Sidney I. Ploss, *Conflict and Decision-Making in Soviet Russia: A Case Study of Agricultural Policy, 1953–1963* (Princeton University Press 1965)

Richard A. Falk and Richard J. Barnet, eds., *Security in Disarmament* (Princeton University Press 1965)

Karl von Vorys, *Political Development in Pakistan* (Princeton University Press 1965)

Harold and Margaret Sprout, *The Ecological Perspective on Human Affairs, With Special Reference to International Politics* (Princeton University Press 1965)

Klaus Knorr, *On the Uses of Military Power in the Nuclear Age* (Princeton University Press 1966)

Harry Eckstein, *Division and Cohesion in Democracy: A Study of Norway* (Princeton University Press 1966)

Cyril E. Black, *The Dynamics of Modernization: A Study in Comparative History* (Harper and Row 1966)

Peter Kunstadter, ed., *Southeast Asian Tribes, Minorities, and Nations* (Princeton University Press 1967)
E. Victor Wolfenstein, *The Revolutionary Personality: Lenin, Trotsky, Gandhi* (Princeton University Press 1967)
Leon Gordenker, *The UN Secretary-General and the Maintenance of Peace* (Columbia University Press 1967)
Oran R. Young, *The Intermediaries: Third Parties in International Crises* (Princeton University Press 1967)
James N. Rosenau, ed., *Domestic Sources of Foreign Policy* (Free Press 1967)
Richard F. Hamilton, *Affluence and the French Worker in the Fourth Republic* (Princeton University Press 1967)
Linda B. Miller, *World Order and Local Disorder: The United Nations and Internal Conflicts* (Princeton University Press 1967)
Henry Bienen, *Tanzania: Party Transformation and Economic Development* (Princeton University Press 1967)
Wolfram F. Hanrieder, *West German Foreign Policy, 1949–1963: International Pressures and Domestic Response* (Stanford University Press 1967)
Richard H. Ullman, *Britain and the Russian Civil War: November 1918–February 1920* (Princeton University Press 1968)
Robert Gilpin, *France in the Age of the Scientific State* (Princeton University Press 1968)
William B. Bader, *The United States and the Spread of Nuclear Weapons* (Pegasus 1968)
Richard A. Falk, *Legal Order in a Violent World* (Princeton University Press 1968)
Cyril E. Black, Richard A. Falk, Klaus Knorr and Oran R. Young, *Neutralization and World Politics* (Princeton University Press 1968)
Oran R. Young, *The Politics of Force: Bargaining During International Crises* (Princeton University Press 1969)
Klaus Knorr and James N. Rosenau, eds., *Contending Approaches to International Politics* (Princeton University Press 1969)
James N. Rosenau, ed., *Linkage Politics: Essays on the Convergence of National and International Systems* (Free Press 1969)
John T. McAlister, Jr., *Viet Nam: The Origins of Revolution* (Knopf 1969)
Jean Edward Smith, *Germany Beyond the Wall: People, Politics and Prosperity* (Little, Brown 1969)
James Barros, *Betrayal from Within: Joseph Avenol, Secretary-General of the League of Nations, 1933–1940* (Yale University Press 1969)
Charles Hermann, *Crises in Foreign Policy: A Simulation Analysis* (Bobbs-Merrill 1969)
Robert C. Tucker, *The Marxian Revolutionary Idea: Essays on Marxist Thought and Its Impact on Radical Movements* (W. W. Norton 1969)

Harvey Waterman, *Political Change in Contemporary France: The Politics of an Industrial Democracy* (Charles E. Merrill 1969)

Cyril E. Black and Richard A. Falk, eds., *The Future of the International Legal Order*. Vol. I: *Trends and Patterns* (Princeton University Press 1969)

Ted Robert Gurr, *Why Men Rebel* (Princeton University Press 1969)

C. Sylvester Whitaker, *The Politics of Tradition: Continuity and Change in Northern Nigeria 1946–1966* (Princeton University Press 1970)

Richard A. Falk, *The Status of Law in International Society* (Princeton University Press 1970)

Klaus Knorr, *Military Power and Potential* (D. C. Heath 1970)

Cyril E. Black and Richard A. Falk, eds., *The Future of the International Legal Order*. Vol. II: *Wealth and Resources* (Princeton University Press 1970)

Leon Gordenker, ed., *The United Nations in International Politics* (Princeton University Press 1971)

Cyril E. Black and Richard A. Falk, eds., *The Future of the International Legal Order*. Vol. III: *Conflict Management* (Princeton University Press 1971)

Francine R. Frankel, *India's Green Revolution: Political Costs of Economic Growth* (Princeton University Press 1971)

Harold and Margaret Sprout, *Toward a Politics of the Planet Earth* (Van Nostrand Reinhold 1971)

Cyril E. Black and Richard A. Falk, eds., *The Future of the International Legal Order*. Vol. IV: *The Structure of the International Environment* (Princeton University Press 1972)

Gerald Garvey, *Energy, Ecology, Economy* (W. W. Norton 1972)

Richard Ullman, *The Anglo-Soviet Accord* (Princeton University Press 1973)

Klaus Knorr, *Power and Wealth: The Political Economy of International Power* (Basic Books 1973)

Anton Bebler, *Military Rule in Africa: Dahomey, Ghana, Sierra Leone, and Mali* (Praeger Publishers 1973)

Robert C. Tucker, *Stalin as Revolutionary 1879–1929: A Study in History and Personality* (W. W. Norton 1973)

Edward L. Morse, *Foreign Policy and Interdependence in Gaullist France* (Princeton University Press 1973)

Henry Bienen, *Kenya: The Politics of Participation and Control* (Princeton University Press 1974)

Gregory J. Massell, *The Surrogate Proletariat: Moslem Women and Revolutionary Strategies in Soviet Central Asia, 1919–1929* (Princeton University Press 1974)

James N. Rosenau, *Citizenship Between Elections: An Inquiry Into The Mobilizable American* (Free Press 1974)

Ervin Laszlo, *A Strategy For The Future: The Systems Approach To World Order* (Braziller 1974)

John R. Vincent, *Nonintervention and International Order* (Princeton University Press 1974)
Jan H. Kalicki, *The Pattern of Sino-American Crises: Political-Military Interactions in the 1950s* (Cambridge University Press 1975)
Klaus Knorr, *The Power of Nations: The Political Economy of International Relations* (Basic Books 1975)
James P. Sewell, *UNESCO and World Politics: Engaging in International Relations* (Princeton University Press 1975)
Richard A. Falk, *A Global Approach to National Policy* (Harvard University Press 1975)
Harry Eckstein and Ted Robert Gurr, *Patterns of Authority: A Structural Basis for Political Inquiry* (John Wiley & Sons 1975)
Cyril E. Black, Marius B. Jansen, Herbert S. Levine, Marion J. Levy, Jr., Henry Rosovsky, Gilbert Rozman, Henry D. Smith, II, and S. Frederick Starr, *The Modernization of Japan and Russia* (Free Press 1975)
James P. Sewell, *UNESCO and World Politics: Engaging in International Relations* (Princeton University Press 1975)
Richard A. Falk, *A Global Approach to National Policy* (Harvard University Press 1975)
Harry Eckstein and Ted Robert Gurr, *Patterns of Authority: A Structural Basis for Political Inquiry* (John Wiley & Sons 1975)
Cyril E. Black, Marius B. Jansen, Herbert S. Levine, Marion J. Levy, Jr., Henry Rosovsky, Gilbert Rozman, Henry D. Smith, II, and S. Frederick Starr, *The Modernization of Japan and Russia* (Free Press 1975)
Leon Gordenker, *International Aid and National Decisions: Development Programs in Malawi, Tanzania, and Zambia* (Princeton University Press 1976)
Carl Von Clausewitz, *On War*, edited and translated by Michael Howard and Peter Paret (Princeton University Press 1976; indexed edition 1984)
Gerald Garvey and Lou Ann Garvey, eds., *International Resource Flows* (D.C. Heath 1977)
Walter F. Murphy and Joseph Tanenhaus, *Comparitive Constitutional Law Cases and Commentaries* (St. Martin's Press 1977)
Gerald Garvey, *Nuclear Power and Social Planning: The City of the Second Sun* (D.C. Heath 1977)
Richard E. Bissell, *Apartheid and International Organizations* (Westview Press 1977)
David P. Forsythe, *Humanitarian Politics: The International Committee of the Red Cross* (Johns Hopkins University Press 1977)
Paul E. Sigmund, *The Overthrow of Allende and the Politics of Chile, 1964-1976* (University of Pittsburgh Press 1977)
Henry S. Bienen, *Armies and Parties in Africa* (Holmes and Meier 1978)
Harold and Margaret Sprout, *The Context of Environmental Politics* (The University Press of Kentucky 1978)
Samuel S. Kim, *China, the United Nations, and World Order* (Princeton University Press 1979)
S. Basheer Ahmed, *Nuclear Fuel and Energy Policy* (D.C. Heath 1979)

Robert C. Johansen, *The National Interest and the Human Interest: An Analysis of U.S. Foreign Policy* (Princeton University Press 1980)

Richard A. Falk and Samuel S. Kim, eds., *The War System: An Interdisciplinary Approach* (Westview Press 1980)

James H. Billington, *Fire in the Minds of Men: Origins of the Revolutionary Faith* (Basic Books, Inc. 1980)

Bennett Ramberg, *Destruction of Nuclear Energy Facilities in War: The Problem and the Implications* (D. C. Heath 1980)

Gregory T. Kruglak, *The Politics of United States Decision-Making in United Nations Specialized Agencies: The Case of the International Labor Organization* (University Press of America 1980)

W. P. Davison and Leon Gordenker, eds., *Resolving Nationality Conflicts: The Role of Public Opinion Research* (Praeger Publishers 1980)

James C. Hsiung and Samuel S. Kim, eds., *China in the Global Community* (Praeger Publishers 1980)

Douglas Kinnard, *The Secretary of Defense* (The University Press of Kentucky 1980)

Richard Falk, *Human Rights and State Sovereignty* (Holmes & Meier Publishers, Inc. 1981)

James H. Mittelman, *Underdevelopment and the Transition to Socialism: Mozambique and Tanzania* (Academic Press 1981)

Gilbert Rozman, ed., *The Modernization of China* (Free Press 1981; paperback edition 1982)

Robert C. Tucker, *Politics as Leadership*. The Paul Anthony Brick Lectures. Eleventh Series (University of Missouri Press 1981)

Robert Gilpin, *War and Change in World Politics* (Cambridge University Press 1981)

Nicholas G. Onuf, ed., *Law-Making in the Global Community* (Carolina Academic Press 1982)

Ali E. Hillal Dessouki, ed., *Islamic Resurgence in the Arab World* (Praeger Publishers 1982)

Richard Falk, *The End of World Order* (Holmes & Meier Publishers 1983)

Klaus Knorr, ed., *Power, Strategy and Security: A World Politics Reader* (Princeton University Press 1983)

Samuel S. Kim, *The Quest for a Just World Order* (Westview Press 1984)

Gerald Garvey, *Strategy and the Defense Dilemma* (D.C. Heath and Company 1984)